APPLICATIONS OF MOLECULAR SPECTROSCOPY TO CURRENT RESEARCH IN THE CHEMICAL AND BIOLOGICAL SCIENCES

Edited by **Mark T. Stauffer**

Applications of Molecular Spectroscopy to Current Research in the Chemical and Biological Sciences

http://dx.doi.org/10.5772/61896
Edited by Mark T. Stauffer

Contributors

Andrzej Łapiński, Chunrui Wang, Lin Sun, Lingcong Shi, Pedro Prezas, Metin Aydin, Daniel L. Akins, Lien Smeesters, Wendy Meulebroeck, Hugo Thienpont, Laura J. Bonales, José A. Oteo, Lara García-Álvarez, Jesús H. Busto, Jesús M. Peregrina, Alberto A Avenoza, Yanying Zhao, Xin Liu, Shuang Meng, Tingting You, Feng Xu, Yang Song, Chandrabhas Narayana, Partha Kundu, Ping Chen, Lu Sun, Lie Lin, Ivana Durickovic, Philip Westergaard, Peiqiang Yu, Jana Sádecká, Michaela Jakubíková, Veronika Uríčková, Miroslav Dramicanin, Tatjana Dramicanin, Joao Henrique Zimnoch Dos Santos, Larissa Brentano Capeletti

CBS, Edition **2017**

Published by InTech

Janeza Trdine 9, 51000 Rijeka, Croatia

Publishing Process Manager Edi Lipovic

Technical Editor SPi Global

Cover Designer InTech Design Team

Additional hard copies can be obtained from orders@intechopen.com

Applications of Molecular Spectroscopy to Current Research in the Chemical and Biological Sciences
Edited by Mark T. Stauffer
p. cm.
Print ISBN 978-953-51-2680-5
Online ISBN 978-953-51-2681-2

INTECH

open science | open minds

Contents

Preface

The term *spectroscopy* is defined as the study of the interaction of matter with electromagnetic radiation (i.e., "light"). The effects of this relationship were known to the ancient Romans by a rainbow that was produced when light passed through a crystal. In the mid-seventeenth century, Sir Isaac Newton demonstrated that the production of a rainbow when sunlight fell on a clear crystal was due to the components of the sunlight itself and not any component of the crystal. Newton gave this rainbow a name for any diagram of light signal as a function of wavelength or frequency—a *spectrum* . Two centuries later, during the heyday of observational astronomy, astronomers utilized the emerging field of spectroscopy to characterize the stars of deep space via identification of their component elements. In this same timeframe, around 1859, two German scientists, Robert Bunsen (a chemist) and Gustav Kirchhoff (a physicist), discovered two main types of spectra, *absorption* (spectra yielding dark lines) and emission (spectra yielding bright lines), based on the processes of absorption of light by elements (and their atoms) and compounds (and their molecules or formula units) and the emission or discharge of radiation by these same entities. They also demonstrated how both absorption and emission spectra could be used for identification of elements within compounds, based on their observation that every element, no matter the compound within which it is incorporated, produces its own unique spectrum with the same spectral features. Thus the foundation was laid for what would become, over the next century and a half or so, the development and utilization of spectroscopic methods and techniques for identification, characterization, and quantitation of components of samples of myriad.

Spectroscopy is truly a multidisciplinary field in every aspect. The fundamental principles of spectroscopy, whether based on absorption, emission, or scattering of light, are rooted in the laws of physics, which concerns itself with the study of light as waves and particles (quantum theory). By definition, spectroscopy is the study of the interaction of light with matter; thus, spectroscopy is also firmly rooted in chemistry—the study of the composition, properties, and changes of matter. Matter in its various forms, in turn, makes up organic and inorganic compounds that figure prominently in the composition of various forms of life—the realm of biology—and the Earth, the domain of geology. Spectroscopy is used, then, to identify, characterize, and quantify the very elements and compounds that define the composition of the Earth and all life on it.

Despite its multidisciplinary nature, spectroscopy can be considered to reside well within the boundaries of chemistry. Spectroscopy may be used for elucidation of the structures of synthesized molecules (organic and inorganic chemistry), for qualitative identification or quantitative determination of a component of an interesting sample (analytical chemistry), for the study of physical processes and properties of elements and compounds (physical and theoretical chemistry), for characterization of various materials (organic, inorganic, and nuclear chemistry), and in combination with electrochemical and chromatographic analytical methods (analytical chemistry).

Absorption and emission spectra are obtained for atoms of elements and molecules of compounds via methods and techniques that are atom specific or molecule specific. The spectra may be a series of discrete lines, as in atomic spectra, or narrow to broad peaks due to intra- and intermolecular forces, plus any sample/solution matrix effects, as in molecular spectra. Such spectra, which are usually plots of absorption, emission, or scattering signals as a function of wavelength or frequency, can yield a wealth of information about the atoms or molecules being studied. Such information may be qualitative and useful for identification of functional groups as well as elucidation of the structure of a newly synthesized molecule. Information obtained from a spectrum may also be quantitative, via the relationship between the intensity of a signal at a selected wavelength and the concentration of the analyte in a sample.

As the focus of this book is on *molecular* spectroscopic methods and techniques and their applications to a variety of research questions and issues, selected methods and techniques associated with this rather expansive branch of spectroscopy will be presented in a series of original research and review chapters. Such well-known methods as UV-visible spectrophotometry, fluorescence spectrometry, infrared (IR) spectroscopy and its Fourier transform counterpart (FT-IR), Raman and FT-Raman spectroscopy, and nuclear magnetic resonance (NMR) spectroscopy, along with X-ray diffraction (XRD) and photoelectron spectroscopy (PES), are widely used in research in many venues and are covered to varying extents in this text. Additionally, one cannot leave out the many computational and statistical methods used in conjunction with spectroscopic methods for analysis of data and results leading to characterization of samples and quantitation of analytes within samples. Such methods include density functional theory (DFT) for quantum mechanical computations and chemometric methods for pattern recognition and classification, for example, principal component analysis (PCA) and hierarchical cluster analysis (HCA).

The goal of this book is to present an overview of applications of molecular spectroscopy to investigations of organic and inorganic materials, foodstuffs, biosamples and biomedicine, and novel characterization and quantitation methods. This text is a compilation of selected research articles and reviews covering current efforts in various applications of molecular spectroscopy. Sections 1 and 2 deal, respectively, with spectroscopic studies of inorganic and organic materials. Section 3 provides applications of molecular spectroscopy to biosamples and biomedicine. Section 4 explores spectroscopic characterization and quantitation of foods and beverages. Lastly, Section 5 presents research on novel spectroscopic methodologies. Overall, this book should be a great source of scientific information for anyone involved in characterization, quantitation, and method development.

I am most grateful to Mr. Edi Lipović, the Publishing Process Manager who supervised and organized publishing of all materials; assisted me and the authors in completion of our work in an easy, timely manner; and provided helpful advice and guidance throughout the project. I thank the authors for their great contributions to this book. I express many thanks to the technical editor who prepared these manuscripts for publication.Finally, I thank my wife, Resa, who is also an analytical chemist, for her advice and support, the University of Pittsburgh at Greensburg for their support, and all the spectroscopists I have known throughout my career, for all they taught me about this fascinating field.

Mark T. Stauffer, PhD
University of Pittsburgh at Greensburg,
Greensburg, Pennsylvania, United States of America

Section 1

Applications of Molecular Spectroscopy to Inorganic Materials

Chapter 1

Fourier Transform Infrared and Raman Characterization of Silica-Based Materials

Larissa Brentano Capeletti and
João Henrique Zimnoch

Additional information is available at the end of the chapter

http://dx.doi.org/10.5772/64477

Abstract

Fourier Transform Infrared and Raman are powerful techniques to evaluate silica and hybrid silica structure. It is possible to evaluate the silica network formation along the hydrolysis and condensation reactions in terms of siloxane rings formation and Si–O(–Si) angle deformation due to the introduction of organic groups, the employed synthetic route or encapsulated species interaction. The siloxane four- or six-membered rings imply in a more rigid or flexible network, respectively, in order to accommodate the organic groups. A structural analysis of the materials is of high importance, since interactions between the encapsulated molecules and the matrix are critical for the device performance, such as sensors. This type of device needs the permeation of an analyte to activate the encapsulated receptor molecules inside the silica structure. Fourier transform infrared spectrometry can be also used to determine parameters of the silica network as a function of the hydrophilicity/hydrophobicity degree and the siloxane ring structure with respect to thin film porosity. This silica structural analysis is reviewed along the text in a tentative of better exploring the data resulting from these powerful techniques. In addition, the functionalization of silica structures by the use of organoalkoxysilanes, which is important to the creation of high-specific materials, can be well described by these two complementary techniques. The Si–C bonds and the maintenance of the organic substituents such as methyl, octyl, octadecyl, vinyl, phenyl, aminopropyl, mercaptopropyl, isocyanatopropyl, iodopropyl, chloropropyl and glicydoxypropyl could be evaluated after the sol-gel synthesis process. The literature regarding silica vibrational spectroscopy is also explored creating a data bank of wave numbers for the most important bonds for different types of silica and hybrid silica materials obtained by different synthetic routes.

Keywords: hybrid silica, molecular imprinting, silica-based materials, FTIR, Raman

1. Introduction on silica-based materials

Silica-based materials have a wide field of applications nowadays, since it is very flexible in terms of material characteristics and fabrication methods [1]. Different types of devices such as catalysts, chromatographic phases, sorbents, sensors, coatings, etc. can be produced with tuned properties to enhance activity and/or robustness.

In terms of catalysts, several different reactions can take advantage of the silica surface usage, resulting in heterogeneous processes. The hydrogen production, for example, needs high surface area supports, open porosity, nanostructure with uniform morphology, highly and relatively uniform dispersed active phase, which can be achieved by a silica matrix [2]. In addition, very complex structures can be designed as rattle-type magnetic silica composite with nonporous silica-coating magnetic iron oxide encapsulated in mesoporous silica hollow sphere which can also contain active metallic nanoparticles (**Figure 1A**), resulting in a Pt-based catalyst for hydrogenation that exhibits high activity, selectivity and excellent reusability [3]. Complex hierarchical structures can be also obtained with independent functionalization of macropore and mesopore networks on the basis of chemical and/or size specificity affords control over the reaction sequence in catalytic cascades [4]. The catalyst preparation strategies also include the addition of other metal oxides to the silica network: in desulfurization process, the prepared mixed oxide would take advantage of both titania, probably as the main active component, and silica, for its high thermal stability, excellent mechanical strength and high surface area [5] and the same approach can be employed for photocatalytic practices [6]. The mixed oxides can also be used as polymerization catalysts, where the presence of oxides such as WO_3, CrO_3 and MoO_3 in the silica network decreased the necessary cocatalyst amount, suggesting that the support nature has a considerable influence on the process [7].

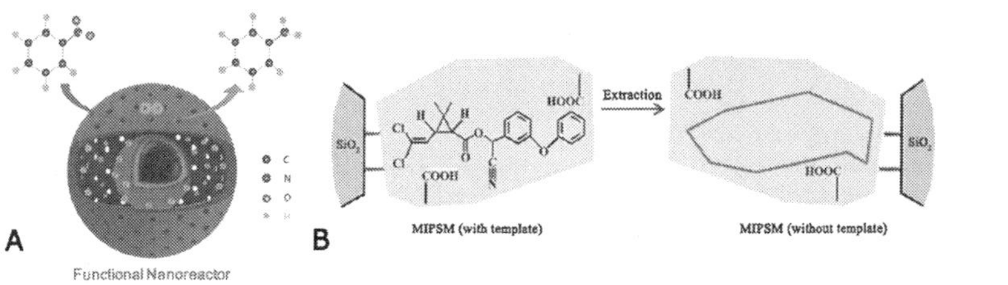

Figure 1. A) Rattle-type magnetic silica composite with nonporous silica-coating magnetic iron oxide encapsulated in mesoporous silica hollow sphere which can also contain active metallic nanoparticles [3]. B) Molecular imprinting process adapted with permission from Zhao et al. [9].

For chromatographic phases and sorbents, the main maneuvers are the so-called molecular imprinting or the silica functionalization with groups that retain the analytes. Taking into consideration the second approach, it is possible to design hybrid silica monoliths functionalized with, for example, aminopropyl or cyanopropyl groups and utilize them as selective stationary phase for microextraction by packed sorbent (MEPS). This method could determine drugs, such as antipsychotics in combination with antidepressants, anticonvulsants and anxiolytics in plasma samples from schizophrenic patients through liquid chromatography-tandem mass spectrometry (LC-MS/MS) in the multiple reactions monitoring (MRM) mode [8]. On the other hand, the molecular imprinting is a methodology which creates cavities for the analytes encapsulating the analyte itself or a template within the silica network that is followed by an extraction process (**Figure 1B**) [9]. The resulting material is of high specificity, increasing the selectivity and performance of the sorbent/phase. This methodology has been employed to extract important compounds such as the β-N-methylamino-L-alanine amino acid from cyanobacteria which is hypothesized to be linked to amyotrophic lateral sclerosis and Parkinson dementia complex from people living in Guam island [10] and to pretreat, detect and analyze trace levels of toxic pyrethroid insecticides in soils [9].

Different types of sensors can also be prepared taking advantages of silica materials' flexibility. Optical sensors use to employ the methodology of a receptor element encapsulation within the silica network and the structural properties and addition of organic groups can be correlated with the device performance [11]. They frequently include the encapsulation of an organic dye which can change color when in contact with the analyte [12]. The introduction of mixed oxides and organic groups can be of high importance in this field to avoid leaching of these dyes during usage [13]. It is also possible to manufacture different devices' configuration such as nanosensors for pH measurement [13], optical fibers for volatile organic compounds' detection [14], electrochemical [15], electrochemiluminescence [16] and biosensors [17]. Furthermore, silica is commonly used to protect or give special features to surfaces. In terms of protection, organosilanes are well known as corrosion protectors for metallic surfaces such as steel and aluminum alloys [18], where other metal such as cerium [19], metallic nanoparticles such as $NiFe_2O_4$ [20] and other compounds such as phosphonic acid [21] can also be added to the coating to enhance the protection. The surface characteristics are another feature that are able to be tuned by the silica coatings. In this field, it is possible to mention the superhydrophobicity that is widely explored with the possibility of self-cleaning surface creation [22, 23] and also the addition of antimicrobial properties is of high importance [24, 25].

Thus, the chapter shall be structured according to the following subitems:

1. *Introduction on silica-based materials.* A panorama on the different applications of such devices (catalysts, chromatographic phases, sorbents, coatings…) should be provided, illustrating recent examples from the literature (2014–2015). Some comments on the general aspects of their production (synthetic routes based on sol-gel routes, grafting reactions, encapsulation via nonhydrolytic sol-gel processes, molecular imprinting).

2. *Organic groups on silica surface.* Recent examples of the use of FTIR and FT-Raman spectroscopies in the monitoring of surface reactions between silanol groups and ligands for organic and organometallic compounds. The possibility of distinguishing liquid-like and crystalline configuration for long-chain alkyl groups from the C–H stretching vibrations position.

3. *Molecules within bulk silica.* The use of FTIR in monitoring the encapsulation of molecules within silica network by deconvolution of Si–O stretching region. The correlation between siloxane four- or six-membered rings and device characteristics/properties.

4. *Silica molecular imprinting*. The use of FTIR and Raman in the monitoring of cavity interaction between templates/target molecules with functional groups from silica pores.

5. *FTIR and Raman modes*. Discussion on the complementary information provided by sampling accessories and detection modes in the characterization/evaluation of hybrid silica materials, namely: attenuated total reflectance (ATR), DRIFTS, photoacoustic spectroscopy (PAS), infrared emission spectroscopy (IRES), micro-FTIR and Raman.

2. Organic groups on silica surface

Recent examples of the use of FTIR and FT-Raman spectroscopies in the monitoring of surface reactions between silanol groups and ligands for organic and organometallic compounds. The possibility of distinguishing liquid-like and crystalline configuration for long-chain alkyl groups from the C–H stretching vibrations position.

The organic groups' presence and interactions are monitored in different types of materials, as the ones discussed above. In terms of corrosion protection coatings, the organosilanes are widely known that due to their efficient properties as coupling agents, representing an interesting and environmentally friendly alternative in the field of surface treatments [26]. One of the employed organosilanes is the glycidyloxypropyltrimethoxysilane (GPS) that when mixed with methyltriethoxysilane can improve coating resistance, charge transfer resistance and present low-frequency impedance parameters. FTIR was employed by Foroozan et al. [18] to investigate the reaction between glycidyl groups of GPS molecules with silanol groups. It was possible to identify bands reflecting the epoxy ring breathing around ~910 and 840 cm^{-1} and also a new band appeared near 1730 cm^{-1} for C=O stretching that could be associated with the oxidation of the epoxide ring. The spectra confirmed a strong network reticulation as a result of the reaction between glycidyl groups of GPS molecules with hydrolyzed silanes. However, the complimentary results of water contact angles decreased probably due to higher amount of –CH–OH, produced in the reaction between glycidyl and silanol groups, so they reached an optimum point at which a more reticulated structure overcomes the silane layer hydrophilicity.

Silanol groups are also employed to investigate the grafting of catalytic compounds at the silica surface. Ochędzan-Siodłak et al. describes a catalytic system where metallocenes and postmetallocene compounds are immobilized in an ionic liquid modified silica surface. The modification process could be followed by the Si–OH stretching vibration, at 3700 cm^{-1}, disappearing after the ionic liquid modified silane reaction with the silanol groups (**Figure 2A**) [27]. This strategy is becoming popular nowadays, where different organosilanes are first used to modify the surface with specific chemical groups with which is possible to graft the catalysts species or precursors itself [28, 29]. Similarly, the Si–OH vibration can be used to follow direct grafting of catalyst in the silica surface, where the silanol groups can react with a metallic center producing chemical bond between the catalyst compound and the silica surface. Li et al. [30] describes the grafting of a nickel complex in silica and alumina surfaces by following a band at 3745 cm^{-1}, which is assigned to isolated surface hydroxyl groups drops gradually with reaction time and almost completely vanishes after 24 h and correspondingly the absorption bands at 3070–2877 cm^{-1} related to C–H stretching vibrations of the catalyst allyl groups steadily raise in intensity (**Figure 2B**). Capel-Sanchez et al. [31] also investigated the grafting by silanol groups and in the development of a single site titanium on an amorphous silica surface, they report that the titanium precursor is preferentially anchored over the silica surface by the bridging hydroxyl groups (broad band around 3500 and 3700 cm^{-1}) over the isolated ones (~3700 cm^{-1}).

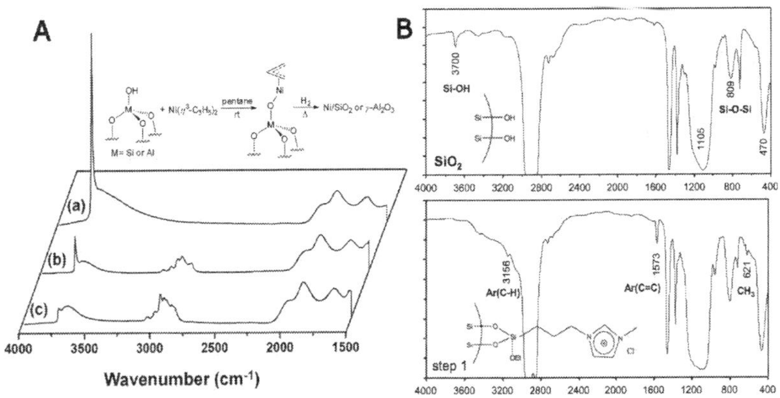

Figure 2. Monitoring of Si–OH stretching vibration, at 3700 cm^{-1}, disappearing after (A) the grafting of a nickel complex in silica and alumina surfaces [27] and (B) the ionic liquid modified silane reaction with the silanol groups [30].

By using Raman spectroscopy, there is also the possibility of final conformation studies in the case of hybrid silica with long alkyl chains at the silica surface. Structure and order information about alkane-based systems can be obtained from multiple indicators in their Raman spectra, especially in the ν(C–H) region between 2750 and 3050 cm^{-1}. Also, significant conformational order information exists for alkane systems in the ν(C–C) and δ(C–H) regions between 900 and 1500 cm^{-1}. However, it is necessary some care about fluorescence phenomena interferences in Raman around this region [32]. Using this methodology, Brambilla et al. evaluated the gauche and trans conformation of hybrid silica with different octadecyl groups (octadecylsilane [ODS]) content by Raman spectroscopy. The two bands centered at 1080 and 1062 cm^{-1} are assigned, respectively, to *v*(C–C) for *gauche* e *trans* conformation of the ODS alkyl chains. The ratio in intensity of these two bands was evaluated in order to monitor the influence of the TEOS/ODS molar ratio in the organic groups' behavior. For all ratios, the intensity between the two bands laid above 1, meaning there is a predominance of *trans* conformation in comparison to *gauche* one, indicating therefore an intense molecular organization in the hybrid silica prepared by the sol-gel method. In addition, it was found that the organization degree of alkyl chains decreases with the ODS content increase, with data from 29Si-NMR and FTIR detected in attenuated total reflectance (ATR) mode, corroborating to the findings [33].

3. Molecules within bulk silica

The FTIR technique can also be employed to evaluate silica network characteristics such as hydrophilicity/hydrophobicity degree and siloxane ring structure regarding thin film porosity [34, 35]. In terms of sensors, these features may impact the analyte interaction and access to the encapsulated molecules, known as receptor elements, within the silica matrix. If the encapsulated molecules' interaction with the silica network occurs through their active sites, it is possible that these active sites are not available to further interact with the analyte. Reduced sensor performance can also occur if the silica network is nonpermeable. In this case, the silica network itself limits the pathways the analyte can travel to reach the encapsulated receptor molecules hindering the reaction. Consequently, both cases could reduce the sensor performance or completely disable it [36].

Silica materials have a prominent band corresponding to the Si–O–Si bond asymmetric stretching in the region from 1300 to 1000 cm^{-1}. Literature reports that the maximum centers and relative intensities of the longitudinal optic (LO) and transversal optic (TO) modes of this bond are shifted with the introduction of chemical groups or organic molecules in the silica network [35]. So, a complete analysis of its components can be conducted, including the deconvolution of the LO and TO modes in their relative main contributions: the four-membered $(SiO)_4$ and six-membered $(SiO)_6$ siloxane rings (**Figure 3A**), resulting in a total of four components (LO_6, LO_4, TO_6 and TO_4) (**Figure 3B**). Usually, materials with higher content of chemical groups or organic molecules use to present higher formation of less stressed six-membered ring, thereby allowing a better accommodation of the nonreactive organic groups [37]. Furthermore, there is a correlation between the formation of six-membered rings and an increase in the relative degree of crystallinity, as well as with the long-range organization in hybrid silica materials that is normally observed with an increase in the degree of matrix alkylation [38].

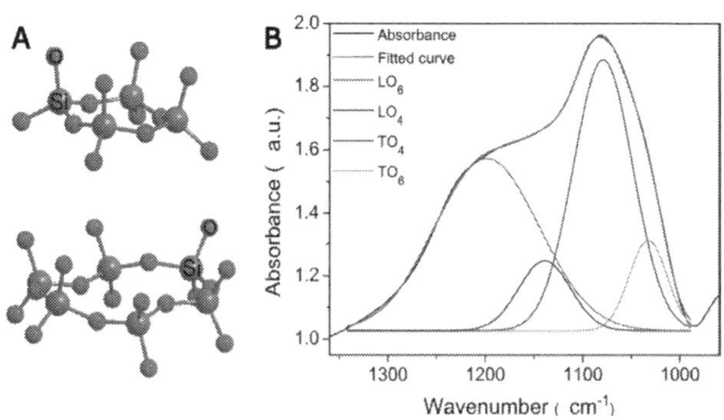

Figure 3. A) Two of the most common cyclical arrangements of SiO_4 structural units in xerogels: four-membered siloxane ring $(SiO)_4$ above and six-membered siloxane ring $(SiO)_6$ below. B) Band deconvolution to asymmetric stretching ν(Si–O(–Si)) bond [39].

Using this approach, a series of silica-based acid-base optical sensors prepared by encapsulating pH indicators using three different sol-gel routes was investigated [36]. The employed routes were: nonhydrolytic, acid-catalyzed and base-catalyzed and the pH indicators were alizarin red, brilliant yellow and acridine. The FTIR spectra were performed for all the materials and the peak corresponding to the Si–O–Si asymmetric stretching was deconvoluted and their respective components analyzed and **Figure 4** assembles the results. For the acidic and nonhydrolytic routes, a positive correlation between the pH indicator content and the increase in $(SiO)_6$ percentage was established, thereby indicating the silica network structure rearrangement in order to accommodate the indicator molecules. Using a basic route, the reached indicator contents were notably low and so this relationship was not observed. In addition, no relationship between the $(SiO)_6$ percentage and the response time could be established in spite of less dense networks, with bigger rings, might render easier the analyte permeation. This behavior may indicate that the analyte probably accessed the receptor elements through the passages between the siloxane rings and not through the siloxane rings themselves [36].

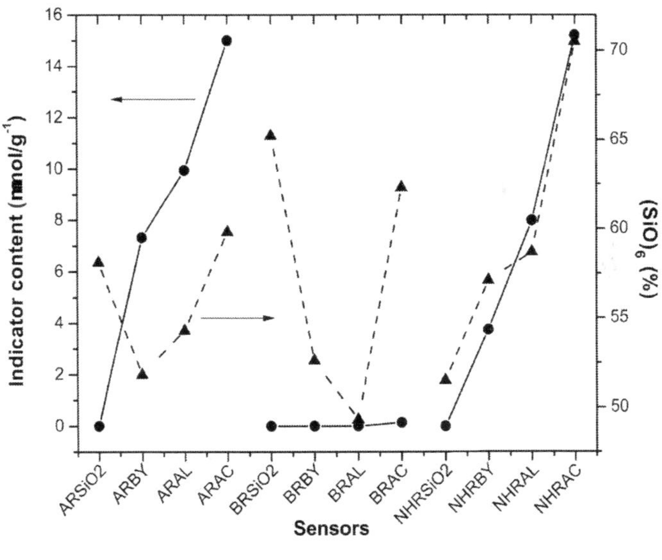

Figure 4. Comparison of $(SiO)_6$ percentage () and encapsulated indicator content () in each sample [39].

It is remarkable that depending on the sol-gel route, the rate between hydrolysis and condensation reactions will change with the pH medium and can explain the behavior of the basic-catalyzed sensor material. Under basic conditions, the condensation reactions of silanol groups are strongly accelerated, and the particles are rapidly formed. Thus, the probability of rearrangement as a function of the presence of other molecules, as pH indicators for example, during the synthesis decreases, since the tridimensional network is quickly formed. When the condensation reactions are slower, such occurs in acid pH, the network can be more influenced by the addition of molecules to be encapsulated [39].

4. Silica molecular imprinting

The molecular imprinting methodology involves a molecular recognition process where the analyte can recognize and preferentially bind to specific sites built by using a template of the target molecule during the matrix network formation. After an extraction process in order to remove the template, the resulting material is bulk silica with cavities that are morphologically and stereochemically compatible with the analyte. Therefore, an option of analytical technique to follow this procedure is FTIR. It is possible to track the template interaction with the silica network by the template bands appearance and/or SiO_2 vibrations change.

With this approach, Morais et al. [40] describe the interactions of different drugs such as fluoxetine, gentamicin, lidocaine, morphine, nifedipine, paracetamol and tetracycline with silica matrix during the preparation of molecular imprinting materials. All these drugs present nitrogen atoms as primary, secondary or tertiary amines that can interact with silanol groups. The investigation was performed comparing the vibrations of bare drug with the encapsulated drug, before removal by extraction. The silica spectrum presents intense bands in the region of ~3500 cm^{-1} assigned to the O–H vibrations of silanol groups and adsorbed water; and in the region of 1200–800 cm^{-1}, where the Si–O stretchings are observed. As a result of these strong bands and the frequently low concentrations of the encapsulated compounds, it is common to observe an overlapping of their signals by the silica ones avoiding this type of evaluation [36].

As an example, **Figure 5A** illustrates the spectra of lidocaine with its main bands at 1675 cm^{-1} is assigned to the ν(C=O) of the amide chemical group, 1655 and 1546 cm^{-1} attributed to δ(C–N–H) of the amine group and the band at 1476 cm^{-1} to the δ(C–CH_3) of the methyl group; and **Figure 5B** shows the spectra of the lidocaine-silica composite where the main bands of the drug can still be observed. However, the bands assigned to the carbonyl stretching and amino bending modes were shifted to 1683 and 1639 cm^{-1}, respectively, suggesting a potential medicine-silica network interaction through these chemical groups. Similar behavior was observed for the other drugs, considering the machine resolution of 4 cm^{-1}. Most of the pharmaceutical presented infrared band shifts toward higher wave numbers (bathochromic shift) when encapsulated, indicating they are interacting with the silica structure, resulting in a rearrangement of the chemical groups, which was confirmed by the rotational isomerism of the molecule. In addition, some of the bands were shifted for lower wave numbers (hypsochromic shift) reflecting a tension increase in the molecule rotational conformation, since the encapsulation process may incur in difficulty of the functional groups vibrational movements demanding more energy. Most of the nitrogen-related bands, for all samples, had their wave numbers shifted. These results denote possibility of a hydrogen bonding interaction through electron donation between these groups and silica network, as illustrated by **Figure 6**. Among the studied drugs, the exception was tetracycline which presented a shift in the OH group deformation band and so indicating an interaction by this group [40].

Chapter 2

Investigations of Phonons in Zinc Blende and Wurtzite by Raman Spectroscopy

Lin Sun, Lingcong Shi and Chunrui Wang

Additional information is available at the end of the chapter

http://dx.doi.org/10.5772/64194

Abstract

The importance of phonons and their interactions in bulk materials is well known to those working in the fields of solid-state physics, solid-state electronics, optoelectronics, heat transport, quantum electronic, and superconductivity. Phonons in nanostructures may act as a guide to research on dimensionally confined phonons and lead to phonon effects in nanostructures and phonon engineering. In this chapter, we introduce phonons in zinc blende and wurtzite nanocrystals. First, the basic structure of zinc blende and wurtzite is described. Then, phase transformation between zinc blende and wurtzite is presented. The linear chain model of a one-dimensional diatomic crystal and macroscopic models are also discussed. Basic properties of phonons in wurtzite structure will be considered as well as Raman mode in zinc blende and wurtzite structure. Finally, phonons in ZnSe, Ge, SnS_2, MoS_2, and Cu_2ZnSnS_4 nanocrystals are discussed on the basis of the above theory.

Keywords: phonons, zinc blende, wurtzite, Raman spectroscopy, molecular vibration

1. Zinc blende and wurtzite structure

Crystals with cubic/hexagonal structure are of major importance in the fields of electronics and optoelectronics. Zinc blende is typical face-centered cubic structure, such as Si, Ge, GaAs, and ZnSe. Wurtzite is typical hexagonal close packed structure, such as GaN and ZnSe. In particular, II–VI or III–V group semiconductor nanowires always coexist two structures, one cubic form with zinc blend (ZB) and another hexagonal form with wurtzite (WZ) structure. Sometimes, this coexistence between zinc blende and wurtzite structure leads to form twinning crystal during the phase transformation between zinc blende and wurtzite [1, 2].

1.1. Basic structure of zinc blende and wurtzite

The crystal structure of zinc selenide in the zinc blende structures is shown in **Figure 1**, which is regarded as two face-centered cubic (fcc) lattices displaced relative to each other by a vector $\frac{a}{4}\vec{i} + \frac{a}{4}\vec{j} + \frac{a}{4}\vec{k}$, where a is lattice constant. Close-packed planes of zinc blende are {111} along <111>, and the stacking is ...ABCABCA...; the adjacent plane separation is $\sqrt{3}/3a$. Along <100>, the sacking is ...ABABAB...; the adjacent plane separation is $a/2$. Along <110>, the sacking is ...ABABABA...; the adjacent plane separation is $\sqrt{2}/4a$. Zinc blende structures have eight atoms per unit cell.

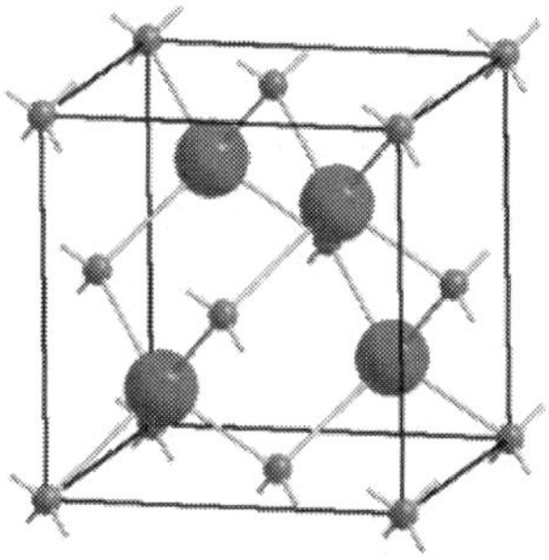

Figure 1. Zinc blende crystal structure.

Figure 2 is wurtzite structure of zinc selenium. Close-packed planes of wurtzite are {0001} along <0001>, and the stacking is ...ABABA.... Adjacent plane spacing is c/2. Wurtzite structures have four atoms per unit cell. In zinc blende, the bonding is tetrahedral. The wurtzite structure may be generated from zinc blende by rotating adjacent tetrahedra about their common bonding axis by an angle of 60° with respect to each other.

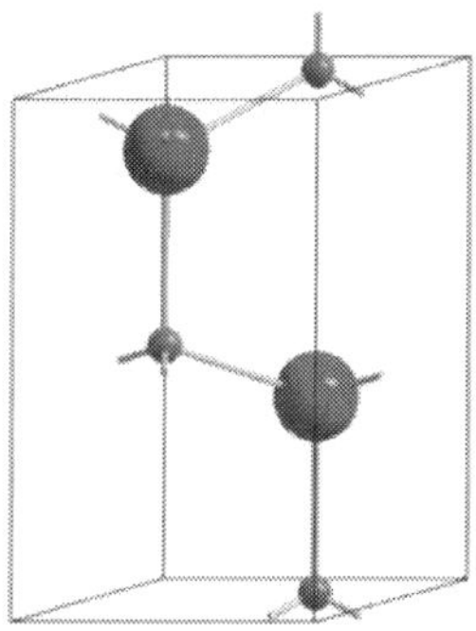

Figure 2. Wurtzite crystal structure.

http://dx.doi.org/10.5772/64194

As was apparent above, polar optical phonon vibrations produce electric fields and electric polarization fields that may be described in terms of Maxwell's equations and the driven-oscillator equations. Assume that the mass of the ions are M_+, M_-, the charges are $\pm Z_e$, displacements are $u_\pm$, the force constant is *k*,

$$M_+\ddot{\vec{u}}_+ = -k\vec{u} + Ze\vec{E}_e \tag{2-5}$$

$$M_-\ddot{\vec{u}}_- = k\vec{u} - Ze\vec{E}_e \tag{2-6}$$

where $\overrightarrow{E_e}$ is the effect electric field, $\vec{u} = \vec{u}_+ - \vec{u}_-$, then

$$\bar{M}\ddot{\vec{u}} = -k\vec{u} + Ze\vec{E}_e \tag{2-7}$$

where $\overline{M} = \dfrac{M_+ M_-}{M_+ + M_-}$ is reduced mass.

The lattice vibration is associated with the electric dipole moment generation, which can be described as follows,

$$\vec{P} = \frac{1}{\Omega}(Ze\vec{u} + \alpha\vec{E}_e) \tag{2-8}$$

where Ω is the volume of the primitive cell, and α is the electron polarization. Under the effective field approximation, the effective field can be described as follows,

$$\vec{E}_e = \vec{E} + \frac{\vec{P}}{3\varepsilon_0} \tag{2-9}$$

Replace the value of $\vec{p}$ in formula (2-9) with (2-8),

$$\vec{E}_e = \frac{3\varepsilon_0\Omega\vec{E} + Ze\vec{u}}{3\varepsilon_0\Omega - \alpha} \tag{2-10}$$

Then, take formula (2-7) and (2-9) into (2-10),

$$\ddot{\vec{u}} = A\vec{u} + B\vec{E} \tag{2-11}$$

$$\vec{P} = C\vec{E} + D\vec{u} \tag{2-12}$$

where

$$A = -\frac{k}{\bar{M}} + \frac{Z^2e^2}{\bar{M}(3\varepsilon_0\Omega - \alpha)} \tag{2-13}$$

$$B = \frac{3\varepsilon_0\Omega Ze}{\bar{M}(3\varepsilon_0\Omega - \alpha)} \tag{2-14}$$

$$C = \frac{3\varepsilon_0\alpha}{3\varepsilon_0\Omega - \alpha} \tag{2-15}$$

$$D = \frac{3\varepsilon_0 Ze}{3\varepsilon_0\Omega - \alpha} \tag{2-16}$$

formula (2-11) and (2-12) are the Huang equations, which are the basic equations of describing the vibrations of long wave in the polar crystals. From the formula (2-14) and (2-16), one can find that,

$$B = \frac{\Omega}{\bar{M}}D \tag{2-17}$$

When the system is under the high-frequency electric field, formula (2-12) reduces to

$$\vec{P} = C\vec{E} \tag{2-18}$$

For $(\infty) = 1 + \frac{\vec{P}}{\varepsilon_0\vec{E}}$, formula (2-18) can be written as follows,

$$C = \varepsilon_0[\varepsilon(\infty) - 1] \tag{2-19}$$

Compute the curl of formula (2-11) and solve the simultaneous equations of (2-12) and electrostatic equations $\nabla \times \vec{E} = 0$,

$$A = -\omega_0^2 \tag{2-20}$$

http://dx.doi.org/10.5772/64194

When the system is under the static electric field, ü = 0, and formula (2-11) reduces to

$$\vec{u} = -\frac{B}{A}\vec{E} \tag{2-21}$$

Take formula (2-21) into (2-12),

$$\vec{P} = (C - \frac{BD}{A})\vec{E} \tag{2-22}$$

Replace the electrostatic equation,

$$\vec{P} = [\varepsilon(0) - 1]\varepsilon_0\vec{E} \tag{2-23}$$

And take formula (2-23) and (2-20) into (2-22),

$$BD = [\varepsilon(0) - \varepsilon(\infty)]\varepsilon_0\omega_0^2 \tag{2-24}$$

Solve the simultaneous equations of (2-17) and (2-24) can obtain

$$B = (\frac{\Omega}{\bar{M}})^{\frac{1}{2}}\{[\varepsilon(0) - \varepsilon(\infty)]\varepsilon_0\}^{\frac{1}{2}}\omega_0 \tag{2-25}$$

$$D = (\frac{\bar{M}}{\Omega})^{\frac{1}{2}}\{[\varepsilon(0) - \varepsilon(\infty)]\varepsilon_0\}^{\frac{1}{2}}\omega_0 \tag{2-26}$$

Solve two simultaneous Maxwell and Huang equations,

$$\nabla \times \vec{E} = -\mu_0\frac{\partial \vec{H}}{\partial t} \tag{2-27a}$$

$$\nabla \times \vec{H} = \frac{\partial}{\partial t}(\varepsilon_0\vec{E} + \vec{P}) \tag{2-27b}$$

$$\nabla \cdot \vec{D} = 0 \tag{2-27c}$$

$$\nabla \cdot \vec{H} = 0 \tag{2-27d}$$

Assume the solution forms are

$$\vec{u} = \vec{u}_0 e^{i(q\cdot r-\omega t)} \quad \text{(2-28a)}$$

$$\vec{P} = \vec{P}_0 e^{i(q\cdot r-\omega t)} \quad \text{(2-28b)}$$

$$\vec{E} = \vec{E}_0 e^{i(q\cdot r-\omega t)} \quad \text{(2-28c)}$$

$$\vec{H} = \vec{H}_0 e^{i(q\cdot r-\omega t)} \quad \text{(2-28d)}$$

Take (2-28) into the Huang and Maxwell equations,

$$\vec{P}_0 = [-\frac{BD}{A+\omega^2} + C]\vec{E}_0 \quad \text{(2-29)}$$

$$(\vec{q}\cdot\vec{E}_0)[\varepsilon_0 + C - \frac{BD}{A+\omega^2}] = 0 \quad \text{(2-30)}$$

To the longitudinal wave, $\vec{q}\cdot\vec{E}_0 \neq 0$, (2-30) reduces to

$$\varepsilon_0 + C - \frac{BD}{A+\omega^2} = 0 \quad \text{(2-31)}$$

Take (2-19) (2-20) (2-25) (2-26) into (2-31)

$$\omega_{LO}^2 = \frac{\varepsilon(0)}{\varepsilon(\infty)}\omega_0^2 \quad \text{(2-32)}$$

Equation (2-23) is the dispersion relations of longitudinal wave, which is commonly called Lyddane-Sachs-Teller (LST) relationship. LST relation indicates that the frequency of longitudinal wave is a constant and independent on the wave vector.

Similarly, to the transverse wave, $\vec{q}\cdot\vec{E}_0 = 0$, solve the simultaneous equations of Maxwell and Huang equations,

$$\frac{q^2}{\mu_0\omega} = \omega\left(\varepsilon_0 + C - \frac{BD}{A+\omega^2}\right) \quad \text{(2-33)}$$

Replace the values of *A, B, C,* and *D* into (2-33),

$$\frac{c^2}{\omega^2}q^2 = \varepsilon(\infty) + \frac{\varepsilon(0)-\varepsilon(\infty)}{\omega_0^2 - \omega^2}\omega_0^2 \qquad (2\text{-}34)$$

Equation (2-34) is the dispersion relations of transverse wave. One can find that the frequency of transverse is dependent on the value of wave vector $\vec{q}$, but independent on its direction [4, 5].

2.2. Dispersion relations

One-dimensional diatomic model can be regarded as the simple double lattice. In the simple linear chain model, it is assumed that only nearest neighbors are coupled, and that the interaction between these atoms is described by Hooke's law; the spring constant α is taken to be that of a harmonic oscillator. Thus, the kinematical equations are established,

$$m\ddot{u}_{2n} = -\beta(2\ddot{u}_{2n} - \ddot{u}_{2n+1} - \ddot{u}_{2n-1}) \qquad (2\text{-}35a)$$

$$M\ddot{u}_{2n+1} = -\beta(2\vec{u}_{2n+1} - \vec{u}_{2n+2} - \vec{u}_{2n}) \qquad (2\text{-}35b)$$

where m and M are the mass of the adjacent atoms *and* $\vec{u}_{2n-1}$, $\vec{u}_{2n}$, $\vec{u}_{2n+1}$, *and* , $\vec{u}_{2n+2}$ are the displacements of the atoms at the position of $2n$-1, $2n$, $2n + 1$, and $2n + 2$, respectively. β is the force constant. The solution forms of (2-35) can be written as follows

$$\vec{u}_{2n} = A_1 e^{i[(\omega t-(2n)\vec{a}\cdot\vec{q}]} \qquad (2\text{-}36a)$$

$$\vec{u}_{2n+1} = A_2 e^{i[(\omega t-(2n+1)\vec{a}\cdot\vec{q}]} \qquad (2\text{-}36b)$$

where $\vec{q}$ is the phonon wave vector and ω is its frequency. Take formulas (2-36a) and (2-36b) into formulas (2-35a) and (2-35b),

$$-m\omega^2 A_1 = \beta(e^{-i\vec{a}\cdot\vec{q}} + e^{i\vec{a}\cdot\vec{q}})A_2 - 2\beta A_1 \qquad (2\text{-}37)$$

$$-M\omega^2 A_2 = \beta(e^{-i\vec{a}\cdot\vec{q}} + e^{i\vec{a}\cdot\vec{q}})A_1 - 2\beta A_2 \qquad (2\text{-}38)$$

Eliminating A_1 and A_2,

$$\omega^2 = \beta \frac{M+m}{Mm} \left\{ 1 \pm [1 - \frac{4Mm}{(M+m)^2} \sin^2(\vec{a} \cdot \vec{q})]^{1/2} \right\} \tag{2-39}$$

The relationship between frequency and wave vector is commonly called dispersion relation [5].

3. Basic properties of phonons in wurtzite structure

In this section, we discuss the phonon effects in wurtzite structure. The crystalline structure of a wurtzite material is depicted in **Figure 2**. There are four atoms in the unit cell. Thus, the total number of optical modes in the long-wavelength limit is nine: three longitudinal optic (LO) and six transverse optic (TO). In these optical modes, there are only three polar optical vibration modes. According to the group theory, the wurtzite crystal structure belongs to the space group C_{6v}^4, and the phonon modes at Γ point of the Brillouin zone are represented by the following irreducible representations:

$$\Gamma = 2A_1 + 2B + 2E_1 + 2E_2$$

Due to the anisotropy of wurtzite structure, the vibrational frequency of oscillates parallel and perpendicular to the optical axis is denoted by ω_{eT} and ω_{oT}, and the corresponding dielectric constants are denoted by ε_{es}, $\varepsilon_{e\infty}$ and ε_{os}, $\varepsilon_{o\infty}$. The corresponding components can be written as the form of Huang equations, and the dispersion relation can be obtained by solving two simultaneous equations of Maxwell and Huang equations.

$$\frac{q^2c^2}{\omega^2} = \varepsilon_0 = \frac{\omega_{oT}^2 \varepsilon_{os} - \omega^2 \varepsilon_{o\infty}}{\omega_{oT}^2 - \omega^2} \tag{3-1}$$

$$\frac{q^2c^2}{\omega^2} = \varepsilon_\theta = \frac{(\frac{\omega_{eT}^2 \varepsilon_{es} - \omega^2 \varepsilon_{e\infty}}{\omega_{oT}^2 - \omega^2})(\frac{\omega_{oT}^2 \varepsilon_{os} - \omega^2 \varepsilon_{o\infty}}{\omega_{oT}^2 - \omega^2})}{(\frac{\omega_{eT}^2 \varepsilon_{es} - \omega^2 \varepsilon_{e\infty}}{\omega_{eT}^2 - \omega^2})\cos^2\theta + (\frac{\omega_{oT}^2 \varepsilon_{os} - \omega^2 \varepsilon_{o\infty}}{\omega_{oT}^2 - \omega^2})\sin^2\theta} \tag{3-2}$$

where ε_0 and ε_θ is the dielectric constants of ordinary and extraordinary wave, is the included angle between wave vector and optical axis.

When the wave vector is parallel to the optical axis, θ = 0, formula (3-2) reduce to

http://dx.doi.org/10.5772/64194

$$\varepsilon_\theta = \frac{\omega_{oT}^2 \varepsilon_{os} - \omega^2 \varepsilon_{o\infty}}{\omega_{oT}^2 - \omega^2} \tag{3-3}$$

which is the same as formula (3-1). When the wave vector is perpendicular to the optical axis, $\theta = 90$, formula (3-2) reduces to

$$\varepsilon_\theta = \frac{\omega_{eT}^2 \varepsilon_{es} - \omega^2 \varepsilon_{o\infty}}{\omega_{eT}^2 - \omega^2} \tag{3-4}$$

Formula (3-4) indicates that the extraordinary wave is transverse wave when the wave vector is perpendicular to the optical axis.

When $q \gg \omega/c$, formulas (3-1) and (3-2) can be rewritten as follows,

$$\omega = \omega_{oT} \tag{3-5}$$

and

$$\left(\frac{\omega_{eT}^2 \varepsilon_{es} - \omega^2 \varepsilon_{e\infty}}{\omega_{eT}^2 - \omega^2}\right)\cos^2\theta + \left(\frac{\omega_{oT}^2 \varepsilon_{os} - \omega^2 \varepsilon_{o\infty}}{\omega_{oT}^2 - \omega^2}\right)\sin^2\theta = 0 \tag{3-6}$$

Formula (3-5) indicates that frequency of ordinary phonon is independent on the wave vector q. Formula (3-6) indicates that the frequency of extraordinary phonon is dependent on the orientation of the wave vector, but independent on its value.

It is most convenient to divide uniaxial crystals into two categories: (a) the electrostatic forces dominate over the anisotropy of the interatomic forces and (b) the short-range interatomic forces are much greater than the electrostatic forces. It has been turned out that crystals with the wurtzite symmetry fall into the first category. In this case, $|\omega_{eT} - \omega_{oT}| \ll |\omega_{eL} - \omega_{oT}|$ and $|\omega_{oL} - \omega_{oT}|$, $\varepsilon_{e\infty} \approx \varepsilon_{o\infty} = \varepsilon_\infty$, formula (3-5) reduces to

$$\left(\frac{\omega_{eL}^2 - \omega^2}{\omega_{eT}^2 - \omega^2}\right)\cos^2\theta + \left(\frac{\omega_{oL}^2 - \omega^2}{\omega_{oT}^2 - \omega^2}\right)\sin^2\theta = 0 \tag{3-7}$$

thus,

$$\omega^2 \approx \omega_{eT}^2 \sin^2\theta + \omega_{oT}^2 \cos^2\theta \tag{3-8}$$

and

$$\omega^2 \approx \omega_{0L}^2 \sin^2\theta + \omega_{eT}^2 \cos^2\theta \tag{3-9}$$

4. Raman mode in zinc blende and wurtzite structure

Raman spectroscopy is a non-destructive technical tool used to gain information about the phonon behavior of the crystal lattice through the frequency shift of the inelastically scattered light from the near surface of the sample. It is well known that different crystal phases have different vibrational behaviors, so the measured Raman shifts of different phases are mostly unique and can be seen as fingerprints for the respective phases. This provides the possibility of detecting different phases in a sample. It has been developed to be a versatile tool for the characterization of semiconductors leading to detailed information on crystal structure, phonon dispersion, electronic states, composition, strain, and so on of semiconductor nano-structures.

In a zinc blende structure, the space group of the cubic unit cell is $F43m(\mathrm{T}_\mathrm{d}^2)$ containing four formula units. The primitive unit cell contains only one formula per unit cell, and hence, there are three optical branches to the phonon dispersion curves. As there is no center of inversion in the unit cell, the zone-center transverse optic (TO) and longitudinal optic (LO) optic modes are Raman active. The optic mode is polar so that the macroscopic field lifts the degeneracy, producing a non-degenerate longitudinal mode that is at a higher frequency than the two transverse modes.

The wurtzite crystal structure belongs to the space group $\mathrm{C}_{6\mathrm{v}}^4$ and group theory predicts zone-center optical modes are A_1, $2B_1$, E_1, and $2E_2$. The A_1 and E_1 modes and the two E_2 modes are Raman active, whereas the B modes are silent. The A and E modes are polar, resulting in a splitting of the LO and the TO modes [6].

5. Phonons in ZnSe, Ge, SnS_2, MoS_2, and Cu_2ZnSnS_4 nanocrystals

In addition to the attached references, this chapter is primarily written on the basis of our research works. Here, we select ZnSe, Ge nanowires and CdSe/Ge-based nanowire hetero-structures, two-dimensional semiconductors SnS_2 and MoS_2, and candidate absorber materials of thin-film solar cells Cu_2ZnSnS_4. These examples will help us to understand the phonons behaviors in nanostructures.

It is well known that ZnSe has two structures: cubic zinc blende (ZB) and hexagonal wurtzite (WZ) due to the difference of the stacking sequence of successive layers, whereas Ge has diamond structure. SnS_2 and MoS_2 belong to the wide family of compounds with layered structures. SnS_2 crystal is isostructural to the hexagonal CdI_2-type structure. MoS_2 usually consists of a mixture of two major polytypes of similar structure, 2H (hexagonal) and 3R (rhombohedral), with the former being more abundant. As for quaternary Cu_2ZnSnS_4 (CZTS), the parent binary II-VI semiconductors adopt the cubic zinc blende structure, and the ternary I-III-VI_2 compounds can be generated by mutating the group II atoms into pairs of group I and III atoms. The quaternary CZTS materials are formed by replacing the two In (III) atoms with Zn (II) and Sn (IV), respectively (see **Figure 5**).

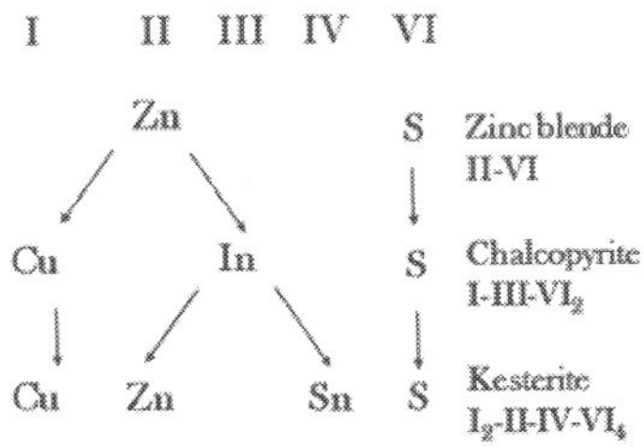

Figure 5. Evolution of multinary compounds.

We use Raman spectroscopy to identify crystal structure of ZnSe one-dimensional material (**Figure 6**). In sample S3, the Raman peaks at 204 and 251 cm^{-1} are attributed to the scatterings of the transverse optic (TO) and longitudinal optic (LO) phonon modes of ZnSe, respectively. A strong peak at 232 cm^{-1}, between the TO and LO phonons, is thought to be surface mode. The Raman peak at ~176 cm^{-1} is attributed to the hexagonal phase E_1(TO) mode of ZnSe, which is inhibited in Raman spectrum (RS) of ZB ZnSe. Compared with S3, Raman peaks at 205.6 (TO mode) and 252 cm^{-1} (LO mode) of S1 show tiny blue-shift. However, in S1, there is no Raman peak corresponding to the surface mode, as well as E_1 (TO) mode, which is suppressed in the

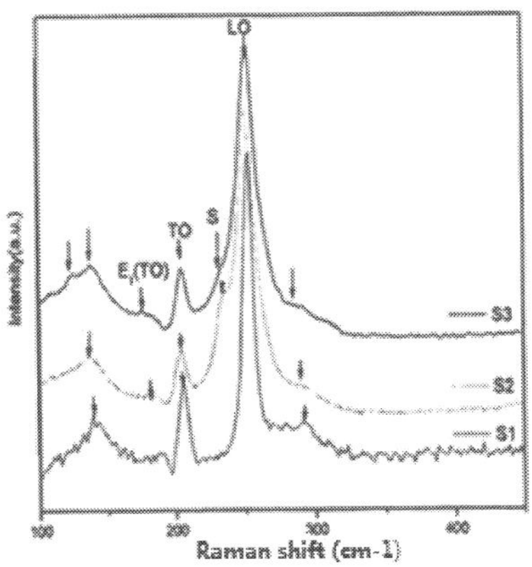

Figure 6. Room temperature Raman spectra of ZnSe. S1, S2, and S3 stand for ZB, coexist of ZB and WZ, WZ ZnSe nanostructure.

ZB phase. This indicates the existence of ZB phase in S1. Thus, structure of the sample can be shown through RS, and we got S1-ZB phase, S3-WZ, S2 the coexist of ZB and WZ [7] (cm^{-1}).

Figure 7 shows the room temperature RS of CdSe/Ge-based nanowires. The LO mode of Ge in CdSe-Ge (or CdSe-Ge-CdSe), -CdSe-Ge core/polycrystalline Ge sheath, and -Ge-GeSe heterostructural nanowires has a downshift by 8, 5, and 2 cm^{-1} in comparison with that of the bulk counterpart Ge (299 cm^{-1}), respectively. With regard to the microstructure of heterostructural nanowires, the downshift of the LO mode may be caused by tensile stress, which affects the Raman line by a downshift. And the different shift scales are attracted by the different sizes of the Ge subnanowires and Ge nanocrystalline [8].

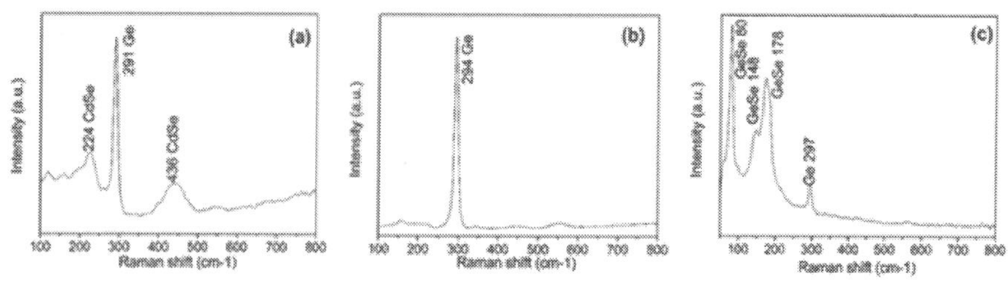

Figure 7. Raman spectrum of (a) CdSe-Ge biaxial nanowires and CdSe-Ge-CdSe triaxial nanowires. (b) CdSe-Ge biaxial nanowire core/polycrystalline Ge sheath heterostructures. (c) Ge-GeSe biaxial heterostructure nanowires.

The individual layer in SnS_2 is known as an S-Sn-S sandwich bonded unit. Each Sn atom is octahedrally coordinated with six nearest neighbor sulfur atoms, while each S atom is nested at the top of a triangle of Sn atoms. The sandwich layers in the elementary cell occur along the *c* axis and bonded together by Vander Waals forces. The normal modes of vibration in SnS_2 are given by the irreducible representations of the D_{3d} point group at the center of the Brillouin zone: $\Gamma = A_{1g} + E_g + 2A_{2u} + 2E_{2u}$. Two Raman-active modes (A_{1g} and E_g) and two IR-active modes (A_{2u} and E_u) are found. In view of the existence of an inversion center, the IR- and Raman-active modes are mutually exclusive. On the other hand, six atoms in the unit cell of SnS_2 extend over two sandwich layers. Eighteen normal vibration modes can be represented by the following irreducible form: $\Gamma = 3A_1 + 3B_1 + 3E_1 + 3E_2$. Based on the analysis above, there are six modes, which are both IR- and Raman-active, belonging to A_1 and E_1, and three Raman-active modes belonging to E_2. The B_1 modes are silent, while the three acoustic modes belong to A_1 and E_1 [9].

The RS of β-SnS_2 nanocrystal is illustrated in our former work [10]. The spectra show one first-order peak at 312 cm^{-1} that corresponding to A_{1g} mode. The RS of as-prepared SnS_2 shows a slight redshift in comparison with that of bulk materials (peak at 317 cm^{-1}). The redshift of phonon peaks is due to spatial confinement of phonon modes. The first-order E_g mode (peak at 208 cm^{-1}) cannot be observed, which likely results from a nanosize effect. A wide peak between 450 and 750 cm^{-1}, which only observed in the bulk materials at lower temperature, may be attributed to second-order effects.

The phonon dispersion of single-layer MoS_2 has three acoustic and six optical branches derivatized from the nine vibrational modes at the Γ point. The three acoustic branches are the in-plane longitudinal acoustic (LA), the transverse acoustic (TA), and the out-of-plane acoustic (ZA) modes. The six optical branches are two in-plane longitudinal optical (LO_1 and LO_2), two in-plane transverse optical (TO_1 and TO_2), and two out-of-plane optical (ZO_1 and ZO_2) branches.

For 2L and bulk MoS_2, there are 18 phonon branches, which are split from nine phonon branches in 1LMoS_2. The phonon dispersions of 1L and bulk MoS_2 are very similar, except for the three new branches below 100 cm^{-1} in bulk because of interlayer vibrations. There are similar optical phonon dispersion curves for 1L, 2L, and bulk MoS_2 because of the weak Vander Waals interlayer interactions in 2L and bulk MoS_2 [11].

Raman spectroscopy is also used to accurately identify the layer number of MoS_2. The frequency difference between out-of-plane A_{1g} and in-plane $E_{2g}{}^1$ mode of MoS_2 is denoted as . $\Delta\omega$. From monolayer to bulk MoS_2, $\Delta\omega$ monotonically increases from 19.57 cm^{-1} to 25.5 cm^{-1}. In our work [12], two strong peak at ~379 cm^{-1} and ~402 cm^{-1} can be assigned as in-plane $E_{2g}{}^1$ mode and out-of-plane A_{1g} mode of MoS_2, respectively, which has a redshift in comparison with that of the bulk MoS_2. The $\Delta\omega$ is about 23 cm^{-1}, indicating that the as-grown MoS_2 contains tri-layer MoS_2.

The phonon dispersion and density-of-states curves along the principal symmetry directions of kesterite CZTS were calculated using a density functional theory by Khare et al. [13]. The phonon states around 50–160 cm^{-1} are mainly composed of vibrations of the three metal cations with some contribution from the sulfur anions. The phonon states around 250–300 cm^{-1} are mainly composed of vibrations of the Zn cations and S anions with some contribution from the Cu cations. The phonon states from 310 to 340 cm^{-1} are mainly a result of vibrations of S anions, whereas those from 340 to 370 cm^{-1} are composed of the vibrations of Sn cations and S anions.

To more exactly confirm secondary phases in Cu_2-II-IV-VI_4 semiconductors, Raman scattering studies have been extensively performed. From the vibrational point of view, the zone-center phonon representation of the kesterite structure space group $I\bar{4}$ is constituted of 21 optical modes: $\Gamma = 3A + 6B + 6E_1 + 6E_2$, where 12$B$, E_1, and E_2 modes are infrared active, whereas 15A, B, E_1, and E_2 modes are Raman active. According to our work [14], the single peak at about 328 cm^{-1} of Raman spectrum of the as-prepared CZTS nanocrystals can be assigned to breathing mode of sulfur atoms around metal ions in CZTS. Moreover, Raman spectrum of CZTS has about 8 cm^{-1} redshifts compared with that of the responding bulk counterpart which may be due to a smaller size effect.

In our work of fabrication of $Cu_2ZnSnS_xSe_{4-x}$ solid solution nanocrystallines [15], RS revealed that vibrating modes were modulated by x-values. The peak position of 170, 189, and 229 cm^{-1} shifted to higher frequency with increasing x-value in CZTSSe, respectively. Those peaks completely disappeared when $x = 4$. Moreover, a wide peak located at about 330 cm^{-1} appeared when $x > 0$ and the relative intensity increased with increasing x-value. Such results indicate that Se elements were gradually replaced by S elements in CZTSSe solid solution system.

Acknowledgements

This work was supported by the National Natural Science Foundation of China under Grant Nos. 11174049 and 61376017.

Author details

Lin Sun, Lingcong Shi and Chunrui Wang*

*Address all correspondence to: crwang@dhu.edu.cn

Department of Applied Physics, Donghua University, Shanghai, China

References

[1] Xu, J., Wang, C., Wu, B., Xu, X., Chen, X., Oh, H., Baek, H., & Yi, G. C. Twinning effect on photoluminescence spectra of ZnSe nanowires. Journal of Applied Physics. 2014;116(17):174303. doi:10.1063/1.4900850.

[2] Xu, J., Wang, C., Lu, A., Wu, B., Chen, X., Oh, H., Baek, H., Yi, G., & Ouyang, L. Photoluminescence of excitons and defects in ZnSe-based longitudinal twinning nanowires. Journal of Physics D: Applied Physics. 2014;47(48):485302. doi: 10.1088/0022-3727/47/48/485302.

[3] Xu, J., Lu, A., Wang, C., Zou, R., Liu, X., Wu, X., Wang, Y., Li, S., Sun, L., Chen, X., Oh, H., Baek, H., Yi, G., & Chu, J. ZnSe-based longitudinal twinning nanowires. Advanced Engineering Materials. 2014;16(4):459–465. doi:10.1002/adem.201300405.

[4] Zhang, G. et al. Lattice Vibration Spectroscopy. Beijing: Higher Education Press; 2001

[5] Huang, K., & Han, R. Solid-State Physics. Beijing: Higher Education Press; 1988.

[6] Stroscio, M. A., & Dutta, M. Phonons in Nanostructures. Cambridge: Cambridge University Press; 2005.

[7] Wang, H. Luminescence and vibrating properties of Zn-based group II-VI nanostructures. Master's thesis. Donghua University. 2012.

[8] Cai, J. Controllable synthesis and vibrating properties of CdSe based heterostructure nanowires. Master's thesis. Donghua University. 2011.

[9] Smith, A. J., Meek, P. E., & Liang, W. Y. Raman scattering studies of SnS_2 and $SnSe_2$. Journal of Physics C: Solid State Physics. 1977;10(8):1321. doi:10.1088/0022-3719/10/8/035.

[10] Wang, C. Synthesis and properties of iodine and sulfide nanomaterials. PhD thesis. University of Science and Technology of China. 2002.

[11] Zhang, X., Qiao, X. F., Shi, W., Wu, J. B., Jiang, D. S., & Tan, P.H. Phonon and Raman scattering of two-dimensional transition metal dichalcogenides from monolayer, multilayer to bulk material. Chemical Society Reviews. 2015;44(9):2757–2785. doi: 10.1039/C4CS00282B.

[12] Fu, Y. Fabrication and properties of $ZnO/CdS/MoS_2$ heterostructure nanorod arrays. Master's thesis. Donghua University. 2016.

[13] Khare, A., Himmetoglu, B., Johnson, M., Norris, D. J., Cococcioni, M., & Aydil, E. S. Calculation of the lattice dynamics and Raman spectra of copper zinc tin chalcogenides and comparison to experiments. Journal of Applied Physics. 2012;111(8):083707. doi: 10.1063/1.4704191.

[14] Wang, C., Cheng, C., Cao, Y., Fang, W., Zhao, L., & Xu, X. Synthesis of Cu_2ZnSnS_4 nanocrystallines by a hydrothermal route. Japanese Journal of Applied Physics. 2011;50(6R):065003. doi:10.1143/JJAP.50.065003.

[15] Cao, Y.. Fabrication and characterization of $Cu_2ZnSnS_xSe_{4-x}$ thin film solar cell absorber layer material. Master's thesis, Donghua University. 2012

1.2. Phase transformation between zinc blende and wurtzite

Research into controlling nanowire crystal structure has intensified. Several reports address the diameter dependency of nanowire crystal structure, with smaller diameter nanowires tending toward a WZ phase and larger diameter nanowires tending toward a ZB phase. Allowing for ZnSe, two phases, zinc blende (ZB) and wurtzite (WZ), exist, and the (111) faces of ZB phase are indistinguishable from and match up with the (001) faces of WZ phase, the subtle structural differences of which lead to the attendant small difference in the internal energies (~5.3 meV/atom for ZnSe). The WZ-ZB phase transformation is considered to be caused by the crystal plane slip. Take the formation of ZnSe longitudinal twinning nanowires, for example [3]. Structurally, the (001) planes of WZ and the (111) planes of ZB are their corresponding close packing planes. ABAB stacking for WZ and ABCABC stacking for ZB are shown in **Figure 3a** and **b**, respectively. It was noteworthy that the arrangement of atoms in A/B packing planes was different in WZ phase. So the phase transition could not be realized until the smaller Zn atoms moved to the interspaces provided by three neighboring bigger Se atoms, within the plane B. In this case, the new layers B' were obtained, and then, the slip occurs between neighboring planes A and B′ by $\frac{1}{3}\vec{a}+\frac{2}{3}\vec{b}$, that is <120> direction, indicated in **Figure 3a**.

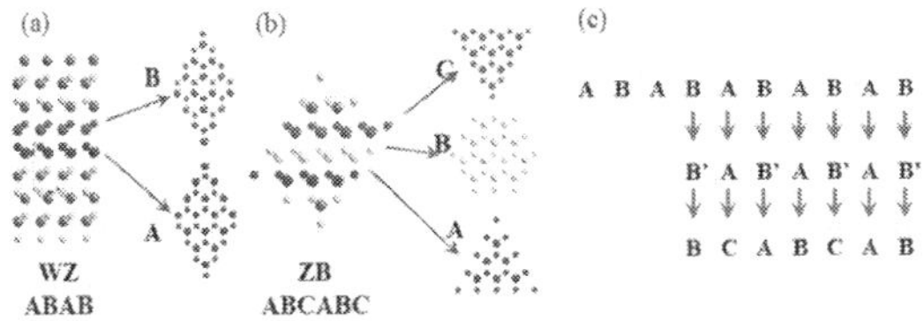

Figure 3. Phase transformation between zinc blende and wurtzite. (a) The arrangement of atoms in WZ phase; (b) The arrangement of atoms in ZB phase; (Se is shown with the bigger sphere and Zn is shown in little one.) (c) The stacking sequence schematic model showing the phase transformation process from WZ phase to ZB phase.

Generally, there are three equivalent directions to realized the slip, which are <120>, <$\overline{21}$0>, and <1$\overline{1}$0>. Such a displacement could be indicated in **Figure 3c**, and the ZB structure could be obtained through the slip between every second close-packed layer in the WZ sequence to form the ABC stacking.

2. Linear-chain model and macroscopic models

To the simple double lattice, lattice vibration can be described by the one-dimensional diatomic model. The linear-chain model of a diatomic crystal is based upon a system of two atoms with masses, m and M, placed along a one-dimensional chain as depicted in **Figure 4**. The separation between the atomics is "a", and the vibration in the vicinity of their equilibrium position is treated as the simple harmonic vibration. The properties of optical phonon can be described based on the macroscopic fields. It is the model based on the Huang and Maxwell equations, which has great utility in describing the phonons in the uniaxial crystals such as wurtzite crystals.

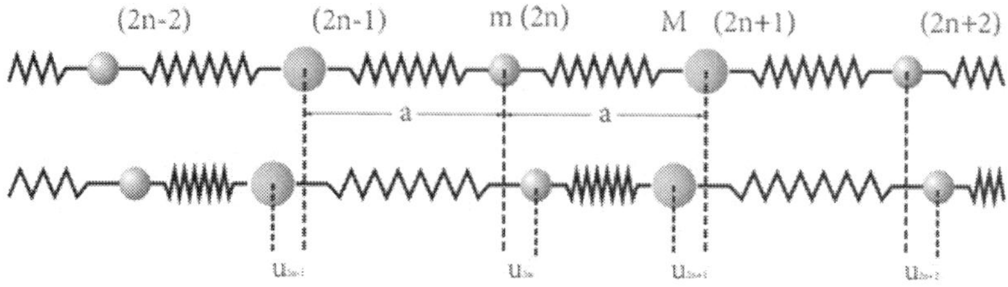

Figure 4. One-dimensional diatomic linear-chain model.

2.1. Polar semiconductors

Polar semiconductor is the crystal that consists of different ions. In polar semiconductor, the lattice vibration is associated with the electric dipole moment and electric field generation. Assume that the vibration frequency is ω, wave vector is $\vec{q}$, then the intensity of polarization can be written as follows,

$$\vec{P} = \vec{P}_0 e^{i(\omega t - \vec{q}\cdot\vec{r})} \tag{2-1}$$

Solve the simultaneous formula (2-1) and Maxwell equations can obtain,

$$\vec{E} = \frac{\omega^2 \vec{P} - \vec{q}c^2(\vec{q}\cdot\vec{P})}{\varepsilon_0(q^2c^2 - \omega^2)} \tag{2-2}$$

To longitudinal polarity lattice mode, $\vec{p}//\vec{q}$, formula (2-2) can be simplified as follows,

$$\vec{E}_L = -\frac{\vec{P}}{\varepsilon_0} \tag{2-3}$$

To transverse polarity lattice mode, $\vec{p} \perp \vec{q}$, formula (2-2) can be simplified as follows,

$$\vec{E}_T = \frac{\omega^2}{\varepsilon_0(q^2c^2 - \omega^2)}\vec{P} \tag{2-4}$$

Liquid Chromatography for the Determination of Trace Pentachlorophenol in Water Samples. Journal of Chromatography A, 1100 (2005) 131–136. DOI: 10.1016/j.chroma.2005.09.035.

[42] A.M. Chrzanowska, A. Poliwoda, P.P. Wieczorek, Surface Molecularly Imprinted Silica for Selective Solid-Phase Extraction of Biochanin A, Daidzein and Genistein from Urine Samples. Journal of Chromatography A, 1392 (2015) 1–9. DOI: 10.1016/j.chroma.2015.03.015.

[43] Z.C. Li, H.T. Fan, Y. Zhang, M.X. Chen, Z.Y. Yu, X.Q. Cao, T. Sun, Cd(II)-Imprinted Polymer Sorbents Prepared by Combination of Surface Imprinting Technique with Hydrothermal Assisted Sol–Gel Process for Selective Removal of Cadmium(II) from Aqueous Solution. Chemical Engineering Journal, 171 (2011) 703–710. DOI: 10.1016/j.cej.2011.05.023.

[44] Y.M. Ren, W.Q. Ma, J. Ma, Q. Wen, J. Wang, F.B. Zhao, Synthesis and Properties of Bisphenol A Molecular Imprinted Particle for Selective Recognition of BPA from Water. Journal of Colloid and Interface Science, 367 (2012) 355–361. DOI: 10.1016/j.jcis.2011.10.009.

[45] L.B. Capeletti, I.M. Baibich, I.S. Butler, J.H.Z. dos Santos, Infrared and Raman Spectroscopic Characterization of Some Organic Substituted Hybrid Silicas. Spectrochimica Acta Part A – Molecular and Biomolecular Spectroscopy, 133 (2014) 619–625. DOI: 10.1016/j.saa.2014.05.072.

[46] A. Fidalgo, L.M. Ilharco, The Defect Structure of Sol–gel-Derived Silica/Polytetrahydrofuran Hybrid Films by FTIR. Journal of Non-Crystalline Solids, 283 (2001) 144–154. DOI: 10.1016/s0022-3093(01)00418-5.

[47] M. Bukleski, V. Ivanovski, E. Hey-Hawkins, A Direct Method of Quantification of Maximal Chemisorption of 3-Aminopropylsilyl Groups on Silica Gel Using Drift Spectroscopy. Spectrochimica Acta Part A – Molecular and Biomolecular Spectroscopy, 149 (2015) 69–74. DOI: 10.1016/j.saa.2015.04.026.

[48] N. Tasinato, D. Moro, P. Stoppa, A.P. Charmet, P. Toninello, S. Giorgianni, Adsorption of F2C=CFCL on TiO2 Nano-Powder: Structures, Energetics and Vibrational Properties from Drift Spectroscopy and Periodic Quantum Chemical Calculations. Applied Surface Science, 353 (2015) 986–994. DOI: 10.1016/j.apsusc.2015.07.006.

[49] V. Ivanovski, M. Bukleski, M. Madalska, E. Hey-Hawkins, Vibrational Spectra of a Ferrocenyl Phosphine Derivative Chemisorbed on 3-Aminopropylsilyl Modified Silica Gel. Vibrational Spectroscopy, 69 (2013) 57–64. DOI: 10.1016/j.vibspec.2013.09.009.

[50] J.F. McClelland, R.W. Jones, S. Luo, L.M. Seaverson, A Practical Guide to FTIR Photoacoustic Spectroscopy, In: P.B. Coleman (Ed.) Practical Sampling Techniques for Infrared Analysis, CRC Press, Boca Raton, 1993, pp. 320.

[51] R. Kizil, J. Irudayaraj, Fourier Transform Infrared Photoacoustic Spectroscopy (FTIR-PAS), In: G.C.K. Roberts (Ed.) Encyclopedia of Biophysics, Springer Berlin Heidelberg, Berlin, Heidelberg, 2013, pp. 840–844.

[52] Y. Gao, N.R. Choudhury, N. Dutta, J. Matisons, M. Reading, L. Delmotte, Organic–inorganic Hybrid from Ionomer Via Sol–gel Reaction. Chemistry of Materials, 13 (2001) 3644–3652. DOI: 10.1021/cm010179s.

[53] P.F. Bernath, 6 Infrared Emission Spectroscopy. Annual Reports Section "C" (Physical Chemistry), 96 (2000) 177–224. DOI: 10.1039/B001200I.

[54] R. Brambilla, J.H.Z. dos Santos, M.S.L. Miranda, R.L. Frost, Thermal Stability of Octadecylsilane Hybrid Silicas Prepared by Grafting and Sol–gel Methods. Thermochimica Acta, 469 (2008) 91–97. DOI: 10.1016/j.tca.2008.01.010.

[55] C. Martinet, V. Martinez, C. Coussa, B. Champagnon, M. Tomozawa, Radial Distribution of the Fictive Temperature in Pure Silica Optical Fibers by Micro-Raman Spectroscopy. Journal of Applied Physics, 103 (2008) 4. DOI: 10.1063/1.2905321.

[56] O. Cristini-Robbe, K. Raulin, F. Dubart, R. Bernard, C. Kinowski, N. Damene, I. El Yazidi, A. Boed, S. Turrell, Porous Silica Supports for Micro-Raman Spectroscopic Studies of Individual Living Cells. Journal of Molecular Structure, 1050 (2013) 232–237. DOI: 10.1016/j.molstruc.2013.06.063.

[57] L. Brigo, V. Auzelyte, K.A. Lister, J. Brugger, G. Brusatin, Phenyl-Bridged Polysilsesquioxane Positive and Negative Resist for Electron Beam Lithography. Nanotechnology, 23 (2012) 7. DOI: 10.1088/0957-4484/23/32/325302.

Silica Nanoparticle Content. Surface & Coatings Technology, 265 (2015) 106–116. DOI: 10.1016/j.surfcoat.2015.01.050.

[20] M. Gharagozlou, R. Naderi, Z. Baradaran, Effect of Synthesized NiFe2O4-Silica Nanocomposite on the Performance of an Ecofriendly Silane Sol–gel Coating. Progress in Organic Coatings, 90 (2016) 407–413. DOI: 10.1016/j.porgcoat.2015.08.009.

[21] V. Dalmoro, J.H.Z. dos Santos, L.M. Baibich, I.S. Butler, E. Armelin, C. Aleman, D.S. Azambuja, Improving the Corrosion Performance of Hybrid Sol–Gel Matrix by Modification with Phosphonic Acid. Progress in Organic Coatings, 80 (2015) 49–58. DOI: 10.1016/j.porgcoat.2014.11.018.

[22] S.A. Mahadik, F. Pedraza, R.S. Vhatkar, Silica Based Superhydrophobic Coating for Long-Term Industrial and Domestic Applications. Journal of Alloys and Compounds, 663 (2016) 487–493. DOI: 10.1016/j.jallcom.2015.12.016.

[23] A.C. Power, A. Barrett, J. Abubakar, L.J. Suarez, L. Ryan, D. Wencel, T. Sullivan, F. Regan, Versatile Self-Cleaning Coating Production through Sol–gel Chemistry. Advanced Engineering Materials, 18 (2016) 76–82. DOI: 10.1002/adem.201500112.

[24] R. Arreche, N. Bellotti, M. Blanco, P. Vazquez, Improved Antimicrobial Activity of Silica-Cu Using a Heteropolyacid and Different Precursors by Sol–Gel: Synthesis and Characterization. Journal of Sol-Gel Science and Technology, 75 (2015) 374–382. DOI: 10.1007/s10971-015-3710-8.

[25] W.L. Storm, J.A. Johnson, B.V. Worley, D.L. Slomberg, M.H. Schoenfisch, Dual Action Antimicrobial Surfaces Via Combined Nitric Oxide and Silver Release. Journal of Biomedical Materials Research, Part A, 103 (2015) 1974–1984. DOI: 10.1002/jbm.a.35331.

[26] M. Fedel, E. Callone, S. Dire, F. Deflorian, M.G. Olivier, M. Poelman, Effect of Na-Montmorillonite Sonication on the Protective Properties of Hybrid Silica Coatings. Electrochimica Acta, 124 (2014) 90–99. DOI: 10.1016/j.electacta.2013.11.006.

[27] W. Ochedzan-Siodlak, K. Dziubek, Metallocenes and Post-Metallocenes Immobilized on Ionic Liquid-Modified Silica as Catalysts for Polymerization of Ethylene. Applied Catalysis A – General, 484 (2014) 134–141. DOI: 10.1016/j.apcata.2014.07.016.

[28] S. Ray, P. Das, A. Bhaumik, A. Dutta, C. Mukhopadhyay, Covalently Anchored Organic Carboxylic Acid on Porous Silica Nano Particle: A Novel Organometallic Catalyst (PSNP-Ca) for the Chromatography-Free Highly Product Selective Synthesis of Tetrasubstituted Imidazoles. Applied Catalysis A – General, 458 (2013) 183–195. DOI: 10.1016/j.apcata.2013.03.024.

[29] M.S. Saraiva, C.I. Fernandes, T.G. Nunes, C.D. Nunes, M.J. Calhorda, New MO(II) Complexes in MCM-41 and Silica: Synthesis and Catalysis. Journal of Organometallic Chemistry, 751 (2014) 443–452. DOI: 10.1016/j.jorganchem.2013.07.081.

[30] L.D. Li, E. Abou-Hamad, D.H. Anjum, L. Zhou, P.V. Laveille, L. Emsley, J.M. Basset, Well-Defined Mono(Eta(3)-Allyl)Nickel Complex MoNi(Eta(3)-C3H5) (M = Si or Al)

Grafted onto Silica or Alumina: A Molecularly Dispersed Nickel Precursor for Syntheses of Supported Small Size Nickel Nanoparticles. Chemical Communications, 50 (2014) 7716–7719. DOI: 10.1039/c4cc02962c.

[31] M.C. Capel-Sanchez, G. Blanco-Brieva, J.M. Campos-Martin, M.P. de Frutos, W. Wen, J.A. Rodriguez, J.L.G. Fierro, Grafting Strategy to Develop Single Site Titanium on an Amorphous Silica Surface. Langmuir, 25 (2009) 7148–7155. DOI: 10.1021/la900578u.

[32] C.J. Orendorff, J.E. Pemberton, Raman Spectroscopic Study of the Conformational Order of Octadecylsilane Stationary Phases: Effects of Electrolyte and pH. Analytical and Bioanalytical Chemistry, 382 (2005) 691–697. DOI: 10.1007/s00216-005-3133-4.

[33] R. Brambilla, G.P. Pires, N.P. da Silveira, J.H.Z. dos Santos, M.S.L. Miranda, R.L. Frost, Spherical and Lamellar Octadecylsilane Hybrid Silicas. Journal of Non-Crystalline Solids, 354 (2008) 5033–5040. DOI: 10.1016/j.jnoncrysol.2008.07.031.

[34] R.M. Almeida, T.A. Guiton, C.G. Pantano, Detection of LO Mode in V-SiO2 by Infrared Diffuse Reflectance Spectroscopy. Journal of Non-Crystalline Solids, 119 (1990) 238–241 DOI: 10.1016/0022-3093(90)90847-f.

[35] A. Fidalgo, L.M. Ilharco, Chemical Tailoring of Porous Silica Xerogels: Local Structure by Vibrational Spectroscopy. Chemistry – A European Journal, 10 (2004) 392–398. DOI: 10.1002/chem.200305079.

[36] L.B. Cappeletti, E. Moncada, J. Poisson, I.S. Butler, J.H.Z. Dos Santos, Determination of the Network Structure of Sensor Materials Prepared by Three Different Sol–gel Routes Using Fourier Transform Infrared Spectroscopy (FT-IR). Applied Spectroscopy, 67 (2013) 441–447. DOI: 10.1366/12-06748.

[37] A. Fidalgo, R. Ciriminna, L.M. Ilharco, M. Pagliaro, Role of the Alkyl-Alkoxide Precursor on the Structure and Catalytic Properties of Hybrid Sol–gel Catalysts. Chemistry of Materials, 17 (2005) 6686–6694. DOI: 10.1021/cm051954x.

[38] S.L.B. Lana, A.B. Seddon, X-Ray Diffraction Studies of Sol–Gel Derived Ormosils Based on Combinations of Tetramethoxysilane and Trimethoxysilane. Journal of Sol-Gel Science and Technology, 13 (1998) 461–466. DOI: 10.1023/a:1008685614559.

[39] L.B. Capeletti, J.H.Z. Dos Santos, E. Moncada, Dual-Target Sensors: The Effect of the Encapsulation Route on pH Measurements and Ammonia Monitoring. Journal of Sol-Gel Science and Technology, 64 (2012) 209–218. DOI: 10.1007/s10971-012-2849-9.

[40] E.C. Morais, G.G. Correa, R. Brambilla, C. Radtke, I.M. Baibich, J.H.Z. dos Santos, The Interaction of Encapsulated Pharmaceutical Drugs with a Silica Matrix. Colloids and Surfaces B – Biointerfaces, 103 (2013) 422–429. DOI: 10.1016/j.colsurfb.2012.10.059.

[41] D.M. Han, G.Z. Fang, X.P. Yan, Preparation and Evaluation of a Molecularly Imprinted Sol–gel Material for On-Line Solid-Phase Extraction Coupled with High Performance

Author details

Larissa Brentano Capeletti and João Henrique Zimnoch*

*Address all correspondence to: jhzds@iq.ufrgs.br

Chemistry Institute, Federal University of Rio Grande do Sul, Porto Alegre, Brazil

References

[1] M. Perchacz, H. Benes, L. Kobera, Z. Walterova, Influence of sol–gel conditions on the final structure of silica-based precursors. Journal of Sol-Gel Science and Technology, 75 (2015) 649–663. DOI: 10.1007/s10971-015-3735-z.

[2] T.Y. Amiri, J. Moghaddas, Cogeled Copper-Silica Aerogel as a Catalyst in Hydrogen Production from Methanol Steam Reforming. International Journal of Hydrogen Energy, 40 (2015) 1472–1480. DOI: 10.1016/j.ijhydene.2014.11.104.

[3] C.Z. Jin, Y.J. Wang, H.L. Tang, K.X. Zhu, X. Liu, J.H. Wang, Versatile Rattle-Type Magnetic Mesoporous Silica Spheres, Working as Adsorbents and Nanocatalyst Containers. Journal of Sol-Gel Science and Technology, 77 (2016) 279–287. DOI: 10.1007/s10971-015-3830-1.

[4] C.M.A. Parlett, M.A. Isaacs, S.K. Beaumont, L.M. Bingham, N.S. Hondow, K. Wilson, A.F. Lee, Spatially Orthogonal Chemical Functionalization of a Hierarchical Pore Network for Catalytic Cascade Reactions. Nature Materials, 15 (2016) 178–182. DOI: 10.1038/nmat4478.

[5] A. Bazyari, A.A. Khodadadi, A.H. Mamaghani, J. Beheshtian, L.T. Thompson, Y. Mortazavi, Microporous Titania-Silica Nanocomposite Catalyst-Adsorbent for Ultra-Deep Oxidative Desulfurization. Applied Catalysis B – Environmental, 180 (2016) 65–77. DOI: 10.1016/j.apcatb.2015.06.011.

[6] B. Moongraksathum, Y.W. Chen, Preparation and Characterization of SiO2-TiO2 Neutral Sol by Peroxo Sol–gel Method and Its Application on Photocatalytic Degradation. Journal of Sol-Gel Science and Technology, 77 (2016) 288–297. DOI: 10.1007/s10971-015-3853-7.

[7] A.A. Bernardes, C. Radtke, M.D.M. Alves, I.M. Baibich, M. Lucchese, J.H.Z. dos Santos, Synthesis and Characterization of SiO2-CrO3, SiO2-MoO3, and SiO2-WO3 Mixed Oxides Produced Using the Non-Hydrolytic Sol–gel Process. Journal of Sol-Gel Science and Technology, 69 (2014) 72–84. DOI: 10.1007/s10971-013-3188-1.

[8] I.D. de Souza, D.S. Domingues, M.E.C. Queiroz, Hybrid Silica Monolith for Microextraction by Packed Sorbent to Determine Drugs from Plasma Samples by Liquid

Chromatography-Tandem Mass Spectrometry. Talanta, 140 (2015) 166–175. DOI: 10.1016/j.talanta.2015.03.032.

[9] M.Y. Zhao, X.D. Ma, F.J. Zhao, H.W. Guo, Molecularly Imprinted Polymer Silica Monolith for the Selective Extraction of Alpha-Cypermethrin from Soil Samples. Journal of Materials Science, 51 (2016) 3440–3447. DOI: 10.1007/s10853-015-9661-1.

[10] P. Svoboda, A. Combes, J. Petit, L. Novakova, V. Pichon, B. Grp, Synthesis of a Molecularly Imprinted Sorbent for Selective Solid-Phase Extraction of Beta-N-Methylamino-L-Alanine. Talanta, 144 (2015) 1021–1029. DOI: 10.1016/j.talanta.2015.07.052.

[11] S. Islam, N. Bidin, S. Riaz, S. Naseem, F.M. Marsin, Correlation between Structural and Optical Properties of Surfactant Assisted Sol–Gel Based Mesoporous SiO2-TiO2 Hybrid Nanoparticles for pH Sensing/Optochemical Sensor. Sensors and Actuators B – Chemical, 225 (2016) 66–73. DOI: 10.1016/j.snb.2015.11.016.

[12] P.C.A. Jeronimo, A.N. Araujo, M. Montenegro. Optical Sensors and Biosensors Based on Sol–gel Films. Talanta, 72 (2007) 13–27.

[13] S. Islam, N. Bidin, S. Riaz, G. Krishnan, S. Naseem, Sol–Gel Based Fiber Optic pH Nanosensor: Structural and Sensing Properties. Sensors and Actuators A – Physical, 238 (2016) 8–18. DOI: 10.1016/j.sna.2015.12.003.

[14] J.C. Echeverria, M. Faustini, J.J. Garrido, Effects of the Porous Texture and Surface Chemistry of Silica Xerogels on the Sensitivity of Fiber-Optic Sensors toward VOCS. Sensors and Actuators B – Chemical, 222 (2016) 1166–1174. DOI: 10.1016/j.snb.2015.08.010.

[15] E.M. Caldas, E.W. de Menezes, T.M. Pizzolato, S.L.P. Dias, T.M.H. Costa, L.T. Arenas, E.V. Benvenutti, Ionic Silsesquioxane Film Immobilized on Silica Applied in the Development of Carbon Paste Electrode for Determination of Methyl Parathion. Journal of Sol-Gel Science and Technology, 72 (2014) 282–289. DOI: 10.1007/s10971-014-3367-8.

[16] A.M. Spehar-Deleze, S. Almadaghi, C.K. O'Sullivan, Development of Solid-State Electrochemiluminescence (ECL) Sensor Based on Ru(bpy)(3)(2+)-Encapsulated Silica Nanoparticles for the Detection of Biogenic Polyamines. Chemosensors, 3 (2015) 178–189. DOI: 10.3390/chemosensors3020178.

[17] A.R. Mukhametshina, S.V. Fedorenko, I.V. Zueva, K.A. Petrov, P. Masson, I.R. Nizameev, A.R. Mustafina, O.G. Sinyashin, Luminescent Silica Nanoparticles for Sensing Acetylcholinesterase-Catalyzed Hydrolysis of Acetylcholine. Biosensors & Bioelectronics, 77 (2016) 871–878. DOI: 10.1016/j.bios.2015.10.059.

[18] A. Foroozan, R. Naderi, Effect of Coating Composition on the Anticorrosion Performance of a Silane Sol–Gel Layer on Mild Steel. RSC Advances, 5 (2015) 106485–106491. DOI: 10.1039/c5ra21744j.

[19] I. Santana, A. Pepe, E. Jimenez-Pique, S. Pellice, I. Milosev, S. Cere, Corrosion Protection of Carbon Steel by Silica-Based Hybrid Coatings Containing Cerium Salts: Effect of

The photoacoustic spectroscopy (PAS) FTIR is a broad-applicable mid-infrared solution when samples present opacity problems [50]. It is a unique extension of IR spectroscopy which combines the utility of interferometry with the standard sample-gas microphone of the photothermal technique for depth-profile analysis of materials. Its signal generation processes automatically and reproducibly isolates a layer extending beneath the sample surface which has suitable optical density for analysis without physically altering the sample. PAS involves measurement of acoustic wave (pressure oscillations) in a hermetically sealed cell fitted with a very sensitive microphone. The microphone signal, when plotted as a function of wavelength, contains a spectrum proportional to the absorption spectrum of the sample. The wave generation follows absorption of light, which is modulated at a frequency in the acoustic range, by the sample. Most FTIR instruments provide modulation frequencies between 50 and 500 Hz in the 400–4000 cm^{-1} wave number span [51]. Therefore, this approach can be employed to evaluate silica samples as described by Gao et al. to widely explore a polyethylene-*co*-Zn-acrylic acid hybrid material prepared by the sol-gel process. They could identify both vibrations for silica and the organic group. By investigating the silica bands it is noted that not only silica but also other forms of silicon groups are formed via the sol-gel reaction, while the existence of the Si–OH group indicates that the condensation reaction was not completely finished. However, analyzing the peaks referring to the organic groups, it was reported that $-CH_2-$ and $-CH_3$ positions in the region of 2800–3000 cm^{-1} remain the same after hybrid material fabrication, indicating no chemical bond between silicon and these two groups forms after the sol-gel reaction. The band at 1700 cm^{-1} is the C=O stretching mode from the –COOH group was also noticed for both materials. This is due to unneutralized acrylic acid. While the transparency of the hybrid is a result of strong organic-inorganic interaction, and a high degree of mixing, no evidence of hydrogen bonding is observed from the FTIR peak position [52].

Infrared emission spectroscopy (IRES) is a method in which a sample is energized by heating and so on, and the infrared light emitted from the sample is measured to obtain a spectrum. It utilizes the contrast between the sample and the base material with possibility of an improved signal-to-noise ratio compared to absorption spectroscopy. Ideally only photons emitted by the sample are detected ("zero background"), free from the noise produced by the continuum lamp in an absorption experiment. This improvement in sensitivity is particularly useful for the spectroscopy of transient molecules because of their intrinsically low concentrations [53]. In the case of hybrid silicas, this methodology can be successfully employed to evaluate the thermal stability of materials. Brambilla et al., for example, describe the behavior of octadecylsilane (ODS) hybrid silicas prepared by grafting and sol-gel methods with a spectra series from 100 to 900°C between 4000 and 500 cm^{-1}. The decrease of physically adsorbed water band and the surface silanol interactions, leading to an increase in the band at 3747 cm^{-1}, attributed to isolated silanol groups was observed. For temperatures higher than 450°C up to 900°C, the band assigned to isolated silanol groups (3747 cm^{-1}) is reduced due to condensation reactions and structural reorganization with generation of siloxane groups. Concerning the ν(C–H) stretching region, the bands at 2963, 2929 and 2855 cm^{-1} are reduced in the range of 250–550°C, as a result of the thermal degradation of ODS chains. **Figure 7** presents the relative band area for this vibration. It was possible to compare the thermal stability of hybrid materials prepared by grafting (GR100) and sol-gel method (SG10A), confirm the higher thermal stability of the first one [54].

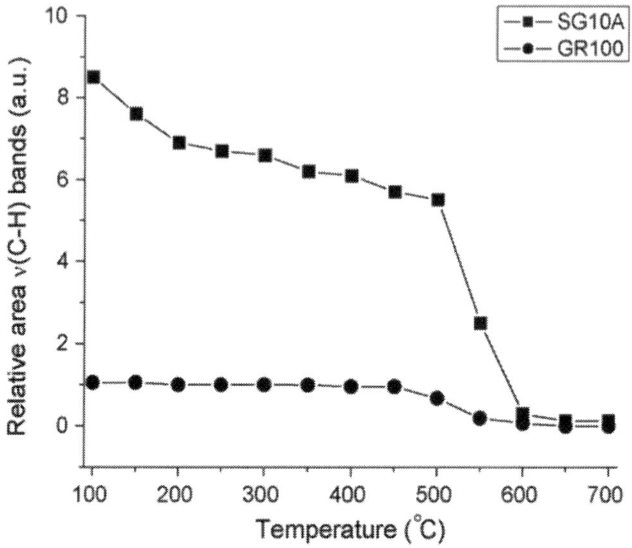

Figure 7. Relative area of ν(C–H) bands versus temperature for grafting prepared (GR100) and sol-gel prepared (SG10) hybrid silicas [54].

Finally, micro-Raman and micro-FTIR modes have less sensibility, however, they are key methods employed when spatial resolution of a few micrometers is necessary. It can be employed, for example, to evaluate radial distribution of the fictive temperature in pure silica optical fibers [55], porous silica supports for individual living cells [56] and phenyl-bridged polysilsesquioxane positive and negative resist for electron beam lithography where the technique helped to propose a description of the tone switching mechanisms [57].

6. Final remarks

Observing all the procedures and results described above, it is noticeable that a vibrational spectrum is not collected only to simply evaluate peak positions anymore. Nowadays, it is possible to obtain deep information about materials' formation, evolution and structure, as well as to acquire spatial resolution spectra or in high-resolution modes with low signal/noise ratios. The reaction chemical processes following methods are getting more and more specific, as collected data are more and more exploited in order to give the maximum results information, bringing fast advance in the materials characterization field.

The silica-related bands presented wave number shifts depending on the employed organic group in the different hybrid silicas. For Si–O(–Si) LO mode, the largest shift occurs from the nonhybrid to the hybrid samples. In this case, shifts to lower wave numbers use to be related to the network deformation in order to accommodate the organic groups within the inorganic silica matrix resulting in larger siloxane rings and greater Si–O–Si angles and longer Si–O bond lengths. LO and TO mode shifts occur mainly near the surface of the material, which can be better detected employing attenuated total reflectance (ATR) mode of FTIR spectroscopy. In addition, the organic groups' introduction can originate from heterogeneous regions that may introduce local deformations in the network resulting in the differences observed for Si–O bond lengths and Si–O–Si angles in the different hybrid materials. Another interesting behavior is reported to Si–O(H) band, considering that, in a general way for this case, the wave number shifts result from hydrogen bond formation with the silanol groups. A significantly higher wave number occurred for C_{18} while the lower ones occurred for Clp and Gp [45]. The very hydrophobic C_{18} organic chain can hinder hydrogen bond formation by comparison with the other samples, thus, the Si–O bond length is decreased and the wave number is shifted to higher values. However, the groups Clp and Gp can facilitate hydrogen bond formation with the silanol groups and, as a consequence, the Si–O bond length is increased and the vibrational wave number is decreased [46].

As mentioned before, the characterization of the organic groups of hybrid network is better done using Raman spectroscopy than FTIR. The last one can observe only some bands, while Raman presents a series of them showing the complementarity of both techniques for hybrid materials' characterization. **Table 1** compiles some important assignments regarding the organic groups and their respective wave numbers detected by both techniques, when applicable [45].

The complementary use of FTIR and Raman spectroscopies can be also employed to deeply investigate the processes taking place during sol-gel process and there are other types of detection modes for these vibrational techniques as DRIFTS, PAS, IRES, micro-FTIR and Raman which can help.

DRIFT spectroscopy is primarily used on samples where most of the reflected radiation is diffused. It is important that the specular reflectance is reduced to a minimum because it distorts the DRIFT spectrum and lowers the band intensities [47]. During the last years, it has become the most effective technique for studying the processes taking place at the gas-solid interface [48]. Ivanovski et al. investigated the region of the OH stretching vibrations of silica gel activated at different temperatures for the purpose of checking the availability of the OH groups for further reaction with 3-aminopropyltrimethoxysilane (APTMS) molecules and whether chemisorption was successful, finding evidence whether chemisorption of APTMS involves all available methoxy groups. It is also possible to investigate the conformation of the aminopropyl groups (APS) backbone on the silica gel surface and the possible proton transfer between the NH_2 groups of APS and OH from silica gel detected by the formation of NH_3^+ and SiO^- which is spectroscopically detectable through the appearance of the $\delta(NH_3^+)$ vibrations at 1668 cm^{-1} [49]. In addition, Bukleski et al. developed a direct quantitative method of quantification of maximal chemisorption of 3-aminopropylsilyl groups on silica gel using

DRIFT spectroscopy, as (APS) modified silica gel plays an important role as a precursor for further modifications, where APS acts as a spacer or bridging molecule. By integrating the spectra in the frequency range of the $\nu(CH_2)/\nu(CH_3)$ vibrations between 3014 and 2808 cm^{-1}, the mass fraction of APTMS of 19.04% was found to correspond to a maximal concentration of APS on silica gel of 2.23 μmol m^{-2}, which was confirmed by elemental analysis for carbon [47].

Assignment	C1	C8	C18	Vy	Ph	SHp	NCOp	Clp	Gp
$C–H_{(3)\ asym}$	2979 R	2955 I	2957 I						
$C–H_{(3)\ sym}$	2916 R	2884 R	2883 R						
$C–H_{(2)\ asym}$		2932 R	2930 R			2928 R	2939 R	2962 R	2926 R
		2926 I	2918 I			2942 I	2945 I	2960 I	2935 I
$C–H_{(2)\ sym}$		2861 R	2850 R			2894 R	2897 R	2901 R	2894 R
		2856 I	2848 I						2880 I
$C–H_{arom}$					3058 R				
Si–C	1412 R				1122 R				
	1280 I								
$CH_{2\ bend}$		1463 R	1460 R						
		1457 I	1468 I						
C–C		1064 R	1062 R	1013 R	999 R				
C=C				1603 R					
$=C–H_{term}$				3072 R					
$=C–H_{term\ bend}$				1412 R					
				1278 R					
(Si)C–H				2991 R					
$C–H_{bend}$					1432 i				
$Ring_{breath}$					737 I				1260 R
$Ring_{def}$					698 I				
S–H						2574 R			
C–S						652 R			
C=N							1553 I		
N=C=O							1449 R		
$H–C–Cl_{def}$								1412 R	
$C–Cl(H)_{trans}$								645 R	
OC–H									1456 R

*asym: asymmetric; sym: symmetric; bend: bending; term: terminal; def: deformation; R: Raman; and I: infrared.

Table 1. Organic group bands detected by the complementary techniques of FTIR and Raman [45].

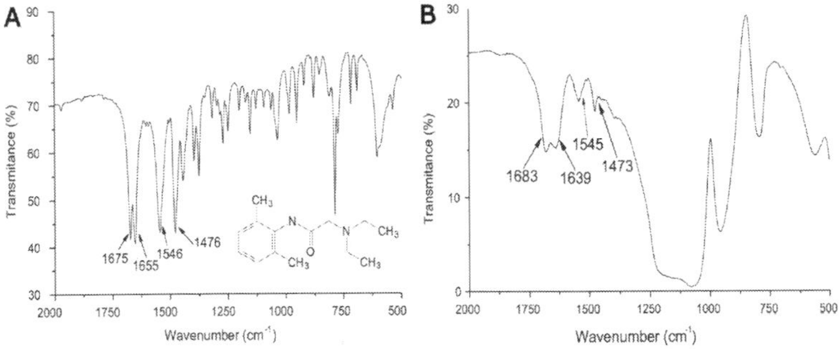

Figure 5. FTIR spectra of bare lidocaine (A) and the respective encapsulated system (B) [40].

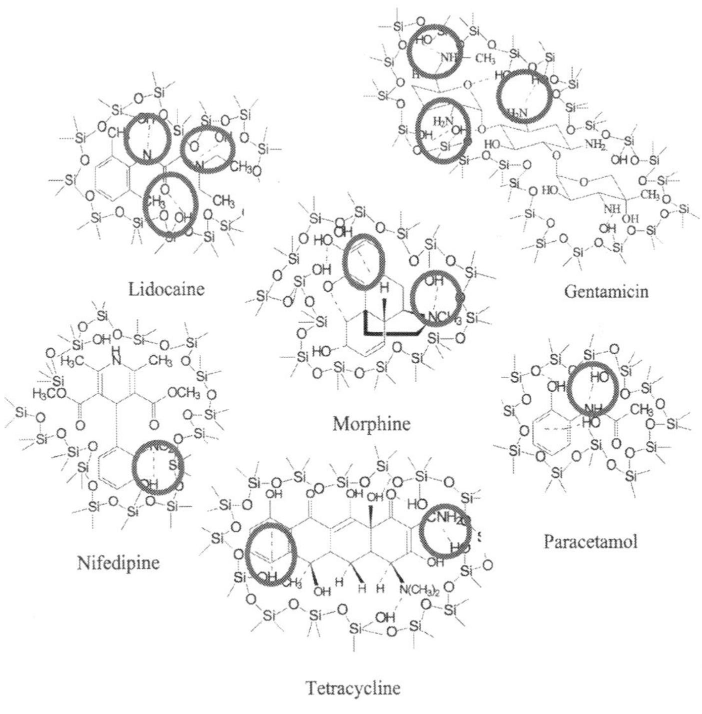

Figure 6. Proposed interactions of the drugs with silica network [40].

In other approaches, hybrid silica networks have been used to improve both the process of molecular imprinting as the following usage of the material as an extraction matrix. Han et al. reports the functionalization of silica with amino groups provided by aminopropyltriethoxysilane to help interaction with the toxic herbicide pentachlorophenol [41]. By using FTIR to monitor this process, they were able to identify the N–H bond around 1560 cm^{-1} and C–H bond around 2935 cm^{-1}, suggesting the –NH_2 grafting onto the activated silica gel surface. In this case, imprinted and nonimprinted sorbents showed similar location and appearance of the major bands, reflecting the already mentioned problem of overlapping bands with the major bands of silica network. Similar behavior was observed by Chrzanowska et al., Ren et al. and Li et al. [42–44]. The first one employed the functionalization of silica nanoparticles surface with aminopropyl groups to promote the encapsulation of biochanin A, producing a selective solid-phase extraction of biochanin A, daidzein and genistein from urine samples [42]. Analogously, Ren et al. employed the same procedure with aminopropyl groups, although the target analyte was bisphenol A [44]. Finally, the later one made use of propylthiocyanate groups to modify the silica surface, creating a selective phase for selective removal of cadmium(II) competing with copper, zinc and lead in aqueous solution [43]. In all the cases, the assisting organic groups' bands were detected; however, the molecular imprinted and nonimprinted spectra were really similar.

5. FTIR and Raman modes

When analyzing hybrid silica materials, sometimes it is necessary to use complementary techniques to better evaluate the materials' characteristics. The same occurs for vibrational spectroscopy methods. A wide investigation was performed with a series of different hybrid silica prepared with tetraethoxysilane (C_0), methyltriethoxysilane (C_1), octyltriethoxysilane (C_8), octadecyltrimethoxysilane (C_{18}), vinyltrimethoxysilane (Vy), phenyltrimethoxysilane (Ph), mercaptopropyltrimethoxysilane (SHp), isocyanatepropyltriethoxysilane (NCOp), chloropropyltrimethoxysilane (Clp) and glycidoxypropyltrimethoxysilane (Gp) [45]. Using FTIR, the main bands of silica were well determined for all the hybrid silicas and they showed shifts depending on the organic group presenting at the network, although the organic groups' bands were barely seen. Using Raman spectroscopy, the organic groups were well described and some of the silica network bands were also observed.

The region around 3600–3000 cm^{-1} is attributed to hydroxyl groups ν(O–H) stretching modes. The shoulder at ~3600 cm^{-1} matches the OAH vibrations associated with alcohols that are a subproduct of sol-gel reaction, while the maximum at ~3425 cm^{-1} is related to surface –OH participating of hydrogen bonds. Water is also described here as a shoulder at ~3230 cm^{-1}, which is also observed at ~1630 cm^{-1}. As mentioned before, silica presents a characteristic region of peaks from 1250 to 700 cm^{-1} that can provide structural characteristics of the network. Specially, when related to the main bands between 1250 and 1000 cm^{-1} corresponding to the asymmetric ν(Si–O–H) and their deconvolution on LO at ~1130 cm^{-1} and TO at 1047 cm^{-1} modes. The Si–O(H) bond stretching appears at slightly different positions in the FTIR (~950 cm^{-1}) and Raman (~980 cm^{-1}) spectra. The symmetric mode of ν(Si–O–Si) band is found at ~791 (FTIR) and ~799 cm^{-1} (Raman), while the Si–O^- rocking mode was observed at ~540 cm^{-1} (IR). Finally, the Raman spectra also show the siloxane ring breathing mode (with 3 or 4 SiO units) located at ~490 cm^{-1} [45].

Chapter 3

Structural Characterization of Lithium Niobate Nanoparticles Prepared by the Sol-Gel Process, Using X-Ray and Raman Spectroscopy and Scanning Electron Microscopy

Pedro R.S. Prezas and Manuel P.F. Graça

Additional information is available at the end of the chapter

http://dx.doi.org/10.5772/64395

Abstract

The widespread use of lithium niobate (LN) in several technological applications, notably in optical and electrooptical systems, is a consequence of its remarkable piezoelectric, electrooptical, photoelastic, acousto-optic, and nonlinear optical coefficients. In this chapter, the structural and electrical characterization of LN nanosized particles synthesized by the Pechini route is discussed. Compared to solid-state reaction processes, wet chemistry processes can be advantageous alternatives for the synthesis of polycrystalline LN, because they require lower processing temperatures, and thus the loss of stoichiometry and formation of secondary phases can be minimized. The powders obtained by drying the gel (base powder) were heat-treated for 4 h at temperatures between 400 and 1000°C, according to the differential thermal analysis (DTA) results. It was found that the powders sintered at 450°C contain only the LN phase, while those heat-treated at 500°C already contain the secondary $LiNb_3O_8$ phase. The structural and electrical characterization of the samples sintered at 450°C, for different times, was performed using X-ray diffraction (XRD) in conjunction with Rietveld refinement, Raman spectroscopy, scanning electron microscopy (SEM), and impedance spectroscopy in the temperature range between 200 and 360 K and in the frequency range between 100 Hz and 1 MHz and by measuring the ac and dc conductivities.

Keywords: lithium niobate, structural properties, X-ray spectroscopy, Raman spectroscopy, electrical properties, lithium triniobate, sol-gel process

1. Introduction

Lithium niobate ($LiNbO_3$, LN) is a well-known artificially synthesized ferroelectric material with considerable technological importance, being in competition with barium titanate ($BaTiO_3$, BTO) in several high-tech applications. In fact, an inspection of the number of publications related with LN and BTO will show that since the mid-1990s, the number of reports on both materials has been following the same increasing trend, with similar number of publications, reinforcing the importance of LN among the scientific communities. **Table 1** displays some of the main physical properties of single-crystalline LN [1].

Melting temperature (°C)	1260
Curie temperature (°C)	1210
Density at RT (g/cm^3)	4.64
Refractive index (ordinary), n_0	2.296
Electrooptical coefficient, r_{33} (m/V)	30×10^{-12}
Transparency window (μm)	0.4–5
Resistivity, ρ (c-axis) (Ω cm)	$\log \rho = (7150/T) - 2.823$ (at RT) $= 10^{21}$
Dielectric constant at RT (ε')—c-axis	80 (100 kHz) >1000 (1 kHz)
Dielectric loss at RT (tan δ)—c-axis	≈0 (100 kHz)
Coercive field (at 1210°C) (V/m)	20
Spontaneous polarization at RT (P_s [$\times 10^{-2}$ C m^{-2}])	70
Piezoelectric coefficient (d_{33} [pC/N])	6
Thermal conductivity at RT (W m^{-1} K^{-1})	3.92

Table 1. Main physical properties of single-crystalline stoichiometric LN. RT stands for room temperature (generally 300 K) [1–3].

LN single crystals display several excellent properties, such as high piezoelectric, electrooptical, photoelastic, acousto-optic, and nonlinear optical coefficients. They are known to have very low acoustic losses, offering a great versatility as a substrate for integrated optic systems: a considerable number of optical devices have been developed based on LN, such as wave guides, surface acoustic wave (SAW) devices, electrooptical wavelength filters and polarization modulators, nonlinear frequency converters (frequency doubling and second harmonic generation), nonvolatile memories, and ultrafast optical processing systems. Its combination of electrooptical and photogalvanic effects makes it photorefractive without the need of applying an external electrical field, thus being able to be applied in holographic data storage. It offers also the possibility of being easily doped, in a controllable way, with optical-active ionic species, using standard techniques such as ion implantation or thermal diffusion.

The physical properties aforementioned are optimized for the case where LN is grown as a single crystal and with stoichiometric composition. Regarding the stoichiometry, LN has a relatively broad composition range, and therefore, it can be labeled as congruent lithium niobate (cLN, 48.35–48.6 mol% Li_2O) and stoichiometric (sLN, ~50 mol% Li_2O). **Figure 1** displays the phase diagram of the Li_2O-Nb_2O_5 binary system, showing the possibility of growing pure LN crystals by using 50% up to ~52% of Nb_2O_5. It also shows the transition of ferroelectric phase to paraelectric by increasing the synthesis temperature and Nb_2O_5 content. The large majority of LN single crystals are grown by the conventional Czochralski method, which yields cLN single crystals. Some competitor methods have been developed for growing stoichiometric crystals, including the vapor transport equilibration (VTE) method, which is a post-grown procedure [4]. The former is more suitable for thin and small samples, because for larger and thicker crystals, very large solid-state diffusion times are required for the Li/Nb ratio equilibration. To attain larger stoichiometric single crystals, more direct growth methods can be applied, such as the double crucible Czochralski method with an automatic power supply [5, 6] or the high-temperature top-seeded solution growth (HTTSSG) method from the K_2O-Li_2O-Nb_2O_5 ternary mixture, which is the one capable of yielding compositions closest to 50 mol% Li_2O [5, 7].

The nonlinear and photorefractive properties will generally degrade with the loss of stoichiometry and consequent increase of defects and impurity density. As a matter of fact, the congruent composition range is regarded as having an intrinsic defect structure which is dominated by lithium vacancies, known as the lithium vacancy model, which translates in Eq. (1) [6]:

$$[Li_{Li}]_{1-5x}[Nb_{Li}]_x[V_{Li}]_{4x}Nb_{Nb}O_3 \tag{1}$$

In this model, as depicted in Eq. (1), for every four lithium vacancies created, a niobium ion occupies a lithium network site, assuring electrical charge neutrality. The cLN is not suitable for high-temperature applications, because degradation processes can start to occur for temperatures starting from 300°C [6, 8]. On the other hand, studies show that sLN can be stable

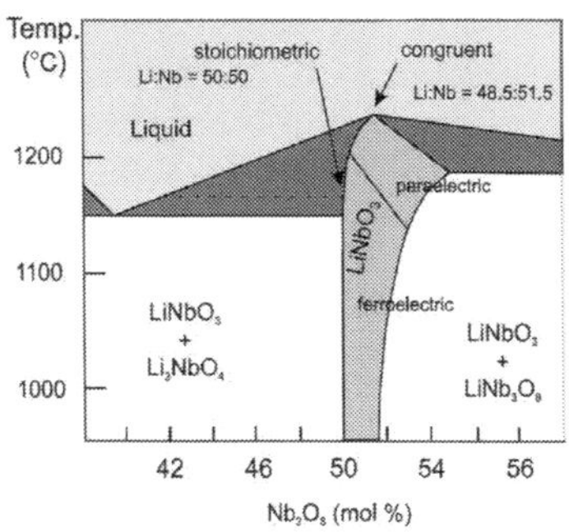

Figure 1. Phase diagram of the Li_2O-Nb_2O_5 binary system [3].

up to temperatures of at least 900°C [6, 9], because some properties like the electrical conductivity do not change under thermal cycling up to such temperatures.

As it was aforementioned, the dominant process in the growth of LN single crystals is the Czochralski method. However, this method is known for its technical and economic drawbacks, as well as being time-consuming. Thus, alternative preparation processes have been researched and explored. Solid-state reaction processes generally require high processing temperatures (>1000°C), which lead to the loss of lithium by evaporation [10]. As a consequence, secondary crystalline phases such as Li_3NbO_4 and $LiNb_3O_8$ can be formed, changing the stoichiometry and deteriorating the properties. Wet chemistry methods, such as sol-gel methodologies and hydrothermal methods, can be good alternatives because they require lower processing temperatures, such as calcination and thermal-sintering treatments, and thus the formation of secondary phases can be minimized [10, 11]. Single crystals in the nanometer and micrometer size ranges can be synthesized at low temperatures, such as 240°C, by these methods [11]. When such methods are applied, typically polycrystalline LN samples are produced, i.e., a material composed by small single crystals in the micrometric or nanometric range randomly distributed with no evident preferential orientation. Polycrystalline LN finds a lot of applications, especially as thin films for integrated optic applications, although generally the properties of polycrystalline materials are not as good as their single-crystalline counterpart. For example, the piezoelectric properties of polycrystalline LN are inferior to the single crystal, and in the best case, they might approach them if all of its domains are perfectly orientated. Further, relative to single-crystalline thin films, the grain boundaries in polycrystalline films may lead to increased light scattering and larger optical losses in wave guides, which may reduce their utility and potentiality in some applications [12]. However, their production is cheaper and easier compared with the growth processes for single crystals, and all these cons and drawbacks have to be considered and well balanced for potential applications.

Amorphous LN is also important for some applications. In an amorphous material, there is no long-range order, and the network can be described as distorted unitary cells randomly oriented. LN single crystals and polycrystals have low electrical conductivity (~10^{-12} S/cm at 500 K for single crystals [13]), and ionic diffusion or mobility is reduced in these materials. However, the amorphous structure is always a more open structure compared with the crystalline composition or, in other words, has a smaller density, which promotes and facilitates the ionic diffusivity, making them suitable for technical application such as solid-state electrolytes for Li-ion batteries. In fact, the reported activation energies for the ionic diffusivity are considerably smaller (half in the 25–150°C temperature range [13]) in amorphous LN.

In this chapter, the structural and electrical characterization of $LiNbO_3$ nanosized particles prepared by the Pechini route, also known in the sol-gel methodologies as the polymeric precursor route, is discussed. The powders obtained by drying the gel (base powder) were heat-treated for 4 h at temperatures between 400 and 1000°C, according to the differential thermal analysis (DTA) results. The sintering temperature revealed to be, as expected, an important parameter in controlling the development of secondary crystalline phases, and it was found that the powders sintered at 450°C contain only the LN phase, while those heat-treated at 500°C already contain the secondary $LiNb_3O_8$ phase. Their structural characterization was performed using X-ray diffraction (XRD) in conjunction with Rietveld refinement, Raman spectroscopy, and scanning electron microscopy (SEM). The grains observed have sizes lower than 100 nm and an approximately spherical geometry. The electrical characterization of pellets made from the base powder heat-treated at 450°C was made by measuring the dc and ac conductivities and measuring the complex impedance (Z^*) in the temperature range between 200 and 360 K and in the frequency range between 100 Hz and 1 MHz. From the measured complex impedance values, the complex permittivity (ε^*) was calculated, since the geometrical characteristics of the pellets legitimate the use of the parallel plate capacitor model. The correlation between the structure and morphology with the electrical and dielectric properties is one of the main topics of the present chapter.

2. Structural, morphologic, and electrical properties

2.1. Structural properties

LN belongs to a group of materials whose crystalline structure is a perovskite. This structure has the typical chemical formula ABO_3, where the cation A usually is too large for an effective complete packaging, causing a distortion in the unit cell and leading to a displacement of the O^{2-} anions from their expected sites. However, in the case of LN, the distortion is related with the small radius of the lithium ion. For temperatures lower than 1415 K, which is the Curie temperature (T_c) of sLN, this material is in its ferroelectric state, and consequently, it exhibits a spontaneous polarization, due to a nonuniform charge distribution of the lithium and niobium ions. In this ferroelectric phase, this material has a trigonal crystalline structure, with threefold rotation symmetry around its c-axis. **Figure 2(a)** and **(c)** shows the atomic model of LN in its ferroelectric crystalline phase. In this configuration, the crystalline structure consists in layers of oxygen atoms parallel to each other with the Li^+ and Nb^{5+} cations lying along the

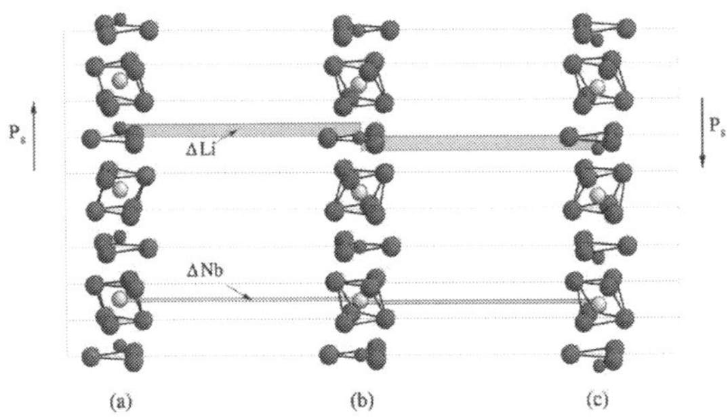

Figure 2. (a) and (c) Atomic model of the LN ferroelectric phase. (b) Atomic model of the LN paraelectric phase. ΔLi indicates the displacement along the c-axis of the Li^+ cations, represented as black spheres, while ΔNb indicates the displacement along the c-axis of the Nb^{5+} cations. Both displacements are represented relatively to the center of the oxygen (red spheres) planes [14].

c-axis, surrounded by oxygen octahedra: in the unitary cell, one-third of the octahedral interstices are occupied by Li^+ cations; another one-third by Nb^{5+} cations and the remaining (one-third) interstices are structural voids [2, 14].

As depicted in **Figure 2**, in the ferroelectric state, the displacement of the Li^+ and Nb^{5+} cations relatively to the center of the oxygen planes originates a spontaneous polarization along the c-axis with a magnitude of 0.7 C/m^2 at 300 K (see **Table 1**). The displacement can be up or down with respect to the oxygen sublattice, and both cations are displaced in the same direction because of the Coulomb repulsion. Above the Curie temperature, due to the thermal expansion of the crystalline lattice axes, the structure is no longer distorted, because the Li^+ and Nb^{5+} cations move to lattice sites lying in the planes of the oxygen layers, as **Figure 2(b)** displays. Thus, the transition to the paraelectric state occurs, and LN ceases to exhibit a permanent spontaneous polarization [2, 14].

As it was stated in Section 1, polycrystalline LN is also of great technological importance. In this form, the structure can be described as composed by small single crystals in the micrometric or nanometric range randomly distributed with no evident preferential orientation. Relatively to the ferroelectricity, it is composed by several ferroelectric domains, which are regions with different orientations of the spontaneous polarization *Ps*. In this case, the thermodynamic potential for describing the ferroelectric phase transition has to account with a nonuniform *Ps* distribution, organized in domains, and therefore the domain depolarization energy W_E and the energy of the domain walls W_W are introduced in the potential [3].

The thermodynamic model for the single-crystal case, an "ideal" ferroelectric, does not need to include the former energy terms W_E and W_W and can be described by Eq. (2) [3]:

$$G(P,T) = G_0 + \frac{\alpha(T)}{2}P^2 + \frac{\beta(T)}{4}P^4 \tag{2}$$

This model is based on the second-order phase transition as described by Landau-Ginsburg; with at least a fourth-order polynomial in *P*, the polarization. *G*(*P*,*T*) is the Gibbs function, and α and β are second- and fourth-order expansion temperature-dependent terms. As it was said, this rather simple form applies for the case when *Ps* is uniform for all the material, as in the case of a single crystal. In the ferroelectric/paraelectric phase transition, the behavior of the plots of the type *G*(*P*,*T*) versus *P*, for different temperatures, is shown in **Figure 3** [3]. As it can be seen, for T_2 and T_0, there is only one minimum, while for T_1, in the ferroelectric phase, two minimum values exist, which correspond to the values of the spontaneous polarization *Ps* (can be positive or negative, according to the direction—**Figure 2**). These values can be determined by solving the differential $(\partial G/\partial P)_{Ps} = 0$, resulting in $Ps = \pm\sqrt{\frac{-\alpha}{\beta}}$ for temperatures lower than T_c and $Ps = 0$ for temperatures higher than T_c [3]. The parameters α and β are related with the dielectric constants of LN, and more specifically, α can be expressed above T_c, in the paraelectric phase, according to the Curie-Weiss law shown in Eq. (3) [3, 15]:

$$\chi = \frac{1}{\alpha} = \frac{C}{T - T_c} \quad (3)$$

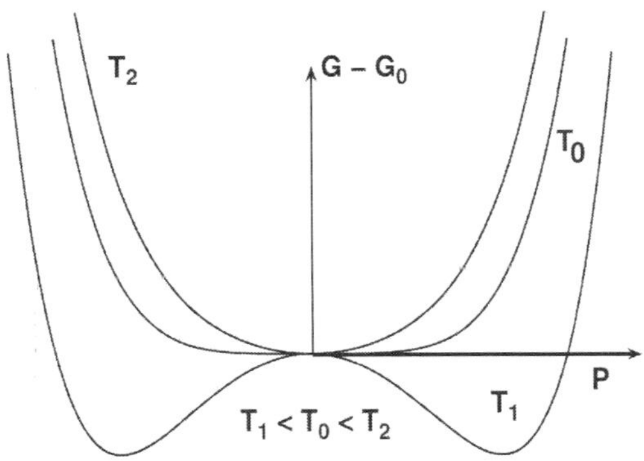

Figure 3. The Gibbs free energy in function of the polarization P. T_0 is equal to T_c, and the quantitative relation between the temperatures is $T_1 < T_0 < T_2$ [3].

where χ is the dielectric susceptibility and C is a material-dependent constant.

The above-considered model fails when predicting quantities such as the coercive field, both for congruent and stoichiometric LN, because the inversion mechanisms of *Ps* occur through the formation of ferroelectric domains and the model does not account with *Ps* discontinuities. However, this thermodynamic model can be improved to better characterize a ferroelectric material containing domains, according to Eq. (4) [3]:

$$G(P,T) = G_0 + \int_V \left[\frac{\alpha}{2}P^2 + \frac{\beta}{4}P^4 + \frac{1}{2}\delta(\nabla P)^2\right] dV + W_E + W_W \quad (4)$$

As it was stated, the terms depolarization energy W_E and the energy of the domain walls W_W are here introduced. The integration volume, W_E and W_W depend on the domain structure and geometry. In [3] the W_E and W_W expressions are described for a simple periodic domain structure model. The manipulation of the structure and geometry of the domain walls in a ferroelectric such as LN was, and still is, an important subject of study, because depending on the technological application, some geometries/shapes may be preferred over others: for example, acoustic and optical frequency conversion devices will benefit with periodic gratings of antiparallel domains [15]. The domain shape will depend on the temperature at which they are created, through the application of external electric fields and also on the crystal stoichiometry/composition. When created at room temperature, they can show different shapes due to small variation of stoichiometric composition [15]. **Figure 4** shows the preferred shapes of domains created at 25 and 125°C, for a congruent LN. It is visible that the domains have a polygonal shape with six sides, known as *y* walls. Curiously, the domain shape as depicted in **Figure 4(a)** is the same for stoichiometric LN, while in other ferroelectric materials such as $LiTaO_3$, the same does not happen [15].

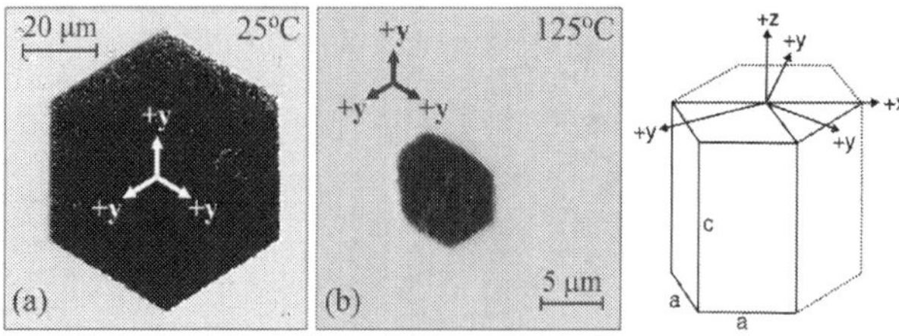

Figure 4. Piezoelectric force microscopy phase contrast images obtained in a congruent LN. The domains were created at 25 and 125°C [15].

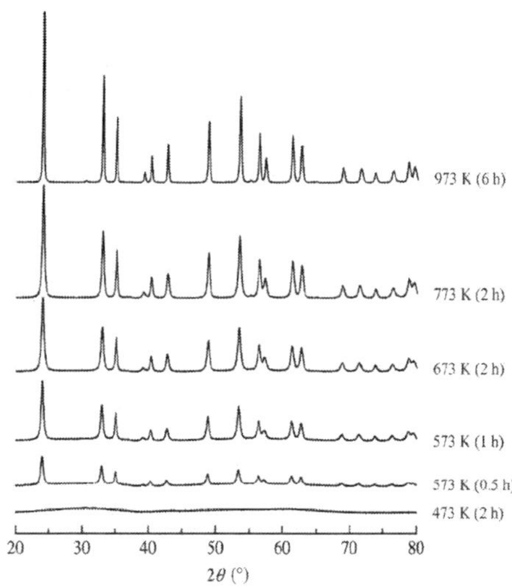

Figure 5. XRD spectra revealing the effect of heat treatments on amorphous LN prepared by complete hydrolysis of the LN double alkoxide [16].

As for the structure of amorphous LN, it was said that the network can be described as distorted unitary cells randomly oriented. **Figure 5** reveals the effect of heat treatments (HTs) on amorphous LN prepared by complete hydrolysis of the LN double alkoxide [16]. The XRD spectra of amorphous LN have the typical form of the spectrum for 473 K displayed in **Figure 5**, with two broad bands around 30 and 50–60°. These broad bands are a trademark of amorphous materials, and they are typically visible for diffraction angles where the crystalline phase has the most intense diffraction peaks, revealing at least a short-range-order preservation. The heat treatments promote the reconfiguration of the amorphous phase to a more thermodynamically stable crystalline phase; although for low treatment temperatures and times, the material may be composed by a heterogeneous mixture of an amorphous and a crystalline phase: for the heat treatment at 573 K for 0.5 h, it is still noticeable the coexistence of a broad band with the diffraction peaks of the LN crystalline phase [16]. The structure of amorphous LN was also described by Kitabatake et al. to be constructed from the network of NbO_6 octahedra which contains a micronetwork similar to crystalline LN [17]. The high dielectric constant and a relaxation mechanism were attributed to the high mobility of the Li^+ ion in the LN structure [17].

Going further on the structural properties, the Raman spectroscopy is a useful nondestructive technique to access about the structure and composition of materials. The Raman spectrum of LN will generally depend on its stoichiometry, i.e., the shape, width, and position of some Raman shifts may change according to the Li/Nb ratio [18]. Furthermore, the Li/Nb ratio of a given LN sample may be determined by analyzing the width of some Raman lines, for a given temperature [18, 19].

Figure 6 exhibits experimental full width at half maximum (FWHM) values of the Raman lines detected at about 153 and 876 cm^{-1}, for samples with different lithium contents (mol%) [19]. The measurements were carried out at room temperature (note that the Raman lines' width also depends on the temperature). The FWHM dependency with the Li content is approximately linear, and hence a calibration line can be obtained. The uncertainty related with the Li content determination by this method was calculated to be 0.05 mol%, with an estimated uncertainty of 0.2 cm^{-1} in the 876 cm^{-1} line FWHM and 0.1 cm^{-1} in the 153 cm^{-1} line [19]. Therefore, this technique can be a simple nondestructive method to estimate the Li/Nb ratio in LN crystals, with an excellent accuracy.

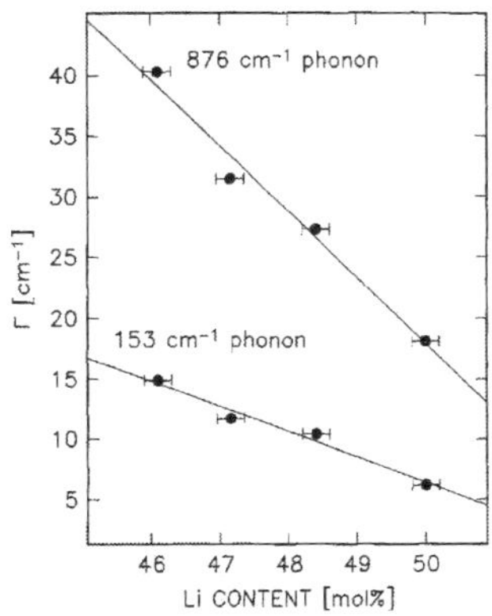

Figure 6. Full width at half maximum of the Raman lines at the wavenumbers 153 and 876 cm^{-1}. The dots are experimental values obtained for samples with different lithium contents (mol%), at room temperature, and the lines are the linear least-squares fits [19].

Figure 7 displays the Raman spectra of a nearly stoichiometric LN crystal, with x_c = 49.7 %, and a congruent LN crystal with x_c = 48.5 %, where x_c is given by $x_C = \frac{[Li]}{[Li] + [Nb]} \times 100\,(\%)$ [18]. According to the group theory, when belonging to the *R3c* spatial group, eighteen vibrational modes are to be expected, which can be reduced in the representation $4A_1 + 9E + 5A_2$ [18]. The A_2 vibrational modes are not active in Raman and FTIR (silent modes), while both A_1 and E modes are active in Raman and FTIR. The A_1 modes are polarized along the Z-axis, while the E modes represent vibrations along the X- and Y-axes (see **Figure 4**). Therefore, in the XYZ coordinate system, as indicated in **Figure 7**, Z-axis lies in the c-axis direction while the X-axis in the *a*-axis crystallographic direction. The Y-axis is perpendicular to Z and X. The notation represented in the same figure is a universally used notation first described by Damen et al. For example, in X(YZ)Y, the symbols inside the parenthesis are, from left to right, the polarization of the incident and scattered light, while the ones outside the parenthesis, from left to right, represent the directions of the incident and scattered light, respectively [20]. As depicted in **Figure 7**, the E(TO) transversal modes can be detected in the X(ZY)Z configuration, while E(TO) and E(LO) can be detected in both X(ZY)Z and X(YZ)Y configurations [18]. The A_1(TO) phonons, represented in **Figure 7**, at right, can be detected in the X(ZZ)Y configuration. The spectra clearly show that there is a broadening of the lines in the congruent composition, relatively to the nearly stoichiometric, i.e., the nearly stoichiometric spectrum lines are more resolved. Furthermore, there are lines that are only clearly visible in the nearly stoichiometric LN, and thus vibrational mode attribution in congruent LN can be an incomplete task [18]. As a final remark, when dealing with polycrystalline LN, the discussion about the different possible configurations to detect different vibrational modes is not applicable, since in the polycrystalline sample, we have nano- or micrometric single crystals randomly oriented, and thus interaction volume of the laser beam with the sample will include all these different orientations. In Section 2.4, the case study, we include the Raman spectra of polycrystalline LN, where a typical overlapping of vibrational modes is visible. The overlapping is due to the fact that, as **Figure 7** shows, some of the E(TO + LO) and A_1(TO) vibration modes are in the same wavenumber range, and consequently, they will overlap in the polycrystalline LN samples.

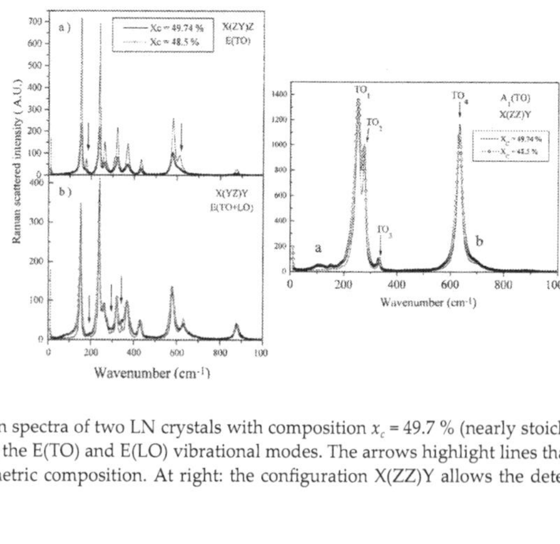

Figure 7. At left: Raman spectra of two LN crystals with composition $x_c = 49.7$ % (nearly stoichiometric) and $x_c = 48.5$ % (congruent), exhibiting the E(TO) and E(LO) vibrational modes. The arrows highlight lines that are more clearly visible in the nearly stoichiometric composition. At right: the configuration X(ZZ)Y allows the detection of the A_1(TO) phonons [18].

2.2. Morphological characteristics

Morphology means "the study of form or pattern," and morphological characterization techniques, such as scanning electron microscopy (SEM) or transmission electron microscopy (TEM), allow to characterize the morphology of a given material. The morphological characteristics of LN will obviously rely upon the synthesis process, stoichiometry, and concentration of intrinsic defects. Depending on the synthesis process and crystal growth conditions, the grains observed for polycrystalline LN can display well-defined symmetries and a different range of sizes.

Figure 8. At left: SEM micrograph of polycrystalline LN prepared by a low-temperature hydrothermal route [21]. At right: SEM micrograph of polycrystalline LN prepared by the reactive molten salt synthesis (RMSS) process [10].

Zhan et al. [21] report in their work a low-temperature hydrothermal route to prepare polycrystalline LN. The XRD characterization revealed the formation of a pure hexagonal single phase of stoichiometric LN. The SEM micrograph of the LN powder, presented in **Figure 8**, shows a regular rhombohedral grain morphology, in agreement with the XRD results, although some imperfections, such as bended surfaces, are visible. The grain size ranges from 300 nm to approximately 1 μm. Kamali et al. [10] prepared polycrystalline LN using a modification of the molten salt synthesis (MSS): in conventional MSS, the mixed powders are heated above the liquids' temperature of the salt mixture, and this molten salt acts as the reaction medium, remaining inert during the synthesis. Salt mixtures such as KCl-NaCl are typically used. In the MSS modification approached by [10], the salt can react with other reagents during the synthesis process, being labeled as reactive molten salt synthesis (RMSS). They heat-treated at 973 K Nb_2Cl_5 powder in a LiCl molten salt, in a water-containing atmosphere, whereby the molten salt is one of the precursors for LN synthesis [10]. Using this RMSS approach, they produced single-phased LN, i.e., the loss by evaporation of Li_2O was avoided. A SEM micrograph of the obtained LN particles is shown in **Figure 8**, on the right side [10]. The revealed morphology shows grains with dimensions ranging from several hundreds of nanometers to a few micrometers. The rhombohedral symmetry is also visible, although cleavages and bends are visible.

In **Figure 9**, it is visible a TEM bright-field micrograph of the polycrystalline LN prepared by the RMSS process. The inset on the right top shows a selected area electron diffraction pattern, being the area marked by the black arrow. The diffraction pattern is consistent with single-crystalline rhombohedral LN. The inset on the left bottom shows a high-resolution micrograph of the area marked by the white arrow. The rhombohedral LN (104) atomic plane patterns are visible, with an interplanar spacing of 0.27 nm [10].

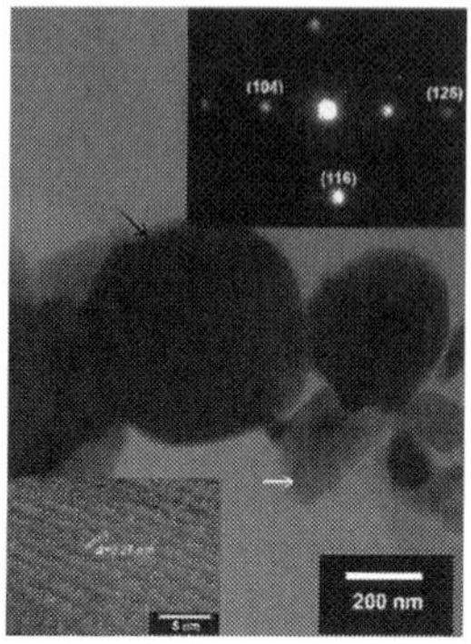

Figure 9. TEM bright-field micrograph of the LN particles prepared by the reactive molten salt synthesis (RMSS). Inset on the top right: selected area electron diffraction pattern of the area marked by the black arrow. Inset on the left bottom: high-resolution micrograph of the area marked by the white arrow [10].

The grain morphology will be determined by the growth conditions in such a way that the final morphology reflects the configuration with the minimum surface energy. In crystalline solid materials, the surface tension will depend on the crystallographic planes and direction, because to create a new surface, it is necessary to break bonds. At a constant pressure and temperature, the work required to create a new portion of surface dA_s in a one-component system is given by Eq. (5) [22]:

$$dW_{T,p} = \gamma dA_s \tag{5}$$

where γ is the surface energy (J/m^2). This represents an excess of energy relatively to the bulk and will depend on the number of bonds of the surface (crystallographic plane) and their bond energy. The change of the Gibbs free energy can be written according to Eq. (6) [22]:

$$dG = -SdT + Vdp + \gamma dA_s \tag{6}$$

where γ is defined as in Eq. (7) [22]:

$$\gamma = \left(\frac{\partial G}{\partial A_s}\right)_{T,P} \tag{7}$$

The formation of new surfaces leads to a positive Gibbs energy contribution, whereby smaller particles will be unstable when compared with larger particles. The equilibrium morphology of the crystal will be determined by the surfaces with lower Gibbs energy, while surfaces of higher energy are sacrificed. Different crystallographic planes will have a different number of bonds per unit of area, and the bond strengths can also change according to the composition. Actually, very often when a surface energy value of a given crystalline material is indicated, it is in fact an average of the surface energy of the different crystalline faces. Taking as example a face-centered cubic lattice, when increasing the Miller indices, typically the atomic density of the planes decreases. The exception is that in the plane family [1 1 1], which contains six nearest neighbors, three bonds for each surface atom have to be broken when cutting the crystal along such direction, while for [1 0 0] and [1 1 0] planes, with lower atomic density, four and six bonds have to be broken, respectively [22]. Therefore, the surface energy of the former planes is larger relatively to the [1 1 1] plane family. Planes with the highest density have a lower surface energy and rate of growth, and therefore the final morphology of the crystal growth will be defined by the high-density atomic planes [21, 22]. However, several studies have indicated that a spherical morphology, which is the configuration that minimizes the surface area, is energetically more favorable for solids at high temperatures, and the difference of surface energy between different crystallographic planes becomes a less important factor [22].

2.3. Electrical and dielectric properties

In this section, we will start to address the electrical properties of LN single crystalline and polycrystalline. Afterward, we will address polycrystalline LN. However, for both cases, it is important to analyze the properties for different temperature ranges and different stoichiometries, since both parameters have influence on the mechanisms of electrical conduction.

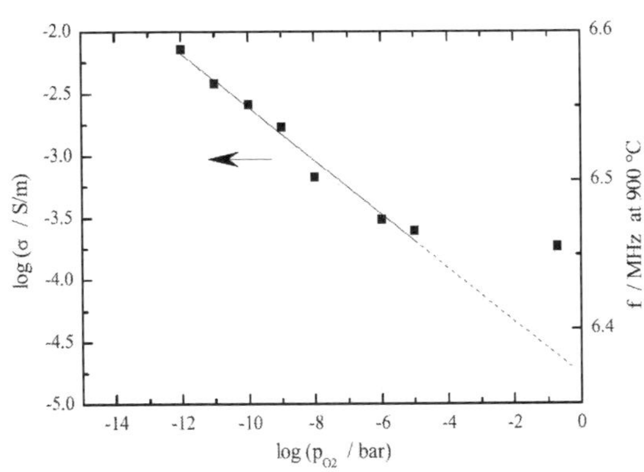

Figure 10. Dependence of a LN stoichiometric sample (sample 3 of **Figure 12**) total electrical conductivity with the oxygen partial pressure p_O, at 1173 K (900°C). The solid line is the linear fit of the experimental values, while the dotted line is the extrapolation [6].

It is well known that the electrical conductivity of LN single crystals, as well as some optical properties, is strongly dependent on the surrounding environmental characteristics, in particular the partial oxygen pressure p_0, as well as the Li/Nb ratio (stoichiometry). It was demonstrated that the electrical conductivity in high temperature ranges, between 600 and 1300 K, has a dependency of the type $p_0^{-1/4}$ for low $p_0 \lesssim 1$ torr (1 torr ≈ 1/760 of a standard atmosphere). **Figure 10** presents the dependence of a sample with [Li]/[Nb] = 1 (sample 3; see inset table of **Figure 12**) total electrical conductivity with p_0, at 1173 K (900°C) [6]. The solid line, which represents the linear fit of the experimental values, has a slope of approximately ¼ [6].

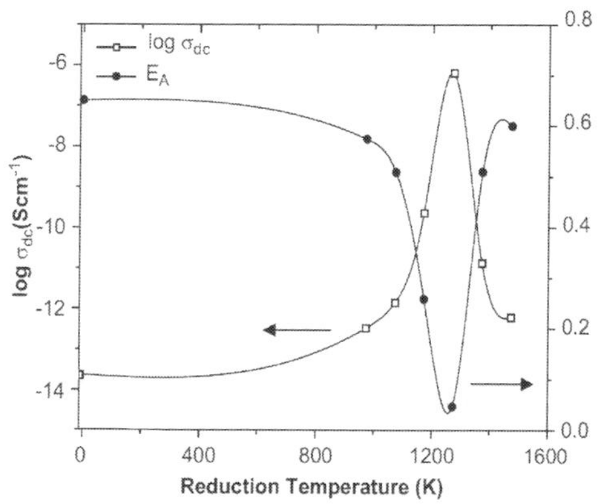

Figure 11. Dependence of σ_{dc} and E_A at room temperature with reduction temperature for different reduced congruent LN single crystals [23].

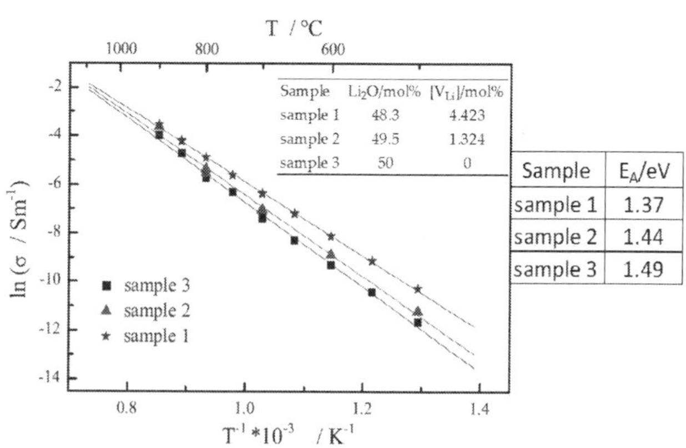

Figure 12. Arrhenius representation of the electrical conductivity of single-crystalline LN samples with different lithium contents, in the temperature range between 500 and 900°C. $[V_{Li}]$ represents the mol% of lithium vacancies, calculated through the lithium vacancy model (Eq. (1)). The activation energies (E_A) for the different samples are also indicated (adapted with permission from [6]).

The electrical properties of LN are conditioned by the oxidation or reduction atmosphere, during thermal annealing: in a low p_0 ($\lesssim$1 torr) environment, it will consist of a reducing atmosphere. The effects of a reducing atmosphere on LN crystals are typically referred in the literature to originate the following modifications: the first is loss of oxygens from the structure, which leads to the release of electrons which are trapped by Nb^{5+} cations, consequently originating Nb^{4+} cations; the second is that the reducing atmosphere leads to the diffusion and loss of lithium cations, creating more lithium vacancies and an excess of niobium cations relatively to lithium, thus leading to the occupation of Li^{+} lattice sites by the Nb^{5+} species (the so-called anti-site niobium defects Nb_{Li}), according to the lithium vacancy model, presented in Eq. (1) [6, 23]. Dhar et al. studied the low temperature (77–373 K) dependency of dc electrical conduction in reduced congruent LN single crystals [23]. The samples had different levels of oxygen reduction according to the temperature of reduction, in a vacuum of approximately 10^{-5} mbar. **Figure 11** shows the dependence of the dc conductivity (σ_{dc}) and activation energy (E_A) at room temperature with the reduction temperature for different samples [23].

The presence of a maximum in σ_{dc} and a minimum in E_A can be explained by the Mott's variable range hopping (VRH) mechanism: the oxygens released during reduction can produce free electrons according to Eqs. (8) and (9) [23]:

$$2Nb^{5+} + O^{2-} \leftrightarrow 2Nb^{4+} + O_v^{2-} + 1/2O_2 \tag{8}$$

and

$$Nb^{4+} \leftrightarrow Nb^{5+} + e^{-} \tag{9}$$

and therefore the release of oxygen during reduction originates free electrons that can get trapped in niobium ions, and the conduction mechanism is assigned to polaronic hopping between Nb^{5+} and Nb^{4+} cations [23]. The maximum and minimum observed in **Figure 11** can be explained according to the ratio of Nb^{4+}/Nb^{5+} states: for low reduction temperatures, few Nb^{4+} states will be created, while for high reduction temperatures, Nb^{4+} will predominate. However, for intermediate reduction temperature, there will be a case where we will get $Nb^{4+}/Nb^{5+} = 0.5$. In that case, a maximum in the conductivity and a minimum in E_A are to be expected, because the different oxidation states through which the polaronic hopping occurs are closer to each other [23]. They also concluded that actually the reduction annealing in low p_0 does not lead to a significant loss of lithium ions, as it was aforementioned (and is frequently mentioned in the literature), because they have shown that when reheating the reduced samples in an oxygen-rich atmosphere, without any content of lithium vapor, the conductivity decreases and regains practically the same value as unreduced samples, which imply that annealing at low p_0 does not have an important role in lithium loss [23]. For a polaronic VRH, the conductivity has a dependency as expressed in Eq. (10):

$$\sigma_{dc} = \sigma_0 exp\left[-\left(\frac{T_0}{T}\right)^{s}\right] \tag{10}$$

where σ_0 is the pre-exponential factor, T_0 is Mott's characteristic temperature, and s is the exponent for the VRH model. For reduced LN single crystals, the exponent s = ¼ is the one which better describes the temperature dependency of σ_{dc} (check on [23] to see a log(σ_{dc}) vs $T^{-1/4}$ plot).

For higher temperature ranges, in reduced LN single crystals, the VRH of polarons continues to be one of the mechanisms for the electrical conduction. However, and especially for congruent samples, lithium diffusion starts to be thermally activated, and when increasing the temperature, the main contribution for the total electrical conductivity may become ionic, assigned to lithium diffusion through lithium vacancies in the network (once again, we recall the lithium vacancy model) [6]. **Figure 12** shows the Arrhenius representation of the total electrical conductivity of LN single-crystalline samples with different lithium oxide molar percentages, in the temperature range between 773 and 1173 K [6]. The conductivity increases with the decrease of the Li_2O content, indicating the influence of lithium diffusion through lithium vacancies. The ionic conductivity will be larger for congruent LN, because the density of lithium vacancies is larger.

With respect to the frequency dependency of the electrical properties, typically studied by means of impedance spectroscopy (IS), some results of the work done by Mansingh and Dhar will be addressed, namely, those related with the ac electrical conductivity (σ_{ac}) and the dielectric constant (ε') of congruent LN single crystals [24]. This work was published in 1985; however, more recent papers reporting dielectric studies as a function of frequency and temperature for LN single crystals are surprisingly not that easy to find, because most of them deal with polycrystalline LN or with LN doped with other elements.

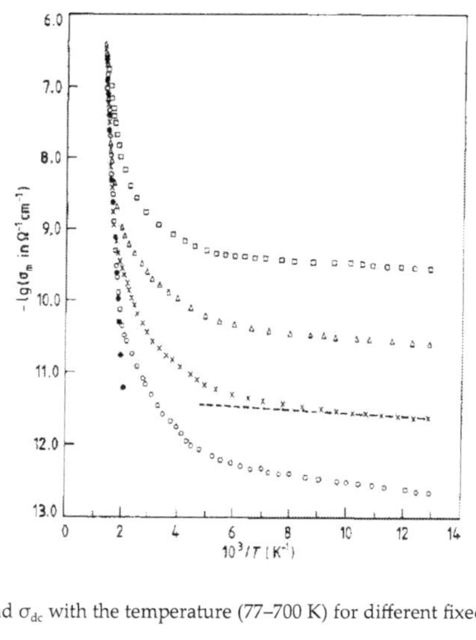

Figure 13. Dependence of σ_{ac} and σ_{dc} with the temperature (77–700 K) for different fixed frequencies (kHz): (•) DC, (○) 0.1, (x) 1, (Δ) 10, (□) 100 [24].

Figure 13 shows for a congruent LN single crystal the dependence of σ_{ac} and also σ_{dc} (although in a more limited temperature range relatively to σ_{ac}) with the temperature (77–700 K) for different frequencies, between 100 Hz and 100 kHz. It is visible that for lower temperatures σ_{ac} presents high-frequency dispersion, and it is considerably higher than σ_{dc}, while for higher temperatures it becomes practically frequency-independent and strongly temperature-dependent. Moreover, the temperature at which σ_{ac} starts to have the same value as σ_{dc} increases with the increase of the frequency. The mechanism for lower temperatures was found to be well described by a hopping-over-the-barrier (HOB) mechanism, and it was correlated with electron hopping between different valence states of the niobium, because of the reduction of Nb^{5+} due to oxygen deficiencies (as it was referred before) [24].

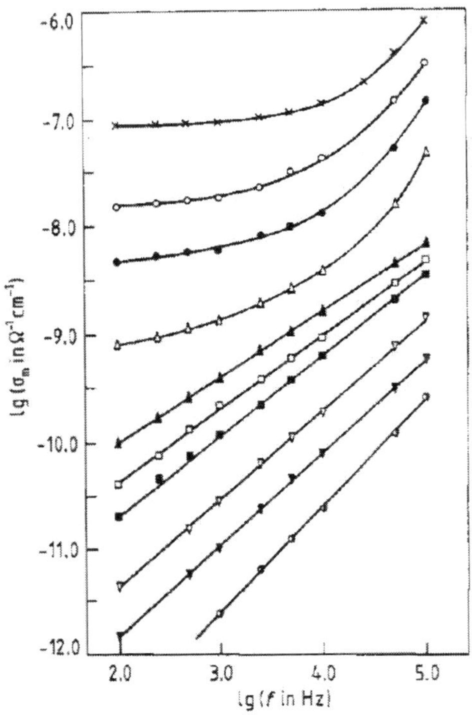

Figure 14. Frequency dependence of σ_{ac} for some fixed temperatures (K): (◑) 77, (▼) 220, (▽) 320, (■) 415, (□) 475, (▲) 530, (Δ) 580, (•) 625, (○) 650, (x) 680 [24].

Figure 14 shows the frequency dependence of σ_{ac} for some fixed temperatures. For lower temperatures, up to 530 K, the dependency of σ_{ac} with the frequency can be expressed by the relation presented in Eq. (11) [24]:

$$\sigma(\omega) = A\omega^s \tag{11}$$

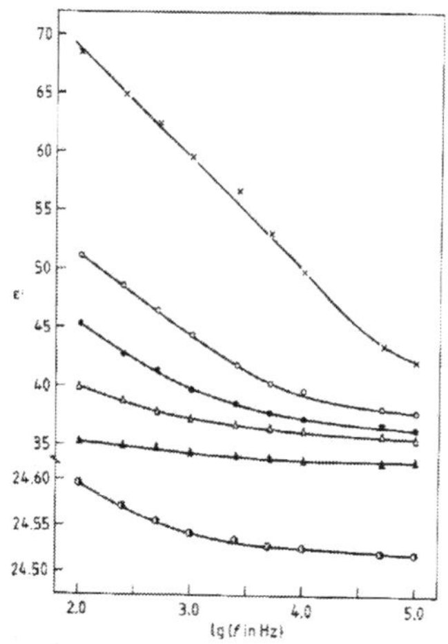

Figure 15. Frequency dependency of the dielectric constant (ε′) for different fixed temperatures (K): (◑) 77, (▲) 530, (Δ) 580, (•) 625, (○) 650, (x) 680 [24].

This is the well-known relation found by Mott and Davis which describes the frequency dependency of σ_{ac} for many amorphous and crystalline materials. The HOB mechanism presents a frequency dependency which can be described by Eq. (11). Furthermore, the HOB mechanism predicts a decrease of the frequency exponent s with the increase of temperature, and the values of s calculated by Mansingh and Dhar (~1 for 77 K and ~0.6 for 530 K) agree satisfactorily with the HOB model [24]. For higher temperatures, as it was aforementioned, σ_{ac} becomes practically frequency-independent and with a magnitude close to σ_{dc}. At the same time, for the same high temperature range (relatively to σ_{ac}, see **Figure 14**), ε' is characterized by a strong frequency dispersion, as shown in **Figure 15**.

The temperature dependence of ε' for different fixed frequencies, between 100 Hz and 100 kHz, is also shown in **Figure 16**.

It can be noted in **Figure 16** that the temperature from which ε' shows a sharp increase increases with the increase of frequency. So, from the presented plots of σ_{ac} and ε', it is evident that the temperature dependences show evidence of two distinct mechanisms for the conductivity. At low temperatures, the mechanism was already identified, the HOB mechanism with a high distribution of relaxation times. For higher temperatures, the strong ε' dispersion is probably associated with the dc conduction mechanisms, while both ac and dc conductivities are determined by the same mechanism, long-range hopping of charge carriers [24]. A sample with thickness reduced to half, keeping the electrode surface area constant, was included in **Figure 16** to demonstrate that the sharp increase of ε' is not related with the electrode barriers and spatial charge accumulation at the electrode/sample interface [24].

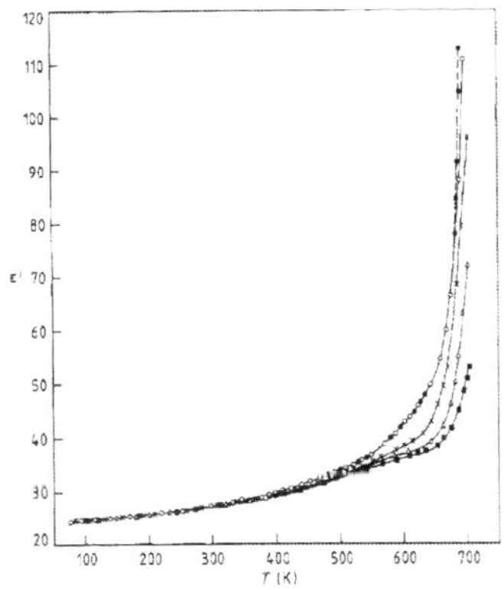

Figure 16. Temperature dependency of the dielectric constant (ε') for different fixed frequencies (kHz): (○) 0.1, (x) 1, (Δ) 10, (■) 100, (•) Sample with half of the thickness, keeping the same area [24].

We will end this section by briefly addressing the electrical and dielectric properties of polycrystalline LN. In such case, the behavior of the referred properties can reflect the presence of grain boundaries in the material. This effect of grain boundaries can be more clearly seen in IS measurements, because the characteristic frequencies (or times) at which grain boundary processes occur are different from those that occur in the bulk of the grains, and therefore Nyquist diagrams [plot of the negative of the imaginary part of the impedance, −Im(Z), in the y-axis over the real part of the impedance, Re(Z), in the x-axis] originate successive semicircles, where each point of the semicircle corresponds to a different frequency value which increases counterclockwise. Lanfredi and Rodrigues report in their work IS studies of the electrical conductivity and dielectric constant of polycrystalline LN [25]. **Figure 17** presents Nyquist diagrams for different temperatures of two polycrystalline LN samples *a* and *b*. Sample *b* has approximately half of the thickness to electrode surface area ratio (*l*/A) relatively to sample *a*.

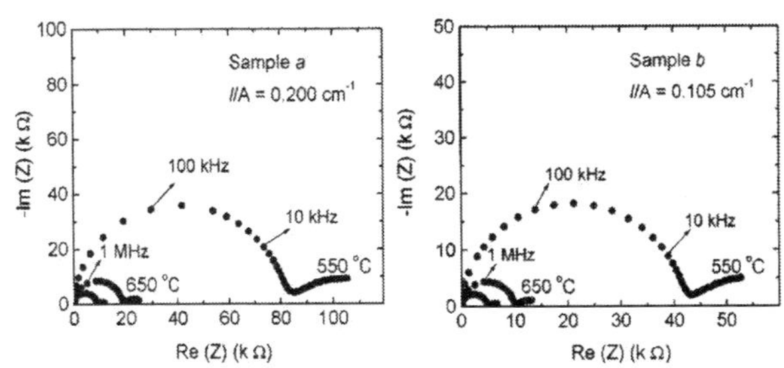

Figure 17. At left: Nyquist diagram for different temperatures for a polycrystalline LN sample with a thickness to electrode surface area ratio *l*/A = 0.200 cm^{-1} (sample *a*). At right: Nyquist diagram for different temperatures for sample *a* polycrystalline LN sample with a thickness to surface area ratio l/A = 0.105 cm^{-1} (sample *b*), approximately half of sample *a*. For both samples, the smallest semicircle corresponds to a measurement performed at 700°C [25].

Comparing both diagrams, it is visible that for the same temperature, the real part of the complex impedance Re(Z) of sample *b* is approximately half of sample *a*, thus confirming that the high-frequency semicircle is the bulk response of the samples, since the resistance is directly proportional to the path length [25]. Furthermore, the frequency value distribution in the bulk-response semicircle is the same for both samples, indicating the homogeneity of the bulk response, and that the relaxation frequency, given by the peak of the bulk semicircle (in the peak, the relation $2\pi f_0 R_b C_b = 1$ is fulfilled, where R_b and C_b are bulk resistance and capacitance, respectively), is an intrinsic property of the material and does not depend on geometrical factors [25]. As expected, the bulk resistance decreases with the increase of the temperature. The low-frequency semicircle is assigned to the response of grain boundaries, and its depressed shape is an indicator of a nonhomogeneous electrical behavior of grain boundaries [25]. This nonhomogeneous behavior can be related with an existence of a distribution of relaxation times.

The complex impedance semicircle for the bulk response can be well fitted by simple $R_b C_b$ equivalent circuit. A bulk electrical conductivity σ_b can be defined according to Eq. (12):

$$\sigma_b = \frac{1}{R_b}\frac{l}{A} \tag{12}$$

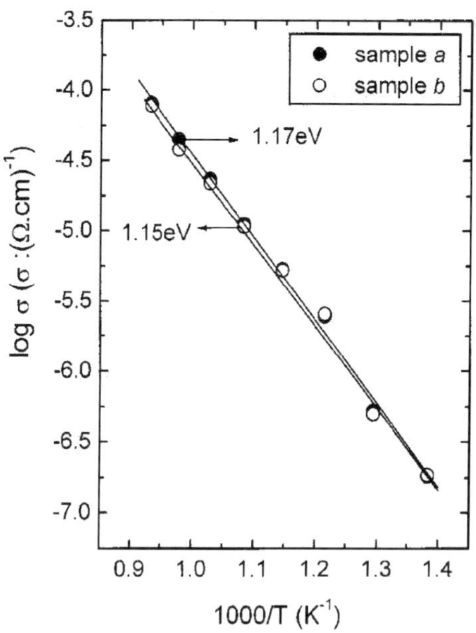

Figure 18. Arrhenius representation of the bulk electrical conductivity σ_b of the polycrystalline LN samples *a* and *b*, in the temperature range between 450 and 800°C [25].

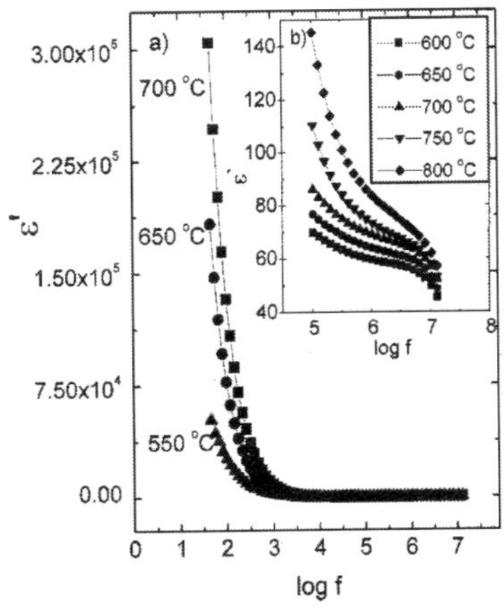

Figure 19. Frequency dependency, in the range of 5–10^7 Hz, of ε' for some fixed high temperatures, for sample *b* [25].

R_b can be determined through the Nyquist plots by the second interception (just before the grain boundary response) of the bulk semicircle with the real axis. **Figure 18** displays for both samples *a* and *b* the Arrhenius representation of σ_b in the temperature range between 450 and 800°C [25]. The activation energies are very similar.

To conclude this section, it is included in **Figure 19** the frequency dependency, between 5 and 10^7 Hz, of ε' for some fixed high temperatures, for sample *b*.

For lower frequencies, a strong dispersion of ε' is observed. This is due to spatial charge accumulation at the grain boundaries, and the charge accumulation at interface electrode/sample may also contribute to the sharp increase of ε' for lower frequencies. This behavior is often observed for polycrystalline materials.

2.4. Case study: preparation and characterization of polycrystalline LN by the Pechini method

2.4.1. Preparation process: the Pechini route

The sol-gel process is a well-known route for the synthesis of different types of materials, and in its basic description, it can be referred as a technique that synthesizes a solid compound through a chemical reaction in solution at low temperatures. There are different sol-gel methodologies according to the type of precursors used and the chemical reactions leading to the formation of the gel: there is the sol-gel methodology based on the hydrolysis-condensation of metal alkoxides, the "chelate-gel" route, involving aqueous solutions containing metal chelates, and the Pechini route. The sol-gel methodologies have general advantages such as the very good control of the stoichiometry and purity of the final material, low processing temperatures, possibility, and good flexibility in developing thin films as well as the possibility to have control over some important characteristics, such as the size and shape of the particles and homogeneity.

The Pechini route takes its name on its developer, Maggio Pechini in 1967 [26]. In particular, it was developed to include metals which are not suitable for traditional sol-gel reactions due to their unfavorable hydrolysis equilibria, and thus this method has the advantage of not requiring that the metallic species involved form stable hydroxo-complexes. This method is known for its use on the synthesis of multicomponent metal oxide materials [26], and basically this method uses an R-hydroxycarboxylic acid, such as citric acid (CA), to lead the formation of stable metal complex, i.e., the metallic cations of interest form stable complexes known as chelates. After this step, a polyalcohol, such as ethylene glycol (EG), is used to promote the polyesterification of the chelates, leading to the formation of a polymeric resin, where the metallic cations are trapped in the organic polymeric network. In other words, the polyalcohol is able to create links between the chelates by polyesterification reactions. The formation of the polymeric resin results in the formation of the gel. The subsequent drying process leads to the pyrolysis of the organic compounds, resulting in the formation of multicomponent metal oxide [26].

In this case study, the precursors lithium nitrate ($LiNO_3$) and niobium chloride ($NbCl_5$) (purity > 99%) were chosen. A molar ratio of 1:1 between $LiNO_3$ and $NbCl_5$ was established in order to enhance the formation of the $LiNbO_3$ stoichiometric phase. Firstly, the $LiNO_3$ and $NbCl_5$ were dissolved in deionized water and in a hydrogen peroxide solution (H_2O_2, 3%, V/V), respectively. For each gram of $NbCl_5$, 3.2 ml of H_2O_2 was used, originating a yellow transparent and clear solution. Both precursor solutions were mixed with citric acid (CA), fixing a molar ratio of 1:1 between the CA and the metallic cations, in order to form the metal complexes (chelates). The mixing was performed using a magnetic stirrer for 30 min at room temperature. After the mixing process, ethylene glycol (EG) was added to promote the polyesterification of the chelates. A mass ratio of 2:3 was established between the CA and EG to determine the quantity of EG to use. The final solution was mixed again with a magnetic stirrer for about 3 h, and the final gel was yellow and transparent, maintaining its macroscopic appearance for a long time period (>1 month). **Figure 20** outlines the entire preparation process.

The base powder was obtained after drying the gel at 300°C for 1 h, with a heating ramp of 5°C/min, yielding a black/grayish powder.

2.4.2. *Thermal, structural, and morphological properties*

The base powder was subjected to several heat treatments (HTs) at temperatures between 400 and 1000°C. These temperatures were chosen according to the differential thermal analysis (DTA) results, presented in **Figure 21**. This thermal technique was performed between room temperature and 1200 °C using a Linseis 63A apparatus. The heating rate was 20 °C/min and Al_2O_3 powder as used as reference. In **Figure 21**, the thermogram shows the presence of three exothermic thermal processes at the temperatures of 490, 790, and 1050°C, approximately. Consequently, HTs were performed at 450, 500, 800, and 1000°C, for 4 h.

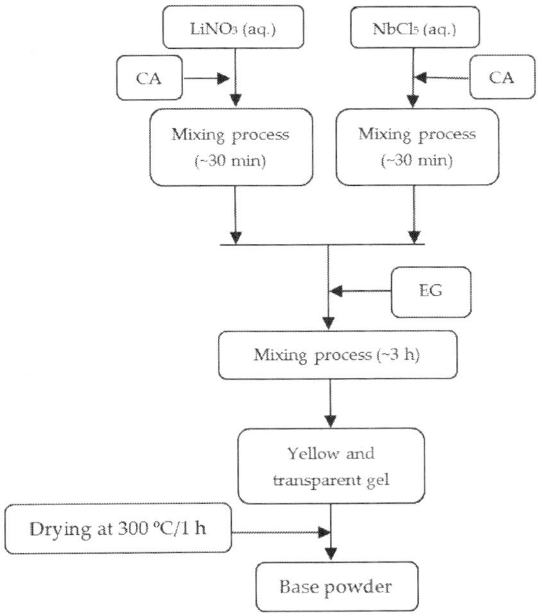

Figure 20. Diagram of the Pechini method used for the synthesis of the LN base powder. CA, citric acid; EG, ethylene glycol.

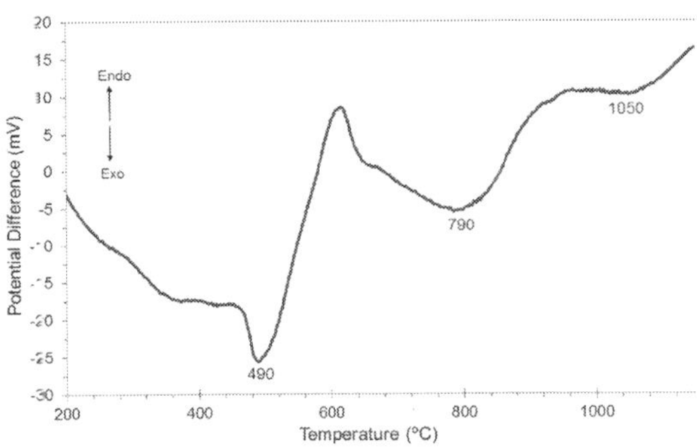

Figure 21. DTA thermogram of the base powder synthesized by the Pechini method. This thermal analysis was performed between room temperature and 1200 °C, with a heating rate of 20 °C/min.

The XRD measurements performed on these powders (not shown here) revealed that for the HT at 500°C, the LN and $LiNb_3O_8$ crystalline phases are present and that the increase of the HT temperature promotes the development of the $LiNb_3O_8$ phase. However, the powder HT at 450°C only contained the $LiNbO_3$ phase, whereby this powder was used to prepare pellets which were then sintered at 450°C for 4, 12, 24, 48, and 96 h. Thirty milligrams of powder was used for the preparation of the 10-mm diameter pellets, resulting in a thickness of about 1 mm when applying a uniaxial pressure of 1.5 tons. Hereafter, we will refer to these pellets as samples.

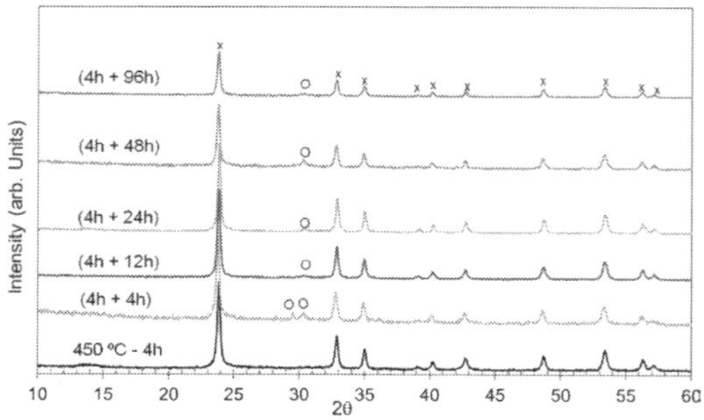

Figure 22. XRD patterns of the base powder HT at 450°C and of the samples sintered for 4, 12, 24, 48, and 96 h (× $LiNbO_3$; O $LiNb_3O_8$).

Figure 22 depicts the XRD patterns of the samples sintered during the aforementioned time intervals. The patterns show that the sintering process activated the formation of the $LiNb_3O_8$ phase. The XRD technique was performed on a Philips X'Pert MPD (CuK_α radiation, λ = 1.54056 Å), with a step 0.02° in 1 s, in the 2θ angle range of 10–60°. The identification of the crystalline phases was made using the database of the Joint Committee on Powder Diffraction Standards–International Center for Diffraction Data. To get a further insight about the contents of each phase in the samples as well as to calculate the crystallite sizes associated with each phase, a Rietveld refinement was performed for all the diffraction patterns shown in **Figure 22**, using the PowderCell software. In **Figure 23**, the Rietveld fits of the samples 4 h and 4 h + 96 h are presented.

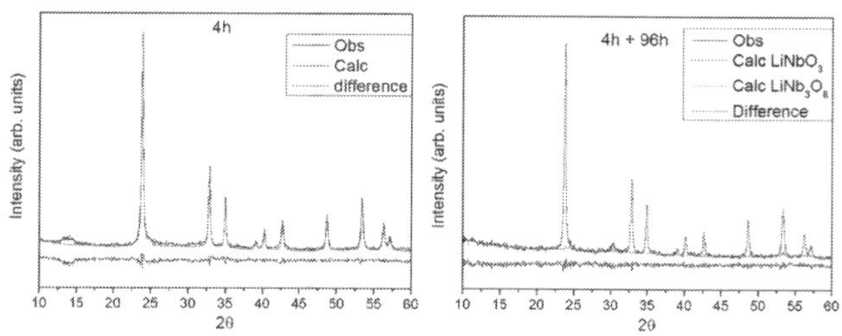

Figure 23. The Rietveld fits of the XRD patterns for the powder HT at 450°C, containing only the LN crystalline phase and the sample sintered for 96 h, containing both LN and $LiNb_3O_8$.

The Rietveld fit parameters indicate a good fit of the structural models to the experimental data (**Table 2**). Although the XRD patterns shown in **Figure 22** may suggest that the $LiNb_3O_8$ phase is present in small amounts, the mass percentages shown in **Table 2** show that the presence of this phase is relevant, reaching the maximum value for the sample 4 h + 96 h. In fact, both LN and $LiNb_3O_8$ have reflections lying in close diffraction angles, and some of the observed peaks contain a contribution of the $LiNb_3O_8$ phase, besides the LN phase, explaining the relatively high mass percentages. The crystallite size of both phases stands approximately constant for the different samples, especially for the $LiNb_3O_8$ phase, which has larger sizes relatively to the LN phase.

Sample	R_{wp}	R_{exp}	χ^2	Crystallite size (nm)		Mass%		ε'	tan δ
				$LiNbO_3$	$LiNb_3O_8$	$LiNbO_3$	$LiNb_3O_8$		
4 h	8.78	6.20	2.01	43.90	–	100	–	–	–
4 h + 4 h	10.40	8.72	1.42	45.52	61.94	72.48	27.52	8.64	0.06
4 h + 12 h	6.26	4.77	1.72	48.30	61.32	66.48	33.52	9.37	0.05
4 h + 24 h	6.31	4.49	1.97	50.23	61.03	66.94	33.06	16.85	0.09
4 h + 48 h	9.72	8.23	1.39	44.84	62.28	75.90	24.10	16.36	0.13
4 h + 96 h	9.37	7.70	1.48	46.65	59.50	61.76	38.24	13.06	0.18

Table 2. The initial three parameters are the weighted profile *R*-factor (R_{wp}), the expected *R*-factor (R_{exp}), and the "chi squared" χ^2. The crystallite size and mass percentage of each crystalline phase in all the samples are also indicated. The last two columns show the dielectric constant (ε') and loss tangent (tan δ) at 10 kHz and room temperature (300 K).

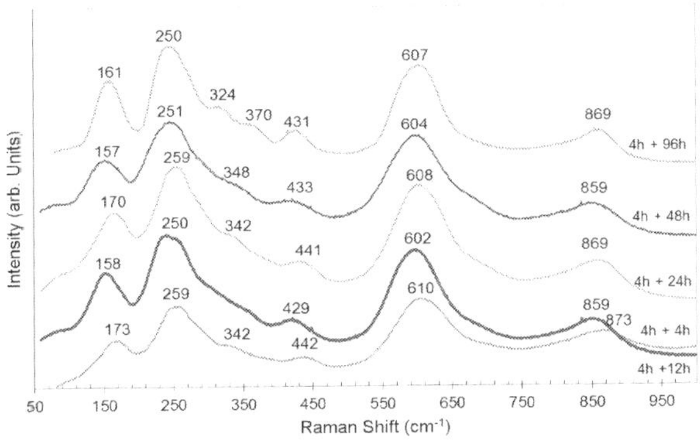

Figure 24. The Raman spectra of the samples sintered for different times between the time range 4 and 96 h, performed at room temperature.

In **Figure 24**, the Raman spectra of the sintered samples are presented. The spectra show the presence of vibrational bands which are the result of an overlapping of vibrational models of LN and also of the $LiNb_3O_8$ crystalline phase, as a consequence of the polycrystalline structure of the prepared samples. For a further analysis of the LN and $LiNb_3O_8$ vibrational modes, the authors suggest the reading of Bartasyte et al. report [27]. The room-temperature Raman spectroscopy was performed in backscattering geometry using a T64000 Jobin-Yvon spectrometer. A microscope objective (50×) focused the exciting light (Ar^+ laser, λ = 532 nm) onto the sample (spot diameter <0.8 μm).

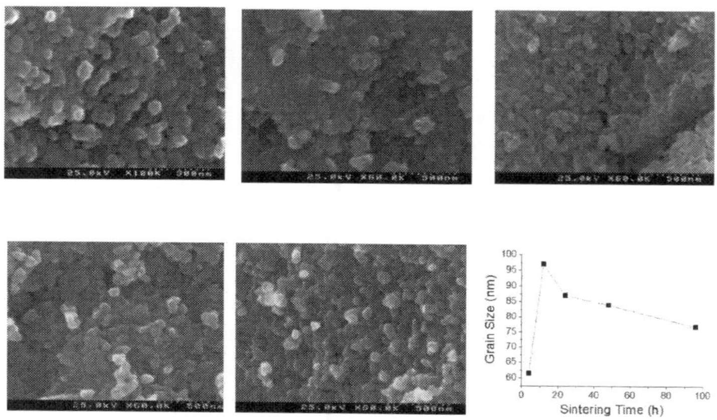

Figure 25. SEM micrographs of the LN polycrystalline samples prepared by the Pechini method, sintered at 450°C for 4, 12, 24, 48, and 96 h.

In **Figure 25**, SEM micrographs of the samples are shown, as well as the variation of the grain size with the sintering time. The grain size increases significantly with the sintering time from 4 to 12 h and then decreases slightly up to 96 h. As for the morphology, the grains show an approximate spherical geometry. The SEM was executed on a HITACHI S4100-1. All samples were sputtered with carbon before the analysis.

2.4.3. Electrical and dielectric properties

The electrical analysis of these types of samples is not evident. **Figure 26** shows the temperature dependence of the dc conductivity (σ_{dc}) for all samples, revealing a decrease of the conductivity with the increase of the sintering time. This suggests a probable decrease in the number of charge carriers.

However, **Figure 27** shows that the activation energy, calculated in the high-temperature region (marked by the Arrhenius regression line in **Figure 26**), decreases substantially from sample sintered at 4 h to the sample sintered at 96 h. This profile indicates that the height of the potential barriers, from which the charge carriers must pass through, decreases, which should promote an increase in their mobility. Therefore, this analysis indicates a decrease in the number of charge carriers and also an increase in their mobility.

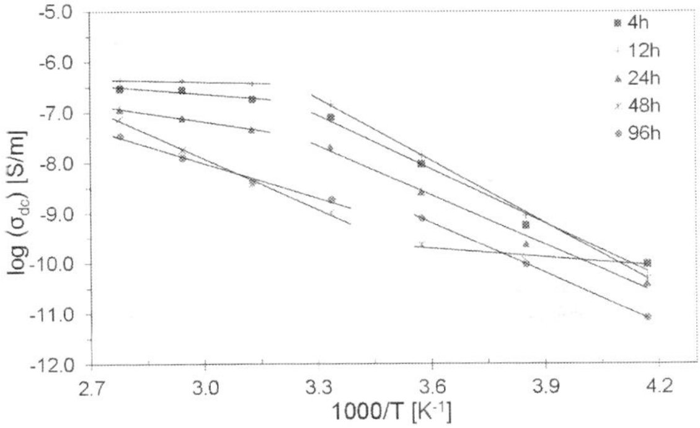

Figure 26. Arrhenius representation of the dc conductivity (σ_{dc}) for the samples sintered at 450°C for 4, 12, 24, 48, and 96 h.

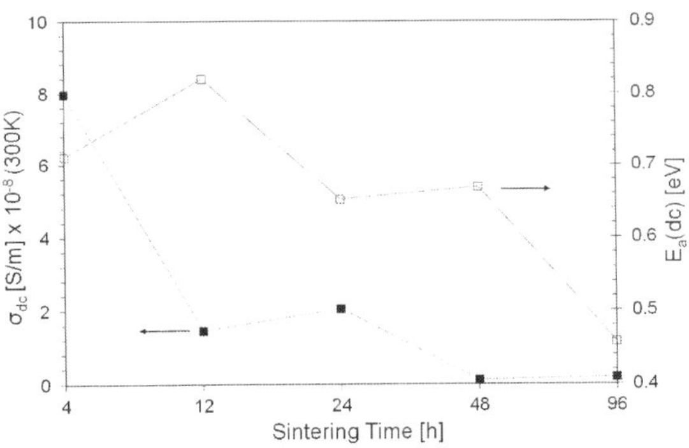

Figure 27. Dc conductivity (σ_{dc}) and activation energy (E_a(dc)) of the samples sintered at 450°C for 4, 12, 24, 48, and 96 h, at room temperature (300 K).

As discussed in Section 2.3, the electrical conduction in crystalline LN can be due to oxygen vacancies, in low-pressure and/or high-temperature conditions (polaron-hopping mechanism), and due to lithium ions in high-pressure conditions (ionic conduction). In this case, the conduction will depend on the ratio between Li/Nb both for high and low pressures. Therefore, in this case context, the conduction related with the lithium-ion (Li^+) hypothesis seems to be more relevant. The existence of the $LiNb_3O_8$ secondary phase, which is a reach niobium LN phase, implies an increase of the number of Li^+ ions not connected to a crystalline structure, which allied to the decrease of the activation energy should improve the σ_{dc}, which is not observed. This fact indicates that the number of Li^+ ions inserted in the crystalline structure increases and is explained by two hypotheses: there is a possible existence of an amorphous phase in all samples and thus the number of lithium ions inserted in this amorphous phase should decrease with the increase of the sintering time, stimulating an increase of the crystalline amount. The second hypothesis is associated with the low-melting point of lithium (~454 K). The increase of the sintering time must raise the possibility of some Li^+ ions to be released during sintering. Density measurements revealed that all samples have a density between 3.53 and 3.69 g/cm^3, which are values lower than the characteristic ones for LN (4.65 g/cm^3) and $LiNb_3O_8$ (3.87 g/cm^3). This fact increases the probabilities of the first hypothesis.

Both dc and ac conductivities (**Figures 26–29**) increase with the rise of the measurement temperature, showing that both mechanisms are thermally activated. However, with the increase in the sintering time, the σ_{ac} shows the opposite trend of σ_{dc} (it increases). This indicates that the conduction mechanism and dominant charge carriers are not the same in both processes. In our opinion, the σ_{ac} increase should be related with the decrease of the grain size (**Figure 25**) observed from sample sintered at 12 h up to the sample sintered at 96 h.

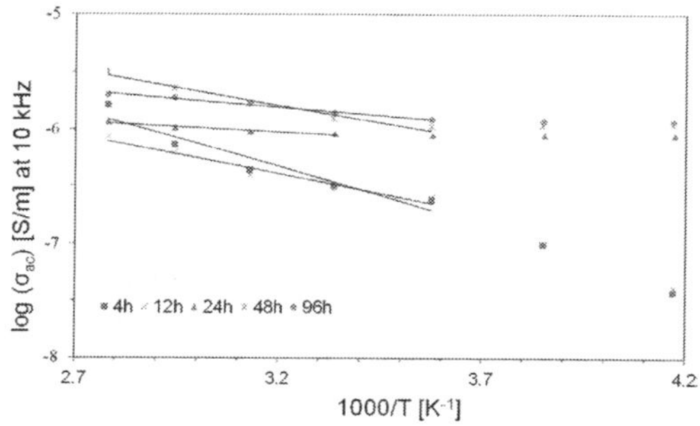

Figure 28. Arrhenius representation of the ac conductivity (σ_{ac}) for the samples sintered at 450°C for 4, 12, 24, 48, and 96 h.

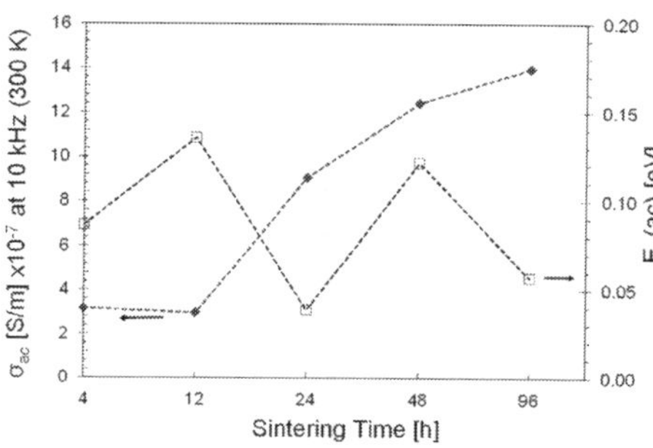

Figure 29. Ac conductivity (σ_{dc}) and activation energy (E_a(ac)) of the samples sintered at 450°C for 4, 12, 24, 48, and 96 h, at 10 kHz and room temperature (300 K).

http://dx.doi.org/10.5772/64395

The dielectric constant is maximum for the sample sintered at 24 h (**Table 2**), measured at room temperature and 10 kHz, which is in accordance with the Rietveld refinement results (**Table 2**), by presenting the highest crystallite size. The sample sintered at 48 h shows a dielectric constant value very near the maximum, which should be related with the amount of LN phase present in the sample and not the size of their grains (**Table 2**). It must be noted that the values presented in **Table 2** were obtained without assuming the existence of the hypothetical amorphous phase. These two samples present a dielectric loss below 0.15 at room temperature and 10 kHz (**Table 2**), being almost constant for temperatures between 200 and 300 K.

The 3D plots (**Figures 30** and **31**) show the presence of a dielectric relaxation phenomenon practically independent on the measurement temperature. However, the frequency of the maximum observed in the loss tangent diagram decreases with the increase of the sintering time, which can be related to the suggested decrease of the amorphous phase content and consequent increase of the LN in crystalline form which is a material where the dielectric depolarization is very difficult, i.e., high-relaxation time characteristic.

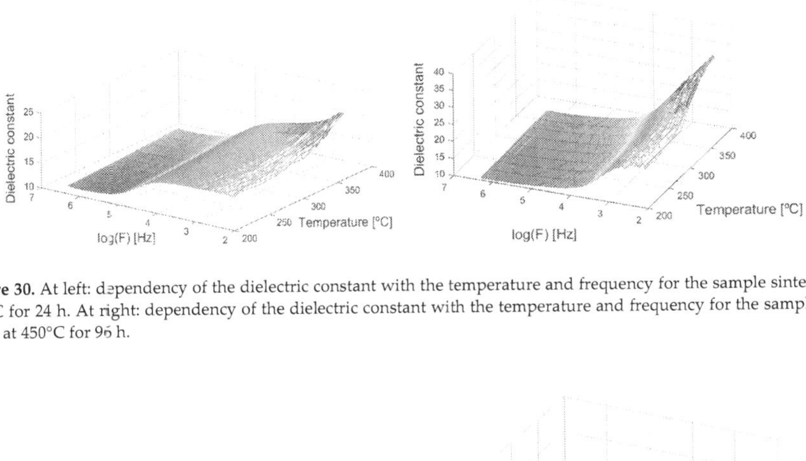

Figure 30. At left: dependency of the dielectric constant with the temperature and frequency for the sample sintered at 450°C for 24 h. At right: dependency of the dielectric constant with the temperature and frequency for the sample sintered at 450°C for 96 h.

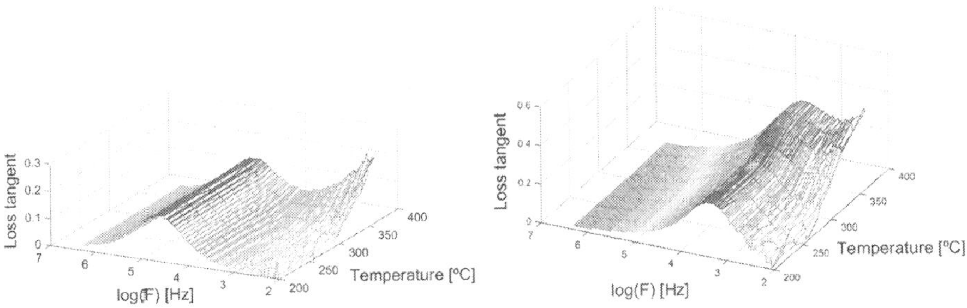

Figure 31. At left: dependency of the loss tangent with the temperature and frequency for the sample sintered at 450°C for 24 h. At right: dependency of the loss tangent with the temperature and frequency for the sample sintered at 450°C for 96 h.

Author details

Pedro R.S. Prezas* and Manuel P.F. Graça

*Address all correspondence to: pedro.rafael@ua.pt

University of Aveiro/I3N—Physics Department, Aveiro, Portugal

References

[1] Abouelleil M.M. and Leonberger F.J., *Waveguides in lithium niobate,* J. Am. Ceram. Soc., 1989, 72(8):1311-1321

[2] Graça M.P.F. and Valente M.A. *Glass ceramics with para, anti or ferroelectric active phases,* Advances in Ceramics – Electric and Magnetic Ceramics, Bioceramics, Ceramics and Environment, Prof. Sikalidis C. (Ed.), InTech (2011). Janeza Trdine 9, 51000 Rijeka, Croatia, ISBN: 978-953-307-350-7.

[3] Volk T. and Wohlecke (2008), Lithium Niobate: Defects, Photorefraction and Ferroelectric Switching, Springer-Verlag, Berlin, Heidelberg, ISBN: 978-3-540-70765-3.

[4] Bordui P.F., Norwood R.G., Jundt D.H., Fejer M.M., *Preparation and characterization of off-congruent lithium niobate crystals,* J. Appl. Phys., 1992, 71(2):875-879

[5] Kitamura K., Yamamoto J.K., Iyi N., Kimura S., Hayashi T., *Stoichiometric $LiNbO_3$ single crystal growth by double crucible Czochralski method using automatic powder supply system,* J. Cryst. Growth, 1992, 116:327-332

[6] Weidenfelder A., Shi J., Fielitz P., Borchardt G., Becker K.D., Fritze H., *Electrical and electromechanical properties of stoichiometric lithium niobate at high-temperatures,* Solid State Ionics, 2012, 225:26-29

[7] Szaller Zs., Péter Á., Polgár K., Szabó Gy., *High temperature top seeded solution growth of stoichiometric lithium niobate $LiNbO_3$ (sLN) with planar interface,* J. Cryst. Growth, 2012, 360:181-184

[8] Damjanovic D., *Materials for high temperature piezoelectric transducers,* Curr. Opin. Solid State Mater. Sci., 1998, 3(5):469-473

[9] Ohlendorf G., Richter D., Sauerwald J., Fritze H., *High-temperature electrical conductivity and electromechanical properties of stoichiometric lithium niobate,* Diff. Fundam., 2008, 8:6.1-6.7

[10] Kamali A.R. and Fray D.J., *Preparation of lithium niobate particles via reactive molten salt synthesis method,* Ceram. Int., 2014, 40:1835-1841

[11] Graça M.P.F., Prezas P.R., Costa M.M., Valente M.A., *Structural and dielectric characterization of $LiNbO_3$ nano-size powders obtained by Pechini method*, J. Sol-Gel Sci. Technol., 2012, 64:78-85

[12] Howlader M.M.R., Suga T., Kim M.J., *Room temperature bonding of silicon and lithium niobate*, Appl. Phys. Lett., 2006, 89:031914

[13] Rahn J., Hüger E., Dörrer L., Ruprecht B., Heitijans P., Schmidt H., *Self-diffusion of lithium in amorphous lithium niobate layers*, Z. Phys. Chem., 2012, 226(5-6):439-448

[14] Sanna S. and Schmidt W.G., *Ferroelectric phase transition in $LiNbO_3$: insights from molecular dynamics*, IEEE Ultrason. Ferr., 2012, 59(9):1925-1928

[15] Scrymgeour D.A., Gopalan V., Itagi A., Saxena A., Swart P.J., *Phenomenological theory of a single domain wall in uniaxial trigonal ferroelectrics: lithium niobate and lithium tantalite*, Phys. Rev. B, 2005, 71(18):184110-1/13

[16] Heitjans P., Masoud M., Feldhoff A., Wilkening M., *NMR and impedance studies of nanocrystalline and amorphous ion conductors: lithium niobate as a model system*, Faraday Discuss., 2007, 134:67-82

[17] Kitabatake M., Mitsuyu T., Wasa K., *Structure and dielectric properties of amorphous $LiNbO_3$ thin films prepared by a sputtering deposition*, J. Appl. Phys., 1984, 56:1780-1784

[18] Ridah A., Bourson P., Fontana M.D., Malovichko G., *The composition dependence of the Raman spectrum and new assignment of the phonons in $LiNbO_3$*, J. Phys.: Condens. Matter., 1997, 9(44):9687-9693

[19] Schlarb U., Klauer S., Wesselmann M., Betzler K., Wöhlecke M., *Determination of the Li/Nb ratio in lithium niobate by means of birefringence and Raman measurements*, Appl. Phys. A, 1993, 56(4):311-315

[20] Damen T.C., Porto S.P.S., Tell B., *Raman effect in zinc oxide*, Phys. Rev., 1966, 142(2): 570-574

[21] Zhan J., Liu D., Du W., Wang Z., Wang P., Cheng H., et al., *Synthesis and characterization of high crystallinity, well-defined morphology stoichiometric lithium niobate nanocrystalline*, J. Cryst. Growth, 2011, 318(1):1121-1124

[22] Stølen S. and Grande T. (2005), *Chemical Thermodynamics of Materials: Macroscopic and Microscopic Aspects*, Wiley, Hoboken, NJ, ISBN: 978-0-471-49230-6

[23] Dhar A., Singh N., Singh R.K., Singh R., *Low temperature dc electrical conduction in reduced lithium niobate single crystals*, J. Phys. Chem. Solids, 2013, 74(1):146-151

[24] Mansingh A. and Dhar A., *The ac conductivity and dielectric constant of lithium niobate single crystals*, J. Phys. D: Appl. Phys., 1985, 18(10):2059-2071

[25] Lanfredi S. and Rodrigues A.C.M., *Impedance spectroscopy study of the electrical conductivity and dielectric constant of polycrystalline $LiNbO_3$*, J. Appl. Phys., 1999, 86(4):2215-2219

[26] Roque-Malherbe R.M.A. (2009), *The Physical Chemistry of Materials: Energy and Environmental Applications*, CRC Press, 6000 Broken Sound Parkway, NW Suite 300 Boca Raton, FL 33487, United States, ISBN: 9781420082722.

[27] Bartasyte A., Plausinaitiene V., Abrutis A., Stanionyte S., Margueron S., Boulet P., et al., *Identification of $LiNbO_3$, $LiNb_3O_8$ and Li_3NbO_4 phases in thin films synthesized with different deposition techniques by means of XRD and Raman spectroscopy*, J. Phys.: Condens. Matter., 2013, 25(20):205901.

Chapter 4

Raman Spectroscopy, a Useful Tool to Study Nuclear Materials

Laura J. Bonales, Jone M. Elorrieta,
Álvaro Lobato and Joaquin Cobos

Additional information is available at the end of the chapter

http://dx.doi.org/10.5772/64436

Abstract

The use of the Raman technique is nowadays being widely spread in many scientific and industrial disciplines. The rise of this spectroscopy is due to the technology developed in some of its main components (the laser, charge-coupled device (CCD) sensors, gratings, filters, etc.), what reduces the cost of the equipment. Characterization by Raman spectroscopy has long and well-established tradition in fields such as condensed matter physics and chemistry. In nuclear sciences, by contrast, it is far from being extensively applied, even though this technique can be especially useful. It is a fact that only a scarce number of Raman laboratories dealing with nuclear materials exist, and therefore a limited database related to these materials. In such a context, this chapter is devoted to the practical use of Raman spectroscopy for nuclear materials characterization.

Keywords: Raman, nuclear, uranium oxides, U-secondary phase

1. Introduction

Nuclear power is a controversial subject that generates a debate in today's society; nevertheless, the existing 438 nuclear reactors [1] produce approximately 15% of the world's electricity [2], making it a major world energy source.

Proponents of nuclear energy describe nuclear power production as a low running costs and a nature friend since it can be considered a low carbon technology compared with fossil fuels such as coal, gas, and oil [3]. Opponents of nuclear power argue that this energy has very high initial costs, complex nuclear waste management, and high plant decommissioning costs, and highlight the environmental and public risks and dangers associated with it [4, 5]. This debate has been mirrored by the number of nuclear power plant (NPP) constructions. For example [6], before the accident in Fukushima Daiichi in 2011, the number of NPPs was increasing due to the concerns over greenhouse gas emissions to avoid further global warming. Then, the accident impacted public acceptance of nuclear power and had an effect of decreasing the number of nuclear reactors [7]. In recent years, the number of constructions commencing are on the rise again [8], and the main concern is related to the nuclear waste management, mainly spent nuclear fuel (SNF).

The most common nuclear fuel consists of uranium dioxide, UO_2, enriched from 0.7 to 3–5% of ^{235}U depending on which reactor it will be loaded into [9]. After its irradiation in the reactor core, the fuel is composed of a matrix of UO_2 (>95%) doped with fission products and trans-uranium elements [10], which are found as bubbles (Xe, Kr), metallic precipitates (Mo, Tc, Ru, Rh, and Pd), oxide precipitates (Rb, Cs, Ba, and Zr), solid solutions, and transuranium elements dissolved by U substitution in the UO_2 matrix [11]. These elements are not distributed homogeneously as a consequence of the thermal gradient within the UO_2 pellet (temperature as high as 1700°C at the center of the pellet and decreasing to 400°C outwards) [12, 13]. Besides, the spent nuclear fuel suffers substantial microstructural modifications from the initial fresh fuel such as coarsening of the grains and extensive microcracking. Thus, SNF can be described as a complex, hot, and radioactive waste and therefore extremely dangerous. Bruno et al. [14] provide a particularly suitable example to demonstrate how hazardous the SNF is, that is, ..."*One year after discharge from a reactor...a person exposed to this level of radioactivity at a distance of one meter would receive a lethal dose in less than one minute....*"

After several thousands of years, the total radioactivity of the SNF equals the radioactivity of natural uranium [15]. Therefore, within the management of spent nuclear fuel, the safe storage of this radioactive waste from the discharge of the reactor until the decay reaches natural uranium radioactivity is considered. Countries adopt different stages of these nuclear storages according to their internal policies [16], but usually the following steps are suitable: (i) spent fuel pools [17], (ii) intermediate storage or reprocessing [18], and (iii) final storage [19].

After SNF is removed from the reactor, it is stored for the first few years on-side in water containers or pools, located close to the reactor in order to allow the spent fuel to decay, both radioactively and thermally. Then, it can be transported to a reprocessing facility or to a definitive storage facility. However, since final repositories for spent fuel do not exist for the moment, interim storage is required. NPPs use the spent fuel dry-cask storages, which are steel and concrete containers filled with an inert gas as a first step for interim storage. Although no ultimate storage in operation exists, the deep geological repository is internationally accepted as the best solution [20].

The performance of the mentioned repositories requires knowledge of the SNF stability at different storage conditions. Thereby, the studies of the spent fuel behavior can be mainly divided into dry and wet conditions, given the different evolution observed in each case. The studies of spent nuclear fuel under dry conditions are mainly focused on the oxidation of both the UO_2 matrix and the AnO_2 [21] present in the SNF.[1] In case of shielding failure, the oxidation of AnO_2 and UO_2 takes place owing to its contact with the atmospheric oxygen and the high temperatures present (up to 400°C) [15]. This oxidation occurs via oxygen incorporation into the fluorite structure (fcc) of the stoichiometric oxide, for example, some actinides as plutonium and uranium can oxidize to AnO_{2+x} ($x < 0.25$) maintaining the fcc structure [22]. Further oxidation to higher-oxidation states (V and VI) leads to different structures. For example, the transformation of UO_2 into U_3O_8 via the two-step reaction [23] $UO_2 \rightarrow U_4O_9/U_3O_7 \rightarrow U_3O_8$ entails an increase in the volume of around 36% and, consequently, it might cause the loss of the UO_2 matrix integrity.

On the other hand, the studies of spent nuclear fuel under wet conditions are focused on the corrosion process of this waste. This might happen in case the SNF shielding fails while stored in pools or in the deep geological repository at timescales of the order of some thousands of years [24] when it is assumed that the barriers that protect the waste will be breached and SNF will be in contact with water [25]. The UO_2 matrix of the spent nuclear fuel might dissolve with water and then the release to the biosphere of the SNF radioactive contents might occur [26–28].

This corrosion process is primarily described by the oxidation of uranium, U(IV) → U(VI), and then the alteration products formation, usually containing UO_2^{2+} in their crystal structures [29] U(VI) → UO_2^{2+} (s).

Great effort has been performed to analyze the reaction mechanism and to establish the key parameter that controls the corrosion of the SNF such as leaching/dissolution experiments [30–32] and studies of the uraninite, a natural analog of the spent nuclear fuel matrix [33, 34]. These stability studies require the characterization of the SNF and its reaction products, with O_2 and/or water, which is a great challenge not only because these materials are very complicated (containing almost the entire periodic table) [11] but also because intense radiation field inherently associated to these materials makes it difficult to examine them in safe conditions.

In order to minimize radiation doses and the release of radioactive material, the working procedure employed to study these materials must fulfill the ALARA principle (acronym for "as low as reasonably achievable") [35]. Such a reliable procedure must, hence, minimize the time that radioactive materials are handled and maximize the distance to them. Raman spectroscopy is an analyzing technique that has been established in recent years as a useful tool since it fulfills the mentioned safe principles, as shown by the increase in the number of publications dealing with the characterization of nuclear materials by this technique [36–44]. This is due to some of its features as (1) that it does not require any special preparation of the sample, (2) it allows the analysis of a very small amount of sample, and (3) it is a nondestructive technique.

Besides these safety principles, the confinement of the whole apparatus in a glove box or a hot cell is also very common, which obviously complicates the measurements [45]. Despite the advantages mentioned above, the characterization of these SNF and related nuclear materials is far from being well established. Existing databases must be improved and new methods

1 A meaning minor actinides such as Np, Pu, and Am.

must be developed. Due to the hazardous feature of nuclear materials, both the development of new protocols and Raman spectra acquisition (for the purpose of extending the databases) are usually performed first by analyzing the behavior of different SNF analogs and, once the method is feasible, by applying it to the real SNF. Such analogs can be divided into two kinds: synthetic analogs such as uranium dioxide (UO_2) [46, 47] or SIMFUEL (simulated fuel) [48, 49], and natural analogs such as uraninite.

In this context, this chapter is structured as follows: In the first part, Raman spectroscopy is described. First, the theoretical aspects on an introductory level are explained. Second, the main components of the Raman spectrometers are presented and, as an example, the *LabRam HR Evolution* spectrometer is described in more detail. The "Results" section has been divided into two, corresponding to dry and wet conditions; in each part, the developed method and the results found for analogs of the SNF are shown. Namely, the materials studied in this section are the different uranium oxides, UO_{2+x} ($0 < x < 0.25$), U_4O_9/U_3O_7, and U_3O_8, and several secondary phases such as rutherfordine, soddyite, uranophane alpha, or kasolite.

2. Raman spectroscopy technique

2.1. Description of the Raman phenomena

Raman effect owes its name to the Indian physicist Chandrasekhara Venkata Raman [50] who won the Nobel Prize for its discovery. In his Nobel lecture, given on December 11, 1930, Sir C.V. Raman said…"*The frequency differences determined from the spectra, the width and character of the lines appearing in them, and the intensity and state of polarization of the scattered radiations enable us to obtain an insight into the ultimate structure of the scattering substance […]. It follows that the new field of spectroscopy has practically unrestricted scope in the study of problems related to the structure of matter*" [51].

As other molecular spectroscopy techniques, Raman scattering is based on the analysis of light-matter interaction [52], that is, absorption, emission, or scattering of a photon. Two interpretations of this phenomenon can be considered: the quantum mechanical method and the classical interpretation. In the purely classical interpretation, the radiation is considered as an electromagnetic wave, and the matter as an assembly of independent classical rotors and vibrators. This model can explain satisfactorily the main features of the light scattering such as the frequency dependence and some key aspect related to their selection rules.

Raman effect is described as an inelastical scattering of light. From a macroscopic point of view, light scattering consists in a deviation of light from its straight trajectory (original direction of incident light). Molecules scatter light because the electric field of the incident light wave forces the electrons within the molecule to oscillate (see **Figure 1**), producing oscillating electric moments leading to the reemission of radiation in all directions [53].

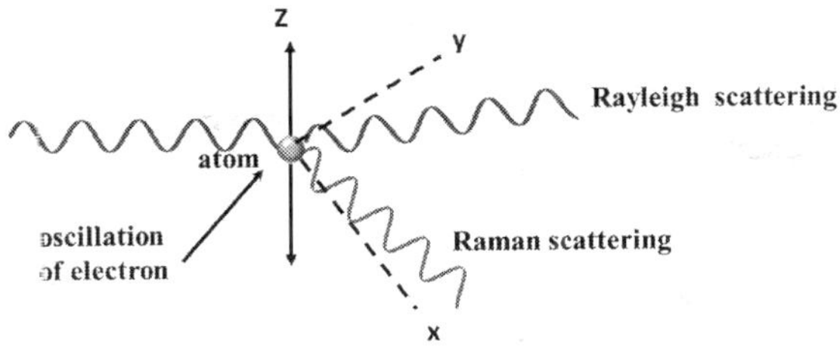

Figure 1. Light scattering produced by the interaction of the incident light's electric field and the molecule electrons.

Such process produces two types of radiation, Rayleigh radiation, which has the same frequency that the incident light (ν_0), and Raman radiation, which consists in a new set of frequencies with more or less energy than the incident radiation ($\nu_0 \pm \nu_1$), where ν_1 is typically related to the rotational, vibrational, and electronic levels of the molecule. In **Figure 2**, a general scheme of the scattering process and its difference with the absorption process from the point of view of the photons and the energy levels of the molecule are represented.

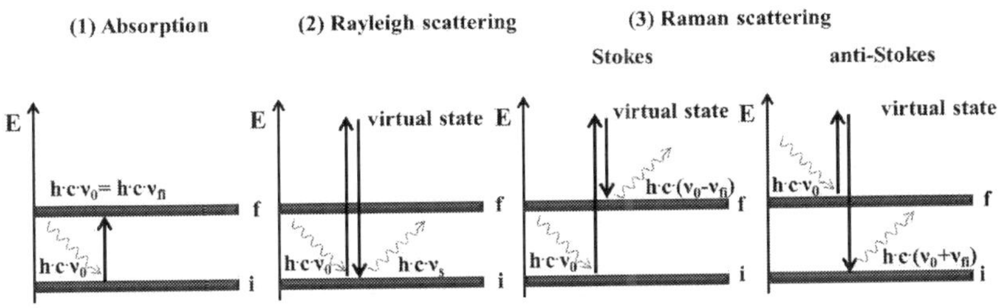

Figure 2. Energy level diagrams describing the physical phenomenon of (1) IR absorption, (2) Rayleigh scattering, and (3) Raman scattering.

Before the interaction of the radiation with the system, there are N photons of energy $hc\nu_0$. In the case of the absorption process, the interaction of the radiation with the system leads to the excitation of the molecule to a higher energy state resulting in a radiation which consists in N − 1 photons of energy $hc\nu_0$. This process can occur, if and only if the incoming photon has the same energy as the difference between the initial and final state of the molecule, $E_f - E_i = hc\nu_{fi} = hc\nu_0$, fulfilling the condition of energy conservation.

Let us now consider the scattering process, the interaction between the incident radiation and the system produces the annihilation of a photon of energy $hc\nu_0$, and simultaneously the creation of a new photon with energy E_s. Now the radiation consists in N − 1 photons of energy $hc\nu_0$, a new photon of energy $hc\nu_s$, and the transition to the molecule to a final state with energy E_f. In the overall process, the energy must be conserved so, $hc\nu_0 = hc\nu_s + E_f$.

This two-photon process can be visualized as two simultaneous stages. First, the annihilation stage which leads the molecule to a virtual high-energy state. Virtual states are created when the laser interacts with the molecule and causes polarization; hence, their energy is determined by the frequency of the incident light source used (ν_0). At this stage, there is no energy conservation implying that the role of the incident radiation in the scattering is to perturb the molecule given the possibility to allow different spectroscopic transitions rather than the absorption process. Second, the creation stage, where the molecule reaches its final state with energy E_f producing the new photon. At this point, we can consider two types of scattered light: (1) the Rayleigh scattering, or elastic scattering, where the final state of the molecule is its own initial state, $E_f = E_i$, and, correspondingly, the energy of the scattered light corresponds to the initial frequency value, $\nu_s = \nu_0$. (2) The Raman scattering, or inelastic scattering, in which the molecule reaches a final state different from its initial state; hence, the energy of the scattered light has a different frequency value from the incident radiation, $\nu_s = \nu_0 \pm \nu_{fi}$. This process is much less probable than Rayleigh scattering (only $10^{-5} - 10^{-8}$ of the incident beam intensity). If the final state has a higher energy than the initial state, $E_f > E_i$, the scattered photon loses energy, $\nu_s = \nu_0 - \nu_{fi}$. This radiation is known as Stokes Raman scattering. By contrast, if the final state has a lower energy than the initial state, $E_f < E_i$, the scattered photon increases its energy, $\nu_s = \nu_0 + \nu_{fi}$, giving the anti-Stokes Raman scattering. Relative probabilities of Stokes and anti-Stokes radiation depend on the population of the molecule states, f and i, and therefore on temperature according to the Maxwell-Boltzmann distribution. As both give the same information, it is customary to measure only the "Stokes" side of the spectrum.

Even though this general scheme describes scattering phenomena in a qualitative way, it highlights some key aspect of the Raman spectroscopy and its differences with the absorption process. Nevertheless, it is worth to describe the classical treatment of the Raman scattering in order to provide a deeper insight in the frequency dependence and the microscopic origin of the scattered light. Classical wave interpretation [54, 55] of the Raman effect is based on the time-dependent polarizability of the molecules. Consider one of the simplest scattering systems, a vibrating diatomic AB molecule.

Such a system can be modeled, at first approximation, as two balls attached by a spring (**Figure 3**). According to Hook´s law, its relative movement can be described by the second Newton law as follows:

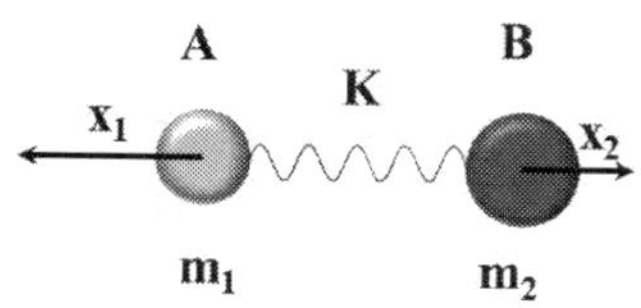

Figure 3. Simplified model of a diatomic AB molecule.

$$\mu\left(\frac{d^2x_1}{dt^2}+\frac{d^2x_2}{dt^2}\right)=K\left(x_1+x_2\right) \tag{1}$$

http://dx.doi.org/10.5772/64436

where μ represents the reduced mass of the molecule, x represents the displacement, and K represents the bond strength. For small vibrations, the harmonic approximation holds, and then the normal coordinates $q(t)$ of the vibrating molecule can be expressed as

$$q = q_0 cos\left(2\pi v_m t\right) \tag{2}$$

where q_0 is the amplitude and v_m is the natural vibration frequency which is defined in terms of its bond strength as

$$v_m = \frac{1}{2\pi}\sqrt{\frac{K}{\mu}} \tag{3}$$

When incident light interacts with a molecule, induces a dipole moment, P, equal to the product of the polarizability of the molecule, α, and the electric field of the incident light source E

$$P = \alpha\, E_0 \cos\left(2\pi v_0 t\right), \tag{4}$$

where E_0 and v_o are the electric field amplitude and frequency, respectively. As far as the molecule is vibrating, its polarizability varies according to the relative displacement of these atoms and therefore we can express α as a power series q

$$\alpha = \alpha_0 + q\left(\frac{\partial\alpha}{\partial q}\right)_0 + \ldots \tag{5}$$

which when combined with Eqs. (3) and (5) results in,

$$P = \alpha_0 E_0 \cos\left(2\pi v_0 t\right) + q_0 \cos\left(2\pi v_m t\right) E_0 \cos\left(2\pi v_0 t\right)\left(\frac{\partial\alpha}{\partial q}\right)_0 \tag{6}$$

$$P = \alpha_0 E_0 \cos\left(2\pi v_0 t\right) + \left(\frac{\partial\alpha}{\partial q}\right)_0 q_0 E_0\left[\cos\left(2\pi\left\{v_0 - v_m\right\}t\right) + \cos\left(2\pi\left\{v_0 + v_m\right\}t\right)\right] \tag{7}$$

From Eq. (7), it is evident that the induced electric dipole is formed by three different terms. The first one gives rise to an oscillating moment at the same frequency of the incident light, the Rayleigh scattering, and two additional terms which accounts for the Stokes and anti-stokes Raman scattering. Therefore, Rayleigh scattering arises from an electric dipole which oscillates at the same frequency induced in the molecule by the electric field of the incident radiation, whereas Raman scattering arises from the modulation of the electric dipole with the natural frequency of the vibrating molecule. This modulation is produced by the electrons of the molecule, whose rearrangement produces a coupling between the nuclear motion and the electric field of the radiation.

2.2. Dispersive Raman spectrometer

From the basis of the Raman effect described above, it is easy to deduce that in a conventional (or dispersive) Raman spectrometer,[2] the main difficulty lies in separating the intense stray light of the Rayleigh scattering from the much weaker Raman-scattered light. Besides, as Raman scattering has low efficiency, the optimization of each of the instrumental components becomes critically important.

The main components of a Raman setup are as follows:

(1) Excitation source

(2) Sample illumination systems and collection optics

(3) Wavelength selectors and separators

(4) Detector

(5) Recording device

1. *Excitation source:* Traditionally, mercury arc lamps were used as light sources until being replaced by laser sources. Laser beams are highly monochromatic, present small diameter and, with the help of different optic devices, can be focused on small samples. Different lasers can be used as the light source in Raman spectrometry, as the ones shown in **Table 1** [56].

Laser	Wavelength (nm)
Ar ion	530.9/647.1
He-Ne	632.8
Near IR diode	785/830
Nd-YAG	1064
Frequency-doubled Nd:YAG	532
Nd:YVO_4 diode	532

Table 1. Lasers used as light source in Raman spectroscopy.

In addition, in order to enhance the laser quality it is possible to employ a pass-band filter, designed to pass only a certain band of frequencies while attenuating all signals outside this band. This component is commonly known as interferometric filter.

2. *Sample illumination system and collection optics*: The collimation and focusing optics of the exciting radiation onto the sample depends on the experimental setup. In principle, excitation

2 Raman systems are subdivided into two principals according to the spectral analysis of the Raman light, namely Fourier-transform (FT) systems using an interferometer, and dispersive systems.

http://dx.doi.org/10.5772/64436

and collection from the sample can be accomplished in any geometry, although 90 and 180°C (backscattering) are more frequently employed. The use of fiber optics helps to make the spectrometers more versatile.

3. *Wavelength selectors and/or separators:* The separation or removal of the intense Rayleigh scattering can be achieved by using two different types of filters: notch and edge filters. Notch filters allow the acquisition of the anti-Stokes and Stokes Raman spectra down to ~30 cm^{-1}, but their use is expensive since they must be replaced very frequently (~2 years). For this reason, the use of edge filters is widespread. These are wide pass-band filters, which imply that the anti- Stokes Raman spectrum cannot be obtained and typical minimum wavenumbers are ~50 cm^{-1}. After the removal or suppression of the Rayleigh radiation, the separation of the different Raman radiations scattered by the sample should be performed. The first Raman spectrometers used prisms, but nowadays these are replaced by gratings that are typically holographically produced. It is worth noting that filters can be neglected if coupling of two or three monochromators is set in a series. This configuration allows not only to separate the Raman lines but also to remove the Rayleigh scatter.

4. *Detectors:* Just like in other spectrometers, the former detectors, that is, photographic films, were substituted first by photodiode array detectors and then by charge transfer devices (CTDs) such as charge-coupled devices (CCDs). CCDs are silicon-based semiconductors arranged as an array of photosensitive elements, each one generating photoelectrons and storing them as an electrical charge. Charges are stored on each individual pixel as a function of the number of photons striking that pixel and then read by an analog-to-digital converter [54].

A schematic representation of a modern micro-Raman spectrometer is shown in **Figure 4**.

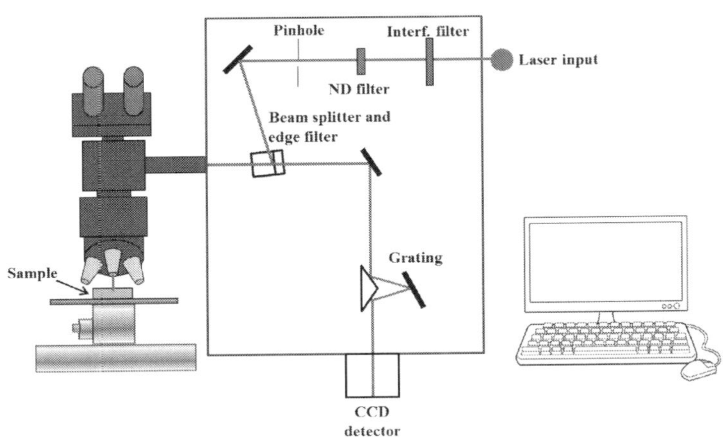

Figure 4. Descriptive scheme of the main components of a Raman microspectrometer.

In the micro-Raman technique, a microscope is integrated in a conventional Raman spectrometer, enabling both visual and spectroscopic measurements. As can be seen in **Figure 4**, in these types of equipment the focusing and collection optics of the scattered radiation are identical. In addition to the analysis of a single point, these spectrometers allow mapping and imaging measurements.

3. Results

As explained before, this section has been divided into two parts corresponding to dry and wet conditions; into the corresponding part the developed methods and the results obtained for analogs of the spent nuclear fuel matrix and its alteration products are shown. Namely, the materials studied in this section are the different uranium oxides, UO_{2+x} ($0 < x < 0.25$), U_4O_9/U_3O_7, and U_3O_8, and several secondary phases such as rutherfordine, soddyite, uranophane alpha, and kasolite.

The results shown in this section have been obtained by using the LabRaman HR Evolution (Horiba Jobin Yvon Technology, i.e., a dispersive spectrometer equipped with a microscope that enables the unification of both focusing and collection optics. It is possible to couple any laser to the spectrometer optical system as the excitation source. We specifically use the internal HeNe laser of 20-mW nominal power and an excitation wavelength of 632.8 nm (red). The laser beam is focused on the sample through a confocal microscope with different magnifications (5×, 20×, 50×, and 100×). Scattered radiation is then collected by the microscope on its way back (180° scattering) and the Rayleigh contribution removed by an edge filter that cuts at <50 cm^{-1}. Thereafter, the Raman-scattered radiation is registered in a Peltier-cooled CCD of 1024 × 256 pixels after crossing a diffraction grating which disperses the signal into its constituent parts. Three gratings of 600, 1800, and 2400 grooves/mm can be selected. The spectral resolution of each grating is ~1, 0.5, and 0.25 cm^{-1}/pixel, respectively. The microscope holds a motorized platen in order to carry out both point-to-point and image-scanning spectra, with a 0.1-µm resolution.

In addition to the abovementioned components, different lenses and mirrors can be found throughout the whole optical path, whose function is to correctly direct the beam within the optical system. Due to the amount of components that the laser beam encounters along its way before reaching the sample, a consequent power reduction of around 50% takes place. Nevertheless, the power attained at the sample surface is still sufficient to properly carry out the experiments.

The sample optical image and the spectra acquisition/visualization are performed by means of specific software designed by Horiba. The acquisition parameters and data processing of each spectrum depend on the sample and, above all, on the aim of the study.

3.1. Characterization of different uranium oxides

Here, a methodology to characterize uranium oxide powder with different oxidation degrees is described, and the spectra of stoichiometric UO_2, hyperstoichiometric UO_{2+x}, ($0 < x < 0.25$),

U_4O_9/U_3O_7, and finally U_3O_8 have been shown. The sample preparation for obtaining these uranium oxides with different stoichiometry is based on the TGA technique and described in detail elsewhere.

The protocol used to analyze any uranium oxide powder can be summarized in the following steps:

1. *Sample visualization*: A small amount of uranium oxide is spread over the surface of a microscope slice, which is housed in the stage of the microscope. By using different objectives, we visualize the particles that compound the uranium oxide powder. The microscope digital camera allows us to get optical image of the sample, such as the images shown in **Figure 5**. As can be seen, the powder is finely divided in particles of a size of ~10–20 μm.

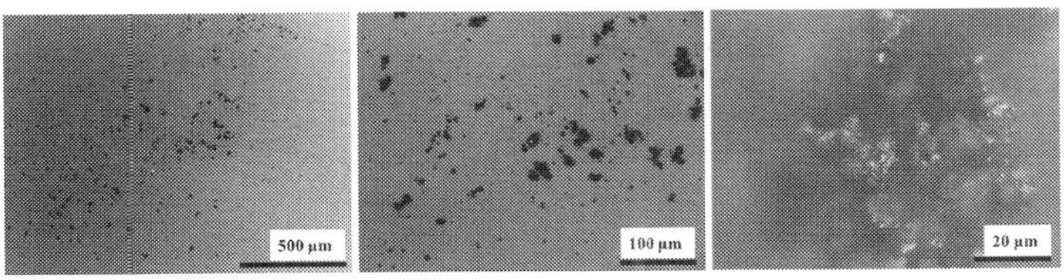

Figure 5. Optical image of UO_2 powder, acquired with the 5×, 20×, and 100× microscope objectives (from left to right).

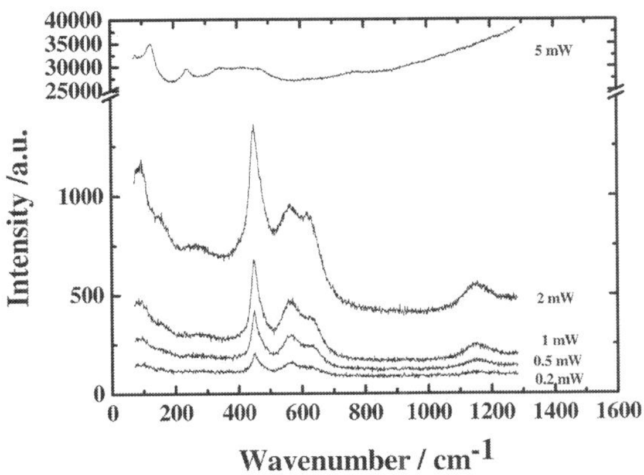

Figure 6. Raman spectra acquired as laser power is increased (for the same acquisition times).

2. *Setting the acquisition conditions*: All spectra are acquired using the laser with an excitation wavelength of 632.8 nm and the 600-grooves/mm grating, thus obtaining a spectral resolution of ~1 cm^{-1}/pixel (see *Micro-Raman spectrometer set-up* section). The laser beam is focused on the sample through the 100× magnification objective. Due to the fact that lasers can induce a local temperature increase by up to several hundred degrees if they are focused on a small spot, it

is crucial to check the stability of the sample at high temperature; otherwise, the laser can damage it. In order to do that, a previous study of the sample behavior at different laser powers needs to be done. In this way, **Figure 6** shows the spectra obtained as increasing the laser power, for an acquisition time of 10 min. As can be observed, the sample is stable up to 2 mW for such acquisition time, and then it is burnt from that power on. Therefore, when carrying out the uranium oxide powder characterization experiments, the laser power is minimized to 1 mW in order to ensure that there is no sample alteration during the measurements.

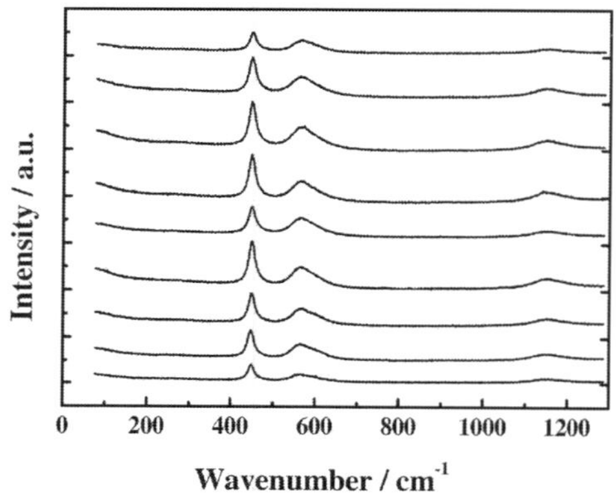

Figure 7. Raman spectra acquired by the multipoint sampling method, in order to check homogeneity of the sample.

3. Checking the sample homogeneity: Since the Raman laser excites only a very small portion of the sample (few microns), its homogeneity becomes a critical issue. Therefore, in order to check the sample homogeneity several spectra of different particles are acquired and compared. For obtaining these spectra, the multipoint sampling method is employed. This method uses sequential sample movement and spectrum acquisition, repeated as many times as desired, that is, the motorized *XY* microscope stage is moved to the position in which each spectrum is acquired. **Figure 7** shows the Raman spectra of different particles of a uranium dioxide sample. It can be appreciated that all spectra are similar, indicating that the sample is very homogeneous.

4. Spectra acquisition: Once the sample homogeneity is checked, and in order to enhance the intensity/noise ratio all spectra are added. In **Figure 8**, the sum spectrum is shown.

5. Spectra analysis: The aim of the Raman spectra analysis is to obtain information about the frequency, intensity, width, and area of their bands; for this purpose, it is necessary to previously know the number of bands or contributions in the spectrum. One way to detect such contributions is to calculate the spectrum second derivative. This method allows us, first, to determine the number of contributions, since each will lead to a minimum, and, second, to accurately approximate the Raman frequency band center from the position of this minimum.

As an example, the second derivative of the uranium oxide sum spectrum is shown in **Figure 10**. Once the number of contributions of each band is estimated, a Lorentzian fit is conducted using the obtained frequency values as fixed parameters. **Figure 8** also shows the profile analysis of the sum spectrum, where four bands are detected at ~445, 560, 630, and 1150 cm^{-1}.

UO_2 presents a fluorite-type structure (fcc), where uranium cations are located at the cubic-coordinated sites and oxygen anions at the tetrahedral-coordinated ones. As oxidation takes place, excess oxygen occupies interstitial positions thus leading to hyperstoichiometric UO_{2+x} [57]. The crystal system remains cubic up to $x = 0.25$, known as U_4O_9 [58]. Further oxidation causes the transformation from cubic to tetragonal structure (U_3O_7) [59] and finally to orthorhombic (U_3O_8).

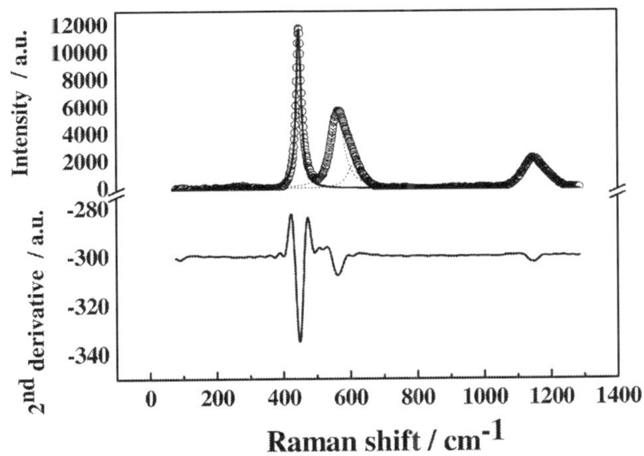

Figure 8. Raman spectrum of uranium oxide powder, corresponding to the sum of several spectra acquired at distinct points of the sample, and its profile analysis (up), and second derivative of such sum Raman spectrum (down).

Since the space group corresponding to uranium dioxide is *Fm3m* [60], group theory predicts two vibrational modes for UO_2: a Raman active mode (T_{2g}) and an infrared active mode (T_{1u}). In such a way, the Raman spectrum of stoichiometric UO_2 presents a band at ~445 cm^{-1} assigned to the mentioned T_{2g} vibrational mode [61]. Likewise, another band at ~1150 cm^{-1} is observed, which has been attributed to the 2LO phonon [62]. With regard to the characteristic spectrum of UO_{2+x}, it features the same two bands as the stoichiometric oxide, but also an additional broad band located at 500–700 cm^{-1} [63]. The same broad band is detected for U_4O_9 and U_3O_7, with a much greater contribution; besides, the 1150 cm^{-1} band completely disappears in these oxide spectra [64]. Since U_4O_9 and U_3O_7 spectra are similar, it is difficult to distinguish one from each other. Moreover, due to the change to the orthorhombic structure, a widely different spectroscopic profile is obtained for U_3O_8 [65].

By following the protocol described above, spectra corresponding to uranium oxide powder samples with different sequential oxidation degrees have been obtained: UO_{2+x} ($0 < x < 0.25$), U_4O_9/U_3O_7, and U_3O_8. Some of these spectra are shown in **Figure 9**. Therefore, the behavior of the uranium dioxide Raman spectrum as the oxidation degree increases can be described by the next features: (1) the apparition of different contribution bands at ~500–700 cm^{-1}, typical of the hyperstoichiometric oxides UO_{2+x} ($0<x<0.25$), which are due to the incorporation of oxygen into the cubic structure of UO_2 [66]; (2) the disappearance of the 1150 cm^{-1} band, characteristic of U_4O_9/U_3O_7 [67]; and (3) the typical three bands at ~375, 450, and 515 cm^{-1} and a band at ~812 cm^{-1} corresponding to the structure of U_3O_8 [68, 69].

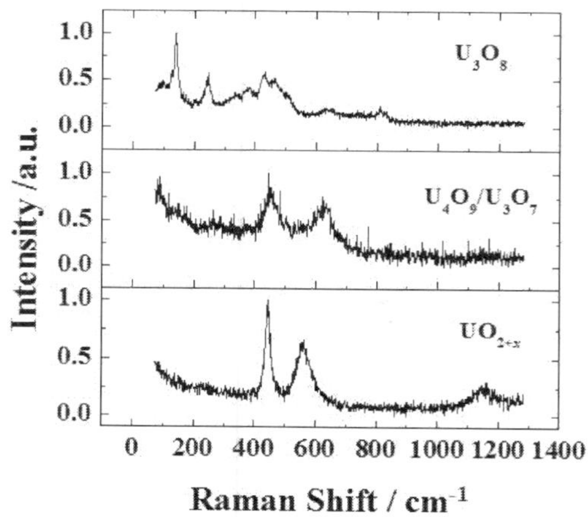

Figure 9. Raman spectra of different uranium oxides, where the bottom spectrum corresponds to the hyperstoichiometric UO_{2+x}, the middle one to U_4O_9/U_3O_7, and the top one corresponds to the final U_3O_8.

3.2. Characterization of secondary phases in natural samples

In this section, we present a method based on Raman spectroscopy that allows us an easy and fast identification of secondary phases formed at nature. Secondary phases are present in nature as rims of corrosion products (typically of one or two centimeters wide) found on weathered uraninite[3] crystals. These structures are known as *gummites* because of the difficulties in distinguishing individual phases. The typical alteration rim around an oxidized uraninite crystal, as was described by Frondel [70, 71], is schematically shown in **Figure 10**.

The ore is composed by the uraninite, usually brown to dark brown depending on its oxidation state. Zone 1 contains domains of uranyl oxide hydrates: fourmarierite vandendriesscheite, wölsendorfite, calciouranoite, clarkeite, becquerelite, curite, and schoepite, whereas zone 2

3 Mineral composed by uranium dioxide, UO_2, sometimes with small amounts of thorium, therefore with variable formula $(U,Th)O_2$ (http://www.minerals.net/mineral/uraninite.aspx#sthash.aTasdt1U.dpuf).

consists most commonly of uranyl silicates: uranophane, kasolite, sklodowskite, and soddyite. Anyhow, as expected, the specific alteration products depend on local conditions [72].

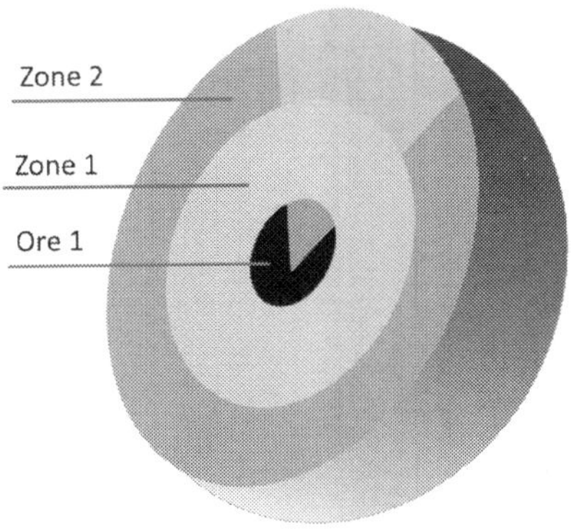

Figure 10. Scheme corresponding to the alteration rim of an oxidized uraninite crystal, as described by Frondel.

As an example of how to use Raman spectroscopy to analyze this kind of samples, we show below the characterization of a gummite sample collected in 1960 from Sierra Albarrana (Córdoba, Spain). More details of this study can be found in Ref. [73]. The surface of a polished section of the sample was analyzed by acquiring different spectra along a 10-mm line from the center outwards, in order to know the alteration products sequence (see **Figure 11**).

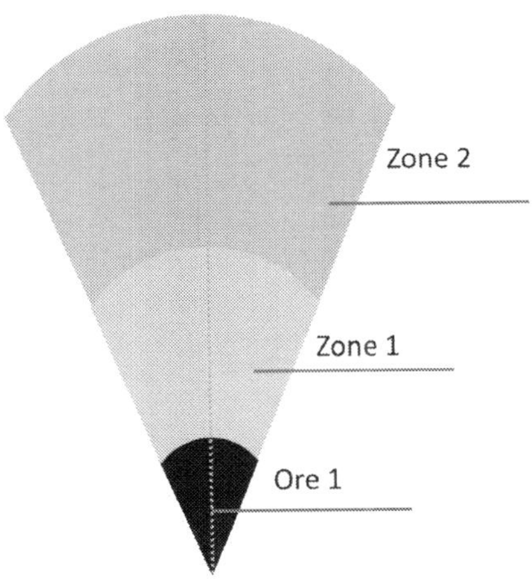

Figure 11. Scheme of the 10-mm line along which different spectra were acquired from the center of the oxidized uraninite crystal outwards.

A combination of the line-mapping and step-by-step procedures can be used to acquire the spectra in this kind of samples. Specifically, in this study a line mapping is performed using the automatized line-scanning tool. This tool allows Raman spectra acquisition of different sample points along a line by automatically moving the stage in one or two directions (*X-Y*). The microscope objective used with a magnification of 20× allows the visualization of a maximum 500 × 70-μm area. Therefore, in order to analyze the whole sample (10 mm), 20 lines with five equidistant points each have been measured, thus acquiring 100 spectra. This was performed with the step-by-step procedure, in which the motorized stage is moved 500 μm (the line-mapping length) in the *x*-direction to allow the analysis of the next part of the sample. The acquisition time for each spectrum was 100 s on an extended shift of 100–1200 cm^{-1}. A typical spectrum acquired in this way is shown in **Figure 12**. This spectrum is the characteristic of a mixture of U-minerals that contain uranyl groups in their structure.

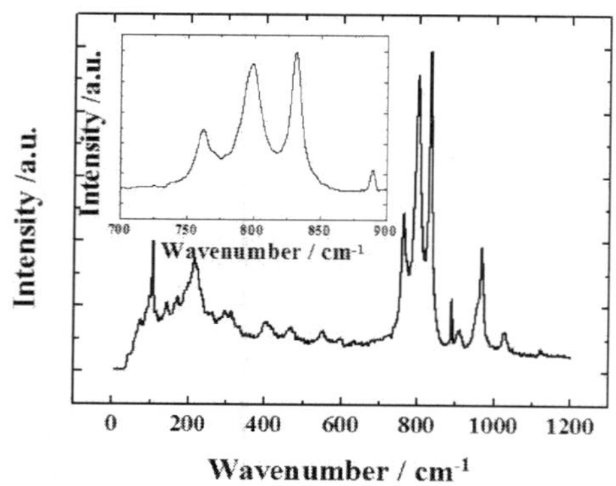

Figure 12. Typical Raman spectrum acquired during the step-by-step and line-mapping combination procedure. The inset shows the frequency range corresponding to the ν_1 symmetric stretch.

It has been demonstrated that the symmetrical-stretching vibration of UO_2^{2+} can be used as a fingerprint to identify each U-mineral phase [74]. UO_2^{2+} presents a linear symmetry which corresponds to the punctual group $D_{\infty h}$. It has four normal modes (3 *N* – 5, *N* = number of atoms) and three fundamental vibrations: the symmetric-stretching vibration ν_1, the doubly degenerate bending vibration $\nu_2(\delta)$, and the anti-symmetric-stretching vibration ν_3 (see **Table 2**).

Fundamental mode	Vibration	Activity	Infrared frequency (cm^{-1})
ν_1	Symmetric stretch	R	700–900
$\nu_2(\delta)$	Bending	IR	200–300
ν_3	Anti-symmetric stretch	IR	850–1000

Table 2. Normal modes of UO_2^{2+}.

Although the perfect linear structure has only a ν_1 Raman active vibration mode, symmetry lowering ($C_{\infty h} \rightarrow C_{\infty v} \rightarrow C_{2v} \rightarrow C_s$) leads to Raman activation of the two IR bands, as well as their overtones and combination vibrations. Moreover, the frequency of each active band is sensible to the environment in which the uranyl group is housed; therefore, each U-mineral has a characteristic ν_1 symmetric-stretching vibration frequency, which can be used as a fingerprint.

ν_1 (cm^{-1})	Phase
768	Kasolite
798	Uranophane
832	Soddyite
889	Rutherfordine

Table 3. Characteristic frequencies of the found U-minerals.

In the inset of **Figure 12**, we show the frequency range corresponding to the ν_1 symmetric stretch, from 700 to 900 cm^{-1}. As can be seen, there are four bands in this region. This must be due to a mixture of four different phases, if we understand the Raman spectra for mixtures as the direct sum of the individual spectrum of each component in the mixture (as long as these components do not interact with each other). Therefore, in this kind of mixtures the vibration bands do not undergo any displacement, and the band profile of the mixture spectrum results in the spectra of the different components or vice versa.

Taking this into account, the characteristic frequency bands observed in **Figure 12** correspond to the following U-minerals: rutherfordine, $UO_2(CO_3)$, uranophane alpha, $Ca(UO_2)_2(SiO_3OH)_2$ $5H_2O$, soddyite, $(UO_2)_2SiO_2$ $2H_2O$, and kasolite, $PbUO_2SiO_4H_2O$ (see **Table 3**). See Refs. [75–78] for the assignments.

Position (mm)	Min. at 768 cm^{-1} (kasolite)	Min. at 798 cm^{-1} (uranophane)	Min. at 832 cm^{-1} (soddyite)	Min. at 889 cm^{-1} (rutherfordine)
0	No	No	No	No
0.9 mm	No	No	Yes	Yes
3.7 μm	No	Yes	Yes	No
9.1 mm	Yes	Yes	No	No

Length from the center to the sample.

Table 4. Result data matrix.

In order to perform a semi-quantitative analysis of the sample with the aim of detection of different phases along the sample, next spectra processing has been carried out:

(i) Second-derivative calculation in the ν_1 region (700–900 cm^{-1}).

(ii) Verification of the existence of a minimum at each characteristic frequency.

(iii) Construction of a data matrix of 0 and 1, where 0 means there is no minimum at the characteristic frequency of the mineral and 1 means there is such a minimum (see **Table 4**).

This data matrix enables constructing different diagrams. As an example, in **Figure 13** we present a scheme where the existence or absence of each phase at each position can be appreciated only by looking at the minimum of the second-derivative spectrum.

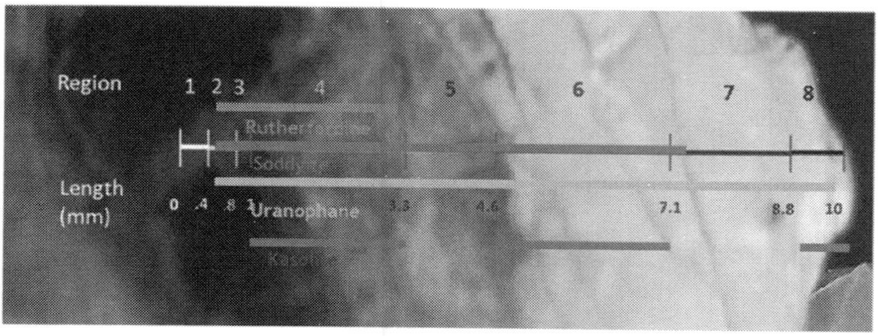

Figure 13. Sequence of the urananite alteration products.

As can be seen, the sequence of alteration products obtained was as follows: (1) uraninite constitutes the unaltered core of the sample, 0–0.4 mm. (2) Rutherfordine appears in the inner part, 0.4–3.3 mm, in contact with the uraninite core. (3) Then, a mixture of uranyl silicates, soddyite, uranophane alpha, and kasolite is found. Soddyite prevails in the inner part, 0.4–7.1 mm; uranophane alpha predominates in the outer part of the sample, 7.1–10 mm, and kasolite appears intermittently (1.0–3.3, 4.6–7.1, and 8.8–10 mm).

Author details

Laura J. Bonales*, Jone M. Elorrieta[1], Álvaro Lobato[2] and Joaquin Cobos[1]

*Address all correspondence to: laura.jimenez@ciemat.es

1 High-Level Radioactive Waste Unit CIEMAT, Madrid, Spain

2 Department of Physical Chemistry, School of Chemistry, University Complutense of Madrid, Madrid, Spain

References

[1] Nuclear Safety Review 2015, IAEA/NSR/2015, IAEA, Austria, 2015.

[2] Organisation for economic co-operation and development (OECD) annual report, 2002. DOI: 10.1787/annrep-2002-en.

[3] Rodriguez Villagra N. Fuel stability of Th / Pu / MOX at disposal conditions [thesis]. Universidad Complutense de Madrid; 2015. ISBN: 978-84-7834-739-1.

[4] Zakaria M. Atoms for Peace? Nuclear Energy and Peace. Bachelor thesis (C-essay) Peace and Conflict Studies (PACS) [thesis]. Faculty of International Migration and Ethnic Relations (IMER); 2006.

[5] de Groot JI, Steg L, Poortinga W. Risk Analysis. 2013;33:2:307–317. DOI: 10.1111/j.1539.

[6] Ahokas J. The role of nuclear power in the future energy system [thesis]. University of LUT School of Technology; 2015.

[7] Jung-Chun H, Chiao-Tzu P, Shu-Fen K, Ruey-Yu C, Leonga M, Hung-Lun C, Wan-Hua H, Chun-Chiao T, Cheng-Fung L, Suei-Loong L. Environment International. 2014;73:295–303.

[8] OECD/IEA Technology Roadmap: Nuclear Energy. [Internet]. 2015. Available from: http://www.iea.org/t&c/termsandconditions/

[9] Choppin G, Liljenzin JO, Rydberg J. Radiochemistry and Nuclear Chemistry, Reed Educational and Professional Publishing Ltd., Oxford; 1995.

[10] Astudillo Pastor J. El almacenamiento geológico profundo de los residuos radiactivos de alta actividad: principios básicos y tecnología. Editorial: Madrid: ENRESA, D.L; 2001.

[11] Kleykamp H. Journal of Nuclear Materials. 2004;131:221–246.

[12] Buck EC, Hanson BD, McNamara BK. The Geological Society of London Special Publication 2004;236:65–88.

[13] Shoesmith DW. Journal of Nuclear Materials. 2000;282:1–31.

[14] Bruno J, Ewing RC. Elements. 2006;2:343–349.

[15] Hedin A. Spent Nuclear Fuel—How Dangerous Is It? SKB Technical Report 97-13, Swedish Nuclear Fuel and Waste Management Co.; 1997, 60 pp.

[16] de Saillan C. Harvard Environmental Law Review. 2010;34:461–519.

[17] Nimander F. Investigation of Spent Nuclear Fuel Pool Coolability, Division of Nuclear Reactor Technology Royal Institute of Technology, Stockholm; 2011.

[18] McEachern RJ, Taylor P. Journal of Nuclear Materials. 1998;254:87–121.

[19] SKB 91. Final Disposal of Spent Nuclear Fuel. Importance of the Bedrock for Safety, SKB Report 92-20; May 1992.

[20] Geological Disposal of Radioactive Waste: Technological Implications for Retrievability. IAEA TR No. NW-T-1.19, 2009. ISBN 978-92-0-100109-2.

[21] Olander DR. Fundamental Aspects of Nuclear Reactor Fuel Elements, Technical Information Center, Office of Public Affairs-Energy Research and Development Administration; 1976, ISBN 0870790315, 9780870790317, 613 pp.

[22] Meredith S., Coonley IM-1, David L. Clark, Gordon D. Jarvinen. Eds. Actinide Oxides, Actinide Research Quarterly, Los Alamos Laboratory, Annual Report 2004. Nuclear Materials Research and Technology.

[23] Belle J, Ed. Uranium Dioxide: Properties and Nuclear Applications, Naval Reactors, Division of Reactor Development, USAEC; 1961.

[24] SKB-91. Final Disposal of Spent Nuclear Fuel. Importance of the Bedrock for Safety. Report 92-20; May 1992.

[25] Ewing RC. Nature Materials. 2015;14:252–257.

[26] Shoesmith DW. Journal of Nuclear Materials. 2000;282:1–31.

[27] Sunder S. Nuclear Technology. 1998;122:211–221.

[28] Christensen H, Sunder S. Journal of Nuclear Materials. 1996;238:70–77.

[29] Roth O, Jonsson M. Central European Journal of Chemistry. 2008;6:1–14.

[30] Amme M, Renker B, Schmid B, Feth MP, Bertagnolli H, Döbelin W. Journal of Nuclear Materials. 2002;306:202–212.

[31] Rey A, Giménez J, Casas I, Clarens F, Pablo J. de. Applied Geochemistry. 2008;23:2249–2255.

[32] Lousada CM, Trummer M, Jonsson M. Journal of Nuclear Materials. 2013;434:434–439.

[33] Frondel C. American Mineralogy. 1956;4:539–568.

[34] Plasil J. Journal of Geoscience. 2014;59:99–114.

[35] Musolino SV, DeFranco J, Schlueck R. The Radiation Safety Journal. Health Physics. 2008;94(2):109–111. DOI: 10.1097/01.HP.0000285801.87304.3f.

[36] Guimbretière G, Desgranges L, Jegou C, Canizarès A, Simon P, Caraballo R, Raimboux N, Barthe MF, Ammar MR, Maslova OA, Duval F, Omnée R. IEEE Transactions on Nuclear Science. 2014;61:2045.

[37] Guimbretière G, Canizares A, Simon P, Tobon-Correa YA, Ammar MR, Corbel C, Barthe MF. Spectroscopic Letters. 2011;44:570.

[38] Jégou C, Caraballo R, Bonfils J de, Broudic V, Peuget S, Vercouter T, Roudil D. Journal of Nuclear Materials. 2010;399:68.

[39] Jégou C, Caraballo R, Peuget S, Roudil D, Desgranges L, Magnin M. Journal of Nuclear Materials. 2010;405:235.

[40] Sarsfield MJ, Taylor RJ, Puxley C, Steele HM. Journal of Nuclear Materials. 2012;427:333.

[41] Desgranges L, Pontillon Y, Matheron P, Marcet M, Simon P, Guimbretière G, Porcher F. Inorganic Chemistry. 2012;51:9147.

[42] Böhler R, Çakır P, Beneš O, Hein H, Konings RJM, Manara D. The Journal of Chemical Thermodynamics. 2015;81:245.

[43] Böhler R, Quaini A, Capriotti L, Çakır P, Beneš O, Boboridis K, Guiot A, Luzzi L, Konings RJM, Manara D. Journal of Alloys and Compounds. 2014;616:5.

[44] Böhler R, Welland MJ, Prieur D, Cakir P, Vitova T, Pruessmann T, Pidchenko I, Hennig C, Guéneau C. Konings RJM, Manara D. Journal of Nuclear Materials. 2014:448:330.

[45] Naji M, Colle JY, Beneš O, Sierig M, Rautio J, Lajarge P, Manara D. Journal of Raman Spectroscopy. 2015;46,750–756.

[46] Casas I, Giménez J, Martí V, Torrero ME, de Pablo J. Radiochimica Acta. 1994;66:23–27.

[47] Fors P, Carbol P, Van Winckel S, Spahiu K. Journal of Nuclear Materials. 2009;394:1–8.

[48] Bruno J, Casas I, Sandino MCA. Journal of Nuclear Materials. 1992;190:61–69.

[49] Ollila K. Journal of Nuclear Materials. 1992;190:70–77.

[50] Raman CV, Krishnan KS. Nature. 1928;121:501–502.

[51] Beran A, Lobowitzky U. EMU Notes in Mineralogy. Eötvös University Press; 2004.

[52] Hollas JM. Modern Spectroscopy. John Wiley & Son Ltd.; 2004.

[53] Long DA, Raman Spectroscopy, McGraw-Hill International Book Company; 1977.

[54] Willes W, Merlin R. Eds. Raman Scattering in Materials. Springer Series in Materials Science, Vol. 42. Springer-Verlag, Berlin; 2000.

[55] Lewis IR, Edwards H. Handbook of Raman Spectroscopy: From the Research Laboratory to the Process Line. Practical Spectroscopy Series, Vol. 28, CRC Press, Boca Raton, FL, USA; 2001.

[56] Egyptian Journal of Forensic Sciences. 2015. Available from: http://dx.doi.org/10.1016/j.ejfs.2015.06.001

[57] Willis BTM. Acta Crystallographic.1978;34:88–90.

[58] Hering H, Perio P. Bulletin de la Societe Chimique. 1952;351.

[59] Jolibois P. Comptes Rendus de l'Academie des Sciences, Paris. 1947; 224:1395.

[60] Graves PR. Applied Spectroscopy. 1990;144:1665.

[61] Marlow PG, Russell JP, Hardy JR. Philosophical Magazine. 1966;14:409.

[62] Livneh T, Sterer E. Physical Review B. 2006;73:085118–085119.

[63] He H, Shoesmith D. Physical Chemistry Chemical Physics. 2010;12:8108.

[64] Manara D, Renker B. Journal of Nuclear Materials. 2003;321:233–237.

[65] Allen C, Butler S, Tuan NA. Journal of Nuclear Materials. 1987;144:17–19.

[66] Manara D, Renker B. Journal of Nuclear Materials. 2003;321:233–237.

[67] Desgranges L, Baldinozzi G, Simon P, Guimbretière G, Canizares A. Journal of Raman Spectroscopy. 2012;43:455–458.

[68] Allen GC, Butler IS, Tuan NA. Journal of Nuclear Materials. 1987;144:17–19.

[69] Palacios ML, Taylor SH. Applied Spectroscopy. 2000;54:1372.

[70] Frondel C. US Geological Survey Bulletin. 1958;1064.

[71] Frondel C. American Mineralogy. 1956;41:539–568.

[72] Robert F, Rodney E. Uraninite alteration in an oxidizing environment and its relevance to the disposal of spent nuclear fuel; SKB Technical Report 91-15; 1990.

[73] Bonales LJ, Menor-Salván C, Cobos J. Journal of Nuclear Materials. 2015;462:296–303.

[74] Frost RL, Čejka J, Weier M, Wayde M. Journal of Raman Spectroscopy. 2006;37(9):879–891.

[75] Frost RL, Jiri C. Journal of Raman Spectroscopy. 2009;40:1096–1103.

[76] Rost RL, Cejka J, Weier ML, Martens W. Journal of Raman Spectroscopy. 2006;37:538–551.

[77] Frost RL, Weier ML, Martens W, Kloprogge T, Cejka J. Spectrochimica Acta. 2006;63:305–312.

[78] Biwer BM, Ebert WL, Bates JK. Journal of Nuclear Materials. 1990;175:188–193.

Chapter 5

Infrared Spectra and Density Functional Theoretical Calculation of Transition Metal Oxide Reaction with Monochloromethane

Yanying Zhao, Xin Liu and Shuang Meng

Additional information is available at the end of the chapter

http://dx.doi.org/10.5772/64437

Abstract

In this chapter, we presented a short review of past and present experimental and theoretical work on the reactions of the transition metal monoxide and dioxide molecules with monochloromethane in excess argon matrices. A series of infrared absorption spectra combining with density functional theoretical (DFT) calculation characterized that the transition metal monoxide molecules produced by laser-ablated higher oxides activated C—H and C—Cl bonds of CH_3Cl to first form the weakly bound $MO(CH_3Cl)$ (M = Sc, Y, Nb, Ta, Ti, Zr, Mn, Fe) complexes, which further photoisomerized to the more stable chlorine-transfer (Cl-transfer) CH_3OMCl (M = Sc, Y), $CH_3M(O)Cl$ (M = Ti, Zr), CH_3MOCl (M = Mn, Fe), and agostic hydrogen-transfer (H-transfer) CH_2ClMOH (M = Sc, Y, Nb, Ta) products upon limited light excitation. Transition metal dioxides reaction with CH_3Cl also formed $MO_2(CH_3Cl)$ (M = Ti, Zr, Nb, Ta) complexes, which were further rearranged to the more stable Cl-transfer $CH_3OM(O)Cl$ (M = Ti, Zr) and agostic H-transfer $CH_2ClM(O)OH$ (M = Nb, Ta) molecules between the metal center atom and the chlorine atom upon ultraviolet light irradiation. Their different reactivity was interpreted according to the different valence electrons of metal center.

Keywords: monochloromethane, chlorine transfer, hydrogen transfer, transition metal oxides, agostic interaction

1. Introduction

Monochloromethane, as the one of the simplest halohydrocarbons, also called methyl chloride, plays an important role in the industrial, synthetic, materials chemistry. It is always regarded that monochloromethane is the largest natural source of ozone-depleting chlorine compounds and accounts for about 15% of the present atmospheric chlorine content as one kind of chlorinated volatile organic compounds (CVOCs). At present, monochloromethane is observed in the dry leaf with the content of 0.1–0.3 μg/g/h, and large emissions of monochloromethane are observed from some common certain types of ferns and dipterocarpaceae [1, 2]. Monochloromethane is also industrially produced by the oxidation and chlorination reaction of methane in the presence of metal chloride catalyst, and drying monochloromethane conversed to gasoline and olefins on the methanol to gasoline (MTG) and the methanol to olefins (MTO) catalysts [3, 4]. The conversion of methyl chloride to hydrocarbons has been investigated since the mid-1980s [5]. The product distribution of methyl chloride to hydrocarbons is strikingly similar to methanol conversion over the same topology [6]. Recently several ZSM-5 zeolites and SAPO sieves catalysts were reported the high performances on the catalytic conversion of monochloromethane to light olefins [7–9]. Modified SAPO-34 catalysts were also chosen to enhance its catalytic performance for the conversion of chloromethane to light olefins [10–13]. The oxidation addition of metal into carbon-halogen bonds is a key step in many stoichiometric and catalytic reactions. Activation of compounds containing C—X (X = Cl, Br, I) bonds attracts widespread interest due to the underactive organic functional group and the inherent chemical properties. Predominantly alkyl and aryl halides are extensively applied as electrophiles in the transition metal-catalyzed cross-coupling reactions [14–16]. It has a far-reaching significance on carbon-chlorine (C—Cl) bond catalytic oxidation on the conversion of monochloromethane to gasoline and olefins.

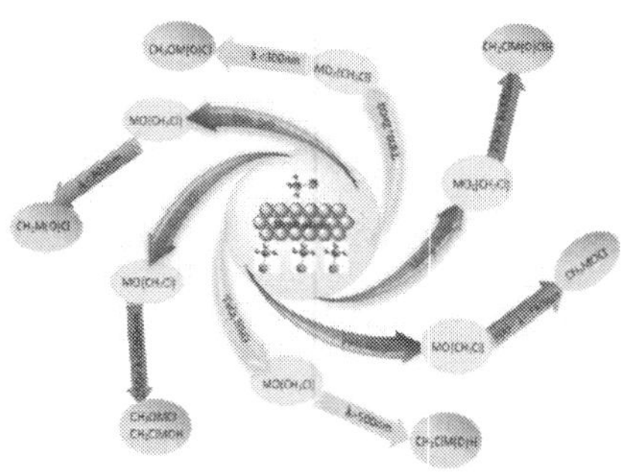

Scheme 1. The reactivity of transition metal monoxide and dioxide with monochloromethane in argon from Refs. [27–30].

The reactions of transition metal centers with chloromethane may serve as the simplest model for understanding the intrinsic mechanism of the organic halides catalytic oxidation processes. The reactions on transition metal atoms with monochloromethane have been intensively studied in solid noble gas matrices. Investigations have reported that C—X bond of CH_3X (X = F, Cl, Br, I) are activated by transition metal atoms [17–22]. The higher valence of group 6 metals can form the methylidyne complexes $CH = MH_2X$ (M = Mo, W, X = H, F, Cl, Br) [23–26]. In this chapter, the reactions of simple transition-metal oxide molecules with monochloromethane in solid argon were reviewed using matrix infrared absorption spectroscopy and density functional theoretical (DFT) calculations. As shown in **Scheme 1**, the ground-state transition metal monoxide molecules activated carbon-hydrogen (C—H) and C—Cl bond of CH_3Cl upon a certain wavelength excitation in argon matrices. The weakly bound $MO(CH_3Cl)$ (x = 1, 2; M = Sc, Y, Nb, Ta, Ti, Zr, Mn, Fe) complexes were initially formed and then isomerized to the more stable Cl-transfer CH_3OMCl (M = Sc, Y) and $CH_3M(O)Cl$ (M = Ti, Zr, Nb, Ta, Mn, Fe), and agostic H-transfer CH_2ClMOH (M = Sc, Y, Nb, Ta) isomers upon limited visible light excitation. The $MO_2(CH_3Cl)$ (M = Ti, Zr, Nb, Ta), which were formed from the reactions on MO_2 with CH_3Cl, were further rearranged to the more stable Cl-transfer $CH_3OM(O)Cl$ (M = Ti, Zr) and H-transfer $CH_2ClM(O)OH$ (M = Nb, Ta) molecules with agostic interactions between the chlorine and the metal center under ultraviolet light irradiation.

2. Experimental and computational methods

The experimental setup for pulsed laser-ablated and matrix isolation Fourier transform infrared (FTIR) spectroscopic technique has been previously described in detail [31]. Briefly, the 1064 nm Nd:YAG laser fundamental (Spectra Physics, DCR 150, 20 Hz repetition rate, and 8 ns pulse width) was focused onto the rotating bulk metal oxide targets, which were prepared by sintered metal oxide powders. Laser-evaporation of bulk higher metal oxide targets has been proved to be an extensively available technique to prepare pure metal oxides in noble gas matrices [32–34]. Using standard manometric technique, the CH_3Cl/Ar samples were mixed at a proper proportion in a stainless steel vacuum line. The CH_3Cl sample was subjected to several freeze-pump-thaw cycles at 77 K before use. The laser-evaporated species were co-deposited with chloromethane in excess argon onto a CsI window cooled normally to 6 K by a closed-cycle helium refrigerator (ARS, 202N). The matrix samples were deposited at a rate of approximately 5 mmol/h for 1–2 h. Isotopic-labeled $^{13}CH_3Cl$ and CD_3Cl (ISOTEC, 99%) were used without further purification. Infrared spectra between 450 and 4000 cm^{-1} were recorded on a Bruker IFS 66v/s spectrometer using HgCdTe (MCT) detector cooled by liquid N_2 at 0.5 cm^{-1} resolution. Samples were annealed to different temperatures and cooled back to 6 K to acquire the spectra, and selected samples were subjected to visible or broadband irradiation using a 250 W high-pressure mercury arc lamp with selected wavelength glass filters.

Density functional theoretical calculations were performed by using Gaussian 03 programs [35] to identify the experimental assignments. The three-parameter hybrid functional, according to Becke with additional correlation corrections from Lee, Yang, and Parr (B3LYP), was utilized [36, 37] to optimize ground geometries, calculate frequencies, and derive the zero-point vibrational energies. Transition-state optimizations were performed with the Berny geometry optimization algorithm at the B3LYP level. The 6-311++G(d, p) basis set was used for the H, C, O, Cl, Sc, Ti, Mn, and Fe atoms [38, 39], DGDZVP basis set for Y, Zr, and Nb atoms [40, 41], and the scalar-relativistic SDD pseudopotential and basis set for Ta atom [42, 43]. In addition, the CCSD(T) method was also applied to accurately calculate the single-point energies of the B3LYP-optimized structures with the same basis sets [44].

3. Transition metal monoxides reaction with CH_3Cl

Reaction of transition metal monoxides (ScO, YO, TiO, ZrO, NbO, TaO, MnO, FeO) with monochloromethane was investigated in solid argon by infrared absorption spectroscopy, combining with isotopic substituted experiments and theoretical calculations. The initial reaction step is the formation of the $MO(CH_3Cl)$ (M = Sc, Y, Ti, Zr, Nb, Ta, Mn, Fe) complex with metal atom bound with chlorine atom and/or oxygen atom with H atoms on annealing. Upon a certain wavelength photolysis, the $MO(CH_3Cl)$ complex was isomerized by the insertion of the M═O to C—H and/or C—Cl/Cl—C bond. Selected region of infrared spectra is illustrated in **Figures 1–4**.

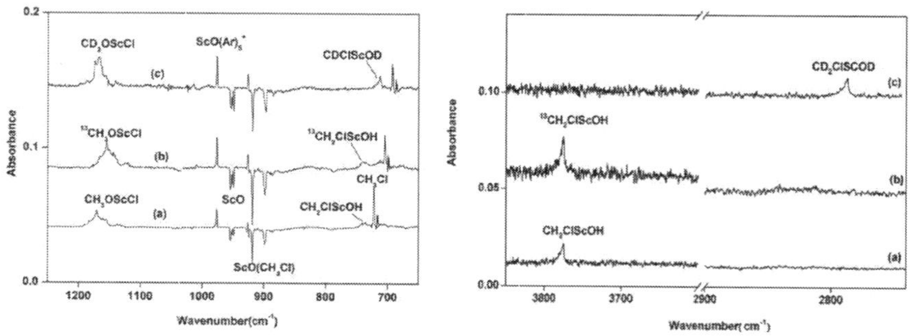

Figure 1. Difference spectra in the selected regions scandium monoxide with isotopic substituted chloromethane in excess argon. (Spectrum taken after 15 min of broadband irradiation minus spectrum taken after 25 K annealing). (a) 0.5% CH_3Cl, (b) 0.5% $^{13}CH_3Cl$, and (c) 0.5% CD_3Cl. (Reprinted with the permission from Ref. [27]. Copyright 2013 American Chemical Society).

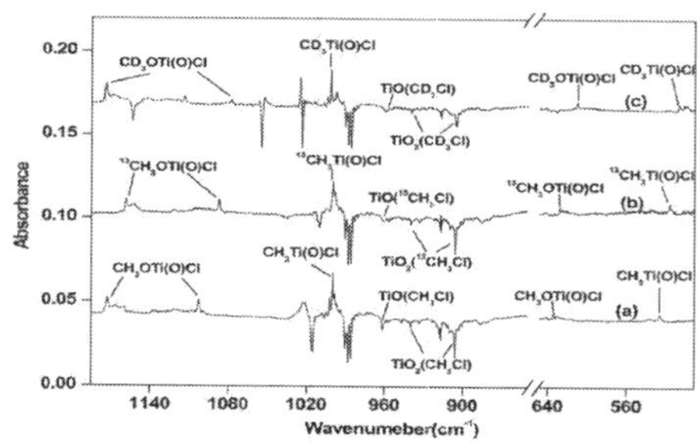

Figure 2. Difference spectra in the selected regions from co-deposition of a laser-ablated TiO_2 target in excess argon. Spectrum taken after 15 min of full-arc broadband photolysis irradiation (λ < 300 nm) followed by the 25 K annealing minus spectrum taken after sample annealing at 25 K. (a) 0.5% CH_3Cl, (b) 0.5% $^{13}CH_3Cl$, and (c) 0.5% CD_3Cl. (Reprinted with the permission from Ref. [28]. Copyright 2013 American Chemical Society).

http://dx.doi.org/10.5772/64437

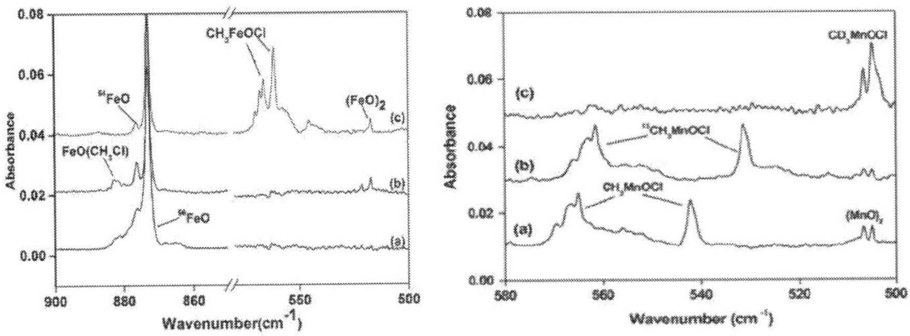

Figure 3. Infrared spectra in the selected region from co-deposition of laser-ablated MnO_2 and Fe_2O_3 target with isotopically substituted CH_3Cl in excess argon. Spectra were taken after 1.5 h of sample deposition followed by 25 K annealing and 15 min of irradiation (300 < λ < 580 nm) and 25 K annealing. (a) 0.5% CH_3Cl, (b) 0.5% $^{13}CH_3Cl$, and (c) 0.5% CD_3Cl. (Reprinted with permission from Ref. [30]. Copyright 2013, with permission from Ref. [28] from Elsevier).

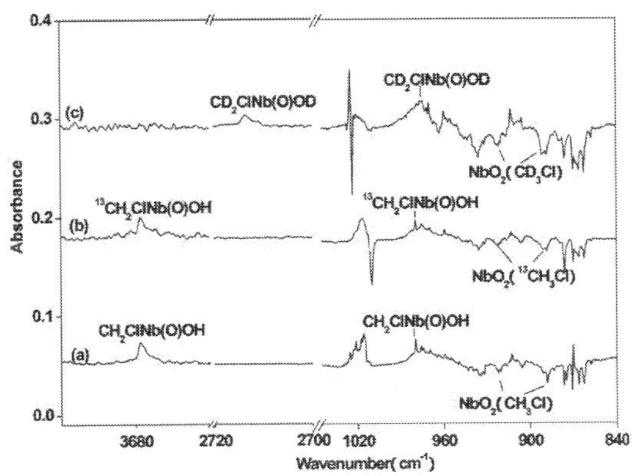

Figure 4. Difference spectra in the selected regions from co-deposition of laser-evaporated niobium oxides with isotopic-substituted monochloromethane in excess argon. Spectrum taken after 15 min of full-arc photolysis minus spectrum taken after sample annealing at 25 K. (a) 0.5% CH_3Cl, (b) 0.5% $^{13}CH_3Cl$, and (c) 0.5% CD_3Cl. (Reprinted with the permission from Ref. [29]. Copyright 2013 American Chemical Society).

In both the scandium and yttrium experiments, two $MO(CH_3Cl)$ (M = Sc, Y) complex isomers were formed spontaneously on annealing [27]. These absorptions of $MO(CH_3Cl)$ (M = Sc, Y) complex are observed at 898.4 and 919.1 cm^{-1} for Sc, and 1050.9, 805.9, and 784.8 cm^{-1} for Y, as shown in **Table 1**, which are corresponding to the Sc—O and Y—O vibration frequencies. The CH_3OMCl and CH_2ClMOH (M = Sc, Y) molecules were produced from the weakly bound $MO(CH_3Cl)$ complexes through photoinduced isomerization reactions on 250–300 nm wavelength irradiation, as shown in **Figure 1**. The CH_3OMCl (M = Sc, Y) isomer observed at

1171.5 and 565.6 cm^{-1} for Sc and 1149.2 and 490.9 cm^{-1} for Y can be regarded as being formed through the addition of the C—Cl bond to the O═M bond, whereas the CH_2ClMOH (M = Sc, Y) isomer observed at 3775.0 and 738.4 for Sc, and 3774.2 and 627.6 for Y is formed through the addition of the C—H bond to the M═O bond. On the basis of DFT calculations, the $MO(CH_3Cl)$ (M = Sc, Y) complex with C_s structure is more stable than the complex with C_{3v} structure by 25.5 (Sc) or 24.0 (Y) kJ/mol. Both CH_3OMCl and CH_2ClMOH (M = Sc, Y) molecules are more stable than the $MO(CH_3Cl)$ complex isomers. The CH_3OMCl (M = Sc, Y) molecule was predicted to proceed through a transition state with an energy barrier of 17.7 for Sc and 8.4 kJ/mol for Y from the $MO(CH_3Cl)$ complex, whereas the CH_2ClMOH isomer also proceeded through a transition state with a much higher energy barrier of 160.1 for Sc and 178.5 kJ/mol for Y from the $MO(CH_3Cl)$ complex. The CH_3OMCl (M = Sc, Y) structure is about 173.0 for Sc and 180.6 kJ/mol for Y lower in energy than the $CH_2ClScOH$ and CH_2ClYOH isomer. The CH_2ClMOH (M = Sc, Y) molecule was also calculated to involve agostic interaction observed between the metal atom and the chlorine atom due to short bond distances of 2.598 Å for Sc—Cl and 2.821 Å for Y—Cl. Such interaction is quite similar to the agostic interactions generally defined to characterize the distortion of an organometallic moiety, which brings an appended C—H bond into close proximity with the metal center [17, 21, 45].

Molecule	Ground state	Point group	Vibrational frequency[b]	Binding energy[c]	Ref.
$ScO(CH_3Cl)$	$^2A'$	C_s	898.4	−34.9	[27]
	$^2A_1$	C_{3v}	919.1	−9.4	
$YO(CH_3Cl)$	$^2A'$	C_s	1050.9, 783.5	−37.0	[27]
	$^2A_1$	C_{3v}	805.9	−13.0	
$TiO(CH_3Cl)$	$^3A'$	C_s	961.8	−41.4	[28]
$ZrO(CH_3Cl)$	^3A	C_1	898.2	−40.5	[28]
$NbO(CH_3Cl)$	$^4A'$	C_s	935.6	−29.5	[29]
$TaO(CH_3Cl)$	^2A	C_1	991.5	−5.0	[29]
$MnO(CH_3Cl)$	$^6A'$	C_s	843.4	−38.9	[30]
$FeO(CH_3Cl)$	$^5A'$	C_s	882.7	−97.4	[30]

[a] Only the values for the most abundant metal isotope are listed.
[b] The mode assignments of the experimental vibrational frequencies are discussed in the cited literature.
[c] Relative to the energy sum of ground metal oxide and CH_3Cl.

Table 1. Ground electronic states, symmetry point groups, vibrational frequencies (cm^{-1}) and binding energies (kJ/mol) for the $MO(CH_3Cl)$ species in solid argon[a].

For IVB metal monoxides, the ground-state $MO(CH_3Cl)$ (M = Ti, Zr) complexes correlate to the ground-state TiO ($^3\Delta$) and ZrO ($^1\Sigma^-$). The binding energies are predicted to be 20.5 (Ti) and 12.2 kcal/mol (Zr), which are larger than the corresponding values of $TiO(CH_4)$ and $ZrO(CH_4)$ [45, 46]. The $MO(CH_3Cl)$ (M = Ti, Zr) complexes can rearrange to the $CH_3M(O)Cl$ isomers by metal terminal insertion to C—Cl bond upon UV light irradiation (λ < 300 nm), which are observed at 999.5 and 526.2 cm^{-1} for Ti, and 915.2 and 488.3 cm^{-1} for Zr, as shown in **Figure 2**. Theoretical calculations also indicated that the electronic state crossings exist from the MO (M = Ti, Zr) + CH_3Cl reaction to the more stable $CH_3M(O)Cl$ molecules through the $MO(CH_3Cl)$ complexes traverse their corresponding transition states. The $CH_3M(O)Cl$ (M = Ti, Zr) molecule was predicted to have a singlet ground state without symmetry. According to CCSD(T) single-point calculations on B3LYP optimization geometry, the singlet ground state is 44.3 kcal/mol for $CH_3Ti(O)Cl$ and 52.2 kcal/mol for $CH_3Zr(O)Cl$ lower in energy than its corresponding triplet state. The triplet $MO(CH_3Cl)$ (M = Ti, Zr) isomerized to the singlet $CH_3M(O)Cl$ (M = Ti, Zr) molecule through their corresponding transition states, which indicated that these reactions related to the spin crossing under UV light irradiation.

Molecule	Ground state	Point group	Vibrational frequency[b]	Binding energy[c]	Ref.
CH_3OScCl	^2A	C_1	1171.5, 565.6	−268.8	[27]
$CH_2ClScOH$	^2A	C_1	3775.0, 738.4	−95.8	[27]
CH_3OYCl	^2A	C_1	1149.2, 490.9	−300.3	[27]
CH_2ClYOH	^2A	C_1	3774.2, 627.6	−119.7	[27]
$CH_3Ti(O)Cl$	^1A	C_1	999.5, 526.2	−324.1	[28]
$CH_3Zr(O)Cl$	^1A	C_1	915.2, 488.3	−349.0	[28]
$CH_2ClNb(O)H$	^2A	C_1	1698.0, 985.0	−136.7	[29]
$CH_2ClTa(O)H$	^2A	C_1	1760.0, 984.8	−182.7	[29]
CH_3MnOCl	^6A	C_1	569.6, 542.2	−55.2	[30]
CH_3FeOCl	^5A	C_1	570.4, 561.5	−50.2	[30]

[a] Only the values for the most abundant metal isotope are listed.
[b] The mode assignments of the experimental vibrational frequencies are discussed in the cited literature.
[c] Relative to the energy sum of ground metal oxide and CH_3Cl.

Table 2. Ground electronic states, symmetry point groups, vibrational frequencies (cm^{-1}), and binding energies (kJ/mol) for the isomers of $MO(CH_3Cl)$ in solid argon[a].

The ground-state $NbO(CH_3Cl)$ and $TaO(CH_3Cl)$ molecules are related to the ground-state NbO ($^4\Sigma$) and TaO ($^2\Delta$). The predicted binding energies of 29.5 (Nb) and 5.0 kJ/mol (Ta) are larger than the corresponding values of $NbO(CH_4)$ and $TaO(CH_4)$ complexes [46], which were predicted to be very weakly interaction with the metal atom being bound to three hydrogen atoms of CH_4. The $MO(CH_3Cl)$ (M = Nb, Ta) complexes rearranged to the more stable doublet $CH_2ClM(O)H$ isomer upon visible light excitation, as shown in **Table 2**. Thus, some excited states may be involved during the reaction process. The $CH_2ClM(O)H$ molecules were predicted to involve agostic interactions between the chlorine atom and the metal center. It is quite interesting to note that the $CH_2ClM(O)H$ (M = Nb, Ta) molecules involve agostic interactions between the chlorine atom and the metal atom. It is notable that the group 5 metal methylidene complexes are more agostically distorted than the group 4 metal complexes.

Taking $CH_2ClNb(O)H$ as an example, the ∠ClCNb was predicted to be only 80.4° with a Cl—Nb distance of 2.624 Å. Agostic distortion interaction is a universal phenomenon in the structures of the early transition metal alkylidene complexes and even more popular in the structures of the small methylidene complexes, in which agostic interactions are also observed between the group 4–6 transition metal atom and one of the R-hydrogen atoms.

The reactions of FeO and MnO with CH_3Cl first formed the $MO(CH_3Cl)$ (M = Mn, Fe) complexes when annealing, which can isomerize to CH_3MOCl (M = Mn, Fe) upon $300 < \lambda < 580$ nm irradiation. The products were characterized by isotopic IR studies with CD_3Cl and $^{13}CH_3Cl$ and density functional calculations, as shown in **Figure 3**. Based on theoretical calculations, the $MO(CH_3Cl)$ (M = Mn, Fe) complexes have $^5A'$ for Fe and $^6A'$ ground state for Mn with C_s symmetry, respectively, as listed in **Table 1**. The binding energies of $MO(CH_3Cl)$ (M = Mn, Fe) are 9.3 and 23.3 kcal/mol lower than MO + CH_3Cl, which are higher in energy than $MO(CH_4)$ and MO(Ng) (Ng = Ar, Kr, Xe) at the same calculation level [46–48]. The accurate CCSD(T) single-point calculations illustrate the CH_3MOCl isomerism are 13.8 and 3.1 kcal/mol lower in energy than the $MO(CH_3Cl)$ (M = Mn, Fe) complexes.

The different reactivity of metal monoxide with CH_3Cl can be rationalized in terms of changes in valence electron structures accompanied by electronic spin state crossing. In the scandium and yttrium reactions, the ground ScO and YO molecules reacted with CH_3Cl to form two isomeric $MO(CH_3Cl)$ (M = Sc, Y) complexes spontaneously on annealing. Broad-band irradiation produced either the addition of the C—Cl bond to the O═M (M = Sc, Y) bond to form the CH_3OMCl (M = Sc, Y) molecules with +II oxidation state of center metal or the addition of the C—H bond to the M═O bond to give the CH_2ClMOH isomer with the valence of metal remaining in +II oxidation state. The CH_2ClMOH (M = Sc, Y) include one α-chlorine atom to form agostic molecules between chlorine atom and metal center atom with less than 90° of ∠ClCM and short Cl---M (M = Sc, Y) distances. No α-H and/or α-Cl atom for the $MO(CH_3Cl)$ complex exist, so no agostic interaction is observed. Sc and Y have only three valence electrons, and hence they are not able to form high oxidation state structures. However, Mn and Fe have five and six valence electrons. Because their d orbitals are fully half-filled and hence are not easily lost, upon $300 < k < 580$ nm irradiation the $MO(CH_3Cl)$ (M = Mn, Fe) complexes triggered the addition of the C—Cl bond to the M═O bond to form the CH_3MOCl molecules with +II valence state. The Ti and Zr metals have four valence electrons, and their oxidation states increase from +II to +IV during the addition of MO insertion into the C—Cl bond to the metal to form $CH_3M(O)Cl$ molecules. For Nb and Ta, visible light irradiation triggered the H-atom transfer of the $MO(CH_3Cl)$ complexes from CH_3Cl to the metal center to form the more stable $CH_2ClM(O)H$ isomers with the oxidation states of the metal increasing from the +II to +IV. However, the Nb and Ta have five valence electrons, and they cannot form +V oxidation structures, but possessing one valence electron characteristic of the agnostic chlorine effect.

4. Transition metal dioxides reaction with CH_3Cl

The ground-state MO_2 (M = Ti, Zr, Nb, Ta) molecules react with CH_3Cl to first form the weakly bound $MO_2(CH_3Cl)$ complexes with O···H and M···Cl bonds. For Ti and Zr, the $MO_2(CH_3Cl)$

complexes can isomerize to the more stable $CH_3OM(O)Cl$ molecules with the addition of the C—Cl bond of CH_3Cl to one of the O=M bond of MO_2 on annealing after broadband light irradiation ($\lambda < 300$ nm), as shown in **Figures 2** and **4**. And the reaction potential energy profile interpreted the chemical reaction mechanism of C—Cl activation by MO_2 (M = Ti, Zr). The photoisomerization reaction of $MO_2(CH_3Cl)$ (M = Nb, Ta) is quite different from those of $MO_2(CH_3Cl)$ (M = Ti, Zr). The $MO_2(CH_3Cl)$ (M = Nb, Ta) complexes were initiated H-transfer under ultraviolet light irradiation to isomerize the more stable $CH_2ClM(O)OH$ molecules. The $CH_2ClM(O)OH$ (M = Nb, Ta) molecules were predicted to involve agostic interactions between the chlorine atom and the metal center. During the photoisomerization process, no electronic spin state crossings were found, as shown in **Table 3**, different from the reaction of metal monoxides with CH_3Cl.

Molecule	Ground state	Point group	Vibrational frequency[b]	Binding energy[c]	Ref.
$TiO_2(CH_3Cl)$	^1A	C_1	940.3, 906.2	−95.7	[28]
$ZrO_2(CH_3Cl)$	$^1A'$	C_s	874.1, 804.2	−84.4	[28]
$NbO_2(CH_3Cl)$	^2A	C_1	948.1, 890.9	−75.66	[29]
$TaO_2(CH_3Cl)$	$^2A'$	C_s	948.4, 890.8	−52.9	[29]
$CH_3OTi(O)Cl$	^1A	C_1	1173.0, 1102.6, 634.1	−326.0	[28]
$CH_3OZr(O)Cl$	^1A	C_1	1153.1, 901.2	−331.9	[28]
$CH_2ClNb(O)OH$	^2A	C_1	3678.4, 979.4, 712.8	−185.5	[29]
$CH_2ClTa(O)OH$	^2A	C_1	3690.8, 975.9, 605.8, 495.7	−171.8	[29]

[a] Only the values for the most abundant metal isotope are listed.
[b] The mode assignments of the experimental vibrational frequencies are discussed in the cited literature.
[c] Relative to the energy sum of ground metal oxide and CH_3Cl.

Table 3. Ground electronic states, symmetry point groups, vibrational frequencies (cm^{-1}), and binding energies (kJ/mol) for the product from $MO_2 + CH_3Cl$ in solid argon[a].

5. Conclusion and outlook

C—Cl and/or C—H bond of monochloromethane activation by transition metal monoxide and dioxide molecules has been investigated using matrix infrared spectroscopy in excess argon and density functional theoretical calculations. The metal monoxide and dioxide molecules prepared by laser-ablated bulk higher oxide targets reacted with monochloromethane to form the weakly bound $MO(CH_3Cl)$ (x = 1, 2; M = Sc, Y, Nb, Ta, Ti, Zr, Mn, Fe) complexes, which isomerized to the more stable CH_3OMCl (M = Sc, Y), agostic CH_2ClMOH (M = Sc, Y, Nb, Ta) and $CH_3M(O)Cl$ (M = Ti, Zr, Nb, Ta, Mn, Fe) isomers upon limited visible light excitation. Metal dioxides also reacted with CH_3Cl to form $MO_2(CH_3Cl)$ (M = Ti, Zr, Nb, Ta), which was rearranged to the more stable $CH_3OM(O)Cl$ (M = Ti, Zr) and $CH_2ClM(O)OH$ (M = Nb, Ta) molecules under ultraviolet light irradiation. Agostic interactions were observed in CH_2ClMOH (M = Sc, Y, Nb, Ta) and $CH_2ClM(O)OH$ (M = Nb, Ta) between the chlorine atom and the metal center atom.

Acknowledgements

We gratefully acknowledge the financial support from National Natural Science Foundation of China (Grants No. 21273202 and 21473162). Y. Zhao is grateful to the Project Grants 521 Talents Cultivation of Zhejiang Sci-Tech University and China Scholarship Council (CSC) Foundation. This work is also supported by Zhejiang Provincial Top Key Academic Discipline of Chemical Engineering and Technology.

Author details

Yanying Zhao[1,2*], Xin Liu[1] and Shuang Meng[1]

*Address all correspondence to: yyzhao@zstu.edu.cn

1 Department of Chemistry, Zhejiang Sci-Tech University, Hangzhou, China

2 State Key Laboratory of Advanced Textiles Materials and Manufacture Technology, MOE, Zhejiang Sci-Tech University, Hangzhou, China

References

[1] Yokouchi Y, Noijiri Y, Barrie LA, Toom-Sauntry D, Machida T, Inuzuka Y, Akimoto H, Li HJ, Fujinuma Y, Aoki SA. Strong source of methyl chloride to the atmosphere from tropical coastal land. Nature. 2000;403:295–298. DOI: 10.1038/35002049

[2] Yokouchi Y, Ikeda M, Inuzuka Y, Yukawa T. Strong emission of methyl chloride from tropical plants. Nature. 2002;416:163–165. DOI: 10.1038/416163a

[3] Tajima N, Tsuneda T, Toyama F, Hirao K. A new mechanism for the first carbon–carbon bond formation in the MTG process: a theoretical study. J. Am. Chem. Soc. 1998;120:8222–8229. DOI: 10.1021/ja9741483

[4] Hazari N, Lglesia E, Labinger JA, Simonetti, DA. Selective homogeneous and heterogeneous catalytic conversion of methanol/dimethyl ether to triptane. Acc. Chem. Res. 2012;45: 653–662. DOI: 10.1021/ar2002528

[5] Olah GA, Gupta B, Felberg JD, et al. Electrophilic reactions at single bonds. 20. Selective monohalogenation of methane over supported acid or platinum metal catalysts and hydrolysis of methyl halides over γ-alumina-supported metal oxide/hydroxide catalysts. A feasible path for the oxidative conversion of methane into methyl alcohol/dimethyl ether. J. Am. Chem. Soc. 1985;107:7097–7105. DOI: 10.1021/ja00310a057

[6] Olsbye U, Svelle S, Bjørgen M, Beato P, Janssens TVW, Joensen F, Bordiga S, Lillerud KP. Conversion of methanol to hydrocarbons: how zeolite cavity and pore size controls product selectivity. Angew. Chem. Int. Ed. 2012;51:5810–5831. DOI: 10.1002/anie.201103657

[7] Xu T, Zhang Q, Song H, Wang Y. Fluoride-treated H-ZSM-5 as a highly selective and stable catalyst for the production of propylene from methyl halides. J. Catal. 2012;295:232–241. DOI: 10.1016/j.jcat.2012.08.014

[8] Zhang R, Zhang B, Shi Z, Liu N, Chen B. Catalytic behaviors of chloromethane combustion over the metal-modified ZSM-5 zeolites with diverse SiO_2/Al_2O_3 ratios. J. Mol. Catal. A: Chem. 2015;398:223–230. DOI: 10.1016/j.molcata.2014.11.019

[9] Kong L, Shen B, Zhao J, Liu J. Comparative study on the chloromethane to olefins reaction over SAPO-34 and HZSM-22. Ind. Eng. Chem. Res. 2014;53:16324–16331. DOI: 10.1021/ie5028155

[10] Olsbye U, Saure OV, Muddada NB, Bordiga S, Lamberti C, Nilsen M H, Lillerud KP, Svelle S. Methane conversion to light olefins – how does the methyl halide route differ from the methanol to olefins (MTO) route? Catal. Today. 2011;171:211–220. DOI: 10.1016/j.cattod.2011.04.020

[11] Wei Y, Zhang D, Chang F, Xia Q, Su BL, Liu Z. Ultra-short contact time conversion of chloromethane to olefins over precoked SAPO-34: direct insight into the primary conversion with coke deposition. Chem. Commun. 2009;40:5999–60001. DOI: 10.1039/b909218h

[12] Xu L, Du AP, Wei YX. Synthesis of SAPO-34 with only Si(4Al) species: effect of Si contents on Si incorporation mechanism and Si coordination environment of SAPO-34. Micropor. Mesopor. Mater. 2008;115:332–337. DOI: 10.1016/j.micromeso.2008.02.001

[13] Jiang Z, Shen B, Zhao J, Wang L, Kong L, Xiao W. Enhancement of catalytic performances for the conversion of chloromethane to light olefins over SAPO-34 by modification with metal chloride. Ind. Eng. Chem. Res. 2015;54(49):12293–12302. DOI: 10.1021/acs.iecr.5b03586

[14] Yang CT, Zhang ZQ, Liang J, Liu JH, Lu XY, Chen HH, Liu L. Copper-catalyzed cross-coupling of nonactivated secondary alkyl halides and tosylates with

secondary alkyl Grignard reagents. J. Am. Chem. Soc. 2012;134:11124–11127. DOI: 10.1021/ja304848n

[15] Rudolph A, Lautens M. Secondary alkyl halides in transition metal-catalyzed cross-coupling reactions. Angew. Chem. Int. Ed. 2009;48:2656–2670. DOI: 10.1002/anie.200803611

[16] Xiao B, Liu ZJ, Liu L, Fu Y. Palladium-catalyzed C–H activation/cross-coupling of pyridine N-oxides with nonactivated secondary alkyl bromides. J. Am. Chem. Soc. 2013;135:616–619. DOI: 10.1021/ja3113752

[17] Cho HG, Andrews L. Persistent photo-reversible transition-metal methylidene system generated from reaction of methyl fluoride with laser-ablated zirconium atoms and isolated in a solid argon matrix. J. Am. Chem. Soc. 2004;126(33):10485–10492. DOI: 10.1021/ja0486115

[18] Cho HG, Andrews L. Infrared spectra of CH_3–MX and CH_2X–MH prepared in reactions of laser-ablated gold, platinum, palladium, and nickel atoms with CH_3Cl and CH_3Br. Organometallics. 2013;32(9):2753–2759. DOI: 10.1021/om400192v

[19] Cho HG, Andrews L. Infrared spectra of CX_3—MnX and CX_2=MnX_2 (X = H, F, Cl) prepared in reactions of laser-ablated manganese atoms with halomethanes. Organometallics. 2011;30(3):477–486. DOI: 10.1021/om100791h

[20] Cho HG, Andrews L. Infrared spectra of simple methylidyne, methylidene, and insertion complexes generated in reactions of laser-ablated rhodium atoms with halomethanes and ethane. Organometallics. 2010;29(10):2211–2222. DOI: 10.1021/om900902p

[21] Cho HG, Andrews L. Infrared spectra of the CH_3—MX, $CHMH_2X$— complexes formed by reaction of methyl halides with laser-ablated group 5 metal atoms. J. Phys. Chem. A. 2006;110:10063–10077. DOI: 10.1021/jp0629644

[22] Cho HG, Andrews L. Infrared spectra of the CH_3—MX and CH_2—MHX complexes formed by reactions of laser-ablated group 3 metal atoms with methyl halides. J. Phys. Chem. A. 2007;111:2480–2491. DOI: 10.1021/jp067662g

[23] Cho HG, Andrews L. Formation of CH_3TiX, CH_2=TiHX, and $(CH_3)_2TiX_2$ by reaction of methyl chloride and bromide with laser-ablated titanium atoms: photoreversible α-hydrogen migration. Inorg. Chem. 2005;44(4):979–988. DOI: 10.1021/ic048615a

[24] Cho HG, Andrews L. Infrared spectra of methylidynes formed in reactions of Re atoms with methane, methyl halides, methylene halides, and ethane: methylidyne C—H stretching absorptions, bond lengths, and s character. Inorg. Chem. 2008;47:1653–1662. DOI: 10.1021/ic701505w

[25] Cho HG, Andrews L. Infrared spectra of platinum insertion and methylidene complexes prepared in oxidative C—H(X) reactions of laser-ablated pt atoms with methane,

ethane, and halomethanes. Organometallics. 2009;28:1358–1368. DOI: 10.1021/om801077x

[26] Jonathan TL, Cho HG, Andrews L. Methylidyne XC : MX_3 (M = Cr, Mo, W; X = H, F, Cl) diagnostic C–H and C–X stretching absorptions and methylidene CH_2=MX_2 analogues. Organometallics. 2007;26(25):6373–6387. DOI: 10.1021/om700689e

[27] Zhao Y. C–Cl activation by group IV metal oxides in solid argon matrixes: matrix isolation infrared spectroscopy and theoretical investigations of the reactions of MO_x (M = Ti, Zr; x = 1, 2) with CH_3Cl. J. Phys. Chem. A. 2013;117(27):5664–5674.

[28] Zhao Y, Huang Y, Zheng X, Zhou M. Preparation and characterization of the agostic bonding molecules between metal and chlorine from the reactions of niobium and tantalum monoxide and dioxide molecules with monochloromethane in solid argon. J. Phys. Chem. A. 2010;114(18):5779–5786. DOI: 10.1021/jp102199c

[29] Huang Y, Zhao Y, Zheng X, Zhou M. Matrix isolation infrared spectroscopic and density functional theoretical study of the reactions of scandium and yttrium monoxides with monochloromethane. J. Phys. Chem. A. 2010;114(7):1476–2482. DOI: 10.1021/jp9101948

[30] Zhao Y, Fan K, Huang Y, Zheng X. Matrix isolation infrared spectra, assignment and DFT investigation on reactions of iron and manganese monoxides with CH_3Cl. Spectrochim. Acta A Mol. Biomol. Spectrosc. 2013;116:96–101. DOI: 10.1016/j.saa.2013.07.010

[31] Gong Y, Zhou M, Andrews L. Spectroscopic and theoretical studies of transition metal oxides and dioxygen complexes. Chem. Rev. 2009;109(12): 6765–6808. DOI: 10.1021/cr900185x

[32] Zhao Y, Gong Y, Chen M, Zhou M. Noble gas-transition-metal complexes: coordination of VO_2 and VO_4 by Ar and Xe atoms in solid noble gas matrixes. J. Phys. Chem. A. 2006;110(5):1845–1849. DOI: 10.1021/jp056476s

[33] Wang C, Zhuang J, Wang G, Chen M, Zhao Y, Zheng X, Zhou M. Tantalum dioxide complexes with dinitrogen. Formation and characterization of the side-on and end-on bonded $TaO_2(NN)_x$ (x = 1–3) complexes. J. Phys. Chem. A. 2010;114(31):8083–8089. DOI: 10.1021/jp103866r

[34] Zhao Y, Gong Y, Chen M, Ding C, Zhou M. Coordination of ScO^+ and YO^+ by multiple Ar, Kr, and Xe atoms in noble gas matrixes: a matrix isolation infrared spectroscopic and theoretical study. J. Phys. Chem. A. 2005;109(51):11765–11770. DOI: 10.1021/jp054517e

[35] Frisch MJ, Trucks GW, Schlegel HB, Scuseria GE, Robb MA, Cheeseman JR, Montgomery JA, Vreven T, Kudin KN, Burant JC, Millam JM, Iyengar SS, Tomasi J, Barone V, Mennucci B, Cossi M, Scalmani G, Rega N, Petersson GA, Nakatsuji H, Hada M, Ehara M, Toyota K, Fukuda R, Hasegawa J, Ishida M, Nakajima T, Honda Y, Kitao O, Nakai H, Klene M, Li X, Knox JE, Hratchian HP, Cross JB, Adamo C, Jaramillo J, Gomperts R, Stratmann RE, Yazyev O, Austin AJ, Cammi R, Pomelli C, Ochterski JW, Ayala PY,

Morokuma K, Voth GA, Salvador P, Dannenberg JJ, Zakrzewski VG, Dapprich S, Daniels AD, Strain MC, Farkas O, Malick DK, Rabuck AD, Raghavachari K, Foresman JB, Ortiz JV, Cui Q, Baboul AG, Clifford S, Cioslowski J, Stefanov BB, Liu G, Liashenko A, Piskorz P, Komaromi I, Martin RL, Fox DJ, Keith T, Al-Laham MA, Peng CY, Nanayakkara A, Challacombe M, Gill PMW, Johnson B, Chen W, Wong MW, Gonzalez C, Pople JA. Gaussian 03, Revision B.05, Gaussian. Inc.: Pittsburgh, PA, 2003.

[36] Becke AD. Density-functional thermochemistry. III. The role of exact exchange. J. Chem. Phys. 1993;98:5648–5652. DOI: 10.1063/1.464913

[37] Lee C, Yang W, Parr RG. Development of the Colle–Salvetti correlation-energy formula into a functional of the electron density. Phys. Rev. B. 1988;B37:785–789. DOI: 10.1103/PhysRevB.37.785

[38] McLean AD, Chandler GS. Contracted Gaussian basis sets for molecular calculations. I. Second row atoms, Z = 11 – 18. J. Chem. Phys. 1980;72:5639–5648. DOI: 10.1063/1.438980

[39] Krishnan R, Binkley JS, Seeger R, Pople JA. Self-consistent molecular orbital methods. XX. A basis set for correlated wave functions. J. Chem. Phys. 1980;72:650–654. DOI: 10.1063/1.438980

[40] Godbout N, Salahub DR, Andzelm J, Wimmer E. Optimization of Gaussian-type basis sets for local spin density functional calculations. Part I. Boron through Neon, optimization technique and validation. Can. J. Chem. 1992;70:560–571. DOI: 10.1139/v92-079

[41] Sosa C, Andzelm BC, Wimmer E, Dobbs KD, Dixon DA. A local density functional study of the structure and vibrational frequencies of molecular transition-metal compounds. J. Phys. Chem. 1992;96:6630–6636. DOI: 10.1021/j100195a022

[42] Dolg M, Stoll H, Preuss H. Energy-adjusted abinitio pseudopotentials for the rare earth elements. J. Chem. Phys. 1989;90(3):1730–1734. DOI: 10.1063/1.456066

[43] Andrae D, Haeussermann U, Dolg M, Stoll H, Preuss H. Energy-adjusted *ab initio* pseudopotentials for the second and third row transition elements. J. Theor. Chim. Acta, 1990;77(2):123–141. DOI:10.1007/BF01114537

[44] Pople JA, Head-Gordon M, Raghavachari K. Quadratic configuration interaction. a general technique for determining electron correlation energies. J. Chem. Phys. 1987;87:5968–5975. DOI: 10.1063/1.453520

[45] Wang G, Gong Y, Chen M, Zhou M. Methane activation by titanium monoxide molecules: a matrix isolation infrared spectroscopic and theoretical study. J. Am. Chem. Soc. 2006;128(17):5974–5980. DOI: 10.1021/ja0604010

[46] Wang G, Zhou M. Probing the intermediates in the MO + $CH_4 \leftrightarrow$ M + CH_3OH reactions by matrix isolation infrared spectroscopy. Int. Rev. Phys. Chem. 2008;27(1):1–25. DOI: 10.1080/01442350701685946

[47] Zhao Y, Gong Y, Zhou M. Matrix isolation infrared spectroscopic and theoretical study of NgMO (Ng = Ar, Kr, Xe; M = Cr, Mn, Fe, Co, Ni) complexes. J. Phys. Chem. A. 2006;110(37):10777–10782. DOI: 10.1021/jp064100o

[48] Zhao Y, Zhou M. Are matrix isolated species really "isolated"? Infrared spectroscopic and theoretical studies of noble gas-transition metal oxide complexes. Sci. China Chem. 2010;53(2):327–336. DOI: 10.1007/s11426-010-0044-9

Section 2

Molecular Spectroscopic Studies of Organic Materials

Chapter 6

Vibrational and Electronic Structure, Electron-Electron and Electron-Phonon Interactions in Organic Conductors Investigated by Optical Spectroscopy

Andrzej Łapiński

Additional information is available at the end of the chapter

http://dx.doi.org/10.5772/63242

Abstract

The physical properties of organic conductors formed by tetrathiafulvalene (TTF) derivatives have been discussed in this chapter. The results were obtained using spectroscopic methods including infrared (IR), Raman, and UV-Vis. The experimental data were supported by theoretical DFT and TD-DFT calculations. Special attention has been paid to the description of electronic and vibrational structures, electron-electron and electron-phonon interactions and determination of transport parameters.

Keywords: organic conductors, electron-electron and electron-phonon interactions, vibrational end electronic structures, infrared and Raman spectroscopy, DFT calculations

1. Introduction

Research on the development of organic metals and superconductors was stimulated by finding the first organic metal—charge-transfer (CT) complex composed of tetrathiafulvalene (TTF) and tetracyanoquinodimethane (TCNQ) [1–2] and by the discovery of superconductivity in CT salts of tetramethyltetraselenafulvalene (TMTSF) [3].

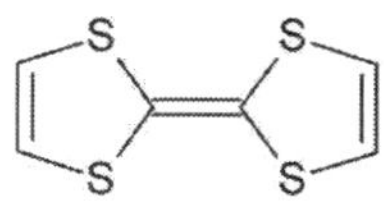

TTF

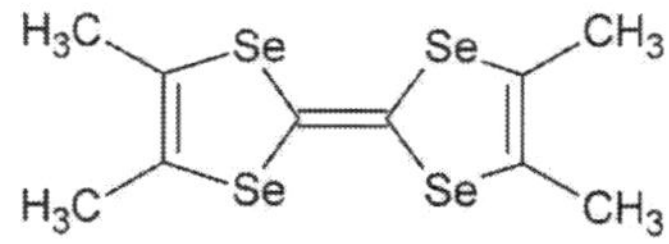

TMTSF

At the beginning, one of the most important targets in the field of molecular conductors was to search for high electrical conductivity systems [4]. Nowadays, the organic conductors based on the π-electron donors with various anions show a wealth of structural modifications and variety of physical properties [5–8]. They have attracted broad interest from the experimental and theoretical side because they exhibit a lot of fascinating phenomena related to the electronic states near the Fermi level and to relevant interactions. Superconductivity, metal-insulator phase transitions associated with charge-density wave (CDW) or spin-density wave (SDW) condensates and other cooperative states leading to antiferromagnetic-, charge- and dielectric order [9] are highly topical subjects in this field.

The great interest in this field has been sustained by chemists, who produced a huge number of π-electron donors for organic conductors. In many laboratories molecular conductors formed by the organic donors derived from TTF molecule are intensively studied. In designing π-donors, one can distinguish two main strategies. One has evolved mainly from: (a) the planarity for a facile formation of donor stacking, (b) the extension of π-conjugation for a decrease of on-site Coulombic repulsion involved in the formation of a dicationic species, and (c) the introduction of chalcogen atoms for an increase in dimensionality of the conduction pathway. The second approach to the donors design is totally different and it is based on the following requirements: (a) extension of the σ-bond framework which will lead to the lack of planarity, and (b) reduction of the π-electron system, which will increase the on-site Coulombic repulsion. The latter strategy was proposed by Yamada et al. [10]. They believe if donors are synthesized in accordance to these requirements, then it will be possible to produce superconductors using such donors.

In the last years, there is a great interest in the design and study of new molecular-based materials involving interplay between multiple physical properties. This kind of effects leads either to competition, coexistence or cooperation between the desired properties. A possible approach to reach this goal consists of building up hybrid solids formed by two molecular networks. Among these hybrid materials, the highly conducting CT salts formed by TTF-derived donors with various inorganic acceptors with permanent magnetic moments are intensively investigated [7, 11]. A characteristic feature of these salts is a spatial segregation of the organic cations and inorganic anions into alternating layers. These materials are of special interest, because their properties are determined by presence of both the system of delocalized π-electron in the fulvalene-derived stacks or layers and by the localized d-electrons of anions. These systems may exist independently or can interact, leading to new physical properties of the conducting material. One of the most interesting phenomena is a possibility of the

http://dx.doi.org/10.5772/63242

interaction between π-electrons within conducting layers and localized magnetic d-electrons in the counterions. Interaction between these two systems may lead to magnetically ordered conducting structures. The problem of magnetic order and electrical conductivity coexistence is highly topical [5, 8, 12, 13]. Magnetic interactions in these compounds are basically explained by the RKKY-type interaction mediated by the π-d coupling between the donor and the magnetic anions [14]. Moreover, Coulomb interactions between organic (cations) and inorganic (anions) subsystems may also lead to charge-ordering phenomena.

Coulomb interactions between electrons play an important role in organic conductors and have a considerable influence on optical, magnetic and conducting properties [15–17]. In many TTF based one- and two-dimensional organic conductors, the long-range Coulomb interactions are responsible for charge-ordering (CO) phenomena. In the field of CT salts, such behavior is of great interest among researches in many laboratories in the world [18–21]. In order to understand the nature of the charge localization Seo and Fukuyama [22] have performed theoretical calculations [23]. They have shown that a stripe-patterned charge ordering is stabilized in the insulating phase owing to intermolecular Coulomb repulsive forces. Moreover, Tajima *et al.* [24] have estimated the charge-ordering patterns in bis(ethylenedithio)tetrathiafulvalene (BEDT-TTF or ET) salts on the base of the spectral analysis and the mean-field calculations.

The most convenient experimental method to investigate charge distribution in conducting layers [25] and effects of electron-molecular vibration coupling [26, 27] is the vibrational spectroscopy, therefore, study of the vibrational structure in many cases is crucial to understand physical properties of organic conductors. Electron-molecular vibration (EMV) coupling constant is one of the parameters in estimating of critical temperature (T_c) of organic superconductors, whereas the degree of ionicity, or average charge per molecule (ρ) is one of the fundamental parameters characterizing the physical properties of CT salts [28]. When Coulomb interaction prevails, the salts may undergo a charge-order instability. For the determination of ρ in TTF derivatives, the C=C stretching modes of TTF framework are mainly taken into account. It was shown that such modes are very sensitive to the ionization degree of the molecule [20, 25, 27]. Nevertheless, one should keep in mind that C=C modes can be coupled to the electronic system [29, 30] and hence their positions in infrared (IR) spectra can shift towards lower frequencies [31] due to this coupling; the frequencies of molecular modes may exhibit a non-linear dependence on ρ [30, 31].

The main purpose of the chapter is to present a description of physical properties of the organic conductors using experimental and theoretical methods of molecular spectroscopy. The special attention will be paid to the description of the electronic structure and determination of transport parameters. It will be shown that the most powerful method to investigate the charge ordering phenomena is the vibrational spectroscopy [32–34]. The results presented in this chapter were obtained by the spectroscopic methods including IR, Raman and UV-Vis. The experimental data were supported by theoretical DFT and TD-DFT calculations. It should be emphasized that the spectroscopic methods are very suitable tools for investigations of crystalline organic conductors because they provide a lot of information about electronic and vibrational structures, electron-electron and electron-phonon interactions [35].

2. Electronic structure investigated by optical spectroscopy

The electronic transitions that appear in the spectra of organic conductors recorded in IR, Vis and UV regions fall into two classes. On the one hand, those at high frequencies generally are a result of localized excitations, which are related to intramolecular transitions in which an electron is excited to a higher level on the same molecule. For characterization of such type of excitations, the time-dependent DFT method (TD-DFT) can be applied for organic conductors [36, 37]. The quantitative understanding of such molecular electronic excited states is important in many domains, including spectroscopy, photochemistry and the design of new optical materials.

On the other hand, transitions at lower frequencies that are along the stacking directions or along to the S···S short contacts correspond to CT excitations between the molecules. The frequencies and oscillator strengths of these CT bands are clearly related to the electronic structure of these compounds, but they are a consequence of three types of interactions among the unpaired electrons occupying the highest molecular orbital (HOMO): (i) the overlap of the electronic wave functions between sites, (ii) the Coulomb repulsion of two electrons on the same or adjacent sites, and (iii) interactions of the electron with phonons. Theoretical models for the electronic structure of organic conductors have shown the importance of one or the other of these interactions, e.g. tight-binding theory [38], the Hubbard model [39] and the Peierls model [40].

Compounds	Charge-transfer bands		Intramolecular excitations	References
	A band	B band		
β''-$(ET)_4NH_4[Cr(C_2O_4)_3]\cdot$DMF	3400	10,200	16,300; 20,600; 30,800	[41]
β''-$(ET)_4K[Cr(C_2O_4)_3]\cdot$DMF	3620	9900	16,000; 21,000; 30,000	[42]
$(ET)_6(Mo_8O_{26})(DMF)_3$	2500	7300	11,200; 21,900; 35,400	[43]
κ-$(ET)_4[Co(CN)_6][N(C_2H_5)_4]\cdot 2H_2O$	2500, 3390	7200	10,000	[44]
τ-(P-*S*,*S*-DMEDT-TTF)$_2$ $(AuBr_2)(AuBr_2)_y$	1000, 5900	12,300	18,100; 22,900; 31,400; 33,500	[45]
β-(EDT-DTDSF)$_4Hg_3I_8$	2800	11,700	26,000; 32,500; 40,000	[46]
(DOEO)$_4HgBr_4\cdot$TCE	3600	11,500	19,500; 21,800; 28,400; 30,600; 32,000; 33,200	[37]

Table 1. Electronic transitions (in cm^{-1}) observed for selected organic conductors.

The CT bands (denoted as A and B) and intramolecular excitations for selected organic conductors are shown in **Table 1**. The band A corresponds to a charge transfer of an electron from an occupied A^+ to a neutral A^0 molecule ($A^++A^0 \rightarrow A^0+A^+$), whereas band B is associated with the CT from one A^+ anion to a neighboring A^+ ($A^++A^+ \rightarrow A^0+A^{2+}$). Within the Hubbard theory the position of A band relates to the Coulomb repulsion energy between two electrons on adjacent molecules (V) and hopping integrals (t). The position of B band depends on the value of the effective Coulomb interaction between two electrons reside on the same site. The energy of its transition is proportional to (U-V); U and V are the Hubbard parameters for on-site and nearest-neighbor Coulomb repulsion, respectively. From the center position of B band and assuming V to be small, we can estimate the on-site Coulomb repulsion. Hubbard has also shown that the electron distribution in the ground state can be periodic and may be considered as a generalization of the classical Wigner lattice. The arrangement of the electrons within the period can be non-uniform and then it will give rise to electric fields that can distort the ordinary lattice; distortions can manifest themselves as satellites in the X-ray diffraction pattern. For example, the ground-state configuration for ϱ=1/2 may be presented by two patterns: ...10101010... and ...11001100... (0 indicates the neutral molecule and 1 corresponds to the monocation). Furthermore, it was also shown that the observed polarizations of optical transitions are consistent with the proposed model [39].

Considerable information about the electronic structure can also be extracted from the oscillator strength sum rule [47]. Such an approach is used very often for organic conductors [41, 42, 46, 48, 49]. The effective number of electrons (N_{eff}) participating in optical transitions for energies less than $\hbar\omega$ is given by

$$\left[\frac{m_0}{m_{eff}}\right] N_{eff}(\omega) = \frac{m_0}{32\pi N_c e^2} \int \sigma(\omega')\, d\omega' \qquad (1)$$

where m_{eff} is the effective mass of the carriers; m_o means the electronic mass; N_c is the number of molecules per unit volume and σ means optical conductivity.

In order to analyze the electronic dispersion for organic metals and semiconductors the least-squares fits to the experimental data of the reflectance calculated from the Drude and Drude-Lorentz dielectric functions can be performed [41–43, 45, 46, 48, 50, 51]. The Drude model is used for describing the intraband transitions of the free charge carriers, whereas interband transitions can be investigated using the Lorentz model. Within the Drude-Lorentz model the complex dielectric constant can be written as [47]:

$$\varepsilon(\omega) = \varepsilon_{core} - \frac{\omega_p^2}{\omega(\omega + i\Gamma)} + \frac{\Omega_p^2}{\omega_0^2 - \omega^2 - i\gamma\omega} \qquad (2)$$

where ω_p is the plasma frequency of the free charge carries, Γ the relaxation constant of the free charge carries (Γ is related to the relaxation time of carriers τ by $\Gamma=1/\tau$), ε_{core} represents all higher frequency contributions to the dielectric function, Ω_p, ω_0 and γ are the oscillator strength, the resonance frequency and the linewidth of the Lorentz oscillators, respectively.

The polarized reflectance and optical conductivity spectra measured for ethylenedithio-dithiadiselenafulvalene (EDT-DTDSF) salt: β-$(EDT\text{-}DTDSF)_4Hg_3I_8$ is presented in the **Figure 1**

[46]. In the bottom panel of this figure there is the inset with sum rule calculations performed for two polarizations of electrical vector. For E_{max}, $(m_o/m_{eff})N_{eff}$ increases rapidly at first and begins to level off at a value near 0.18 and then rises rapidly above 1 eV. The function rises much more slowly and smoothly for the other polarization. For this salt the effective masses of holes (m_{eff}= 1.1m_o for E_{max}andm_{eff}= 3.8m_o for E_{min}) suggest that it belongs to quasi-two-dimensional material with a closed Fermi surface [46].

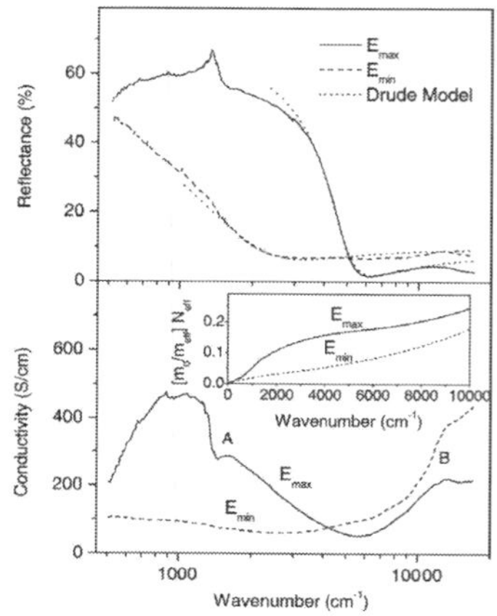

Figure 1. Polarized reflectance (upper panel) and optical conductivity (lower panel) spectra of β-(EDT-DTDSF)$_4$Hg$_3$I$_8$ at room temperature. Least-squares fits to the reflectance assuming a Drude dielectric function (upper panel, dotted line) and sum rule calculations based on the optical conductivity data (in the inset). (Reprinted with permission from Łapiński et al. [46]. Copyright© 2006, Elsevier).

The polarized reflectance and optical conductivity spectra measured for pyrazino-*S,S*-dimethyl-ethylenedithio-tetrathiafulvalene (P-*S,S*-DMEDT-TTF) salt: τ-(P-S,S-DMEDT-TTF)$_2$(AuBr$_2$)(AuBr$_2$)$_y$ are shown in **Figure 2** [45].

Dotted line in the upper panel of this figure was used for display Drude-Lorentz model. The electronic band centered between 5500 and 6050 cm^{-1} observed for this organic conductor is related to the transitions between the lower and upper Hubbard bands. Within the basic Hubbard framework, the optical conductivity spectrum at low-temperature is predicted to show: a Drude-type band centered at zero frequency due to quasi-particle transitions, a resonance at $\omega \approx U/2$ (where U means the on-site Coulomb repulsion) due to transitions between the Hubbard bands and quasi-particle band; and a contribution which appears at $\omega \approx U$ due to transitions between lower and upper Hubbard bands [52].

Among the two-dimensional organic CT salts, the κ-phase ET salts are of particular interest, because the constituting cationic dimers are arranged in an anisotropic lattice [53–55]. To understand the electronic properties of such group of organic conductors Kino and Fukuyama [56–58] considered a triangular lattice. For κ-(ET)$_2$X family the electronic transition observed in mid-infrared has the double nature origin. One can identify intraband transitions within the correlated manifold and interband transitions due to CT within the ET dimer. Two-dimensional metallic properties are related to the carriers which can move between the dimer "lattice sites." In the Mott-insulating state, the strong electronic repulsion limits their mobility and immobilizes them. The other types of carriers are localized on the sites of a triangular lattice. The effective on-site Coulomb interaction is related to the intradimer overlap integrals, whereas the interdimer overlap integrals define the hopping *t* between the sites.

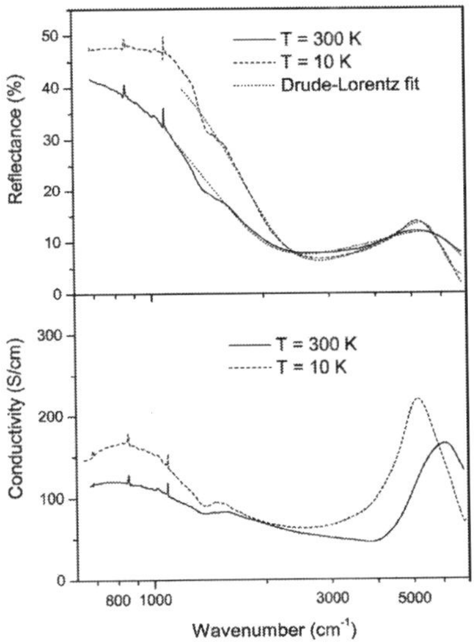

Figure 2. Polarized reflectance spectra of τ-(P-S,S-DMEDT-TTF)$_2$(AuBr$_2$)(AuBr$_2$)$_y$ at room temperature and 10 K (represented by solid and dashed lines, respectively). Least-squares fits to the reflectance assuming a Drude-Lorentz dielectric function (upper panel, dotted line) and the optical conductivity spectra derived from the reflectance spectra by the Kramers-Krönig transformation. (Reprinted with permission from Łapiński et al. [45]. Copyright© 2003, Elsevier).

Figure 3 shows the electronic feature A for κ-(ET)$_4$[Co(CN)$_6$][N(C$_2$H$_5$)$_4$]·2H$_2$O [44] which can be explained by two contributions: CT inside the dimer "lattice sites" (from 3300 to 4700 cm^{-1}) and interdimer CT by correlated charge carriers (from 2500 to 3000 cm^{-1}).

Rice [59], Yartsev and coworkers [60–62] have shown that the electronic feature with a maximum at around 3500 cm^{-1} is related to the CT within a dimer; A$_g$ vibrations of ET molecule can be coupled with such transition. Moreover, the electronic band at about 2700 cm^{-1} reveals transitions between the Hubbard bands formed by the correlated conduction electrons [15, 16, 52, 63].

The family of two-dimensional ET salts with half-filled band are especially interesting for the charge disproportionation phenomena and has been extensively investigated [64–71], with the special focus on the κ-phase salts [72–79]. However, such phenomenon in such group of salts with an effectively half-filled band was very rarely observed. It has been reported only for a few salts, such as for κ-$(ET)_4PtCl_6 \cdot C_6H_5CN$, the triclinic κ-$(ET)_4[M(CN)_6][N(C_2H_5)_4] \cdot 3H_2O$ and the monoclinic κ-$(ET)_4[M(CN)_6][N(C_2H_5)_4]$ $\cdot 2H_2O$ (with M = Co^{III}, Fe^{III} and Cr^{III}). For κ-$(ET)_4[Co(CN)_6][N(C_2H_5)_4] \cdot 2H_2O$ and κ-$(ET)_4[Fe(CN)_6][N(C_2H_5)_4] \cdot 2H_2O$ salts [44] the crystallographic studies have been performed at different temperatures at about 100, 200 and 293 K [78]. At ambient temperature, both salts exhibit rather poor conductivity and large paramagnetic susceptibility, which allows us to assert that they are on the insulator side of the Mott-Hubbard criterion as in κ-$(ET)_2Cu_2(CN)_3$ and κ-$(ET)_2Cu[N(CN)_2]Cl$ [80, 81]. Both investigated salts undergo a charge-ordering phase transitions at T_{CO}=150 K [78]. Above 150 K they are in the Mott insulating state with a uniform charge distribution among ET molecules ($ET_A^{+0.5}ET_A^{+0.5}$, $ET_B^{+0.5}ET_B^{+0.5}$), whereas below this temperature a charge pattern ($ET_A^{+1}ET_A^{+1}$, $ET_B^{0}ET_B^{0}$) is observed.

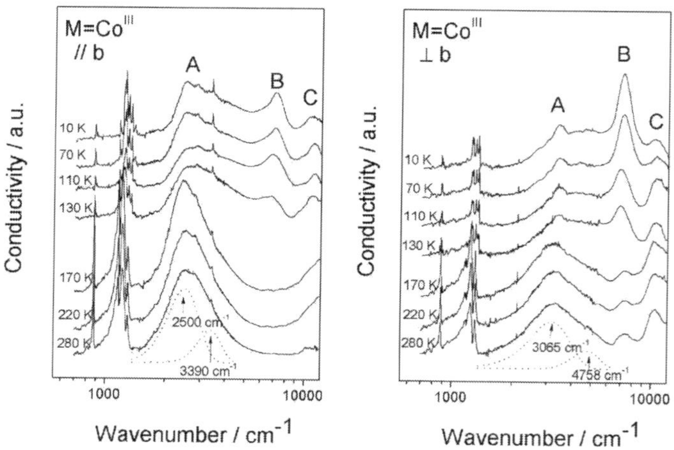

Figure 3. Temperature dependence of the optical conductivity spectra of κ-$(ET)_4[Co(CN)_6][N(C_2H_5)_4]$ $\cdot 2H_2O$. Dotted lines show the deconvolution of the charge-transfer band into two components. (Reprinted with permission from Łapiński et al. [44]. Copyright© 2013, American Chemical Society).

The phase transition at 150 K induces considerable modifications in electronic structures. Due to the charge ordering phase transition, some new bands related to the fully ionized and neutral ET molecules appear in the spectra. One of the most important spectral changes is appearance of the new band at about 7000 cm^{-1} [44] (see **Figure 3**; band B). This electronic band can be attributed to intermolecular electronic transitions between neighboring ET^{+} cations. The appearance of this feature is a consequence of the charge-ordering phenomenon. Below 150 K the intensity of this band increases on cooling down. For polarisation E⊥b we can also find this band above 150 K which means that even at room temperature charge density fluctuations are present and hence we can expect that the relatively short-living dimers $(ET_A)_2^{2+}$ exist in our system.

http://dx.doi.org/10.5772/63242

3. Vibrational structure studied by IR and Raman spectroscopy

The most well-known donor in organic conductors is ET molecule. In the solid state, they are never flat and their symmetry is lowered caused by the deformations of outer CH_2-CH_2 groups. Nevertheless, for the classification of normal modes, a planar D_{2h} molecular symmetry was often taken by many authors, e.g. by Kozlov *et al.* [82, 83] and Eldridge *et al.* [84]. The equilibrium geometry of a neutral ET has a boat conformation (C_2 symmetry). An ionized ET can be either staggered (D_2 molecular symmetry) or eclipsed (C_{2h} symmetry) [85]. Correlations between the spectral predictions based on D_{2h} and on D_2 symmetry can be easily made, the difference being associated with the lack of inversion center (A_g and A_u become A, B_{1g} and B_{1u} become B_1, B_{2g} and B_{2u} become B_2, B_{3g} and B_{3u} become B_3. Moreover, A and B_1 modes in D_2 correlate with A in C_2 and B_2, B_3 in D_2 with B in C_2).

For ET molecule, there are three normal modes related to C=C stretching vibrations, which exhibit the largest ionization frequency shift (120–130 cm^{-1}) [25, 31, 54]. Assuming D_{2h} symmetry for the free ET molecule, these modes are called $\nu_2(A_g)$, $\nu_3(A_g)$ and $\nu_{27}(B_{1u})$ (see **Figure 4**) where the symbols in parentheses denote the symmetry species [82, 83].

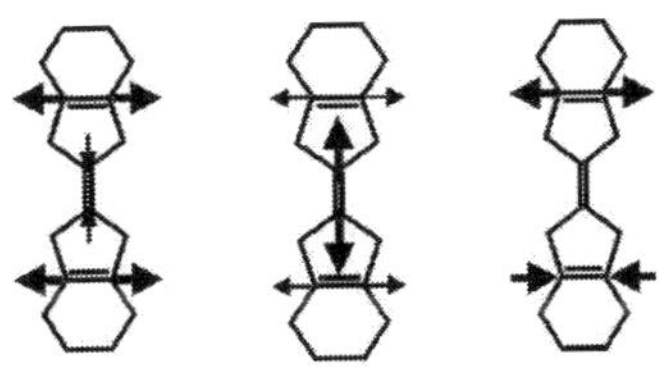

Figure 4. Schematic views of the three C=C stretching modes in ET molecule.

They have been the first ones to be proposed for the determination of ρ in ET [25] and the linear dependence of $\nu_2(A_g)$ and $\nu_{27}(B_{1u})$ modes against charge density has been experimentally verified. The Raman-active ν_2 and ν_3 modes involve the central and symmetric ring C=C stretching vibrations. For the neutral ET molecule, C=C stretching vibrations are mixed almost equally in ν_2 and ν_3 modes, whereas for the ionized ET, they are almost separated; in the latter case the ν_2 and ν_3 modes are mainly assigned to ring C=C stretching and bridge C=C stretching, respectively [86, 87]. The IR-active $\nu_{27}(B_{1u})$ mode is due to the anti-symmetric ring C=C stretching and is thus completely separated from the central C=C stretching.

Figure 5 shows that for ET salts in their IR spectra a group of strong bands related to C=C bonds vibrations can be found within the spectral region from 1100 to 1400 cm^{-1}. The ionization results in the meaningful red shift of these modes 95 and 60 cm^{-1} [20, 25, 88]. For organic metals, where electrical conductivity is relatively high, situation becomes a little more complicated, because instead of well defined group of vibrational features, additional broad maxima can be observed (e.g. β''-$(ET)_4A[M(C_2O_4)_3]\cdot$DMF where $A=NH_4^+$, K^+ and $M=Cr^{III}$, Fe^{III} [41–42],

$(ET)_6(Mo_8O_{26})(DMF)_3$ [43]). Such broad bands are a consequence of the coupling of $\nu_2(A_g)$ and $\nu_3(A_g)$ modes with conducting electrons, for which their proximity leads to a strong mixing.

Nevertheless, it should be emphasized that the C=C modes are often used in the study of the nature of different phase transitions. For example, for the salts κ-$(ET)_4[Co(CN)_6][N(C_2H_5)_4]\cdot 2H_2O$ and κ-$(ET)_4[Fe(CN)_6][N(C_2H_5)_4]\cdot 2H_2O$ where the phase transition from the Mott insulator state to the charge-ordering state is present, the C=C modes have been discussed [44]. The considerable modifications in vibrational structure due to this phase transition at 150 K are presented in **Figure 5a**. The new bands related to the fully ionized and neutral ET molecules appear in the IR spectra. One of the most important spectral changes is appearance of new bands at 1347 cm^{-1} and 1289, 1297 cm^{-1} related to the $\nu_3(A_g)$ and $\nu_5(A_g)$ modes of ET^{+1} cation, which are activated by coupling with the CT transition observed at about 7000 cm^{-1} [44]. It gives an evidence of doubly charged ET_2^{2+} dimers. The appearance of these features is a consequence of the charge-ordering phenomenon and EMV coupling in these dimers.

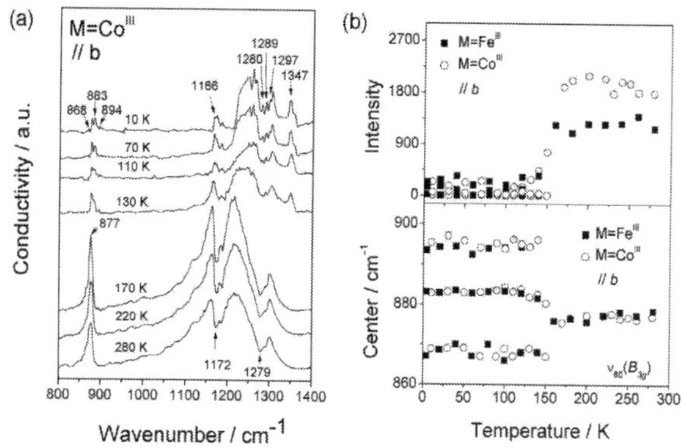

Figure 5. Temperature dependence of the conductivity spectra of κ $(ET)_4[Co(CN)_6][N(C_2H_5)_4]\cdot 2H_2O$ salt (a) and temperature variation of the $\nu_{60}(B_{3g})$ (D_{2h}) mode (ν_{10}(A)for D_2 symmetry) (frequency and intensity) (b). (Reprinted with permission from Łapiński et al. [44]. Copyright© 2013, American Chemical Society).

For the determination of ρ in ET salts, other vibrational modes as$\nu_{29}(B_1)$ and $\nu_{44}(B_2)$ (for D_2 molecular symmetry) can also be used [89]. For example, for κ-$(ET)_4[M(CN)_6][N(C_2H_5)_4]\cdot 2H_2O$ (M= Co^{III} and Fe^{III}) salts [44], the $\nu_{44}(B_2)$ mode is assigned to weak bands observed in the experimental spectra at 863 cm^{-1} (for ET^0), at 875 cm^{-1} (for ET$^{0.5+}$) and at 900 cm^{-1} (for ET$^+$) which are in good agreement with calculated by Girlando values 864, 876 and 903 cm^{-1} for ET^0, ET$^{0.5+}$ and ET$^+$, respectively [89]. The 890 cm^{-1} mode, assigned to $\nu_{60}(B_{3g})$ in the D_{2h} symmetry [84] and to ν_{10}(A) in the D_2 symmetry [89], has attracted a special attention as a spectral feature which is so sensitive to the charge disproportion in ET salts [85]. Moreover, if we consider this mode as a totally symmetric one of a distorted, not a perfectly flat ET molecule (D_{2h} symmetry) then this mode can be coupled to the electronic system [89]. This mode appears in the experimental spectra for the ET$^{0.5+}$ at 878 cm^{-1} [90] and for ET$^+$ at 899 cm^{-1} [91]. In the optical conductivity spectra of κ-$(ET)_4[M(CN)_6][N(C_2H_5)_4]\cdot 2H_2O$ (M= Co^{III} and Fe^{III}) salts this mode at

868 cm^{-1} (for ET^0), at 883 cm^{-1} (for $ET^{0.5+}$) and at 894 cm^{-1} (for ET^+) can be observed [44]. **Figure 5b** shows the temperature dependence of the position of the ν_{10}(A)mode.

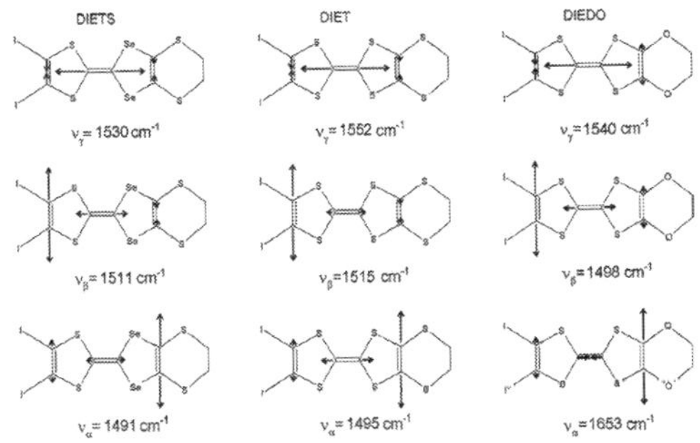

Figure 6. Schematic atomic displacements of C=C in vibrational modes for DIETS, DIET and DIEDO molecules. Note: the position of C=C stretching modes is given for the experimental spectra of these donors. (Reprinted with permission from Łapiński et al. [36]. Copyright© 2010, Elsevier).

In the case of organic unsymmetrical donors derived from TTF molecule: e.g. pyrazino-*S,S*-dimethyl-ethylenedithio-tetrathiafulvalene (P-*S,S*-DMEDT-TTF) [45],2-(4,5-ethylenedithio-1,3-dithiol-2-ylidene)-5-(1,3-dithiolan-2-ylidene)-1,3,4,6-tetrathiapentalene (EDDH-TTP) [48],2,5-bis(1,3-dithiolan-2-ylidene)-1,3,4,6-tetrathiapentalene (BDH-TTP) [48], ethylenedithio-dithiadiselenafulvalene (EDT-DTDSF) [46],diiodoethylenedithio-dithiaselenafulvalene (DIETS) [36, 50], diiodoethylenedithiotetrathiafulvalene (DIET) [50], diiodoethylene-dioxotetrathiafulvalene (DIEDO) [36, 50], (1,4-dioxane-diyl-2,3-ditio) ethylenedioxytetrathiafulvalene (DOEO) [37, 51, 92],dimethyltrimethylenetetrathiafulvalene

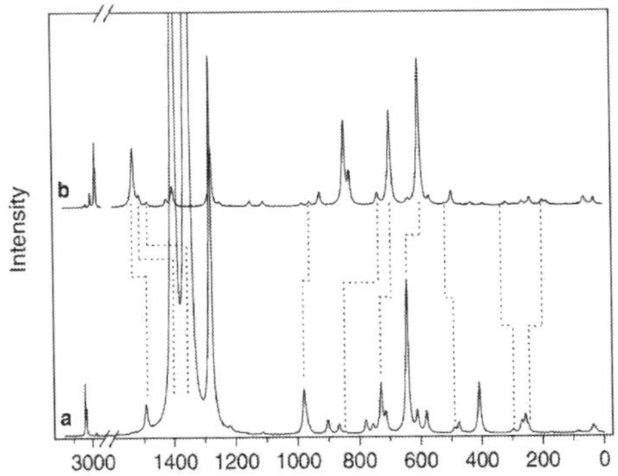

Figure 7. Simulated IR absorption spectra of EDT-DTDSF$^+$(a) and EDT-DTDSF^0(b). (Reprinted with permission from Łapiński et al. [46]. Copyright© 2006, Elsevier).

(DMtTTF) [93], *o* dimethyl-tetrathiafulvalene (*o*-DMTTF) [93] one can find three types of non-equivalent C=C bonds - one central bond and two ring bonds. The schematic picture of atomic displacement of carbon atoms for DIEDO, DIETS, DIET is shown in **Figure 6**. The theoretical calculations show that a strong mixing of all C=C stretching vibrations is present for all modes related to C=C stretching vibrations.

Except the modes related to the C=C stretching vibration, which show the largest low-frequency shifts when the molecule is oxidized, other active modes also exhibit a significant shift due to the ionicity [36, 46, 51], what is schematically illustrated by dashed line for EDT-DTDSF (see **Figure 7**).

For the isolated donor molecules, the optimized geometry depends on whether we have a neutral or charged molecule. The TTF framework for neutral molecules is non-planar deforming to a boat conformation, in contrast with cations where the TTF framework is planar (see **Figure 8**) [36]. The similar situation has also been observed for the other organic donors [37, 51, 94]. In crystal structures of salts derived from TTF, where donors are not isolated molecules, the TTF framework can be flat with small deviations from planarity [95, 96]. For example, for the CT salts derived from DIET, DIEDO and DIETS and the anion $[Fe(bpca)(CN)_3]^-$ the donors are almost flat [97]. In Raman spectra of these salts bands assigned to the C=C stretching vibrations are strongly shifted towards lower frequency and this effect is due to ionization [50].

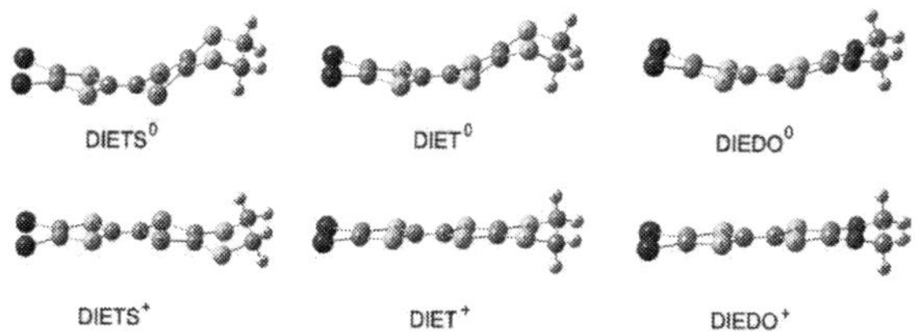

Figure 8. Optimized geometry of the neutral DIETS, DIET, DIEDO molecules and cations. (Reprinted with permission from Łapiński et al. [36]. Copyright© 2010, Elsevier).

Infrared spectroscopy is a powerful method not only in investigating the charge disproportion phenomena but also in the study of interaction between the organic and inorganic layers in organic conductors and its impact on the physical properties. For κ-$(ET)_4[Co(CN)_6]$ $[N(C_2H_5)_4]\cdot 2H_2O$ and κ-$(ET)_4[Fe(CN)_6][N(C_2H_5)_4]\cdot 2H_2O$ salts Ota and co-workers [78] showed that the strong electron–electron correlations and Coulomb interaction between ET and inorganic layers play an important role in the phase transition from the Mott insulator state to the charge-ordering state. The role of such interactions and their contributions to the phase transition has been investigated and discussed in Ref. [44]. The temperature dependence of the modes related to C≡N triple bond vibrations (~2100 cm^{-1}) in $M(CN)_6^{3-}$ (M=Co^{III} and Fe^{III}) anions and C-H stretching in ET molecules (~3400 cm^{-1}) is presented in **Figure 9**. For these modes, the frequency dependences reflect the sensitivity of the CN-CH_2 interaction to the 150 K phase transition. The blue shift of the CN stretching frequency with decreasing temperature proves that the interaction between the hydrogen atom and the CN group of the anions is present. Moreover, the examination of the CH_2 stretching frequencies in κ-$(ET)_4[Co(CN)_6][N(C_2H_5)_4]\cdot 2H_2O$ and κ-$(ET)_4[Fe(CN)_6][N(C_2H_5)_4]\cdot 2H_2O$ salts leads us to the conclusion that in the last salt the degree of donor-anion interaction is slightly smaller.

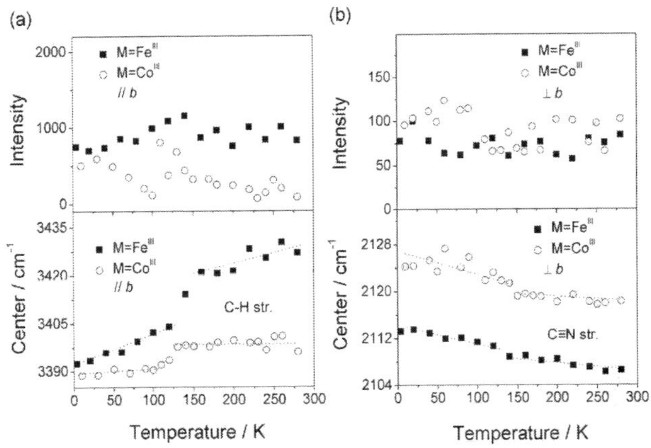

Figure 9. Temperature variation of the C-H (a) and C≡N (b) modes (frequency and intensity). (Reprinted with permission from Łapiński et al. [44]. Copyright© 2013, American Chemical Society).

What should be emphasized at this point is the fact that the temperature variation of bands related to C-H stretching indicate that apart from the long-range Coulomb interactions between electrons within the conducting layers, the anions have an influence on the formation of the charge-ordered state as well [44]. Laversanne and co-workers [98] showed that periodic distribution of the anions could play an important role in the physical properties of organic conductors.

They drew attention that the anion potential effects on an terminal part of the donors. The orientation of the anions along the chain influences on the charge density distribution within conducting layers and the description of the anion potential should be taken into account in calculations [98]. Moldenhower *et al.* [88] investigated the correlation between the T_c temperature and frequencies of the CH_2 stretching modes of ET radical salts with the superconducting phase transition. The position of these modes is evidently not charge dependent but it can reflect the strength of the interaction of the donor molecule with the respective anion. They showed that phases with a higher T_c of their superconducting transition exhibit a smaller red shift of these frequencies, which is due to the hydrogen-bonding like interaction of the donor with the anion, i.e., a less attractive donor-anion interaction [88].

For $(DOEO)_4HgBr_4\cdot$TCE salt it has been shown in [92] that the role of anion layers cannot be neglected. The temperature evolution of vibrational features reveals that only the bands related to the deformation of the outer ring of DOEO molecule (e.g. 1371, 1092 cm^{-1}) are sensitive to the metal-insulator phase transitions (**Figure 10a**), whereas the bands related to the deformation of C-O bonds in inner part of donor molecule observed at 1000 and 902 cm^{-1} do not show such behavior (**Figure 10b**). The systematic analysis of vibrational and electronic structure performed for DOEO salt one can find in [37] and [51].

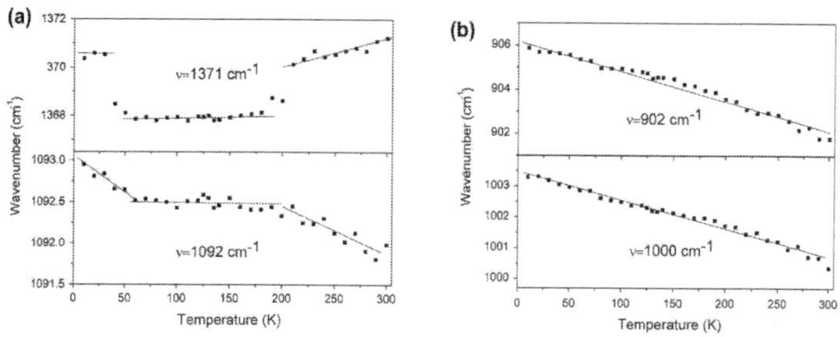

Figure 10. Temperature dependence of the wave number of the selected bands: 1371 and 1092 cm^{-1} (a) and 902 and 1000 cm^{-1} (b). Note: lines are used as a guide for the eyes. (Reprinted with permission from Łapiński et al. [92]. Copyright© 2012, Elsevier).

In the case of organic conductors formed by iodinated TTFs as DIEDO, DIET, DIETS [36], the position of modes related to C-I and C-S vibrations could be also sensitive to strong interaction between the iodo group of donor and the cyano group or halogen of acceptors [95, 97, 99, 100]. In the experimental IR spectra of these neutral donors, one can find the bands related to the simultaneous deformation of C-S and C-I bonds at 693, 817, 903 cm^{-1} (for DIETS), 696, 817, 917 cm^{-1} (for DIET) and 699, 829, 912 cm^{-1} (for DIEDO) [36].

The strong interaction between donors and acceptors should also have an influence on frequencies of acceptor molecules. For example, for the 2:1 salts formed by the iodine substituted donors DIEDO, DIET, DIETS with the anion $[Fe(bpca)(CN)_3]^-$ (where bpca=bis(s-pyridylcarbonyl)amide anion) we have observed the influence of such interactions on frequencies of C≡N and C=O vibrations of acceptor molecule [50].

4. Electron-phonon interaction

Infrared spectra of organic conductors are usually dominated by strong vibrational features related to the coupling between the electrons in the highest occupied molecular orbital (HOMO) and the molecular vibrational modes of molecules [101, 102]. Because of this coupling, the vibrational modes borrow intensity from the nearby CT electronic transition and occur at frequencies lower than the corresponding Raman active modes. These strong bands are polarized perpendicularly to the molecular planes, like the CT transition.

The EMV coupling phenomenon in conducting organic salts can be analyzed in terms of various models depending on structure of the molecular stacks. Microscopic theories for regular stacks or stacks consisting of quasi-isolated dimers, trimers, tetramers or n-mers have been developed [28–30].The above mentioned models were successfully applied to IR spectra of CT salts formed by TTF and its derivatives with various acceptors [62]. More specifically, the dimer model has proven to be especially useful in determining the coupling constants for TTF-containing CT materials. For example, reliable values for the EMV constants were obtained for the salt $(ET)_2[Mo_6O_{19}]$ which contained well-isolated $(ET^+)_2$ dimers [91].

According to symmetry considerations, only the totally symmetric vibrational modes within a specified linear approximation of non-degenerate molecular orbitals can couple with electrons [103]. For other symmetry modes, the electron–vibrational interaction is forbidden by the selection rules and is only correct if the molecules within the dimer have the same symmetry. When considering the dimer, where the molecules are asymmetric with respect to one another, if these molecules are different or inequivalent, their modes are no longer degenerate and couple both in-phase and out-of-phase. In this case, the lack of an inversion center suppresses the mutual exclusion rule leading to all of the modes for the constituent molecules within the dimer becoming both IR and Raman active. Additionally, all of their symmetric modes can then couple to the CT electron [104]. Moreover, if we consider the sufficiently fast charge transfer between the dimer molecules due to electromagnetic radiation, their nuclear configurations do not have time to change in response to the charge transfer and their molecular vibrations arise as the result of the relaxation to each molecule's respective equilibrium configuration. In this case, the symmetry type of the arising vibrational modes depends not only on the final and initial state symmetries, but also on that of the intermediate states from which the transferred charge among the neighboring molecules belongs. These facts could suggest that in some cases non-totally symmetric modes can be also coupled [105].

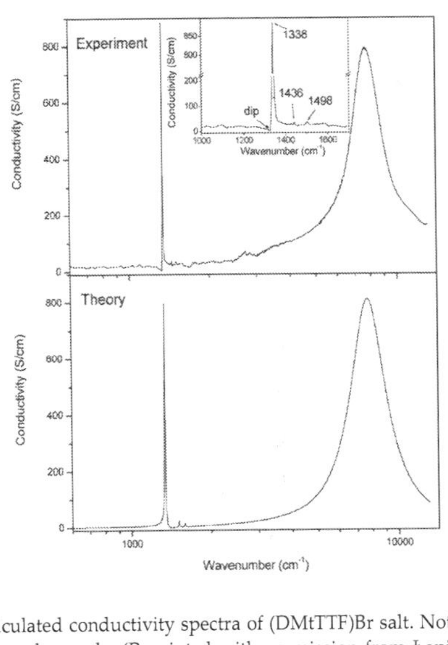

Figure 11. Experimental and calculated conductivity spectra of (DMtTTF)Br salt. Note: electric vector along the stacking axis; the logarithmic wave number scale. (Reprinted with permission from Łapiński et al. [93]. Copyright© 2014, Elsevier).

For salts formed by unsymmetrical TTF derivatives such as *o*-DMTTF, DMtTTF and DOEO in the IR spectra one observes vibrational features suggesting the EMV coupling to intramolecular vibrations of donors, especially those related to C=C stretching [106–109]. In the IR spectra of the salts $(DOEO)_4HgBr_4 \cdot TCE$ [51], (DMtTTF)Br [93] and $(o\ DMTTF)_2[W_6O_{19}]$ [93] a clear evidence of the EMV coupling to C=C stretching modes have been found. The observation of vibrational bands shifted towards lower wave numbers with respect to Raman data and, moreover, with characteristic antiresonance deeps, give an evidence of their interaction with CT transition. This effect is very well visible for polarization parallel to the stacking axis; in particular there are distinct antiresonance dips at 1370 and 1379 cm^{-1} for (DMtTTF)Br and $(o\text{-}DMTTF)_2[W_6O_{19}]$, respectively [93]. Moreover, the bands at 1338 and 1345 cm^{-1} [93] have a characteristical asymmetric Fano-like lineshape [110]. It is observed when the electric field is polarized along the chain and the electronic absorption overlaps with the phonon frequencies; the interaction between electrons and phonons gives rise to characteristically asymmetric Fano lineshape of bands. The analysis of Fano-effect combined with Raman data can lead to quantitative predictions for the various electron phonon couplings [111].

(DMtTTF)Br [93]			$(o\text{-}DMTTF)_2[W_6O_{19}]$ [93]			$(DOEO)_4HgBr_4 \cdot TCE$ [51]			Assignment
ω_α	g_α	λ_α	ω_α	g_α	λ_α	ω_α	g_α	λ_α	
1587	97	0.002	1556	484	0.022	1575	183	0.040	Ring C=C stretch
1507	161	0.006	1483	645	0.042	1494	252	0.080	Ring C=C stretch
1412	597	0.095	1417	805	0.068	1460	233	0.070	Central C=C stretch

Note: ω_α and g_α are given in cm^{-1}.

Table 2. Dimer model parameters by the fit to the polarized conductivity spectrum of (DMtTTF)Br, $(o\text{-}DMTTF)_2[W_6O_{19}]$ and $(DOEO)_4HgBr_4 \cdot TCE$ salts obtained by Kramers-Krönig transformation from the reflectance spectra.

TTF [28–29]		ET [91]		TMTTF [112]		TTP [113]		Assignment
ω	g_α	ω	g_α	ω	g_α	ω	g_α	
1505	42	1460	43	1567	32	1540	8	Ring C=C stretch
						1474	54	Ring C=C stretch
1420	133	1414	71	1418	133	1423	9	Central C=C stretch

Table 3. The EMV coupling constants, g_α (meV) of the C=C stretching modes ω (cm^{-1}) for several symmetrical radical cations based on TTF.

Figure 11 shows the conductivity spectrum of (DMtTTF)Br salt obtained by Kramers-Krönig transformation from the experimental reflectance spectrum (upper panel) and conductivity spectrum calculated within the framework of the dimer model proposed by Rice et al. [103] (lower panel). **Table 2** presents the model parameters for the fit to the conductivity spectra of (DMtTTF)Br [93], $(o\ DMTTF)_2[W_6O_{19}]$ [93] and $(DOEO)_4HgBr_4 \cdot TCE$ [51] salts. Comparing the relevant EMV coupling constants for unsymmetrical DMtTTF, *o*-DMTTF and DOEO donors with the corresponding ones measured for symmetrical TTF-based electron donor molecules: TTF itself [28–29], ET [91], tetramethyl-tetrathiafulvalene (TMTTF) [112], bis-fused TTF (TTP) [113] (see **Table 3**) one can see that the values of coupling constants are comparable and the most strongly coupled modes are assigned to the TTF skeleton vibrations, namely the modes related to the stretching vibrations of both central and ring C=C bonds.

It should be also emphasized that the EMV coupling plays a role in the charge-ordering instabilities, as the modulation of the frontier molecular orbitals, pushing charges back and forth, which may in turn provoke their localization on the molecular sites [89].

5. Conclusions

In this chapter, selected problems of solid state physics of organic conductors have been discussed. The IR, Raman and UV-Vis spectroscopies were used providing the information about vibrational and electronic structures, electron-electron and electron-phonon interactions. The detailed spectral analysis led to a wider recognition and provides the necessary information about physical properties of organic conductors.

It was shown that the role of anion layers cannot be neglected. The periodic distribution of anions could play an important role in the physical properties of organic conductors. On the basis of vibration spectra it was shown that the anion potential influences on a terminal part of the molecules, which are close to anions. Apart from the long-range Coulomb interactions between electrons within layers formed by donors also the anions can have a significant influence on the formation of the charge-ordered state. Moreover, it was shown that in the case of organic unsymmetrical donors derived from TTF molecule the values of coupling constants are comparable with another symmetrical TTF derivatives and the most strongly coupled modes are assigned to the TTF skeleton vibrations, namely the modes related to the stretching vibrations of both central and ring C=C bonds. It was also shown that the special attention should be paid to the impact of the EMV coupling on the charge-ordering instabilities.

For characterization of intramolecular and intermolecular transitions the time-dependent DFT method and Drude-Lorentz models can be successfully applied. Considerable information about the electronic structure can also be extracted from the oscillator strength sum rule.

The degree of ionicity or average charge per molecule is one of the fundamental parameters characterizing the physical properties of CT salts. For the organic donors, this parameter can be studied using spectroscopic methods. It was shown that in the case of organic donors derived from TTF molecule the most convenient for this analysis are the C=C stretching modes of TTF framework which show sensitivity to the ionization degree. It was also shown that spectroscopic methods are very powerful tool in the study of the nature of different phase transitions and the interaction between the organic and inorganic layers in organic conductors and its impact on the physical properties.

Acknowledgements

Author is indebted to his co-workers Barszcz B., Graja A., Ostrowski A.,Olejniczak I., Połomska M., Świetlik R., Waplak S. at Institute of Molecular Physics PAS for their contribution to this work and to Baumer V., Faulques E.,Fourmigué M., Gąsecka A., Golhen S., Golub M., Imakubo T., Jankowski D.,Kotov A.I.,Kravczenko A., Lyubovskaya R.N., Mousdis G.A., Ouahab L.,Papavassiliou G.C.,Prokhorova T.G., Reinheimer E.W.,Setifi F., Starodub V.A., Yamada J., Zhilyaeva E.I. for a fruitful collaboration.

Author details

Andrzej Łapiński

Address all correspondence to: lapinski@ifmpan.poznan.pl

Institute of Molecular Physics, Polish Academy of Sciences, Poznan, Poland

References

[1] Wudl F, Wobschall D, Hufnagel EJ. Electrical conductivity by the bis(1,3-dithiole)-bis(1,3-dithiolium) system. J. Am. Chem. Soc. 1972;94(2):670–672. DOI: http://dx.doi.org/10.1021/ja00757a079

[2] Ferraris J, Cowan DO, Walatka VV, Perlstein JH. Electron transfer in a new highly conducting donor-acceptor complex. J. Am. Chem. Soc. 1973;95(3):948–949. DOI: http://dx.doi.org/10.1021/ja00784a066

[3] Jérome D, Mazaud A, Ribault M, Bechgaard K. Superconductivity in a synthetic organic conductor $(TMTSF)_2PF_6$. J. Physique (Paris) Lett. 1980;41:95–98. DOI: http://dx.doi.org/10.1051/jphyslet:0198000410409500

[4] Little WA. Possibility of Synthesizing an Organic Superconductor. Phys. Rev. 1964;134:A1416–A1424. DOI: http://dx.doi.org/10.1103/PhysRev.134.A1416

[5] Enoki T, Yamaura J, Miyazaki A. Molecular Magnets Based on Organic Charge Transfer Complexes. Bull. Chem. Soc. Jpn. 1997;70:2005–2023. DOI: http://dx.doi.org/10.1246/bcsj.70.2005

[6] Mori TJ. Magnetic Interactions in Insulating ET Salts. Solid State Chem. 2002;168:433–437. DOI: http://dx.doi.org/10.1006/jssc.2002.9694

[7] Coronado E, Day P. Magnetic Molecular Conductors. Chem. Rev. 2004;104:5419–5448. DOI: http://dx.doi.org/10.1021/cr030641n

[8] Enoki T, Miyazaki A. Magnetic TTF-based charge-transfer complexes. Chem. Rev. 2004;104:5449–5477. DOI: http://dx.doi.org/10.1021/cr0306438

[9] Negishi E, Kuwabara T, Komiyama S, Watanabe M, Noda Y, Mori T, Matsui H, Toyota N. Dielectric ordering and colossal magnetodielectricity in the antiferromagnetic insulating state of λ-(BEDT–TSF)$_2$FeCl$_4$. Phys. Rev. B. 2005;71:012416–012419. DOI: http://dx.doi.org/10.1103/PhysRevB.71.012416

[10] Yamada J, Akutsu H,Nishikawa H, Kikuchi K. New trends in the synthesis of π-electron donors for molecular conductors and superconductors. Chem. Rev. 2004;104:5057–5083. DOI: http://dx.doi.org/10.1021/cr0306687

[11] Ouahab L, Enoki T. Multiproperty Molecular Materials: TTF-based conducting and magnetic molecular materials. Eur. J. Inorg. Chem. 2004;2004:933–941. DOI: http://dx.doi.org/10.1002/ejic.200300869

[12] NishijoJ, Miyazaki A, Enoki T, Watanabe R, Kuwatani Y, Iyoda M. d-Electron-induced negative magnetoresistance of a π–d interaction system based on a brominated-TTF donor. Inorg. Chem. 2005;44:2493–2506. DOI: http://dx.doi.org/10.1021/ic048548k

[13] Sugimoto T, Fujiwara H, Noguchi S, Murata K. New aspects of π–d interactions in magnetic molecular conductors. Sci. Technol. Adv. Mater.2009;10:024302–024321. DOI: http://dx.doi.org/10.1088/1468–6996/10/2/024302

[14] Mori T, Katsuhara M. Estimation of π-d interactions in organic conductors including magnetic anions. J. Phys. Soc. Jap. 2002;71:826–844. DOI: http://dx.doi.org/10.1143/JPSJ.71.826

[15] McKenzie RH. A strongly correlated electron model for the layered organic superconductors kappa-(ET)$_2$X. Com. Condens. Matter Phys. 1998;18:309–337.arXiv:cond-mat/9802198

[16] Susaki T, Ito I, Yoneyama N, Kobayashi N, Hanasaki N, Tajima H, Ito T, Iwasa Y. Electronic correlation in the infrared optical properties of the quasi-two-dimensional κ-type BEDT-TTF dimer system. Phys. Rev. B. 2004;69:064508–064514. DOI: http://dx.doi.org/10.1103/PhysRevB.69.064508

[17] Giamarchi T. Theoretical framework for quasi-one dimensional systems. Chem. Rev. 2004;104:5037–5056. DOI: http://dx.doi.org/10.1021/cr030647c

[18] Hiraki K, Kanoda K. Wigner crystal type of charge ordering in an organic conductor with a quarter-filled band: (DI–DCNQI)$_2$Ag. Phys. Rev. Lett. 1998;80:4737–4740. DOI: http://dx.doi.org/10.1103/PhysRevLett.80.4737

[19] Chow DS, Zamborszky F, Alavi B, Tantillo DJ, Baur A, Merlic CA, Brown SE. Charge ordering in the TMTTF family of molecular conductors. Phys. Rev. Lett. 2000;85:1698–1701. DOI: http://dx.doi.org/10.1103/PhysRevLett.85.1698

[20] Yamamoto K, Yakushi K, Miyagawa K, Kanoda K, Kawamoto A. Charge ordering in θ –(BEDT–TTF)$_2$RbZn(SCN)$_4$ studied by vibrational spectroscopy. Phys. Rev. B. 2002;65:085110–085117. DOI: http://dx.doi.org/10.1103/PhysRevB.65.085

[21] Seo H, Merino J, Yoshioka H, Ogata M. Theoretical aspects of charge ordering in molecular conductors. J. Phys. Soc. Jap. 2006;75:051009–051028. DOI: http://dx.doi.org/10.1143/JPSJ.75.051009

[22] Seo H, Fukuyama H. Frustrated Spin System in θ-(BEDT-TTF)$_2$RbZn(SCN)$_4$. J. Phys. Soc. Jpn. 1998;67:1848–1851. DOI: http://dx.doi.org/10.1143/JPSJ.67.1848

[23] Seo H. Charge ordering in organic ET compounds. J. Phys. Soc. Jpn. 2000;69:805–820. DOI: http://dx.doi.org/10.1143/JPSJ.69.805

[24] Tajima H, Kyoden S, Mori H, Tanaka S. Estimation of charge-ordering patterns in θ –ET$_2$MM'(SCN)$_4$ (MM'=RbCo,RbZn,CsZn) by reflection spectroscopy. Phys. Rev. B. 2000;62:9378–9385. DOI: http://dx.doi.org/10.1103/PhysRevB.62.9378

[25] Wang HH, Ferraro JR, Wlliams JM, Geiser U, Schlueter JA. Rapid Raman spectroscopic determination of the stoichiometry of microscopic quantities of BEDT–TTF-based organic conductors and superconductors . J. Chem. Soc. Chem. Commun. 1994;1893–1894. DOI: http://dx.doi.org/10.1039/C39940001893

[26] Ramakumar R, Tanaka Y, Yamaji K. Electron–phonon interactions, pair-transfer processes and superconductivity in TTF[Ni(dmit)$_2$]$_2$. Phys. Rev. B. 1997;56:795–801. DOI: http://dx.doi.org/10.1103/PhysRevB.56.795

[27] Drozdova O, Yamochi H, Yakushi K, Uruichi M, Horiuchi S, Saito G. Determination of the charge on BEDO-TTF in its complexes by Raman spectroscopy. J. Am. Chem. Soc. 2000;122:4436–4442. DOI: http://dx.doi.org/10.1021/ja9927523

[28] Painelli A, Girlando A. Electron–molecular vibration (e–mv) coupling in charge-transfer compounds and its consequences on the optical spectra: A theoretical framework. J. Chem. Phys. 1986;84:5655–5671. DOI: http://dx.doi.org/10.1063/1.449926

[29] Painelli A, Girlando A, Pecile C. Accurate electron-molecular vibration coupling constants from powders optical spectra: TCNQ and TTF. Solid State Commun. 1984;52:801–806. DOI: http://dx.doi.org/10.1016/0038-1098(84)90009-7

[30] Pecile C, Painelli A, Girlando A. Studies of organic semiconductors for 40 years. V. Mol. Cryst. Liq. Cryst. 1989;171:69–87. DOI: http://dx.doi.org/10.1080/00268948908065787

[31] Yamamoto T, Uruichi M, Yamamoto K, Yakushi K, Kawamoto A, Taniguchi H. Examination of the charge-sensitive vibrational modes in bis(ethylenedithio)tetrathiafulvalene. J. Phys. Chem. B. 2005;109:15226–15235. DOI: http://dx.doi.org/10.1021/jp050247o

[32] Guionneau P, Kepert CJ, Chasseau D, Truter MR, Day P. Determining the charge distribution in BEDT-TTF salts. Synth. Met. 1997;86:1973–1974. DOI: http://dx.doi.org/10.1016/S0379-6779(97)80983-6

[33] Hatsui T, Okamoto K, Yokoyama T, Kitajima Y, Tanaka H, Ojima E, Kobayashi A, Ohta T. Sulfur K-edge X-ray absorption spectra for BETS and BEDT-TTF charge transfer salts: a novel probe for the determination of hole concentration. Chem. Phys. Lett. 2000;330:309–314. DOI: http://dx.doi.org/10.1016/S0009-2614(00)01121-0

[34] Ouyang J, Yakushi K, Misaki Y, Tanaka T. Raman spectroscopic evidence for the charge disproportionation in a quasi-two-dimensional organic conductor θ–(BDT–TTP)$_2$Cu(NCS)$_2$. Phys. Rev. B. 2001;63:054301–054306. DOI: http://dx.doi.org/10.1103/PhysRevB.63.054301

[35] Bozio R, Pecile C. Molecular vibration studies of quasi-one-dimensional organic charge-transfer compounds. In: L. Alcácer, editor. The Physics and Chemistry of Low Dimensional Solids. Dordrecht, Holland: Reidel Publishing Company; 1980. 165–186. DOI: http://dx.doi.org/10.1007/978-94-009-9067-8

[36] Łapiński A, Ouahab L, Imakubo T. Spectroscopic and theoretical study on organic donors: DIET, DIEDO and DIETS. Vib. Spectr. 2010;52:22–30. DOI: http://dx.doi.org/10.1016/j.vibspec.2009.09.007

[37] Łapiński A, Kotov AI. TD-DFT investigation of the electronic spectra of (DOEO)$_4$HgBr$_4$·TCE salt. Mol. Phys. 2008;106:33–42. DOI: http://dx.doi.org/10.1080/00268970701823180

[38] Ziman JM. Principles of the Theory of Solids. second ed. Cambridge: Cambridge University Press; 1972. 448 p. ISBN-13: 978-0521083829

[39] Hubbard J. Generalized Wigner lattices in one dimension and some applications to tetracyanoquinodimethane (TCNQ) salts. Phys. Rev. B. 1978;17:494–505. DOI: http://dx.doi.org/10.1103/PhysRevB.17.494

[40] Peierls RE. Quantum Theory of Solids. London, England: Oxford University Press; 2001. 238 p. ISBN: 978-0198507819

[41] Łapiński A, Graja A, Prokhorova TG. Spectral investigation of the new organic metal β″-(BEDT-TTF)$_4$NH$_4$[Cr(C$_2$O$_4$)$_3$]·DMF. J. Mol. Structure. 2004;704:83–87. DOI: http://dx.doi.org/10.1016/j.molstruc.2003.11.058

[42] Łapiński A, Klemt K, Graja A, Prokhorova TG. Spectral investigations of organic metals β″-(BEDT-TTF)$_4$A[M(C$_2$O$_4$)$_3$]·DMF, where A=NH$_4^+$, K$^+$ and M=Cr^{III}, Fe^{III}. Synth. Metals. 2005;151:191–196. DOI: http://dx.doi.org/10.1016/j.synthmet.2005.04.007

[43] Łapiński A, Starodub VA, Golub M, Kravczenko A, Baumer V, Faulques E, Graja A. Characterization and spectral properties of the new organic metal (BEDT-TTF)$_6$(Mo$_8$O$_{26}$)·(DMF)$_3$. Synth. Metals. 2003;138:483–489. DOI: http://dx.doi.org/10.1016/S0379-6779(02)00465-4

[44] Łapiński A, Świetlik R, Ouahab L, Golhen S. Spectroscopic studies of the phase transition from the Mott insulator state to the charge-ordering state ofκ-(ET)$_4$[M(CN)$_6$]

$[N(C_2H_5)_4]\cdot 2H_2O$ (M=Co^{III} and Fe^{III}) salts. J. Phys. Chem. A. 2013;117:5241–5250. DOI: http://dx.doi.org/10.1021/jp402027x

[45] Łapiński A, Graja A, Papavassiliou GC, Mousdis GA. Optical properties of the conducting salt τ-(P-S,S-DMEDT-TTF)$_2$($AuBr_2$)($AuBr_2$)y (y≈0.75). Synth. Metals. 2003;139:405–409. DOI: http://dx.doi.org/10.1016/S0379-6779(03)00189-9

[46] Łapiński A, Lyubovskaya RN, Zhilyaeva EI. Spectral investigations of unsymmetrical π-donor ethylenedithio-dithiadiselenafulvaleneiodomercurate, (EDT-DTDSF)$_4$$Hg_3I_8$. Chem. Phys. 2006;323:161–168. DOI: http://dx.doi.org/10.1016/j.chemphys.2005.08.050

[47] Wooten F. Optical Properties of Solids. New York: Academic Press; 1972. 260 p.

[48] Łapiński A, Świetlik R, Setifi F, Ouahab L, Golhen S, Yamada J. Temperature dependence of the IR reflectance spectra of molecular crystals: κ-(EDDH-TTP)$_3$[Cr(phen)$(NCS)_4$]$\cdot 2CH_2Cl_2$ and κ_{21}-(BDH-TTP)$_5$[Cr(phen)$(NCS)_4$]$_2\cdot 2CH_2Cl_2$. J. Mol. Structure. 2003;651–653:683–687. DOI: http://dx.doi.org/10.1016/S0022-2860(03)00119-4

[49] Łapiński A, Świetlik R. Electronic structure and electron-molecular vibration interaction in dimerized [Pd(dddt)$_2$]$_2$X (where X=$AuBr_2$, SbF_6, $TeCl_x$) salts studied by IR spectroscopy. Phys. Rev. B. 1999;59:15669–15679. DOI: http://dx.doi.org/10.1103/PhysRevB.59.15669

[50] ŁapińskiA, Świetlik R, Połomska M, Ouahab L, Imakubo T. Spectral studies of DIET, DIEDO and DIETS organic donors and their charge-transfer salts with paramagnetic cyano complex. J. Low Temp. Phys. 2006;142:601–604. DOI: http://dx.doi.org/10. 1007/s10909-006-9172-1

[51] Łapiński A, Kotov AI. Optical properties of the conducting salt (DOEO)$_4$$HgBr_4$•TCE. Chem. Phys. 2006;326:551–562. DOI: http://dx.doi.org/10.1016/j.chemphys.2006.03.018

[52] Rozenberg MJ, Kotliar G, Kajueter H. Transfer of spectral weight in spectroscopies of correlated electron systems. Phys. Rev. B. 1996;54:8452–8468. DOI: http://dx.doi.org/10.1103/PhysRevB.54.8452

[53] Elsässer S, Wu D, Dressel M, Schlueter JA. Power-law dependence of the optical conductivity observed in the quantum spin-liquid compound κ-(BEDT-TTF)$_2$$Cu_2(CN)_3$. Phys. Rev. B. 2012;86:155150–155157. DOI: http://dx.doi.org/10.1103/PhysRevB.86.155150

[54] Sedlmeier K, Elsässer S, Neubauer D, Beyer R, Wu D, Ivek T, Tomić S, Schlueter JA, Dressel M. Absence of charge order in the dimerized κ-phase BEDT-TTF salts. Phys. Rev. B. 2012;86:245103–245112. DOI: http://dx.doi.org/10.1103/PhysRevB.86.245103

[55] Dressel M, Faltermeier D, Dumm M, Drichko N, Petrov B, Semkin V,Vlasova R, Méziere C, Batail P. Disentangling the conductivity spectra of two-dimensional organic conductors. Physica B. 2009;404:541–544. DOI: http://dx.doi.org/10.1016/j.physb.2008.11.051

[56] Kino H, Fukuyama H. Electronic States of Conducting Organic κ-(BEDT-TTF)$_2$X. J. Phys. Soc. Jpn. 1995;64:2726–2729. DOI: http://dx.doi.org/10.1143/JPSJ.64.2726

[57] Kino H, Fukuyama H. Interrelationship among Electronic states of α-(ET)$_2$I$_3$, (ET)$_2$MHg(SCN)$_4$ and κ -(ET)$_2$X. J. Phys. Soc. Jpn. 1995;64:4523–4526. DOI: http://dx.doi.org/10.1143/JPSJ.64.4523

[58] Kino H, Fukuyama H. Phase diagram of two-dimensional organic conductors: (BEDT-TTF)$_2$X. J. Phys. Soc. Jpn. 1996;65:2158–2169. DOI: http://dx.doi.org/10.1143/JPSJ.65.2158

[59] Rice MJ. Organic linear conductors as systems for the study of electron–phonon interactions in the organic solid state. Phys. Rev. Lett. 1976;37:36–39. DOI: http://dx.doi.org/10.1103/PhysRevLett.37.36

[60] Yartsev VM, Drozdova OO, Semkin VN, Vlasova RM. Modelling κ - phase organic conductors. J. Phys. I (France). 1996;6:1673–1681. DOI: http://dx.doi.org/10.1051/jp1:1996102

[61] Yartsev VM. Optical properties of low-dimensional molecular crystals and LB Films. In: Kajimura K, Kanoda S, editors. Materials and Measurements in Molecular Electronics. Berlin: Springer; 1996. 189–202. DOI: http://dx.doi.org/10.1007/978-4-431-68470-1

[62] Yartsev VM, Graja A. Electron–intramolecular vibration coupling in charge-transfer salts studied by infrared spectroscopy. Intern. J. Mod. Phys. B. 1998;12:1643–1672. DOI: http://dx.doi.org/10.1142/S0217979298000909

[63] Rozenberg MJ, Kotliar G, Kajueter H, Thomas GA, Rapkine DH, Honig JM, Metcalf P. Optical conductivity in Mott-Hubbard systems. Phys. Rev. Lett. 1995;75:105–108. DOI: http://dx.doi.org/10.1103/PhysRevLett.75.105

[64] Hotta C. Classification of quasi-two dimensional organic conductors based on a new minimal model. J. Phys. Soc. Jpn. 2003;72:840–853. DOI: http://dx.doi.org/10.1143/JPSJ.72.840

[65] Hotta C. Quantum electric dipoles in spin-liquid dimer Mott insulator κ-ET$_2$Cu$_2$(CN)$_3$. Phys. Rev. B. 2010;82:241104–241107. DOI: http://dx.doi.org/10.1103/PhysRevB.82.241104

[66] Naka M, Ishihara S. Electronic ferroelectricity in a dimer mott insulator. J. Phys. Soc. Jpn. 2010;79:063707–063710. DOI: http://dx.doi.org/10.1143/JPSJ.79.063707

[67] Gomi H, Imai T, Takahashi A, Aihara M. Purely electronic terahertz polarization in dimer Mott insulators. Phys. Rev. B. 2010;82:035101–035107. DOI: http://dx.doi.org/10.1103/PhysRevB.82.035101

[68] Li H, Clay RT, Mazumdar S. The paired-electron crystal in the two-dimensional frustrated quarter-filled band. J. Phys.: Condens. Matter. 2010;22:272201–272207. DOI: http://dx.doi.org/10.1088/0953-8984/22/27/272201

[69] Dayal S, Clay RT, Li H, Mazumdar S. Paired electron crystal: Order from frustration in the quarter-filled band. Phys. Rev. B. 2011;83:245106–245117. DOI: http://dx.doi.org/10.1103/PhysRevB.83.245106

[70] Clay RT, Dayal S, Li H, Mazumdar S. Beyond the quantum spin liquid concept in frustrated two dimensional organic superconductors. Phys. Status Solidi B. 2012;249:991–994. DOI: http://dx.doi.org/10.1002/pssb.201100727

[71] Hotta C. Theories on frustrated electrons in two-dimensional organic solids. Crystals. 2012;2:1155–1200. DOI: http://dx.doi.org/10.3390/cryst2031155

[72] Doublet M-L, Canadell E, Shibaeva RP. Concerning the first-order transition in the κ-phase $(BEDT\text{-}TTF)_4PtCl_6C_6H_5CN$. J. Phys. I. (France). 1994;4:1479–1490. DOI: http://dx.doi.org/10.1051/jp1:1994202

[73] Le Maguerès P, Ouahab L, Conan N, Gomez-García CJ, Delhaès P, Even J, Bertault M. Phase transitions in new BEDT-TTF κ-phase salts with hexacyanometalate anions $[M(CN)_6^{3-}$ M=$Co^{(III)}$ and $Fe^{(III)}]$. Solid State Commun. 1996;97:27–32. DOI: http://dx.doi.org/10.1016/0038-1098(95)00598-6

[74] Ouahab L, Golhen S, Le Hoerff T, Guillevic J, Tual L, Hélou N, Amiell J, Delhaès P, Binet L, Fabre JM. Radical cation salts with paramagnetic hexacyanometallate anions. Synth. Met. 1999;102:1642–1645. DOI: http://dx.doi.org/10.1016/S0379-6779(98)00242-2

[75] Świetlik R, Łapiński A, Połomska M, Ouahab L, Guillevic J. Spectroscopic evidence of the charge ordering in κ-$[Et_4N](BEDT\text{-}TTF)_4M(CN)_6\cdot 3H_2O$ (M = Co^{III}, Fe^{III}, Cr^{III}). Synth. Metals. 2003;133–134:273–275. DOI: http://dx.doi.org/10.1016/S0379-6779(02)00401-0

[76] Świetlik R, Łapiński A, Ouahab L, Yakushi K. Charge ordering in the κ-phase BEDT–TTF salts with $Co(CN)_6$ and $Fe(CN)_6$ anions studied by infrared and Raman spectroscopies. C.R. Chimie. 2003;6:395–403. DOI: http://dx.doi.org/10.1016/S1631-0748(03)00049-3

[77] Świetlik R, Ouahab L, Guillevic J, Yakushi K. Infrared and Raman studies of the charge ordering in the organic semiconductor κ-$[Et_4N][Co(CN)_6(ET)_4]\cdot 3H_2O$. Macromol. Symp. 2004;212:219–224. DOI: http://dx.doi.org/10.1002/masy.200450821

[78] Ota A, Ouahab L, Golhen S, Yoshida Y, Maesato M, Saito G, Świetlik R. Phase transition from mott insulating phase into the charge ordering phase with molecular deformation in charge-transfer salts κ-$(ET)_4[M(CN)_6][N(C_2H_5)_4]\cdot 2H_2O$ (M = Co^{III} and Fe^{III}). Chem. Mater. 2007;19:2455–2462. DOI: http://dx.doi.org/10.1021/cm063036k

[79] Yasin S, Rose E, Dumm M, Drichko N, Dressel M, Schlueter JA, Zhilyaeva EI, Torunova S, Lyubovskaya RN. Electronic and magnetic studies of κ-$(BEDT\text{-}TTF)_2Hg(SCN)_2Cl$. Physica B. 2012;407:1689–1691. DOI: http://dx.doi.org/10.1016/j.physb.2012.01.007

[80] Kanoda K. Recent progress in NMR studies on organic conductors. Hyperfine Interact. 1997;104:235–249. DOI: http://dx.doi.org/10.1023/A:1012696314318

[81] Welp U, Fleshler S, Kwok WK, Crabtree GW, Carlson KD, Wang HH, Geiser U, Williams JM, Hitsman VM. Weak ferromagnetism in κ-$(ET)_2Cu[N(CN)_2]Cl$, where (ET) is bis(ethylenedithio)tetrathiafulvalene. Phys. Rev. Lett. 1992;69:840–843. DOI: http://dx.doi.org/10.1103/PhysRevLett.69.840

[82] Kozlov ME, Pokhodnia KI, Yurchenko AA. The assignment of fundamental vibrations of BEDT-TTF and BEDT-TTF-d_8. Spectrochim. Acta A. 1987;43:323–329. DOI: http://dx.doi.org/10.1016/0584-8539(87)80112-5

[83] Kozlov ME, Pokhodnia KI, Yurchenko AA. Electron molecular vibration coupling in vibrational spectra of BEDT-TTF based radical cation salts. Spectrochim Acta A. 1989;45:437–444. DOI: http://dx.doi.org/10.1016/0584-8539(89)80039-X

[84] Eldrige JE, Homes CC, Williams JM, Kini AM, Wang HH. The assignment of the normal modes of the BEDT-TTF electron-donor molecule using the infrared and Raman spectra of several isotopic analogs. Spectrochim Acta A. 1995;51:947–960. DOI: http://dx.doi.org/10.1016/0584-8539(94)00236-5

[85] Musfeldt JL, Świetlik R, Olejniczak I, Eldridge JE, Geiser U. Understanding electron-molecular vibrational coupling in organic molecular solids: Experimental evidence for strong coupling of the 890 cm^{-1} mode in ET-based materials. Phys. Rev. B. 2005;72:014516–014522. DOI: http://dx.doi.org/10.1103/PhysRevB.72.014516

[86] Eldridge JE, Xie Y, Wang HH, Williams JM, Kini AM, Schlueter JA. Two Interesting Features in the Infrared and Raman Spectra of the 12K Organic Superconductor κ-$(ET)_2Cu[N(CN)_2]Br$. Mol. Cryst. Liq. Crys. 1996;284:97–106. DOI: http://dx.doi.org/10.1080/10587259608037914

[87] Eldridge JE, Xie Y, Wang HH, Williams JM, Kini AM, Schlueter JA. Electron–phonon effects in the organic superconductor κ-$(BEDT\text{-}TTF)_2Cu[N(CN)_2]Br$. Spectrochim. Acta A. 1996;52:45–56. DOI: http://dx.doi.org/10.1016/0584-8539(95)01545-0

[88] Moldenhower J, Horn Ch, Pokhodnia KI, Schweitzer D, Heinen I, Keller HJ. FT-IR absorption spectroscopy of BEDT-TTF radical salts: charge transfer and donor-anion interaction. Synth. Metals. 1993;60:31–38. DOI: http://dx.doi.org/10.1016/0379-6779(93)91180-A

[89] Girlando A. Charge Sensitive Vibrations and Electron-Molecular Vibration Coupling in Bis(ethylenedithio)-tetrathiafulvalene (BEDT-TTF). J. Phys. Chem. C. 2011;115:19371–19378. DOI: http://dx.doi.org/10.1021/jp206171r

[90] Faltermeier D, Barz J, Dumm M, Dressel M, Drichko N, Petrov B, Semkin V, Vlasova R, Mezière C, Batail P. Bandwidth-controlled Mott transition in κ–(BEDT–TTF)$_2$Cu[N(CN)$_2$]Br$_x$Cl$_{1-x}$: Optical studies of localized charge excitations. Phys. Rev. B. 2007;76:165113–165124. DOI: http://dx.doi.org/10.1103/PhysRevB.76.165113

[91] Visentini G, Masino M, Bellitto C, Girlando A. Experimental determination of BEDT –TTF+ electron-molecular vibration constants through optical microreflectance. Phys. Rev. B. 1998;58:9460–9467. DOI: http://dx.doi.org/10.1103/PhysRevB.58.9460

[92] Łapiński A, Gąsecka A, Graja A, Waplak S, Ostrowski A, Kotov AI. Spectral investigation of the phase-transition behaviour in $(DOEO)_4HgBr_4{\cdot}TCE$ salt. Opt. Mat. 2012;34:1651–1655. DOI: http://dx.doi.org/10.1016/j.optmat.2012.03.012

[93] Łapiński A, Jankowski D, Świetlik R, Reinheimer EW, Fourmigué M. Electron-molecular vibration coupling in (DMtTTF)Br and $(o\text{-}DMTTF)_2[W_6O_{19}]$ salts studied by vibrational spectroscopy. Synth. Metals. 2014;188:92–99. DOI: http://dx.doi.org/10.1016/j.synthmet.2013.11.004

[94] Demiralp E, Dasgupta S, Goddard III WA. MSX force field and vibrational frequencies for BEDT-TTF (neutral and cation). J. Phys. Chem. A. 1997;101:1975–1981. DOI: http://dx.doi.org/10.1021/jp963004s

[95] Imakubo T, Tajima N, Tamura M, Kato R, Nishio Y, Kajita K. A supramolecular superconductor $\theta\text{-}(DIETS)_2[Au(CN)_4]$. J. Mater. Chem. A. 2002;12:159–161. DOI: http://dx.doi.org/10.1039/B110303B

[96] Imakubo T, Tajima N, Tamura M, Kato R, Nishio Y, Kajita K. Supramolecular organic conductor $\theta\text{-}(DIETS)_2[Au(CN)_4]$: unique crystal structure and superconductivity under uniaxial strain (DIETS=diiodo(ethylenedithio)diselenadithiafulvalene). Synth. Metals. 2003;133–134:181–183. DOI: http://dx.doi.org/10.1016/S0379-6779(02)00383-1

[97] Ouahab L, Setifi F, Golhen S, Imakubo T, Lescouezec R, Lloret F, Julve M, Świetlik R. Charge transfer salts containing a paramagnetic cyano-complex and iodine substituted organic donor involving $I_{(donor)}\cdots N_{(anion)}$ interactions. Comptes Rendus Chimie. 2005;8:1286–1297. DOI: http://dx.doi.org/10.1016/j.crci.2004.11.009

[98] Laversanne R. The role of anions in the electronic localisation of organic conductors. J. Phys. C: Solid State Phys. 1988;21:3081–3089. DOI: http://dx.doi.org/10.1088/0022-3719/21/16/020

[99] Imakubo T, Shirahata T, Harvé K, Ouahab L. Supramolecular organic conductors based on diiodo-TTFs and spherical halide ion X^- (X = Cl, Br). J. Mater. Chem. 2006;16:162–173. DOI: http://dx.doi.org/10.1039/B512036E

[100] Imakubo T, Sawa T, Kato R. Novel molecular conductors, $(DIETS)_4M(CN)_4$(M = Ni, Pd, Pt): highly reticulated donor···anion contacts by –I ··· NC– interaction. J. Chem. Soc. Chem. Commun. 1995:1667–1668. DOI: http://dx.doi.org/10.1039/C39950001667

[101] Bozio R, Pecile C. Charge Transfer Crystals and Molecular Conductors. In: Clark RJH, Hester RE, editors. Spectroscopy of Advanced Materials. Chichester: Wiley; 1991. 1–86. ISBN 0-4719-2981-6

[102] Graja A. Optical properties. In: Fargés J-P, editor. Organic Conductors-Fundamentals and Applications. New York: Dekker; 1994.229–268. ISBN 0-8247-9216-5

[103] Rice MJ, Pietronero L, Brüesch P. Phase phonons and intramolecular electron–phonon coupling in the organic linear chain semiconductor TEA(TCNQ)$_2$. Solid State Commun. 1977;21:757–760. DOI: http://dx.doi.org/10.1016/0038-1098(77)91146-2

[104] Girlando A, Bozio R, Pecile C, Torrance JB. Discovery of vibronic effects in the Raman spectra of mixed-stack charge-transfer crystals. Phys. Rev. B. 1982;26:2306–2309. DOI: http://dx.doi.org/10.1103/PhysRevB.26.2306

[105] Kaplunov MG, Lyubovskaya RN. Infrared optical properties of the new organic conductors family (ET)$_8$[Hg$_2$X$_6$(C$_6$H$_5$Y)]$_2$ (X, Y = Cl, Br). J. Phys. I France. 1992;2:1811–1818. DOI: http://dx.doi.org/10.1051/jp1:1992247

[106] Delhaes P, Garrigou-Lagrange C, Fabre J-M. Electronic and vibrational absorption spectra in DMtTCF and TMTCT salts: Estimation for electronic correlations. Physica B. 1986;143:485–487. DOI: http://dx.doi.org/10.1016/0378-4363(86)90173-7

[107] Drozdova O, Yakushi K, Yamamoto K, Ota A, Yamochi H, Saito G, Tashiro H,Tanner DB. Optical characterization of 2k$_F$ bond-charge-density wave in quasi-one-dimensional 34-filled (EDO–TTF)$_2$X (X=PF$_6$ and AsF$_6$). Phys. Rev. B. 2004;70:075107–075114. DOI: http://dx.doi.org/10.1103/PhysRevB.70.075107

[108] Jankowski D, Świetlik R, Reinheimer EW, Fourmigué M. Involvement of weak C---H···X hydrogen bonds in metal-to-semiconductor regime change in one-dimensional organic conductors (o-DMTTF)$_2$X (X = Cl, Br and I): combined IR and Raman studies. J. Raman Spectr. 2011;42:1518–1527. DOI: http://dx.doi.org/10.1002/jrs.2889

[109] Jankowski D, Świetlik R, Reinheimer EW, Fourmigué M. Spectroscopic studies of charge localization in organic metals (DMtTTF)$_2$X (X = ClO$_4$, ReO$_4$). Phys. Status Solidi C. 2012;9:1164–1166. DOI: http://dx.doi.org/10.1002/pssc.201100770

[110] Fano U. Effects of Configuration Interaction on Intensities and Phase Shifts. Phys. Rev. 1961;124:1866–1878. DOI: http://dx.doi.org/10.1103/PhysRev.124.1866

[111] Horovitz B, Gutfreund H, Weger M. Infrared and Raman activities of organic linear conductors. Phys. Rev. B. 1978;17:2796–2799. DOI: http://dx.doi.org/10.1103/PhysRevB.17.2796

[112] Pedron D, Bozio R, Meneghetti M, Pecile C. Electronic interactions in the organic conductors (TMTSF)$_2$X (X=ClO$_4$ and PF$_6$) and (TMTTF)$_2$X (X=Br and PF$_6$) from their infrared spectra. Phys. Rev. B. 1994;49:10893–10907. DOI: http://dx.doi.org/10.1103/PhysRevB.49.10893

[113] Taniguchi M, Misaki Y, Tanaka K. Theoretical estimation of the electron–molecular vibration coupling in several organic donor molecules. Solid State Commun. 2000;114:75–79. DOI: http://dx.doi.org/10.1016/S0038-1098(00)00008-9

Chapter 7

Infrared and Raman Spectroscopic Characterization of Porphyrin and its Derivatives

Metin Aydin and Daniel L. Akins

Additional information is available at the end of the chapter

http://dx.doi.org/10.5772/64582

Abstract

Density functional theory (DFT) was employed to investigate protonation, deuteration, and substitution effects on the vibrational spectra of porphyrin molecules. The results of the calculations were compared with experimental data. The calculations show that *meso*-substitutions produced a substantial shift in frequencies when the *meso*-carbons within the parent porphine are involved in the vibrational motion of molecules, while protonation of the N atoms leads to a significant blue shift when the H atoms covalent bonded to the N atoms that are substantially involved in the vibrational motion. Deuteration of N atoms at the porphyrin core is found to result not only in a red shift in the frequencies of the corresponding peaks below 1600 cm^{-1}, but also to generate new Raman bands of frequencies in the range of 2565–2595 cm^{-1}, resulting from N-D bond stretching. Also, the deuteration of O atoms within the sulfonato groups ($-SO_3^-$) results in a new peak at near 2642 cm^{-1} due to O-D bond stretching. Calculated IR spectra of the compounds studied here showed similar differences. Finally, we discuss solvent effects on the IR spectrum of TSPP.

Keywords: porphyrins, protonation, Raman, IR, DFT calculation

1. Introduction

Molecular vibrations may be induced through two well-known optical excitation processes. One is the absorption of photons and the other is the inelastic scattering of photons. Excitation of molecular vibration by absorption of photons is achieved by irradiation of a species using radiation containing photons of a frequency equivalent to the frequency difference $\Delta\nu$ between the initial (i) and the final (f) vibrational states of the species; i.e., $\Delta\nu = \nu_f - \nu_i$. Unlike IR spectroscopy, the scattering mechanism for exciting molecular vibrations generally exploits monochromatic radiation. In this latter case, a number of incident photons is scattered inelastically such that the frequency of the scattered photons (ν_S) differs from that of the incident photons (ν_0). And with conversation of the energy, this energy difference is the energy change associated with a transition from the initial (i) vibrational state to the final (f) vibrational state of the scattering species; i.e., $\nu_0 - \nu_S = \nu_f - \nu_i$. Inelastic scattering of the photons was first discovered by the Indian scientist C. V. Raman in 1928 and is referred to as the Raman effect.

In this chapter, we discuss IR and Raman spectra of protonated, deuterated, and *meso*-substituted parent porphyrin using density functional theory (DFT) to calculate the IR and Raman spectra, and, where possible, make comparison to experimental spectra. We also discuss spectra of aggregates involving several of the porphyrin species, using vibrational band assignments to ascertain which motions of the vibrating molecule couple more effectively, with excitonic motion, and as a result, we derive molecular alignment information from the enhancement that certain vibronic bands in the Raman spectrum experience for the various porphyrins.

2. Overview of Raman spectroscopy

In this section, we focus on Raman scattering. It is convenient to define the Raman scattering cross-section for the n → m vibrational transition as $\sigma_{n\to m}$, and to relate it to the scattering intensity as follows:

$$\mathbf{I_{n\to m} = \sigma_{n\to m} I_0} \tag{1}$$

In this equation, $\mathbf{I_0}$ is the intensity of the incident radiation and $\mathbf{I_{n\to m}}$ the intensity of the light scattered by molecules integrated over all scattering angles and polarization directions for randomly oriented molecules. The Raman cross-section is associated with the Raman polarizability by utilizing the fact that the intensity for electric dipole radiation scales as the fourth power of the frequency:

$$\boldsymbol{\sigma_{n\to m} \propto (\nu_0 \mp \nu_k)^4 \sum_{\rho,\sigma} |\alpha_{\rho\sigma}|^2} \tag{2}$$

In this equation, the indices ρ and σ indicate the molecule-fixed directional coordinates. Moreover, for this equation, the scattering tensor $\alpha_{\rho\sigma}$ can be formulated in terms of Kramers -Heisenberg-Dirac dispersion theory, as indicated in Eq. (3) below [1]:

$$\boldsymbol{[\alpha_{\rho\sigma}]_{nm} = \frac{1}{h}\sum_{S,r}\left\{\frac{\langle nG|M_\rho|Sr\rangle\langle rS|M_\sigma|Gm\rangle}{\nu_{Sr}-\nu_k-\nu_0+i\Gamma_S} + \frac{\langle rS|M_\sigma|Gm\rangle\langle nG|M_\rho|Sr\rangle}{\nu_{Sr}-\nu_k+\nu_0+i\Gamma_S}\right\}}, \tag{3}$$

where M_ρ and M_σ represent the electronic transition dipole moment in a molecule-fixed coordinate system (Albrecht [2] and Warshel and Dauber [3]). The symbols ν_0 and ν_k represent the frequencies of the excitation radiation and the normal mode Q_k, respectively, S and r represent the respective electronic and vibrational states of the molecule, and Γ_s is a damping constant, which is associated with the lifetime of the vibroelectronic state Sr. The sum in Eq. (3) indicates that for the Raman transition of all vibronic states must be used, which indicates that the scattering tensor, and thus, the Raman intensity, is controlled by the transition probabilities involving all vibronic states, even though the initial and final states refer to the vibrational ground and excited states of the electronic ground state. Thus, the sum of integrals in Eq. (3) describes the transitions: $|nG\rangle \rightarrow |Sr\rangle$ and $|Sr\rangle \rightarrow |Gn\rangle$. When the excitation frequency ν_0 is in resonance or preresonance with the frequency of an electronic transition, the scattering is referred to as resonance Raman (RR) scattering. In this case, Eq. (3) may be simplified to:

$$\left[\alpha_{\rho\sigma}\right]_{nm} \cong \frac{1}{h}\sum\nolimits_{S,r}\left\{\frac{\langle nG|M_\rho|Sr\rangle\langle rS|M_\sigma|Gm\rangle}{\nu_{Sr}-\nu_k-\nu_0+i\Gamma_S}\right\}, \tag{4}$$

where the summation is now restricted to the vibrational states r of the resonantly excited electronic state (Albrecht [2]; Warshel and Dauber [3]). The wave functions of the integrals in Eq. (4) depend on the electronic and nuclear coordinates and may be separated by taking into account the Born-Oppenheimer approximation:

$$\langle nG|M_\rho|Sr\rangle = \langle G|M_\rho|S\rangle\langle n|r\rangle = M_{GS,\rho}\langle n|r\rangle \tag{5}$$

Here, the $\langle n|r\rangle$ integral represents the Franck-Condon factor, which is the integral over the product of two vibrational wave functions. With this approximation, Eq. (4) becomes:

$$\left[\alpha_{\rho\sigma}\right]_{nm} \cong \frac{1}{h}\sum\nolimits_{r}\left\{\frac{M_{GS,\rho}M_{GS,\sigma}\langle n|r\rangle\langle r|m\rangle}{\nu_{Sr}-\nu_k-\nu_0+i\Gamma_S}\right\}, \tag{6}$$

where $M_{GS,\rho}$ is the electronic transition-dipole moment associated with the electronic transition from the ground state G to the electronically excited state S. Thus, $M_{GS,\rho}$ can be expanded in a Taylor series with respect to the normal coordinates Q_k:

$$M_{GS,\rho}(Q_k) = M_{GS,\rho}\left(Q_k^0\right) + \sum\nolimits_k Q_k\left(\frac{\partial M_{GS,\rho}}{\partial Q_k}\right)_0 + \cdots \tag{7}$$

And, within the harmonic approximation, we neglect higher order terms and combine Eqs. (6) and (7) to obtain the scattering tensor as the sum of two terms, the so-called Albrecht A and B terms:

$$\left[\alpha_{\rho\sigma}\right]_{nm} \cong A_{\rho\sigma} + B_{\rho\sigma} + \ldots \tag{8}$$

$$A_{\rho\sigma} \cong \frac{1}{h}\sum\nolimits_{r}\left\{\frac{M^0_{GS,\rho}M^0_{GS,\sigma}\langle n|r\rangle\langle r|m\rangle}{\nu_{Sr}-\nu_k-\nu_0+i\Gamma_S}\right\} \tag{9}$$

$$\mathbf{B}_{\rho\sigma} \cong \frac{1}{h}\sum_{r}\left\{\frac{M^{0}_{GS,\sigma}\left(\frac{\partial M_{GS,\rho}}{\partial Q_k}\right)_0 \langle n|Q_k|r\rangle\langle r|m\rangle}{v_{Sr}-v_k-v_0+i\Gamma_S} + \frac{M^{0}_{GS,\rho}\left(\frac{\partial M_{GS,\sigma}}{\partial Q_k}\right)_0 \langle r|Q_k|m\rangle\langle n|r\rangle}{v_{Sr}-v_k-v_0+i\Gamma_S}\right\} \quad (10)$$

In the above equations, $M^{0}_{GS,\rho}$ and $M^{0}_{GS,\sigma}$ are the components of transition dipole moment of the vertical electronic transition G → S.

The A and B terms represent different scattering mechanisms, but the dominators are minimized in both terms when the frequency of the excitation v_0 is in preresonance or resonance with the frequency of an electronic transition. In such a case, both the A and the B terms are enhanced, leading to amplified scattering of radiation.

It is to be noted that if the resonant electronic transition exhibits a large oscillator strength, i.e., a large transition dipole moment $\mathbf{M}^{0}_{GS}$, then the A term may be increased substantially more than the B term, and therefore become the more important scattering term. In this case, the enhancement of a normal mode depends on the products of Franck-Condon factors, i.e., the term $\langle n|r\rangle\langle r|m\rangle$.. It is to be noted that whether or not a normal mode is resonance enhanced via the Franck-Condon mechanism depends on the geometry of the resonant excited state.

The intensity of a vibrational band attributable to a normal mode Q of frequency v_Q can be estimated in the double harmonic approximation. For the nonresonant situation (for a normal mode $\mathbf{Q}_K$ of frequency v_{Qk} and excitation frequency v_0), the Raman intensity $\mathbf{I}_{Qk}$ can be computed according to the following equations [1, 4]:

$$\mathbf{I}_{Q_k} = \frac{f(v_0-v_Q)^4}{v_{Q_k}\left[1-\exp\left(-\frac{hv_Q}{kT}\right)\right]}\mathbf{S}_{Q_k} \quad (11)$$

$$\mathbf{S}_{Q_k} \cong \left\{45\left(\frac{\partial\alpha}{\partial Q_k}\right)^2 + 7\left(\frac{\partial\gamma}{\partial Q_k}\right)^2\right\} \quad (12)$$

$$\left(\frac{\partial\alpha}{\partial Q_k}\right)^2 = \left(\frac{1}{3}\right)\left\{\left(\frac{\partial\alpha_X}{\partial Q_k}\right) + \left(\frac{\partial\alpha_Y}{\partial Q_k}\right) + \left(\frac{\partial\alpha_Z}{\partial Q_k}\right)\right\}^2 \quad (13)$$

$$\left(\frac{\partial\gamma}{\partial Q_k}\right)^2 = \left(\frac{1}{2}\right)\left\{\left[\left(\frac{\partial\alpha_X}{\partial Q_k}\right) - \left(\frac{\partial\alpha_Y}{\partial Q_k}\right)\right]^2 + \left[\left(\frac{\partial\alpha_Y}{\partial Q_k}\right) - \left(\frac{\partial\alpha_Z}{\partial Q_k}\right)\right]^2 + \left[\left(\frac{\partial\alpha_Z}{\partial Q_k}\right) - \left(\frac{\partial\alpha_X}{\partial Q_k}\right)\right]^2\right\} \quad (14)$$

In the above equations, $\mathbf{S}_{Q_k}$ is the Raman activity for a normal mode $\mathbf{Q}_k$, $\left(\frac{\partial\alpha}{\partial Q}\right)$ and $\left(\frac{\partial\gamma}{\partial Q}\right)$ are, respectively, the derivatives of the polarizability tensor and the corresponding anisotropy with respect to the normal mode Q, and f is a physical constant that includes the intensity of the incident radiation. We have calculated Raman intensities of the Raman active modes using Eq. (11), which is implemented in Gauss Sum software [5]. The software provides $\mathbf{S}_{Q_k}$ (the Raman activity, Eq. 12) and the frequency v_{Qk} from the output files of the quantum chemical calculation program (specifically, Gaussian 09).

We explore in this chapter the effect of protonation, deuteration, and *meso*-substitutions on the vibronic spectra of porphyrin and some of its derivatives. Specific molecules considered are the following: parent porphyrin (FBP), diprotonated FBP (H_4FBP), deuterated H_4FBP (D_4FBP); *meso*-tetraphenylporphyrin (TPP), diprotonated TPP (H_4TPP or dicationic TPP) deuterated H_4TPP (D_4TPP); *meso*-tetrakis (*p*-sulfonatophenyl) porphyrin (TSPP), diprotonated TSPP (H_4TSPP or dianionic TSPP), deuterated H_4TSPP (D_4TSPP), dicationic TSPP (H_8TSPP), as well as deuterated H_8TSPP (D_8TSPP). We also deal with how molecular aggregation of some of the aforementioned species affects Raman spectra. Density functional theory has been employed to calculate the vibronic structural properties for both IR and Raman spectra.

Our motivation for focusing on the porphyrin monomers and aggregates is that porphyrin monomers and their aggregates play fundamental roles in natural systems and increasingly in artificial photonic devices. As regards aggregates, the primary mechanism through which molecular aggregate structures are formed in both natural and artificial systems is self-assembly through intrinsic intermolecular interactions, without the formation of covalent linkages. Self-assembled molecular aggregates often assume a structure that can be classified as being of J- or H-type, defined by the relative orientations of induced transition dipoles of the constituent molecules, either "head-to-tail" or "head-to-head," respectively [6]. Structural pictures such as those provided by J- and H-aggregates have provided a framework for theoretical analysis of structure and dynamics of aggregated systems.

Moreover, aggregated porphyrin species are model composite structures for gaining insight into the roles that optically induced transient structural changes and photon dynamics play in photosynthesis [7, 8]. And through the study of spectral properties and photodynamic behaviors of aggregated porphyrin structures, an important outcome sought is the translation of the electron transfer specificities and speeds often found for biological reactions to the realm of molecular photonic devices (i.e., biomimetics) or photonic materials; indeed, enormous interest in the applications area has been evidenced [9, 10]. Thus, experimental and quantum chemical calculations of structures and optical dynamics of porphyrin monomers and aggregates have both scientific and technological importance.

We deduce that the observed Raman bands of the TPP, TSPP, H_4TSPP, and aggregated H_4TSPP may most properly be characterized by the vibrations of the pyrrole and pyrroline rings, the sulfonatophenyl groups, and their combinations rather than as vibrations of isolated chemical bonds.

As regards IR spectra, we have found that calculated IR spectra of H_4TSPP can be assigned by comparison with the calculated IR spectra of other porphyrin derivatives and the experimentally measured IR spectra that are obtained from the literature. We further point out that the experimental and theoretical data used in this chapter are taken from prior experimental measurements performed in our laboratories [11–13].

The Raman and IR spectra of porphyrin derivatives in water, used as solvent in the calculations, were calculated at the B3LYP/6-311G (d, p) level of density functional theory.

3. The Raman spectra of porphyrin and derivatives

Figure 1 provides the measured Raman spectra of the TPP (**Figure 1B**) and H_4TSPP (**Figure 1G**) from our previous works [11–13]. Many Raman bands with strong and medium intensity, as well as numerous weak bands are found throughout the spectrum. The Raman spectrum of the H_4TSPP when compared to that of the TPP are quite similar, however, the positions of several bands are substantially shifted in frequency. As examples, in the observed Raman spectrum of the TPP, the strongest band at 1564 cm^{-1} and the bands at 334, 1234, 1327, 1438, 1577, and 1595 cm^{-1} (with relatively weak intensity) are respectively red shifted to 1537 cm^{-1} (the most intense peak), 312, 1229, 1339, 1427, 1562, and 1494 cm^{-1} in the H_4TSPP spectrum. Also, the bands at 201, 334, 962, and 1002 cm^{-1} are respectively blue shifted to 236, 314, 983, and 1014 cm^{-1} in the H_4TSPP spectrum. Additionally, the bands at 1476 and 701 cm^{-1} in the H_4TSPP spectrum are considerably enhanced compared to their corresponding ones in the TPP.

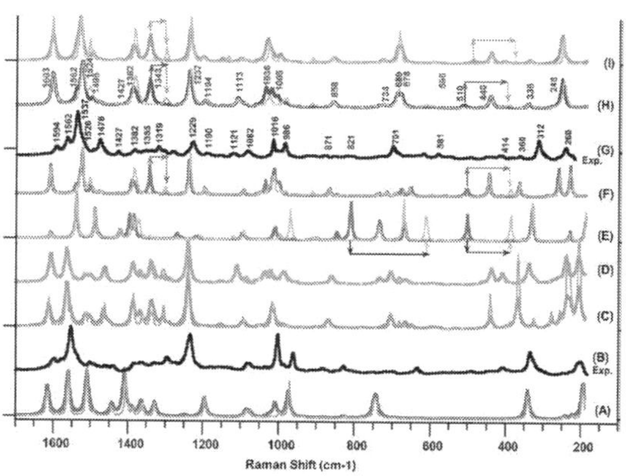

Figure 1. The predicted Raman spectra of porphyrin derivatives: (**A**) free-base porphyrin (FBP) and deuterated FBP (D_2FBP); (B) the experimentally measured Raman spectrum of the TPP; (**C**) *meso*-tetraphenylporphyrin (TPP) and (D_2TPP); (**D**) anionic *meso*-tetrakis(*p*-sulfonatophenyl)porphyrin (TSPP) and deuterated TSPP (D_2TSPP); (**E**) diprotonated FBP (H_4FBP) and deuterated H_4FBP (D_4FBP); (**F**) diprotonated-TPP (H_4TPP) and deuterated H_4TPP (D_4TPP); (**H**) diprotonated TSPP (H_4TSPP) and deuterated-H_4TSPP (D_4TSPP); and (**I**) dicationic TSPP (H_8TSPP) and deuterated H_8TSPP (D_8TSPP). The plotted spectra in the gray color belong to the deuterated molecules. The calculations were carried out in water at B3LYP/6-311G(d, p) level of DFT, and the line arrows show the frequency shift in the deuterated molecule [11].

Also, calculated Raman spectra of the FBP/D_2FBP, H_4FBP/D_4FBP, TPP/D_2TPP, H_4TPP/D_4TPP, TSPP/D_2TSPP, H_4TSPP/D_4TSPP, and H_8TSPP/D_8TSPP in water used as a solvent are given in **Figure 1**, with the observed Raman spectra of the TPP and H_4TSPP for comparison. (It is important to note that the D_8TSPP symbolize the dicationic TSPP where four of eight deuterium atoms (D) covalently bounded to the nitrogen atoms at the core and the other four covalent bonded to one of three oxygen atoms within each of four *meso*-sulfonatophenyl substituted groups.)

TPP						TSPP				H_4TSPP						Assignments
Sym	$\Delta\nu_{sc.}$	S_R	I_R	$\Delta\nu_{exp}$	$I_{R/exp}$	Sym	$\Delta\nu_{sc.}$	S_R	I_R	Sym	$\Delta\nu_{sc.}$	S_R	I_R	$\Delta\nu_{exp}$	$I_{R/exp}$	
A2	1612	25	24	1595	29	A2	1607	44	42	A2	1603	45	40	1593	21	C-C bond stretching within phenyl rings and rocking of their H, no any contribution comes from mac-
A1	1612	29	27			A1	1607	53	50	A1	1603	55	49			rocycle and sulfonato groups (-SO_3^-) (-SO_3).
A1	1588	<1	<1	1577	27	A1	1574	1	1	A1	1568	2	1	1563	42	C-C bond stretching within phenyl rings and rocking of their H, accompanied by relatively weak asymmetric stretching of C_α-C_m-C_β bond stretching, no any contribution comes from sulfonato groups.
A1	1564	100	100	1553	100	A1	1564	100	100	A1	1524	100	100	1537	100	C_β-C_β bond stretching, $\nu(C_\beta$-$C_\beta)$, symmetric stretching of C_α-C_m-C_α bonds, $\nu_s(C_\alpha$-C_m-$C_\alpha)$ that leads to bending deformation of the C-N-C bonds, θ(C-N(H)-C).
				1540	40									1528	42	
A2	1559	8	8			A2	1558	9	9	A2	1540	29	28			$\nu_s(C_\alpha$-C_m-$C_\alpha)$/rocking of C-N(H)-C and H on N atoms, ρ(C-N(H)-C)/ρ(NH)
A1	1555	15	15			A1	1554	8	8	A1	1529	26	25			Asymmetric stretching of (C_α-C_m-C_α) bonds $\nu_a(C_\alpha$-C_m-$C_\alpha)$/θ(C-N(H)-C).
A1	1514	24	25	1502	21	A1	1515	21	23	A1	1477	5	5	1476	38	$\nu(C_\beta$-$C_\beta)$ and rocking of the H on C atoms within macrocycle (not on the phenyl groups), $\rho(C_\beta H)$, and relatively weak θ(C-N(H)-C)
A2	1502	14	15	1491	16	A2	1499	14	15	A1	1496	7	7	1489	15	ρ(CH within phenyl groups only)
A1	1501	4	4			A1	1498	3	4	A2	1495	8	8			
A2	1466	31	36	1461	13	A2	1464	37	44							$\nu(C_m$-$C_\alpha)$/$\rho(C_\beta H)$, and relatively weak ν_a(C-N(H)-C)
A1	1454	3	3	1438	12	A1	1454	2	2	A1	1426	3	3	1428	11	$\nu_s(C_\alpha$-C_m-$C_\alpha)$/ ρ(C-N(H)-C)/ρ(CH)
A2	1387	46	61	1378	20	A2	1387	44	58	A2	1391	31	38	1384	15	$\nu_a(C_\beta$-C_α-N(H))/$\rho(C_\beta H$-$C_\beta H)$
A1	1367	22	30			A1	1367	22	29	A1	1382	20	25	1354	14	$\nu(C_\beta$-$C_\alpha)$/θ(C-C_m-C)/θ(C-N(H)-C), which leading to macrocycle getting a square shape)
A2	1339	19	26	1327	20	A2	1338	42	60	A2	1343	52	70	1340	14	$\nu(C_\varphi$-$C_m)$/$\rho(C_\beta H)$/ρ(NH), and relatively weak ν_a(C-N(H)-C)
A2	1335	26	38			A2	1329	13	19					1319	22	
A1	1306	13	19			A1	1305	7	10	A1	1321	1	1	1304	14	ν_a(C-C-C) within phenyl groups/θ(C-N(H)-C)/ρ(CH).
A1	1291	4	7			A1	1290	4	6	A1	1300	1	2	1283	12	
A1	1239	75	127	1234	84	A1	1239	77	131	A1	1237	60	97	1229	34	$\nu(C_\varphi$-$C_m)$ (primarily)/ν_s(C-N(H)-C)/ρ(CH)/$\nu(C_\beta$-$C_\beta)$ (relatively weak)
A2	1189	0.	0.			A2	1188	1	1	A2	1194	3	6	1190	8	ρ(CH) within phenyl groups.
A1	1189	1	2			A1	1189	1	1	A1	1194	5	9			
A2	1152	4	8	1137	10	A2	1153	3	5							ρ(NH)/$\rho(C_\beta H)$ and relatively weak structural deformation
						A2	1146	<1	<1	A2	1151	<1	<1	1122	9	ν_a(O-S-O) within sulfonato groups
A1	1097	1	2	1080	21	A1	1109	19	42	A1	1108	9	18	1082	14	ν(S-O)/θ(C-C(S)-C) within sulfonato groups.
A1	1091	5	11			A1	1092	3	6	A1	1093	2	4			$\rho(C_\beta H)$
A1	1048	2	6			A1	1036	1	3	A1	1033	<1	1			θ(C-C-C) within the phenyl groups
A2	1013	6	15			A2	1037	1	4	A2	1035	1	3			

TPP						TSPP				H_4TSPP						Assignments
Sym	$\Delta\nu_{sc.}$	S_R	I_R	$\Delta\nu_{exp}$	$I_{R/exp}$	Sym	$\Delta\nu_{sc.}$	S_R	I_R	Sym	$\Delta\nu_{sc.}$	S_R	I_R	$\Delta\nu_{exp}$	$I_{R/exp}$	
A1	1020	4	10	1002	85	A1	1020	2	5	A1	1036	3	7	1016	40	Expansion of the pyrrole/pyrroline groups along N(H)…N(H) direction due to ν(C_α-C_β), leading to macrocycle getting rectangular shape instead of square shape.
A1	983	2	4	962	44	A1	986	2	5	A1	1005	2	5	1002	15	Expansion of the pyrrole/pyrroline groups along N(H)…N(H) direction in the same phase like macrocycle getting square shape or similar to breathing of the macrocycle
A2	990	1	2			A2	988	1	2	A2	992	0.	1			Out of plane wagging of the H on the phenyl rings, w(CH)
A1	899	1	3			A1	899	1	3	A1	904	<1	1	986	32	Bending deformation inside entire molecule.
A1	868	3	13			A1	859	4	14	A1	858	2	8	879	6	w(CH on the macrocycle and phenyl rings)
A1	815	<1	<1			A1	814	<1	<1	A1	820	<1	1	821	5	
A1	768	<1	<1			A1	749	<1	1	A1	751	1	2			Out of plane bending deformation of whole molecule including w(CH/NH)
A1	727	<1	<1			A1	734	3	15	A1	734	1	5	728	5	ν(S-O) and expansion of the phenyl rings along S…C_m direction including w(CH/NH)
										A2	689	5	31	701	27	Out of plane twisting of the macrocycle
A1	665	2	12	636	13	A1	669	1.33	9	A1	678	5	33			
A1	584	<1	2			A1	589	<1	1	A1	596	<1	3	580	10	w(NH and CH on the macrocycle and phenyl rings) and wagging of the macrocycle.
A1	531	0.	3			A1	574	0.	1	A1	566	0.	1	548	5	Wagging of entire molecule
										A1	510	0.	2	494	3	w(NH)
A2	468	<1	1			A2	457	1	13	A2	468	<1	1			In-plane wagging of macrocycle and translational motion of phenyl rings.
A2	437	2	33	408	15	A2	435	2	25	A2	440	2	25	439	5	Out of plane bending of the phenyl rings.
						A1	407	2	47	A1	404	<1	6	414	8	Breathing macrocycle and translational motion of phenyl rings in opposite phase.
A1	365	6	160	334	50	A1	336	3	111	A1	338	1	24	314	41	Breathing of whole molecule.
A2	322	<1	17											363	7	Out of plane wagging of macrocycle.
A1	235	2	127	201	29	A1	237	2	152	A1	248	2	144	242	26	Out of plane wagging of macrocycle.
A1	252	<1	28			A1	257	<1	18	A1	252	<1	12			Out of plane wagging of phenyl rings and relatively weak out of plane wagging macrocycle.

The calculations were obtained in water used as solvent at B3LYP/6-311G(d,p) level. Where $\Delta\nu_{sc}$ symbolizes the scaled vibrational frequencies, $\Delta\nu_{sc.} = 0.96\ (\Delta\nu_{calc}) + 40$, and SR and IR represents, respectively, the predicted Raman scattering activity and intensity; and $\Delta\nu_{exp}$ and $I_{R/exp}$ symbolize the measured Raman frequency and Intensity, respectively [11–13].

Table 1. Predicted and measured Raman active modes of frequencies (in cm^{-1}) of the H_4TSPP (C_{2v}) with the TPP (C_{2v}) and TSPP (C_{2v}).

http://dx.doi.org/10.5772/64582

FBP	H_4FBP	TPP		H_4TPP	TSPP	H_4TSPP		H_8TSPP
Calc.	Calc.	Calc.	Exp.	Calc.	Calc.	Calc.	Exp.	Calc.
1007	1013	1020	1002	1036	1020	1036	1016	1036
972	1010	983	962	1002	985	1005	983	1001
D_2FBP	D_4FBP	D_2TPP	D_4TPP	D_2TSPP	D_4TSPP	D_8TSPP	D_2FBP	D_4FBP
Calc.	Calc.			Calc.	Calc.	Calc.	Exp. [14]	Calc.
996	995	1017		1026	1012	1026	1004	1026
968	968	979		980	977	983	957	979

Table 2. The predicted Raman active bands of frequencies (for the protonated and deuterated porphyrin derivatives) exhibited significant frequency shift in the range of 1040–950 cm^{-1}.

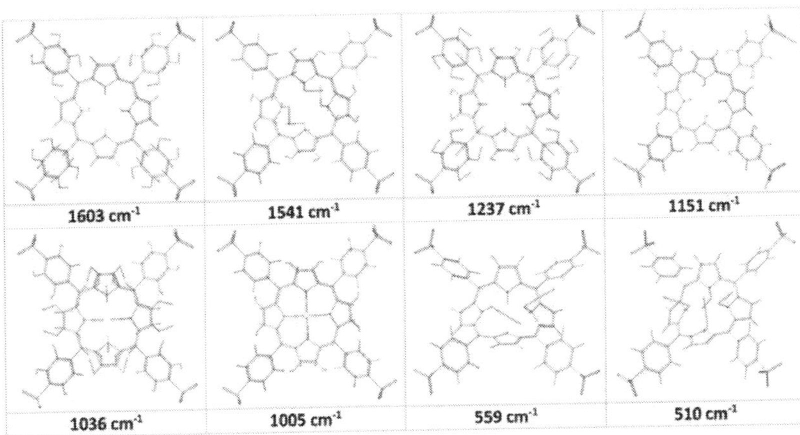

Figure 2. Calculated molecular motions for some vibrational bands of the H_4TSPP, from reference [11].

The assignment of the observed vibrational bands in the Raman spectra of the TPP and H_4TSPP were made based on the density functional prediction at the B3LYP/6-311G (d, p) level and on the atomic displacements visualized by using the GaussView program. The calculated vibrational frequencies coincided with those observed in their Raman spectra. We used the calculated frequencies and, to some degree, the predicted intensity distribution to attribute observed vibrational frequencies and intensities to specific intramolecular motions of the H_4TSPP and TPP. These latter assessments were facilitated by analysis of the calculated nuclear displacements, combined with animation of their vibrations, to identify specific motions as the dominant movements within the molecule. This is not a truly rigorous approach but should provide adequate insight. The assignments of the vibrational mode are provided in **Tables 1** and **3**, whereas **Figure 2** presents the nuclear displacement for several selected vibrational modes.

TPP						TSPP			H_4TSPP				H_8TSPP			Assignments
Sym	$\Delta\nu_{sc}$[a]	$\Delta\nu_{sc}$[b]	I_{IR}	$\Delta\nu_{exp}$ [15]	$\Delta\nu_{exp}$ [16]	Sym	$\Delta\nu_{sc}$[a]	I_{IR}	Sym	$\Delta\nu_{sc}$[a]	I_{IR}	$\Delta\nu_{exp}$ [16]	Sym	$\Delta\nu_{sc}$[a]	I_{IR}	
B2	447	414	6	409	406	B2	431	1	B2	439	3	415	B2	439	1	In-plane rotational motion of the pyrroline rings, including relatively weak out-of-plane twisting deformation of the phenyl rings, but no contributions come from the pyrroline rings
						B2	438	13	B2	427	5	445				Rocking of phenyl rings (ρ(phenyl) and wagging of macrocycle w(macrocycle).
B2	447.	414	6			B2	475	59	B2	439	3	457	B2	469	1	Out-of-plane bending of phenyl groups only.
B1	553	521	10		516	B1	523	2	A1	484	8		A1	484	8	Twisting of phenyl τ(phenyl) and w(macrocycle)
A1	584	553	2						A1	510	8		A1	487	4	w(NH only)
B2	570	539	2	559	558				B1	559	10		B1	567	10	
						A1	540	8	A1	566	14	560				Out of plane twisting of the molecule and θ(O-S-O)/w(CH and NH)
						B1	541	1	B1	567	6	580				
						B2	627	45	B2	624	63	637	B2	582	39	Due to bending deformation of the SO_3^- groups like closing and opening umbrella shape.
	647	618	2	619	618	A1	657	0.6								In plane bending deformation of phenyl rings, including w(NH and C_βH only) and out of plane deformation of the macrocycle.
A1	666	636	4	638	636	A1	668	1.0								
B2	688	659	6	658	657											
B2	728	700	43	699	701											w(CH on phenyl) and relatively weak out of plane deformation of the phenyl rings.
						B2	732	6	B2	722	12	715	B2	725	5	w(CH on phenyl) and out of plane deformation of the phenyl rings and macrocycle.
						B2	747	16	B2	748	19	741	B2	739	6	Primarily due to ν(S-C)/θ(phenyl) and relatively weak w(CH an NH) and out of plane bending (or twisting) deformation of macrocycle,
A1	745	716	61			A1	749	10	A1	751	6		A1	753	19	Primarily due to w(C_βHs an NH) and out of plane bending (or twisting) deformation of macrocycle, relatively weak out of plane deformation of the phenyl.
B2	757	729	32	727	728											w(C_βH an NH) and out of plane bending (or twisting) deformation of macrocycle, relatively weak out of plane deformation of the phenyl.
B1	776	749	37			B1	757									w(CH in phenyl and macrocycle) and out of plane bending (or twisting) deformation of phenyl rings the macrocycle.

TPP						TSPP			H_4TSPP				H_8TSPP			Assignments
Sym	$\Delta\nu_{sc}$[a]	$\Delta\nu_{sc}$[b]	I_{IR}	$\Delta\nu_{exp}$ [15]	$\Delta\nu_{exp}$ [16]	Sym	$\Delta\nu_{sc}$[a]	I_{IR}	Sym	$\Delta\nu_{sc}$[a]	I_{IR}	$\Delta\nu_{exp}$ [16]	Sym	$\Delta\nu_{sc}$[a]	I_{IR}	
B2	776	749	21	746	748	B2	767	1								
													A1	769	32	Mainly due to ν(S-O(H)), including w(C_βHs an NH) and out of plane bending (or twisting) deformation of macrocycle, relatively weak out of plane deformation of the phenyl.
													B2	773	66	
A1	815	788	3	785	788	B1	823	7	B1	825	2	800	B1	826	3	W(C_βHs and NH) and out of plane bending deformation of macrocycle
A1	829	802	100	798	799	A1	829	20	A1	848	23	854	A1	852	15	
B1	894	869	10	875	871	B1	896	1								θ(N-C_α-C_β and N-$C_{\alpha'}$-$C_{\beta'}$ in the same phase)/θ(C_m-C_α-N)/θ(C_α-C_m-C_α)/θ(phenyl)/ρ(C_βH)
B1	985	960	91	964	962	B1	968	5	B1	974	1	966	B1	998	2	W(CH on phenyl)
						B2	980	63	B2	980	100	984				νs(O-S-O)
B2	1001	977	34			B2	1002	5	B2	1004	2	1012	B2	1004	9	θ(N-C_α-C_β and N-$C_{\alpha'}$-$C_{\beta'}$ in the same phase)/θ(C_m-C_α-N)/θ(C_α-C_m-C_α)/θ(phenyl)/ρ(C_βH)
A1	1020	996	0.	979	980	A1	1020	0.	A1	1036	2	1039	A1	1035	1	Expansion of the pyrrole/pyrroline groups along N(H)…N(H) direction due to ν(C_α-C_β), leading to macrocycle getting rectangular shape instead of square shape.
B1	1014	991	11													θ(C-C-C in phenyl)
B2	1016	993	33	999	1002	B2	1016	1								ρ(C_βH)/θ(C-C-C in phenyl)/θ(C_m-C_α-N)
B1	1049	1026	9	1031	1032				B1	1033	<1					ρ(CH in phenyl)/θ(C-C-C in phenyl)
B1	1087	1065	9	1069	1072				B1	1089	1					ρ(C_βH)
A1	1094	1072	10													ρ(C_βH and CH on phenyl).
						B1	1109	21	B1	1108	11		B1	1107	3	ν(C-S)/θ(C-C(S)-C)/ρ(CH on phenyl only)
						A1	1116	14	A1	1125	6	1125	A1	1133	1	ρ(CH on phenyl only)
						B2	1146	64	B1	1151	30					νa(O-S-O)
						A1/B1	1156	100	A1/B1	1160	35	1188a	A1/B1	1151	100	νa(O-S-O)/ρ(CH on phenyl)
B2	1180	1159	38		1155	B2	1180	8								νa(C-N-C)/θ(C-N-C)/ ρ(C_βH)
B2	1189	1168	4	1174	1176	B2	1188	0.	B2	1193	9		B2	1200	3	ρ(CH on phenyl only)
B1	1204	1183	25	1187	1187	B1	1204	4								ρ(NH) and relatively weak θ(whole molecule)
B2	1224	1203	13	1211	1211	B2	1224	2	B2	1220	14	1218	B2	1221	2	νs(C-N-C)/νs(C_φ-C_m-C_α)/ρ(NH and CH)

TPP						TSPP			H$_4$TSPP				H$_8$TSPP			Assignments
Sym	$\Delta\nu_{sc}$-[a]	$\Delta\nu_{sc}$-[b]	I_{IR}	$\Delta\nu_{exp}$ [15]	$\Delta\nu_{exp}$ [16]	Sym	$\Delta\nu_{sc}$-[a]	I_{IR}	Sym	$\Delta\nu_{sc}$-[a]	I_{IR}	$\Delta\nu_{exp}$ [16]	Sym	$\Delta\nu_{sc}$-[a]	I_{IR}	
B1	1235	1215	17	1220	1219	B1	1235	3	B1	1240	12		B1	1241	7	va(C-N-C)/ρ(NH and C$_\beta$H)
B1	1262	1242	9	1247	1251	B1	1262	2	B1	1299	11		B1	1301	2	ν(C$_\beta$-C$_\alpha$)/θ(C-N-C)/ρ(C$_\beta$H and NH).
B1	1355	1337	47	1348	1351	B1	1354	8	B1	1348	1	1350	B1	1387	1	va(C$_\beta$-C$_\beta$-C$_\alpha$)/ θ(C$_\alpha$-N-C$_\alpha$)/va(C$_\varphi$-C$_m$-C$_\alpha$)/θ(C$_\varphi$-C$_m$-C$_\alpha$)/ρ(CH).
B1	1373	1356	24	1359	1358	B1	1373	4	B1	1386	5	1384	B1	1387	1	va(C-N-C)/ρ(C$_\beta$H and NH).
B2	1411	1394	22	1400	1400	B2	1411	4								ν(C$_\beta$-C$_\alpha$)/ν(C$_\alpha$-C$_m$) which also leading to va(C$_\beta$-C$_\alpha$-C$_m$), including ρ(C$_\beta$H).
A1	1451	1435	12	1459	1437	B2	1399	2	B1	1405	9	1395	B1	1407	6	ρ(CH on phenyl), including relatively weak vs(C-C-C$_\varphi$)
B2	1482	1466	82	1471	1471	B2	1482	16	B2	1460	62	1472	B2	1467	17	N(C$_\beta$-C$_\beta$)/ν(C$_\alpha$-C$_m$) that leading to θ(C-N-C)
B1	1492	1486	7	1488	1493											ν(C$_\alpha$-C$_m$)/ν(C$_\beta$-C$_{\beta'}$)/va(C$_\alpha$-NH-C$_\beta$), including ρ(CH) on the macrocycle only.
B1	1568	1554	67		1555	B1	1567	10	B1	1545	1	1554	B1	1545	3	
B1	1588	1574	1	1573	1575				B2	1566	1	1558 1564				va(C-C-C) within phenyl rings and ρ(H on phenyl)
B2	1612	1598	30	1595	1595	B2	1607	2	B1	1603	8	1592 1597	B2	1604	1	ν(C-C)/ρ(CH) within phenyl rings, including θ(C-C-C in phenyl).

Where $\Delta\nu_{sc}$ represents the scaled vibrational frequencies ((a) $\Delta\nu_{sc} = 0.96(\Delta\nu_{calc}) + 40$ as used for the Raman spectra for all compounds studied here) and I_{IR} symbolizes the predicted IR intensity. In the assignments, the symbols ν, θ, ρ, and w represent the bonding stretching, bending deformation, rocking, and wagging, respectively. It is worthy to note that two different scaling factor used for the TPP: (a) $\Delta\nu_{sc} = 0.96(\Delta\nu_{calc}) + 40$ and (b) $\Delta\nu_{sc} = 0.976(\Delta\nu_{calc})$. The latter one, (b), gives best fitting to measured IR spectrum (from ref. [15, 16]) of the TPP only; not for others. However, the scaling factor of $\Delta\nu_{sc} = 0.96(\Delta\nu_{calc}) + 40$ gives the best fitting to measured IR spectrum of H$_4$TSPP (from ref. [16]). The results of calculations were obtained in water used as solvent at the B3LYP/6-311G (d,p) level of the theory.

Table 3. Assigned IR features of the *meso*-substituted porphyrin derivatives: TPP (C_{2v} point group), TSPP (C_{2v}), H$_4$TSPP (C_{2v}), and H$_8$TSPP (C_{2v}).

http://dx.doi.org/10.5772/64582

Our assignments may be summarized as follows:

1. The observed Raman peak at 1593 cm^{-1}: The calculations, as shown in **Figure 1**, produce a peak at around 1600 cm^{-1} in the Raman spectra of all of the molecules studied here. The motions of atoms within the molecules suggest that this calculated band is as a result of principally C-C bond stretching, ν(C-C), within phenyl rings and rocking of their H, ρ(CH). No contribution appears to derive from the macrocycle and sulfonato groups (-SO_3^-) motions. Calculations for the parent porphyrin (FBP) and diprotonated FBP (H_4FBP) also predicted a band at about 1600 cm^{-1}, resulting from asymmetric stretching of the C_α-C_m-C_α (ν_a(C-C_m-C)) and bending deformation of the C-N(H)-C and C-N-C bonds, and rocking of H atoms covalent bonded to *meso*-carbon atoms (C_m), ρ(C_mH); see **Figure 2**. Consequently, we can conclude that the vibrational motion of the substituted phenyl is responsible for the observed Raman band at 1593 cm^{-1} in the observed spectrum of diprotonated TSPP (H_4TSPP). It is to be noted that, in our prior publication [13], the observed and predicted Raman spectra of the TPP showed the similar Raman pattern; these results are also presented in **Table 1** and **Figure 1**.

2. The experimental peak at 1563 cm^{-1}: Even though the measured Raman spectrum of the H_4TSPP displays a relatively weak bands at 1563 cm^{-1}, the calculation indicates only a very weak peak at 1568 cm^{-1}, attributed to C-C bond stretching within the phenyl rings and rocking of their attached H atoms, as well as a relatively weak asymmetric stretching of C_α-C_m-C_β bonds. However, there is no contribution from the sulfonato group. Upon examination of this peak for vibrational motions for the TPP, TSPP, H_4TPP, and H_8TSPP, we find extremely weak peaks at 1588, 1574, 1585, and 1578 cm^{-1}, respectively. As seen in **Figure 1** and **Table 1**, the calculated spectrum of the TSPP produces the most intense band at 1564 cm^{-1}, but its vibrational motions indicated that this mode is shifted to 1524 cm^{-1} in the H_4TSPP spectrum. Therefore, we believe that the DFT calculations might underestimate the intensity of this peak.

3. The observed strong Raman bands at 1553 cm^{-1} (with a shoulder at 1540 cm^{-1}) and at 1537 cm^{-1} (with a shoulder at 1528 cm^{-1}) are, respectively, in the spectra of TPP and H_4TSPP: Their calculated spectra display the strongest Raman band at 1564 and 1524 cm^{-1} for TPP and H_4TSPP, respectively. Both of experimental and calculated Raman spectra show that the strongest band in the observed spectrum of the TPP is significantly red shifted in the H_4TSPP. The large red shift in the peak position for the observed and calculated bands are, respectively, 16 cm^{-1} and ca. 40 cm^{-1}, and is not due to the substitution effect, which is mainly caused by the out-of-plane distortion of the macrocycle resulting from protonation of the N atoms at porphyrin core. Correspondingly, the Raman band associated with the same vibrational motions within FBP, TPP, and TSPP display a similar red shift when compared to diprotonated species (i.e., H_4FBP, H_4TPP, and H_4TSPP). Additionally, the observed shoulders at 1540 cm^{-1} (TPP) and at 1528 cm^{-1} (H_4TSPP) correspond to the calculated peaks at 1555 cm^{-1} and 1529 cm^{-1}, respectively, which results from the ν_a(C_α-C_m-C_α) and θ(C-N(H)-C motions, respectively.

4. Calculated Raman spectra of TPP, H_4TPP, TSPP, H_4TSPP, and H_8TSPP reveal a relatively strong peak around 1238 cm^{-1}, arising from predominantly ν(C_φ-C_m), as well as contribu-

tions from ν_s(C-N(H)-C), ρ(CH) and a comparatively weak $\nu(C_\beta\text{-}C_\beta)$. This vibrational mode is attributed to the measured Raman bands at 1229 cm^{-1} in H_4TSPP spectrum and 1234 cm^{-1} in the TPP. However, in calculated spectra of the unsubstituted free-base porphyrin (FBP and H_4FBP), this peak is respectively red shifted to 1194 and 1217 cm^{-1}, owing to the rocking of the H atom (covalent bounded to *meso*-carbon atom (C_m)), $\rho(C_mH)$, including vibrational bond stretching within the macrocycle (see **Figure 2**). There is a question here we need to answer that while the peak (at 1238 cm^{-1}) is not significantly shifted in the predicted Raman spectra of the diprotonated and/or *meso*-substituted porphyrin molecules, relative to each other, but it is substantially shifted in the FBP and H_4FBP spectra.

This may be explained by the electrostatic repulsive interactions or steric effect between the H atoms covalent bonded to the C_m and C_β atoms in the FBP and H_4FBP structures. The effect decreases with increasing in the distance between the H atoms on the C_β and C_m atoms because of the out-of-plane distortion from planarity in the H_4FBP molecule (diprotonated porphyrin). In the case of *meso*-phenyl or meso-sulfonatophenyl substituted porphyrin molecules, the steric effect between the H atoms on the C_β and C_p (in the *meso*-phenyl substituent) give rise to the rotation of these *meso*-substituted groups about C_m-C_φ bond, in their ground state structure, up to the tilt angle of about 71° and 48° for their unprotonated and protonated structures, respectively. Due to reduced electrostatic repulsion or steric effect by the C_m-C_φ bond rotation, the calculations do not reveal a substantial frequency shift in this peak position (~1238 cm^{-1}) in the *meso*-substituted porphyrin molecules.

(5) In region of 1050–950 cm^{-1}, there are two Raman peaks that are affected by diprotonated and deuterated parent porphyrin molecule. For instance, the observed two peaks at 1002 and 962 cm^{-1} in the TPP spectrum (exc. at 488 nm) are respectively blue shifted to 1016 and 1002 cm^{-1} in the H_4TSPP (exc. at 514 nm). Calculation indicates that the peaks at 1020 and 983 cm^{-1} in the TPP spectrum occurs at 1020 and 986 cm^{-1} in the TSPP. These same bands are blue shifted to 1036 and 1005 cm^{-1} in the calculated spectrum of the protonated-TSPP (H_4TSPP). Our results clearly show that these shifts in the observed peak positions are due to protonation of the porphyrin core that leads to saddle-type distortions of the porphyrin core (i.e., leads to an increase in the degree of freedom of the rocking of the N-H bonds as a consequence of the reduced repulsive interaction or steric effect between these hydrogen atoms). Moreover, GaussView visualization software shows that the peak at 1036 cm^{-1} (in H_4TSPP) is caused by expansion of the pyrrole groups along N(H)…N(H) direction, but in opposite phase (**Figure 2**), as a consequence of $\nu(C_\alpha\text{-}C_\beta)$, which causes the macrocycle to assume a rectangular shape rather than square shape. The band at 1005 cm^{-1} (H_4TSPP) is caused by expansion of the pyrroles along N(H)…N(H) direction likewise macrocycle breathing (or breathing of pyrroles in the same phase) as assigned by Rich and McHale [14].

(6) Another two fundamental Raman bands in the range of low frequency are found at 248 and 338 cm^{-1} in the H_4TSPP spectrum, and at 235 and 365 cm^{-1} in the TPP are respectively attributed to out-of-plane twisting of the macrocycle and breathing of whole molecule, which are in agreement with their experimental values of 242 and 338 cm^{-1} for the H_4TSPP; 235 and 334 cm^{-1} for the TPP. These blue and red shifted bands in the measured and predicted Raman spectrum of the H_4TSPP (dianionic or diprotonated-TSPP) result from the protonation of the

http://dx.doi.org/10.5772/64582

nitrogen atoms at the core, not owing to the *meso*-sulfonato substituted groups. Additional assignments are provided in **Table 1**.

3.1. Isotope effect on the Raman spectrum

The polarized resonance Raman scattering (RRS) spectra, exc. at 488 nm, of the aggregated H_4TSPP (diprotonated TSPP) and deuterated TSPP (D_2TSPP) by Rich and McHale [14] displayed a frequency shifts in the positions of some of the well-known Raman peaks, in addition to changes in the relative intensities of the Raman bands upon deuteration. The authors have reported that the observed Raman bands at 983 and 1013 cm^{-1} in the aggregated H_4TSPP (or diprotonated-TSPP) spectrum are respectively shifted to 957 and 1004 cm^{-1} in the aggregated D_4TSPP spectrum. Additionally, they suggested that these two modes are pyrrole breathing modes and thus these red shifts may be attributed to the substitution of deuterium ions with the labile protons in the porphyrin core [14].

By comparing the spectral positions of these two peaks in the calculated Raman spectra of diprotonated and deuterated porphyrin core with their corresponding nonprotonated ones (see **Table 2**), we see that while the protonated and *meso*-substituted parent porphyrin cause a blue shift in frequency of the two Raman peaks, the deuteration causes a red shift. For example, in the calculated spectrum of the TSPP, while these bands at 1020 and 985 cm^{-1} are blue shifted respectively to 1036 and 1005 cm^{-1} in the H_4TSPP (diprotonated TSPP), they are red shifted to 1012 and 977 cm^{-1} in the D_2TSPP (deuterated TSPP) spectrum, respectively. When all four nitrogen atoms at the porphyrin core are deuterated, these Raman bands are shifted from 1036 and 1005 cm^{-1} (in the H_4TSPP) to 1026 and 983 cm^{-1} in the D_4TSPP (deuterated H_4TSPP), respectively.

For the other moderately intense Raman peaks in the predicted spectrum, the shift in their spectral positions, due to the deuterated nitrogen atoms at the core, is not more than 5 cm^{-1}, which is in agreement with the experimental observation [14]. However, there are several weaker bands in the calculated spectra that displayed a significant shift in frequency (**Figure 1**). The other *meso*-substituted and free-base porphyrin derivatives displayed analogous results, which are in agreement with the experimental findings as argued above. Moreover, the results of calculations suggest that the *meso*-substituted groups do not significantly alter the spectral position of these two Raman bands.

Consequently, the blue shift of the two Raman bands associated with diprotonated nitrogen atoms at the core is not unexpected when considering the steric effect (or electrostatic repulsive effect) between the hydrogen atoms bounded to nitrogen atoms at the core. This effect may be reduced by departing from the planarity of the porphyrin core (or macrocycle) as argued earlier. The red shift also is to be expected because of the isotopic effect since the vibrational frequency is inversely related to the square root of atomic mass that contributes to the vibrational mode. The deuterated nitrogen atoms at the core and one of three oxygen atoms in each of four sulfonato groups (-SO_3D) revealed new Raman peaks in the 2630–2720 cm^{-1} region, which could be an experimental evidence for the presence of the deuterated TSPP (D_4TSPP) or deuterated H_4TSPP (D_8TSPP) in samples.

4. IR spectra of porphyrin and derivatives

We also calculated (at the same level of the DFT) the IR spectra of FBP, TPP, TSPP, H_4FBP, H_4TPP, H_4TSPP, and H_8TSPP, as well as their deuterated structures (D_2FBP, D_2TPP, D_2TSPP, D_4FBP, D_4TPP, D_4TSPP, and D_8TSPP). It is worthy to note that the D_8 represent that the four of eight deuterium atoms covalent bounded to nitrogen at the core and another four bounded to four sulfonato groups (-SO_3D). Calculated spectra exhibit dispersed about the full spectral range many IR features with medium and relatively weak intense, in addition to intense IR bands (see **Figure 3**). The results of the calculations together with their animated motions indicate that the predicted IR vibrational modes are predominantly linked with: (1) symmetric and asymmetric skeletal deformations of the macrocycle and phenyl rings; (2) wagging and rocking of the hydrogen atoms bonded to carbon and nitrogen atoms, CH and NH; and (3) out-of-plane distortion of the phenyl rings and the parent porphyrin or macrocycle. The selected IR bands in the calculated spectra of these compounds studied here are assigned using GaussView animation software. The assigned IR features are given in **Table 3**.

To test the reliability of the calculated IR spectra of the molecules investigated, we compared the IR bands in the calculated spectra of TPP and H_4TSPP (diprotonated TSPP) with experimentally measured IR spectra of TPP [15] and H_4TSPP [16]; as seen in **Table 3**, the spectra correlate quite well. This analysis indicates that the calculated IR spectra of these compounds (FBP/H_4FBP, TPP/H_4TPP, and TSPP/H_8TSPP) are reasonable. We assigned the predicted IR features for the H_4TSPP in connection with the predicted IR spectra of the FBP/H_4FBP, TPP/

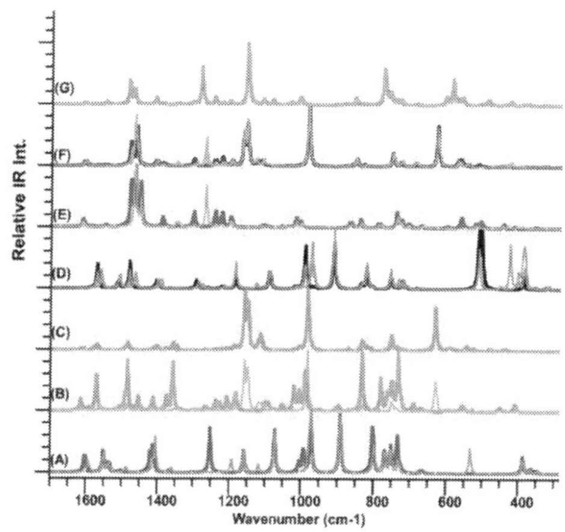

Figure 3. The predicted IR spectra of parent porphyrin and its derivatives: (**A**) free-base porphyrin (FBP) and deuterated FBP (D_2FBP); (**B**) *meso*-tetraphenylporphyrin (TPP) and (D_2TPP); (**C**) anionic *meso*-tetrakis(*p*-sulfonatophenyl)porphyrin (TSPP) and deuterated TSPP (D_2TSPP), and (**D**) diprotonated FBP (H_4FBP) and deuterated H_4FBP (D_4FBP); (**E**) diprotonated TPP (H_4TPP) and deuterated H_4TPP (D_4TPP); (**F**) diprotonated TSPP (H_4TSPP) and deuterated H_4TSPP (D_4TSPP); and (**G**) dicationic TSPP (H_8TSPP) and deuterated H_8TSPP (D_8TSPP). It should be noted that the plotted IR spectra (grey in color) corresponds to that for their deuterated structures. The calculations were carried out in water used as solvent at the B3LYP/6-311G(d, p) level of the theory.

H_4TPP, and TSPP/H_4TSPP/H_8TSPP (in water used as solvent) by taking into account their vibrational motions. Our key conclusions concerning the calculated IR spectra are as following:

1. The predicted IR features at 1603, 1566, 1405, 1193, 1125, 974, and 439 cm^{-1} in the H_4TSPP spectrum are attributable to structural distortion of the *meso*-phenyl substitution, e.g., rocking and/or wagging of hydrogen atoms and bond stretching, no contribution is derived from motions of the sulfonato groups, $-SO_3$.

2. The following IR bands arise from the vibrational motion of the $-SO_3$ groups: the most intense IR band at 980 cm^{-1} in the calculated spectrum of the H_4TSPP, which originates from the symmetric stretching of O-S-O bonds, ν_s(O-S-O); one medium intensity at 1151 cm^{-1}, caused by asymmetric stretching of O-S-O, ν_a(O-S-O); and a moderately intense band at 624 cm^{-1} is as a result of bending distortion of the $-SO_3$ groups, as would describe the closing and opening of an umbrella.

3. A relatively strong IR peak at 1160 cm^{-1} in the spectrum of the H_4TSPP is caused by ν_a(O-S-O) and rocking of CH in phenyl rings, ρ(CH on phenyl).

4. The vibrational motion of the *meso*-sulfonatophenyl rings, ν(S-C), θ(C-C(S)-C) and ρ(CH on phenyl only), produced an IR band with very weak intensity at 1108 cm^{-1}.

5. While the ν(S-C), θ(bending distortion of phenyl), relatively weak wagging of the CH and NH, and twisting of the macrocycle induced an IR peaks at 748 cm^{-1}; twisting of the entire molecule, including bending distortion of the O-S-O bonds and wagging of the CH and NH bonds produced an IR band with very weak intensity at 566 cm^{-1}.

6. The two IR bands with frequencies of ca. 510 and 559 cm^{-1} are indicated as associated with hydrogen atom motions (NH), and are very weak in intensity.

7. A very weak peak at 1089 cm^{-1} is due to the $\rho(C_\beta H)$, which appears essentially at the same spectral position in the calculated IR spectra for the other molecular structures studied here.

8. The calculated IR features, with very weak intensity, at 1386, 1299, 1240, 1036, 848, 825, and 751 cm-1 arise from the bond stretching (ν), rocking (ρ), wagging (w), and bending deformation (θ) of the C and H atoms within macrocycle. These calculated features are found to be spectrally shifted from those in protonated porphyrins, which is in agreement with experimental observation as provided in **Table 3** from references [15] and [16].

9. The peaks at 1004 (weak), 1460 (strong), and 1545 (weak) cm^{-1} are results of symmetric/asymmetric bond stretching, bending deformation, and/or wagging/rocking vibrational motion of the atoms within the porphyrin macrocycle.

10. While in-plane rotational motion of the pyrroline rings, including relatively weak out-of-plane twisting deformation of the phenyl rings, produced weak peak at 439 cm^{-1}; and rocking of phenyl rings and wagging of macrocycle induced a weak IR peak at 427 cm^{-1} in the H_4TSSP spectrum. Another weak one is found at 484 cm^{-1} that originated from the twisting of phenyl and wagging of macrocycle. Complete descriptions for each IR features are provided in **Table 3**.

4.1. Isotopic (or deuteration) effect on the IR spectrum

Calculated IR spectra of the molecules clearly indicate that deuterated porphyrin exhibits relatively intense IR peaks in the range of 2565–2600 cm^{-1}. Such bands are associated with N-D bond stretching. And bands around 2640 cm^{-1} are attributable to O-D bond stretching. The largest frequency shifts are calculated for bands at 540 and 490 cm^{-1} (H_4TSPP, where all N atoms at porphyrin core are protonated) that are shifted to 396 and 366 cm^{-1} in the D_4HTSPP (deuterated-H_4TSPP); in the low-frequency region (below 700 cm^{-1}) bands are attributable to wagging of the N-D bond, w(ND).

In the region of high or mid frequency, where deuterium atom is included in vibrational mode frequency, a red shift in frequency by up to 10 cm^{-1} is shown in the D_2TSPP. Deuteration also has an influence on the intensity of the IR bands; see **Figure 3**. Shift in the region of high frequency of D_2FBP, D_4FBP, and D_4TPP spectra are more significant than those in the spectra of the D_2TSPP and D_4TSPP. These results imply that above the low-frequency region, the frequency shifts as a result of the deuteration decrease with increasing size of the substituent group.

5. Solvent effect on the IR spectrum

We investigated the solvent effect on the IR spectrum of the H_4TSPP by using toluene, dimethyl sulfoxide (DMSO) and water as a solvent. Calculations indicate that below 1100 cm^{-1} there is no significant frequency, shift in peak positions H_4TSPP, but above 1100 cm^{-1} shifts do occur. Specifically, the IR peak centered at around 1200 cm^{-1} is shifted to 1188 cm^{-1} (toluene), 1166 cm^{-1} (DMSO), and 1160 cm^{-1} (water). Also, the IR peaks centered about 1453 and 1477 cm^{-1} in the gas phase spectrum are shifted to 1468 and 1490 cm^{-1} (toluene), 1480 and 1497 cm^{-1} (DMSO), and 1481 and 1499 cm^{-1} (water), respectively. This observation suggests that the IR features, especially in high energy region, of the parent porphyrin and its derivatives, at least for H_4TSPP, are responsive to its surroundings.

6. Resonance Raman spectra of aggregated diprotonated-TSPP

In the section, we will discuss the results vibroelectronic properties of the aggregated-H_4TSPP (known as acidic-, dianionic-, or diprotonated-TSPP). Several structural and spectroscopic studies have shown that TSPP aggregates in acidic aqueous solution. **Figure 4** shows the absorption spectra of free-base TSPP (pH = 12), H_4TSPP (pH = 4.5), and aggregated H_4TSPP (pH = 1.6), with concentration of 5×10^{-5} M in aqueous solution.

While the spectrum of the TSPP [17] exhibited a Soret band at 410 nm and several weak Q bands in the region of 500–640 nm, the monomeric H_4TSPP spectrum exhibited the Soret band at 432 nm, and Q-bands at 594 and 642 nm. **Figure 4** shows the Soret band of the H_4TSPP is split into H- and J-band components in the H_4TSPP aggregate as a sharp and intense absorption

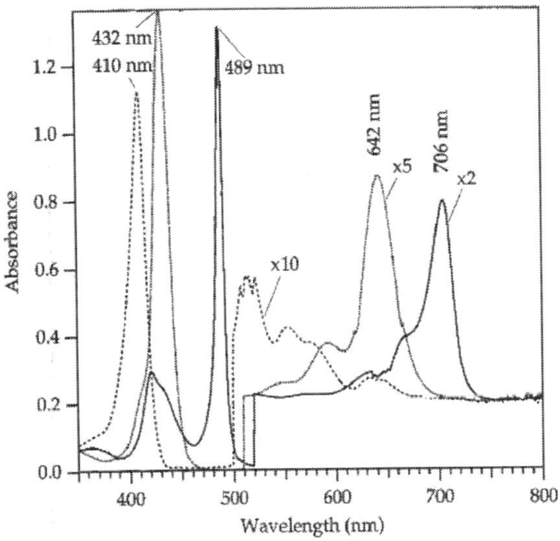

Figure 4. UV-vis spectra of the free-base TSPP (the maximum of the absorption band at 410 nm), monomeric H_4TSPP (the maximum at 432 nm) and aggregated H_4TSPP (the maximum at 489 nm) [20]. The concentration of compounds in each case is 5 × □ 10^{-5} M in the aqueous solution. The pH = 12 for the free-base TSPP; pH = 4.5 for the monomeric H_4TSPP; pH = 1.6 for the aggregated H_4TSPP, and [KCl] = 0.1 M. Spectra above ca. 500 nm have been offset by +0.2 absorbance units and amplified by the indicated factor to aid presentation. The structure on the band at around 500 nm for the free-base TSPP is an artifact attributable to the absorption spectrometer.

band at 489 nm (J-aggregate) and a broad and weak absorption band at 422 nm (H-aggregate) are formed. The Q-bands at 594 and 642 nm (in the monomeric H_4TSPP) is also shifted to ca. 670 and 706 nm in the aggregated H_4TSPP spectrum, respectively. The UV-vis spectra results suggest that aggregation evolves through the formation of diprotonated TSPP (H_4TSPP), and only occurs at a pH below 5. These observations have also reported by other researchers [18, 19].

Figure 5 presents the Raman spectra of free-base (TSPP), monomeric dianion (H_4TSPP) and aggregated H_4TSPP resonantly excited at their respective Soret-band absorption wavelengths (**Figure 4**). Analysis of the resonance Raman (RR) spectra of the H_4TSPP and aggregated H_4TSPP reveal the presence of bands that do not accommodate bands of the free-base porphyrin. However, an in-depth examination indicates that subtle differences bands of the aggregate and the dianionic monomer are correlated. The most important correlation in spectra is found in the low-frequency region, where two bands of dianionic TSPP monomer at 233 and 310 cm^{-1} correlate with two dramatically enhanced aggregate bands at 241 and 317 cm^{-1}, in addition to two weak satellite bands at 205 and 362 cm^{-1}.

Moreover, DFT calculations (at B3LYP/6-311G(d, p) level) show that while the band at 317 cm^{-1} is due to breathing of the whole molecule, the enhanced band at 241 cm^{-1} results from the out-of-plane wagging of the macrocycle. Hence, computationally these two Raman bands of porphyrins originate from out-of-plane modes—in conjunction with bending of the C_m-ph

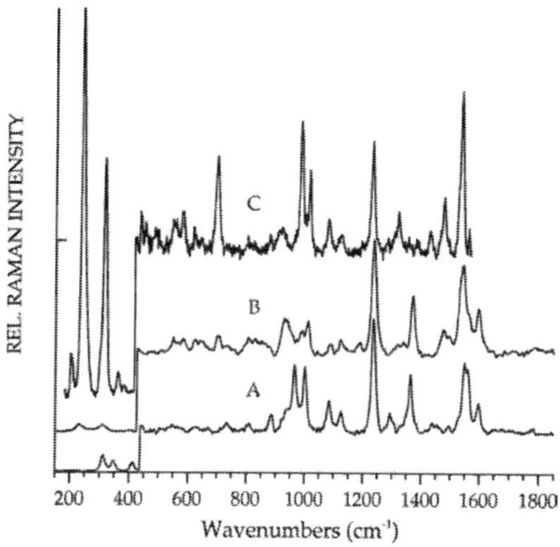

Figure 5. Resonance Raman spectrum (RRS) of the TSPP [20]: (A) free-base TSPP, exc. at = 416 nm; (B) diprotonated TSPP (H_4TSPP), exc. at 432 nm; (C) aggregated H_4TSPP, exc. at 488 nm. Where the solutions used here were the same as used for the measured absorption spectra in **Figure 4**.

bond (ph representing phenyl) and deformation of the core of the porphinato macrocycle caused by pyrrole ring tilt and swivel [20–22]. Moreover, in the case of lanthanide sandwich dimer porphyrins, low-frequency Raman bands have been hypothesized to reflex the degree of intramolecular π-π interaction and to be connected with the intradimer vibration that modulates the separation between the two porphyrin moieties, or owing to symmetrical linear combinations of out-of-plane distortions of the neighboring porphinato macrocycles [23].

Given the enhancement of scattering associated with bands having motions that can strongly couple with excitonic motion, which helps define the aggregate's structure, we report in **Table 1** assignments made through such a scheme and by comparison to the literature. It is to be noted that a theoretical construct known as "aggregation-enhanced Raman scattering (AERS)," has been advanced by one of us (DLA) to explain which bands would experience significant enhancement upon aggregation of the scattering species. And, indeed, the two bands discussed above that show enormous enhancement upon aggregation of H_4TSPP have motions that couple to exciton movement through the aggregate and would be expected to experience significantly enhanced Raman intensities [12].

Acknowledgements

We would like to thank Ömer Andaç (Chemistry Department of Ondokuz Mayıs University) for kindly providing computing facilities and software expertise. We also thank staff at TUBITAK ULAKBIM, the high performance and Grid Computing Center (TR-Grid e-Infra-

http://dx.doi.org/10.5772/64582

structure), for performing the calculations reported herein. Also, we would like to thank the U.S. National Science Foundation (NSF) for support of research efforts under grant No. HRD-08-33180.

Author details

Metin Aydin[1*] and Daniel L. Akins[2]

*Address all correspondence to: aydn123@netscape.net

1 Department of Chemistry, Faculty of Art and Sciences, Ondokuz Mayıs University, Samsun, Turkey

2 Department of Chemistry and Biochemistry, Center for Analysis of Structures and Interfaces (CASI), The City College of the City University of New York, New York, USA

References

[1] Shelnutt, J. A. Thesis: Resonance Raman spectroscopy of manganese (III) Etioporpyrin I: Theory and experiment, Georgia Institute of Technology, November, 1975.

[2] Albrecht, A. C. On the theory of Raman intensities, J. Chem. Phys., 1961, 34, 1476–1484.

[3] Warshel, A.; Dauber, P. Calculations of resonance Raman spectra of conjugated molecules, J. Chem. Phys., 1977, 66, 5477–5488.

[4] Polavarapu, P. L. Ab initio vibrational Raman and Raman optical activity spectra, J. Phys. Chem., 1990, 94, 8106–8112.

[5] O'Boyle, N. M.; Tenderholt, A. L.; Langner, K. M. cclib: A library for package-independent computational chemistry algorithms, J. Comp. Chem., 2008, 29, 839–845.

[6] Kasha, M. Energy transfer mechanisms and the molecular exciton model for molecular aggregates, Radiat. Res. 1963, 20, 55.

[7] Okamura, M. Y.; Feher, G.; Nelson, N. Chapter 5: Reaction centers in *Photosynthesis*, Govindjee, Ed. Academic Press: New York, 1982, pp. 195–272.

[8] Pearlstein, R. M. Amesz, J. (Ed.) Chapter 13: Structure and exciton effects in photosynthesis. Elsevier: Amsterdam, 1987, pp. 299–317.

[9] O'Neil, M. P.; Niemcyzk, M. P.; Svec, W. A.; Gosztola, D.; Gaines III, G. L.; Wasielewski,M. R. Picosecond optical switching based on biphotonic excitation of an electron donor-acceptor-donor molecule, Science, 1992, 257, 63.

[10] Wagner, R. W.; Lindsey, J. S.; Seth, J.; Palaniappan, V.; Bocian, D. F. Molecular optoelectronic Gates, J. Am. Chem. Soc., 1996, 118, 3996.

[11] Aydin, M. Comparative study of the structural and vibroelectronic properties of porphyrin and its derivatives. Molecules, 2014, 19, 20988–21021.

[12] Akins, Daniel L., Enhanced Raman scattering by molecular nanoaggregates (Invited Review Article), nanomater. Nanotechnol, 2014, 4, 4. DOI: 10.5772/58403

[13] Aydin, M. DFT and Raman spectroscopy of porphyrin derivatives: Tetraphenylporphine (TPP). Vib. Spectrosc. 2013, 68, 141–152.

[14] Rich, C. C.; McHale, J. L. Influence of hydrogen bonding on excitonic coupling and hierarchal structure of a light-harvesting porphyrin aggregate. Phys. Chem. Chem. Phys. 2012, 14, 2362–2374.

[15] Mi, Z.; Guo-Hui, L.; Shu-Qin, G.; Zuo-wei, Li. Infrared spectroscopic investigations of J-aggregates of protonated tetraphenylporphine. Chem. Res. Chin. Univ., 2009, 25, 257–260.

[16] Zhang, Y.H.; Chen, D.M.; He, T.; Liu, F.C. Raman and infrared spectral study of meso-sulfonatophenyl substituted porphyrins (TPPSn, n_/1, 2A, 2O, 3, 4). Spectrochim. Acta Part A, 2003, 59, 87–101.

[17] Akins, D. L.; Zhu, H.R.; Guo, C., Absorption and Raman scattering by aggregated meso-tetrakis(p-sulfonatophenyl) porphine. J. Chem. Phys., 1994, 98, 3612–3618.

[18] Pasternack, R. F.; Huber, P. R.; Boyd, Engasser, P.; Francesconi, G.; Gibbs, L.; Fasella, E.; Venturo, P.; Cerio, G.; Hinds, L. deC. On the aggregation of meso-substituted water-soluble porphyrins, J., Am. Chem. Soc., 1972, 94, 4451.

[19] Ohno, O.; Kaizu, Y.; Kobayashi, H. *J*-aggregate formation of a water-soluble porphyrin in acidic aqueous media, J. Chem. Phys., 1993, 99, 4128.

[20] Akins, D. L.; Özçelik, S.; Zhu, H. R.; Guo, C. Fluorescence decay kinetics and structure of aggregated tetrakis(*p*-Sulfonatophenyl)porphyrin, J. Chem. Phys. 1996, 100, 14390–14396.

[21] Choi, S.; Spiro, T. G. Out-of-plane deformation modes in the resonance raman spectra of metalloporphyrins and heme proteins, J. Am. Chem. Soc. 1983, 105, 3683.

[22] Alden, R. G.; Crawford, B. A.; Doolen, R.; Ondrias, M. R.; Shelnutt. J. A. Ruffling of nickel(II) octaethylporphyrin in solution, J. Am. Chem. Soc. 1989, 111, 2070.

[23] Donohoe, R. J.; Duchowski, J. K.; Bocian, D. F. Hole delocalization in oxidized cerium(IV) porphyrin sandwich complexes, J. Am. Chem. Soc. 1988, 110, 6119.

Chapter 8

Novel Pressure-Induced Molecular Transformations Probed by *In Situ* Vibrational Spectroscopy

Yang Song

Additional information is available at the end of the chapter

http://dx.doi.org/10.5772/64617

Abstract

Pressure-induced structural change in molecular systems has demonstrated strong promises to access previously unexplored, novel structures and new properties in molecular materials with practical applications. The *in situ* structural characterization is of fundamental importance to understand the exotic structures and the possible transformation mechanisms. Among all the spectroscopic probes, vibrational spectroscopy that include Raman and Fourier-transform infrared (FTIR) spectroscopy and microscopy allow for highly efficient, sensitive and qualitative characterization of pressure-induced new structures and transformation processes *in situ*. Supported by state-of-the-art, highly customized spectroscopic systems in-house and at synchrotron facilities, molecular structures and materials properties can be probed in a broad pressure-temperature range with very high spectral and spatial resolutions. Complementary to each other, Raman and IR spectroscopy provide valuable information in molecular structures, nature of bonding, lattice dynamics as well as intermolecular interactions. In this chapter, a comprehensive and critical review of examples of pressure-induced molecular transformations in a wide variety of molecules and materials probed by vibrational spectroscopy is provided. The purpose of this chapter is to give readers the most recent advances in high-pressure chemistry and materials research by demonstrating the power of vibrational spectroscopy as a highly effective *in situ* structural characterization tool.

Keywords: high pressure, diamond anvil cells, Raman spectroscopy, FTIR spectroscopy, conformation, hydrogen bonding, phase transitions, guest-host interactions

1. Introduction

As one of three principal thermodynamic variables, pressure (P) plays an important role to alter the interatomic distances and thus the nature of intermolecular interactions, chemical bonding, molecular configurations, crystal structures and stability of materials. Extreme pressure can even induce transformations involving the strongest chemical interactions that exceed 10 eV (965 kJ mol^{-1}) such that chemical bonds and even the well-known properties of atoms and molecules can be completely changed. As a result, investigation of pressure-induced structural transformations and formation of novel functional materials has become a vibrant frontier in chemistry and materials science [1]. On one hand, major advances in high-pressure techniques such as diamond anvil cell have allowed the study of molecules and materials in an unprecedented pressure-temperature (P-T) range. On the other hand, the compatible micro-spectroscopic probes have made possible the characterization of structures and transformational processes *in situ* with great spectral and spatial resolution. The recent advances in high-pressure science and technology and their applications in materials research have been provided in several excellent review articles [2–8].

Among all the available *in situ* structural characterization probes for materials under extreme conditions, in particular, vibrational spectroscopy that include Raman and Infrared (IR) spectroscopy has demonstrated strong sensitivity and accuracy as well as efficiency in monitoring the pressure-induced transformations. Raman and IR spectroscopy, complementary to each other, provide valuable information on molecular structures, nature of bonding, lattice dynamics as well as intermolecular interactions. In this chapter, a comprehensive and critical review of examples of pressure-induced molecular transformations in a wide variety of molecules and materials probed by vibrational spectroscopy is given. The examples include (1) conformational change; (2) pressure-mediated hydrogen bonding; (3) phase and structural transitions; (4) pressure-induced chemical reactions; and (5) porous materials and guest-host interactions. Through these examples, the readers are provided the most recent advances in high-pressure chemistry and materials research by demonstrating the power of vibrational spectroscopy as an effective tool for structural characterizations for materials under extreme conditions.

2. Experimental methods

2.1. The diamond anvil cell

The recent advances in high-pressure technology have enabled the generation of extreme conditions in a broad P-T range with great controllability and accuracy. In particular, diamond anvil cell (DAC) is a fundamental apparatus to achieve static high pressures. Diamonds are known as the hardest material in nature and thus suitable as anvils to generate very high pressure. Moreover, diamonds are transparent to a wide spectral range of electromagnetic radiation from far-IR to hard X-ray. As a result, various analytical probes, including optical spectroscopy, synchrotron and neutron sources, have enabled structural characterization of material under extreme P-T conditions with unprecedented spatial, temporal and spectral resolutions.

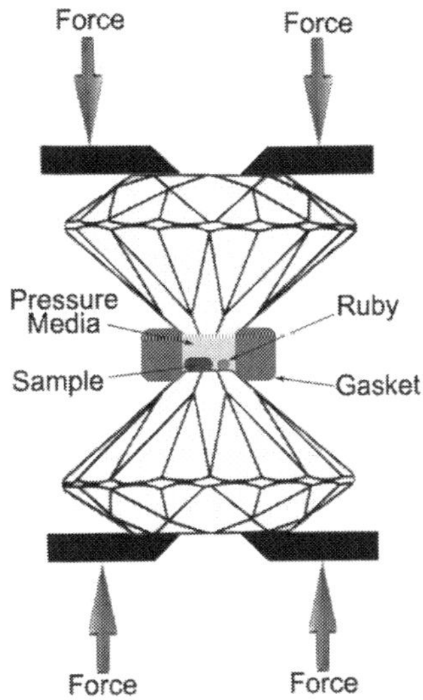

Figure 1. Schematic diagram of a diamond anvil cell.

Figure 1 shows a typical DAC apparatus where two brilliant cut diamonds are used as anvils to exert static pressure up to several million atmospheres (or several hundred GPa) with only moderate force. A metal gasket with a hole drilled at the center serves as the sample chamber. Most of the time the sample to be studied is loaded together with pressure-transmitting medium (PTM) to enhance the hydrostaticity, and a ruby chip which is used for pressure calibration. The extreme pressures can be accurately determined by monitoring ruby fluorescence lines using the following relationship [9]:

$$P = \frac{1904}{B}\left[\left(1 + \frac{\Delta\lambda}{694.24}\right)^{B} - 1\right] \tag{1}$$

where P is the pressure in GPa, $\Delta\lambda$ is the ruby R_1 line shift in nm, and parameter B is 7.665 for quasi-hydrostatic conditions and is 5 for non-hydrostatic conditions. The ruby fluorescence can be conveniently collected using a Raman system such as described below.

To conduct vibrational spectroscopy on materials loaded in DAC effectively, optical transparency is a prime factor in selecting diamond anvils. Two types of diamonds (i.e., type I and type II) are typically used for different spectroscopic probes. Both types have the intense first-order

Raman line at 1332 cm^{-1} (F_{2g} mode of the diamond). The difference between the two types is in the infrared absorption spectrum. Type I anvils (with more nitrogen impurities) have two strong IR absorption regions around 2000 and 1000–1350 cm^{-1}, respectively. In contrast, type II anvils (nitrogen free) have a relatively clean IR window bellow 2000 cm^{-1} allowing effective IR absorption measurements on samples. Therefore, the low-fluorescent type I diamonds are only suitable for Raman spectroscopy, while the type II diamonds are mainly used in IR spectroscopy.

2.2. Raman spectroscopy

Raman spectroscopy is a vibrational spectroscopy based on the inelastic scattering of visible photons (typically from a laser source) by materials, a process with a much smaller cross-section than other spectroscopic processes (e.g., absorption, fluorescence, etc.). Although many commercial Raman microscopy systems are available, they generally have a rigid design that does not allow *in situ* measurements with different DAC configurations. Therefore, a state-of-the-art customized Raman system was constructed to allow the DAC-based measurements in a broad P-T range with multiple excitation laser sources that cover the spectral range from near UV to near IR, such as 488–514 nm lines from an Innova Ar^+ laser (Coherent Inc.), 532 and 782 mn lines from diode-pumped solid-state lasers, as well as 700–1100 nm lines from a Ti: sapphire laser (Spectral Physics) [10]. Using a 20× Mitutoyo objective, the laser can be focused to less than 5 μm on the sample. The combination of a 15× eyepiece and a digital camera allows precise alignment of the focused laser beam on the sample. With backscattering geometry, the Raman signal is collected by the same objective lens. The elastic Rayleigh scattering is removed by either a pair of notch filters or an edge filter that enabled a spectral range above 100 cm^{-1} to be measured before the total scattered photons are focused on the entrance slit of a spectrometer. The scattered light is then dispersed using an imaging spectrograph (SpectroPro-2500i, Acton Research Corporation) that houses a 0.5 m focal distance monochromator equipped with multiple gratings, such as a 1800 lines/mm grating, allowing a spectral resolution of ±0.1 cm^{-1} to be achieved. The Raman signal was recorded using an ultrasensitive back-illuminated, liquid nitrogen cooled, charge-coupled device (CCD) detector from Acton. The Raman system is first calibrated by using a neon lamp giving an uncertainty of ±1 cm^{-1} before each experiment.

2.3. FTIR spectroscopy

Complementary to Raman spectroscopy, IR absorption spectroscopy provides sensitive and fingerprints information on materials loaded in DAC, especially those with high fluorescence that prohibits effective Raman measurements. The IR measurements for the examples demonstrated in this chapter were mostly carried out using a customized IR micro-spectroscopy system constructed in-house [11]. Specifically, a commercial FTIR spectrometer (model Vertex 80v from Bruker Optics Inc.) containing a Globar IR source constitutes the major component of the micro-IR system. The spectrometer is operated under a vacuum of <5 mbar to efficiently remove the absorption by H_2O and CO_2. The IR beam is collimated with varying diameters achieved by using apertures from 0.25 to 8 mm, and then is directed into a relay box through a KBr window. Using the combination of iris optics and 15× reflective objective lens (numerical

http://dx.doi.org/10.5772/64617

aperture of 0.4), the IR beam is then focused onto the sample in the DAC. Using an XYZ precision stage with the aid of an optical microscope equipped with a 20× eyepiece from Edmond Optics and an objective lens of variable magnifications, the sample loaded in DAC can be easily aligned to allow the maximum transmission of the IR beam. Using a series of iris apertures, the size of the IR beam was set to be identical to the entire sample size (e.g., ~200 μm). Another identical reflective objective as the condenser is used to collect the transmitted IR beam, which is subsequently directed to a midband mercury cadmium telluride (MCT) detector. A ZnSe window equipped on the midband MCT detector allows efficient measurements in the spectral range of 600–12,000 cm^{-1}. The combination of 512 scans and a resolution setting of 4 cm^{-1} is typically used for each spectrum measurement that gives an excellent signal-to-noise ratio. The absorption of diamond anvils loaded with KBr but without any sample is used as the reference spectrum, which is divided as background from each sample spectrum to obtain the absorbance.

2.4. Synchrotron-based FTIR spectroscopy

Synchrotron light is a source of electromagnetic radiation produced by a storage ring housing traveling electrons with a near speed of light. Although synchrotron source provides enormous advantages typically in the X-ray region, the infrared synchrotron light has unique applications for DAC-based measurements due to the very intense, very broad and highly focused IR source that allows very high spatial resolution and far-IR measurements. Some examples in this chapter are based on the experiments performed at the U2A beamline at the National Synchrotron Light Source (NSLS) of Brookhaven National Laboratory (BNL). Briefly, the IR beam from the synchrotron storage ring is first extracted through a wedged diamond window from a source with a 40 × 40 mrad solid angle. Then it is collimated to a 1.5" diameter beam and directed into a vacuum FTIR spectrometer (Bruker IFS 66V) equipped with three independent microscope systems. The spectrometer is equipped with a number of combinations of IR beam splitters and detectors (e.g., silicon bolometer and MCT). For mid-IR measurements, a Bruker IR microscope is used to focus the IR beam onto the sample. The absorption spectrum is collected in transmission mode by the MCT detector in the spectral range of 600–4000 cm^{-1}. The far-IR spectra are collected using a customized IR microscope allowing very high collection efficiency and recorded by the bolometer in the spectral region of 100 to 600 cm^{-1}. A resolution of 4 cm^{-1} was used in all IR measurements. For all measurements, mid-IR spectra were collected through a 30 × 30 μm^2 aperture, whereas the effective IR transmission area covered the entire sample (i.e., a circle of about 90 μm in diameter) for the far-IR measurements. The data acquisition, processing and analysis are similar to those obtained using the in-house mid-IR spectroscopy system.

3. Pressure-induced conformational change

Pressure-mediated conformational equilibrium is of particular interests because the reactivity of many organic reagents, product yields, and even reaction pathways are strongly correlated with molecular conformations. Here, two simple halogen substituted alkane molecules, that is, 1,2-dichloroethane (DCE) [12] and chlorocyclohexane (CCH) [13], were investigated under high pressures for conformational and structural changes using *in situ* Raman spectroscopy.

3.1. 1,2-Dichloroethane

As a model molecule, DCE has two representative conformations, that is, *gauche* and *trans* depending on the relative orientations of the halogens attached to the two carbons, making it an interesting candidate for conformational studies under high pressures. At pressures below 0.6 GPa, fluid DCE exhibits two conformations, that is, *gauche* and *trans* in equilibrium. Upon compression, the equilibrium appears to shift toward *gauche* conformation (**Figure 2**). Upon further compression, DCE was found to transform to a solid phase with exclusive *trans* conformation. In fluid phase, all the characteristic Raman shifts remain constant whereas in the solid phase they move to higher frequencies with increasing pressure. At about 4–5 GPa, DCE transforms into a crystalline phase from a possible disordered phase as indicated by the appearance of several new lattice modes and bandwidth narrowing. Dramatic changes in Raman spectra of DCE were observed when compressed to ~8–9 GPa. For instance, the C—C—Cl bending mode at 325 cm^{-1} splits, the inactive internal mode at 684 cm^{-1} becomes observable, and new lattice modes appear. All these observations suggest another pressure-induced phase transformation. Significant changes in pressure dependence of representative Raman modes at the distinctive pressures further confirm the transition and allow the identification of phase boundaries. Although with a likely lower symmetry, the new phase remains crystalline. The transformations are found reversible in the entire pressure region upon decompression. Quantitative analysis on Raman intensities associated with each conformer even allows the determination of the transformation volume of 0.58 ± 0.10 cm^3/mol (**Figure 3**).

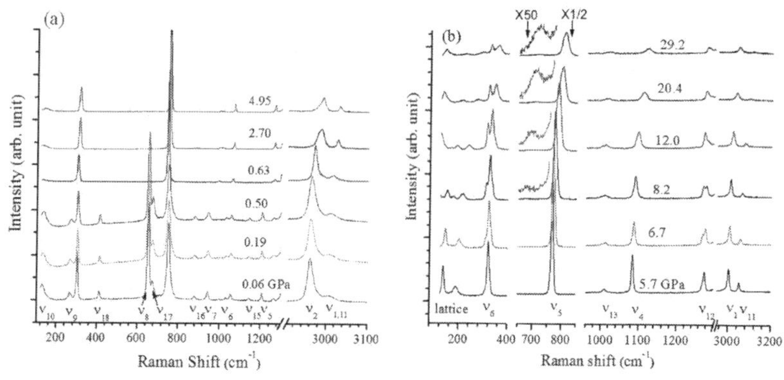

Figure 2. Representative Raman spectra of DCE on compression in the pressure region of (a) 0–5.0 GPa and (b) 5.7–29.2 GPa and the spectral region of 120–1300 cm^{-1} and 2900–3100 cm^{-1}. The assignments of Raman active mode are labeled below for *gauche* (a) and *trans* (b) conformations. Reproduced with permission from reference [12].

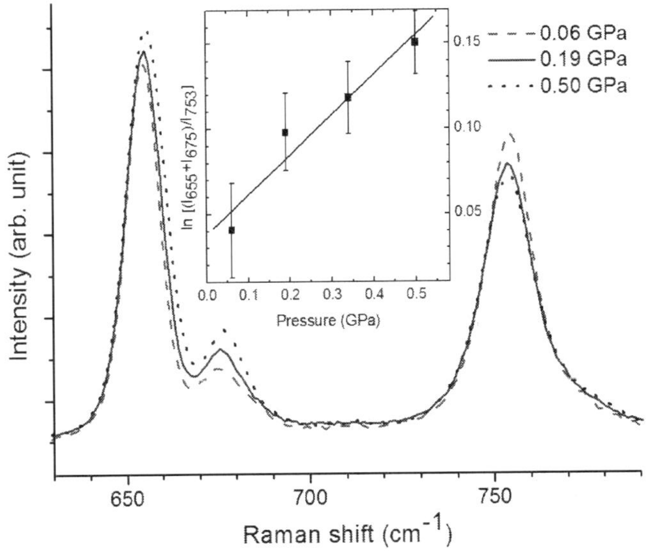

Figure 3. Representative Raman spectra of DCE in the magnified spectral region of 600–800 cm^{-1} for fluid phase. The inset is the plot of logarithm of relative intensities of the first two peaks over the third peak as a function of pressure. Reproduced with permission from reference [12].

3.2. Chlorocyclohexane

In situ Raman measurements on CCH at room temperature and high pressures up to 20 GPa also show interesting pressure-dependent conformational changes [13]. Below 0.7 GPa, CCH exists as a fluid phase with a mixture of axial and equatorial conformations in equilibrium, which is shifted to axial upon compression (**Figure 4a**). The shift was attributed to the smaller volume of axial conformer with a volume difference of −2.2 cm^3 mol^{-1} relative to equatorial conformation, which is consistent with previous studies. When compressed to 2.4 GPa, the depletion of C—H stretching mode at high frequency as well as the splittings of the ν_{22} mode suggest a phase transition (**Figure 4b**). The splittings are further enhanced at 4.8 GPa together with the observation of a new lattice mode, suggesting another phase transition. Upon careful comparison, these high-pressure phases are likely different from the low-temperature phases observed previously. Significant broadening of Raman profiles was observed above 9.5 GPa, indicating that CCH is undergoing gradual disordering at high pressures (**Figure 4b**). Upon releasing of pressure, CCH is fully recoverable indicating that the six-member ring can sustain high pressures up to 20 GPa. The observation of two new modes upon decompression, however, suggests that phase transformation of CCH is partially irreversible above 2.5 GPa. The phase produced by decompression exhibits a contribution from axial conformation of CCH. These pressure-induced hysteresis and partial irreversibility can be attributed to the plastic nature of the CCH crystals.

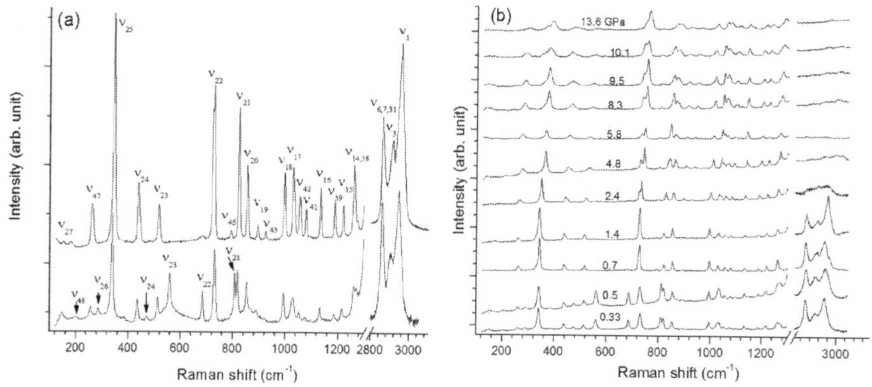

Figure 4. (a) Raman spectra of CCH collected at ambient pressure (top) in comparison with that collected upon slight compression (i.e., at 0.03 GPa, bottom). CCH exists as a mixture of axial and equatorial conformer with the latter dominant at condition and thus the assignment labeled above each Raman modes refers to equatorial conformation for the top spectrum. Axial and equatorial conformers share majority of common Raman modes and thus only those exclusively associated with axial conformer are labeled in the bottom spectrum. (b) Selective Raman spectra of CCH on compression in the pressure region 0–14 GPa in the spectral range of 120–1300 cm^{-1} and 2800–3200 cm^{-1}. Reproduced with permission from reference [13].

4. Pressure-mediated hydrogen bonding

Hydrogen bonding plays an important role in stabilizing a wide range of molecular structures and influences the chemical and physical properties of molecular systems. Typically, the characterizations of hydrogen bonding are inferred from crystal structures or by theoretical modeling. Here two examples are shown to demonstrate that vibrational spectroscopy on materials loaded in DAC can reveal interesting pressure-mediated hydrogen bonding interactions.

4.1. Ethylene glycol

Ethylene glycol (EG) serves as a prototype for understanding hydroxyl group interactions in biological compounds such as sugars and polysaccharides. Using *in situ* high-pressure Raman and infrared absorption spectroscopy, the structural and conformational transformations of EG were found to be substantially influenced by hydrogen bonding interactions under pressure up to 10 GPa [14]. The high-pressure behavior of Raman modes suggests that EG exists as a liquid with a mixture of *trans* and *gauche* conformations up to 3.1 GPa. In the pressure range 4–7 GPa, the solid phase has a varied proportion of *trans* and *gauche* conformations. At pressures above 7 GPa, the EG structure is stabilized to *gauche* conformation and remains stable up to 10 GPa. The increase in the intensity and the large pressure induced red shift of the infrared active OH mode ν_ε suggest that intra-molecular hydrogen bond is formed and strengthened during the stabilization of *gauche* conformation (**Figure 5**). The observed pressure induced changes were found to be completely reversible on decompression to ambient conditions.

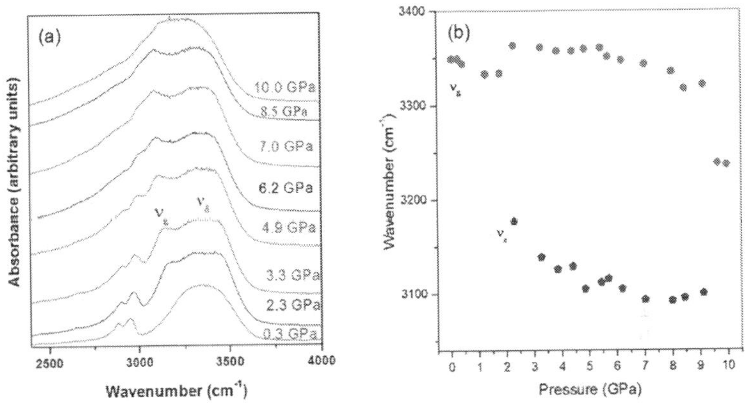

Figure 5. Infrared absorption spectra of ethylene glycol at different pressures under compression in the spectral region of 2500–4000 cm^{-1} (a) and variation of OH stretching infrared active modes of ethylene glycol with pressure (b). Reproduced with permission from reference [14].

4.2. Bis(1H-tetrazol-5-yl)amine monohydrate

Bis(1H-tetrazol-5-yl)amine (BTA) with two tetrazole rings linked by one nitrogen atom that contains 82.5 wt% nitrogen has been considered a promising high energy density material. Moreover, examining the possibility of converting this high nitrogen content precursor to other polymorphs with higher energy density using high pressure is of great interest. *In situ* high-pressure study of $BTA \cdot H_2O$ was carried out up to 25 GPa at room temperature using Raman and IR spectroscopy, X-ray diffraction as well as *ab initio* simulations [15]. Upon compression, both the Raman and IR vibrational bands were found to undergo continuous and gradual broadening without significant change of the profile, indicating pressure-induced structural disordering rather than phase transition. X-ray diffraction patterns confirmed the pressure effect on the structural evolutions of $BTA \cdot H_2O$. Interestingly, in contrast to all other Raman and IR modes of $BTA \cdot H_2O$ which exhibit blue shifts, the N-H stretching mode shows a prominent red shift upon compression to ~8 GPa, strongly suggesting pressure enhanced hydrogen bonding between BTA and H_2O (**Figure 6**). The analysis of X-ray diffraction patterns of $BTA \cdot H_2O$ indicates that the unit cell parameters undergo anisotropic compression rate. The pressure dependence of the unit cell parameters and volumes coincides with the behavior of the hydrogen bonding enhancement (**Figure 7**). Aided with first-principles simulations, these pressure-mediated structural modifications consistently suggest that hydrogen bonding played an important role in the compression behavior and structural stability of $BTA \cdot H_2O$ under high pressures (**Figure 8**).

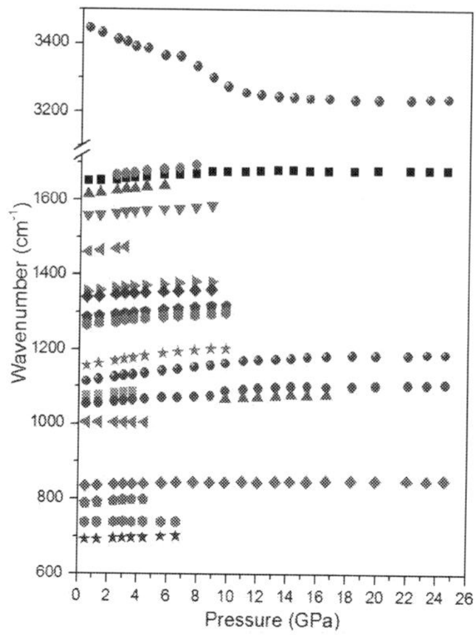

Figure 6. Pressure dependence of IR modes of BTA·H_2O on compression. Reproduced with permission from reference [15].

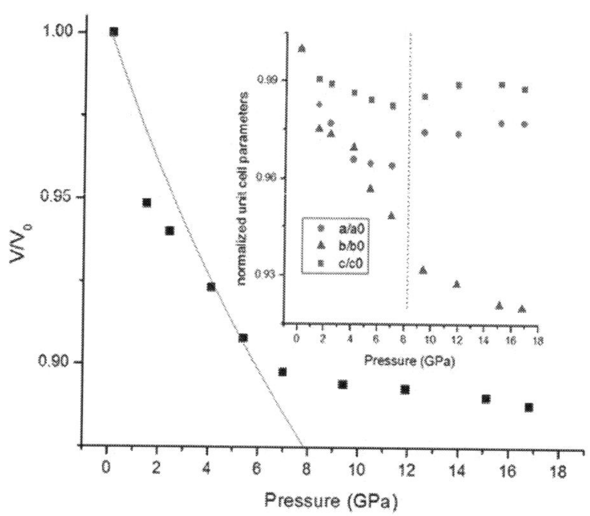

Figure 7. Normalized unit cell volume versus pressure (black squares) for BTA·H_2O on compression and fitted equation of state (red curve) using second-order Birch-Murnaghan equation. The inset shows normalized monoclinic unit cell parameters for *a*, *b* and *c* of BTA·H_2O on compression. The vertical dashed line denotes the pressure at which the monotonic contraction of *a* and *c* axes changed. Reproduced with permission from reference [15].

http://dx.doi.org/10.5772/64617

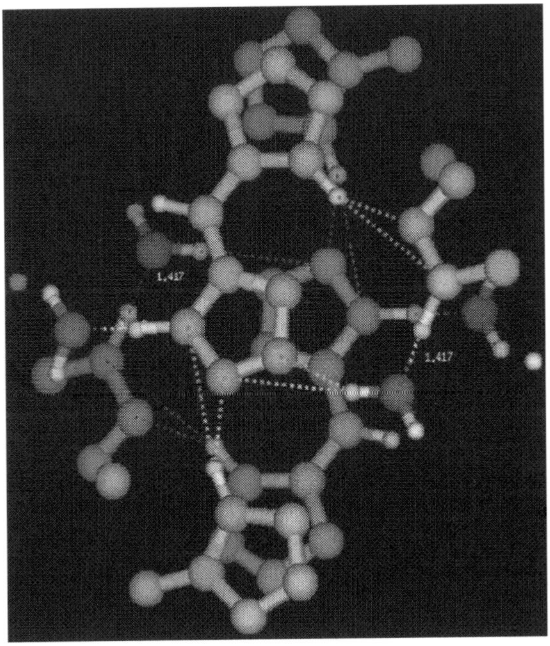

Figure 8. Proton hopping and molecular interactions of BTA·H_2O system based on first-principles simulations.

5. Structural and phase transitions

5.1. Boron nitride nanotubes

Compared to carbon nanotube, boron nitride nanotube (BNNT) has structure-independent wide band gap, enhanced thermal stability, high resistance to oxidation at high temperatures, high thermal conductivity and remarkable yield strength, making it a promising advanced material for a wide range of applications. Multiwalled boron nitride nanotubes (BNNTs) were compressed at room temperature in diamond anvil cells up to 35 GPa followed by decompression and characterized by *in situ* FTIR absorption spectroscopy [11]. Pressure-induced transformations from a hexagonal to a more closely packed wurtzite structure were observed at 11 GPa, which is similar to that reported for bulk BN (**Figure 9**). However, BNNTs exhibit quantitative differences compared to bulk h-BN in terms of transformation completeness and reversibility (**Figure 10**). These findings provide strong evidence that significantly different yield of sp^3 bonding formation originated from different morphologies of the starting BN materials (**Figure 11**). The unique transformation mechanism for BNNTs provides new useful information for developing BNNTs as potential advanced materials with more desirable properties than carbon nanotubes.

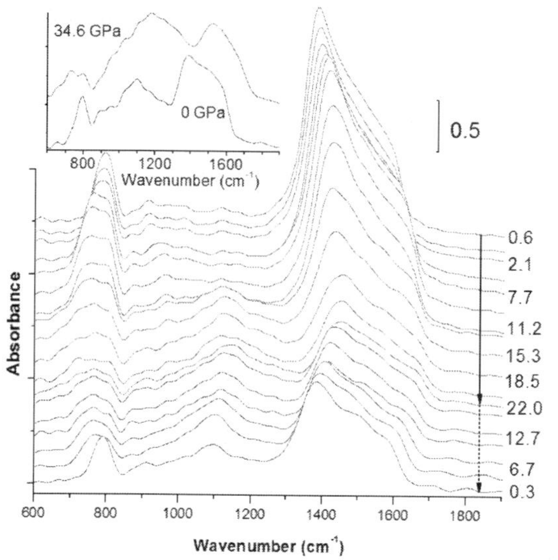

Figure 9. Infrared spectra of BNNTs at selected pressures upon compression (red lines) and decompression (blue lines) in the spectra region of 600–1900 cm^{-1}. The solid and dashed arrows indicate the compression and decompression sequence. The inset shows spectra from another run at a highest pressure of 34.6 GPa on compression (red line) and complete pressure release (blue line). Reproduced with permission from reference [11].

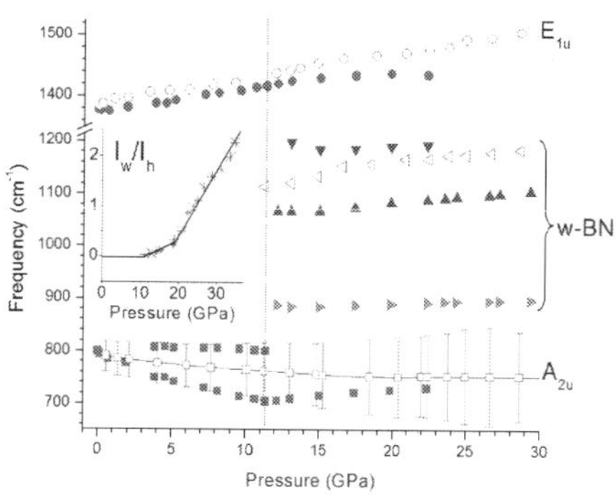

Figure 10. Pressure dependence of representative IR modes of BNNTs (open symbols) and in comparison with those for bulk h-BN (solid symbols) on compression. The squares and circles are the respective A_{2u} and E_{1u} modes of h-BN structure, while other symbols represent IR modes for w-BN structure. The dashed line at around 11 GPa denotes the transition onset for both BNNTs and bulk h-BN. The vertical bars for A_{2u} mode represent the full width at half maximum for BNNTs. The inset shows the ratio of the IR band intensity of the mode at 1125 cm^{-1} for w-BN over the E_{1u} mode for h-BN observed in BNNTs labeled as I_w/I_h. The solid lines are for eye guidance showing three distinctive conversion regions. Reproduced with permission from reference [11].

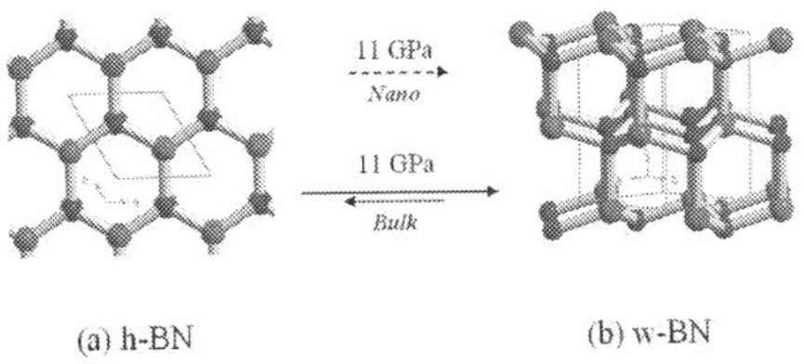

Figure 11. Crystal structures and bonding patterns of (a) h-BN and (b) w-BN with the transformation conditions for BNNTs and bulk h-BN denoted. The red and blue balls represent boron and nitrogen, respectively. The dashed arrow for BNNT indicates incomplete irreversible transformation, while the solid arrows with different length for bulk h-BN indicate partial reversibility. Reproduced with permission from reference [11].

5.2. Aromatic compounds

Aromatic compounds have been investigated under non-ambient conditions over the past few decades due to their great importance in both fundamental and applied science. In particular, they have been widely studied in chemical synthesis under elevated temperatures and pressures as the precursors of technological materials, such as amorphous solids and conjugated polymers [16]. For instance, using *in situ* Raman spectroscopy and infrared absorption spectroscopy, structural transitions of pyridine have been investigated as a function of pressure up to 26 GPa [17]. By monitoring the band profiles in both Raman and IR spectra and especially the Raman shifts in the lattice region, a liquid-to-solid transition at 1 GPa followed by solid-to-solid transitions at 2, 8, 11 and 16 GPa were observed upon compression (**Figure 12**). All these transitions were reversible upon decompression from 22 GPa. When compressed beyond 22 GPa, a further chemical transformation was observed which is evidenced by the substantial and irreversible changes of the Raman and infrared spectra. This transformation could be attributed to the destruction of the ring structure. The high-pressure behavior of pyridine was also compared to that of benzene. The similar transition sequence with well-aligned transition pressures indicates that aromatic compounds with isoelectronic structures may have similar structural stabilities and thus transition behaviors under high pressure.

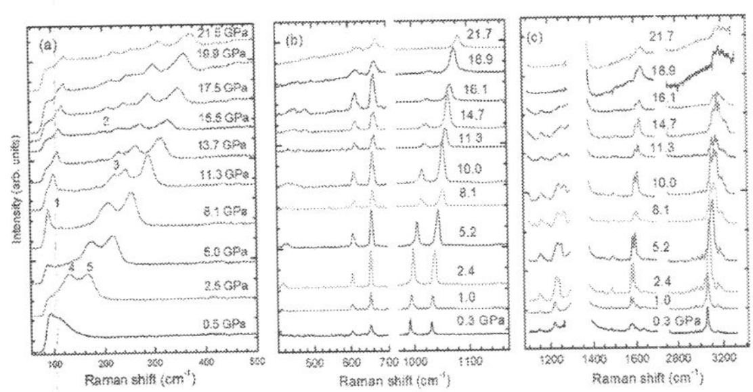

Figure 12. Selected Raman spectra of pyridine on compression in the spectral region of 60–500 cm^{-1} (a), 400–1160 cm^{-1} (b) and 1100–3300 cm^{-1} (c). Reproduced with permission from reference [17].

5.3. Metal and chemical hydrides

High-pressure investigations of potential hydrogen storage materials, especially hydrogen-rich metal and chemical hydrides have received increasing attentions [8]. Not only has pressure demonstrated great promises for producing new structures and materials but also many known hydrogen-rich materials have exhibited new transformations as well as totally different thermodynamic and kinetic behaviors under higher pressures than under ambient conditions. Hydrides in a wide range of different categories, such as calcium borohydride, sodium amide and ammonia borane, have been extensively investigated under high pressures by vibrational spectroscopy, X-ray diffraction and theoretical calculations [18–21]. Here ammonia borane (NH_3BH_3) is chosen as an example to demonstrate that vibrational spectroscopy can be an effective tool to elucidate novel high-pressure structures [18, 21].

Using *in situ* Raman and synchrotron IR spectroscopy, the pressure behavior of ammonia borane complex as a promising hydrogen storage material was investigated up to 14 GPa [18]. In the low-pressure region (<2 GPa), the complex was found to undergo a structural transformation to, an ordered, possibly orthorhombic structure from originally a disordered tetragonal phase. With increasing pressure, the Raman and IR spectra suggest several solid-to-solid transformations at about 2.4, 5.5, 8.5 and 10.4 GPa, as evidenced by the distinctive profiles and the pressure dependences of characteristic modes (**Figure 13**). Upon decompression, these pressure-induced transformations are found completely reversible with intact chemical structure of the NH_3BH_3 complex, but possible modifications to the crystal structures. Analysis of combined Raman and IR measurements, especially the lattice features (**Figure 13a**), suggests that NH_3BH_3 structures below 5.5 GPa resemble a low-pressure orthorhombic structure, while in the higher pressure regions, NH_3BH_3 complexes may undergo transformations to disordered or amorphous structures.

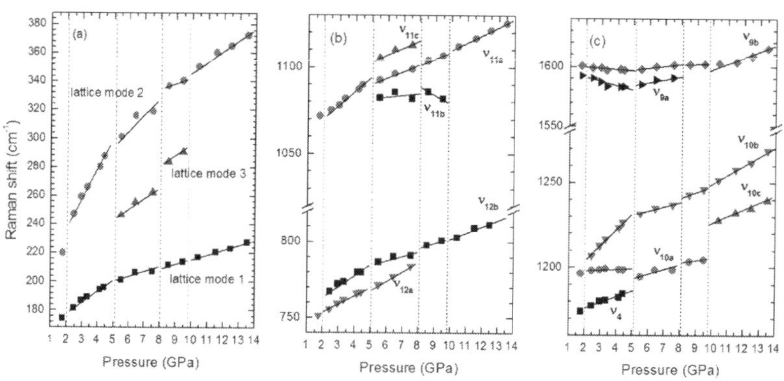

Figure 13. Pressure dependences of Raman shift of NH_3BH_3 on compression for (a) the lattice modes; (b) the ^{11}B-N/^{10}B-N stretch (ν_5/ν_5') modes; and (c) the NBH rock (ν_{12a}, ν_{12b}, ν_{11a}, ν_{11b} and ν_{11c}) modes. The solid lines crossing the solid symbols are based on linear fit. The vertical dashed lines indicate the proposed phase boundaries. Reproduced with permission from reference [18].

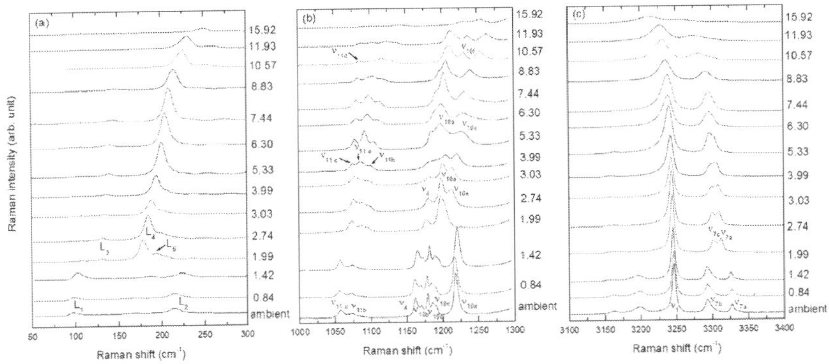

Figure 14. Selected Raman spectra of NH_3BH_3 collected on compression up to 15.92 GPa at 180 K in the region of 50–300 cm^{-1} (a), 1000–1300 cm^{-1} (b), and 3100–3400 cm^{-1} (c). The assignments are labeled for selected Raman mode at selected pressures. Reproduced with permission from reference [21].

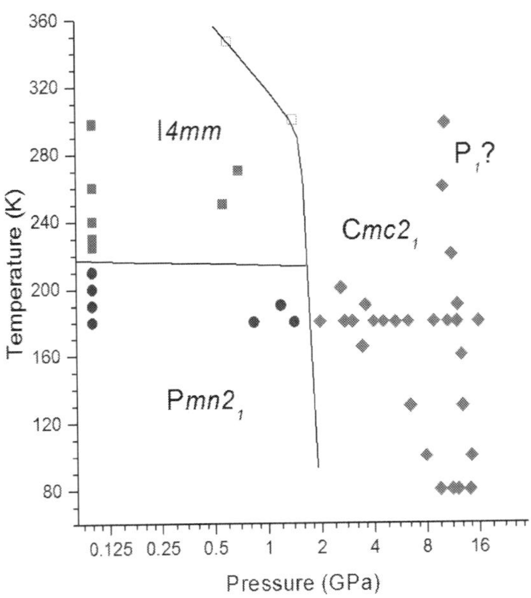

Figure 15. Schematic P-T phase diagram of NH_3BH_3 in the pressure region of 0–15 GPa (in $\log_2$ scale) and temperature region of 80–350 K. Solid symbols are experimental data from this study, with squares for *I4mm* phase, circles for *Pmn*2$_1$ phase and diamonds for *Cmc*2$_1$ phase. The open squares are adopted from reference [22]. The solid lines denote the rough boundaries among the three known phases. The *P*1 phase labeled is considered tentative. Reproduced with permission from reference [21].

Subsequently, ammonia borane was investigated at simultaneous high pressures (up to 15 GPa) and low temperatures (down to 80 K) by *in situ* Raman spectroscopy [21]. Upon cooling to 220 K from room temperature at ambient pressure, ammonia borane transforms from *I4mm* to *Pmn*2_1. Upon isothermal compression to 15 GPa at 180 K, another three pressure-induced structural transformations were observed. These transitions can be evidenced by the change in the Raman profile as well as the pressure dependence of the major Raman modes (**Figure 14**). Upon decompression and warming-up, these P-T-induced transformations are found completely reversible. With the aid of factor group analysis, the phases above 1.5 GPa were found consistent with the crystal structure with space group *Cmc*2_1, and that the transitions at 5 and 8 GPa are second order in nature, which can be interpreted as enhanced inter-molecular interactions within the same or possibly a slightly modified crystal lattice. Further compression above 15 GPa leads to the gradual transformation to an amorphous phase. When combined with previously reported high-pressure and room-temperature data, our Raman measurements from multiple runs covering various P-T paths allowed the significant update of the P-T phase diagram of ammonia borane in the pressure range of 0–15 GPa and the temperature range of 80–350 K (**Figure 15**).

6. Pressure-induced chemical reactions

6.1. Acrylic acid

Pressure-induced polymerization is a chemical process pertaining to green chemistry as the reactions can be carried out in the absence of any solvent or catalyst, which implies a lesser environmental impact. Poly(acrylic acid) is a well-known polymer with a wide variety of industrial applications such as being super absorbent materials, biocompatible polymers, poly-electrolytes and nanopolymers in molecular devices. Therefore, it is very significant in the polymer industry to explore pressure-induced polymerization from this monomer, as the polymer product with improved properties distinct from that obtained using conventional synthetic methods might be obtained. The first pressure-induced structural and polymeric transformations of acrylic acid were studied by *in situ* Raman spectroscopy [23]. Upon compression to 0.3 GPa, a liquid-to-solid transformation was observed, followed by a solid-to-solid transition at ~2.7 GPa. The two new high-pressure crystalline phases are labeled as phase I and II, respectively (**Figure 16a**). Phase I had a possibly similar structure that resembles low-temperature phase reported previously. Phase II can be interpreted as a denser phase with strong intermolecular interactions leading to polymerization or oligomerization ultimately. When compressed to above 8 GPa, acrylic acid transforms into a disordered polymeric phase (**Figure 16b**). Upon decompression to ambient pressure, the retrieved polymeric phase exhibits a significant amount of acrylic acid monomers or oligomers. Comparative Raman measurements on standard commercial poly(acrylic acid) (**Figure 17**) allowed the understanding of possible structures of the polymeric phase of acrylic acid produced in this study. Overall, our analysis suggests that hydrogen bonding played a significant role in the pressure-induced polymerization/oligomerization process.

http://dx.doi.org/10.5772/64617

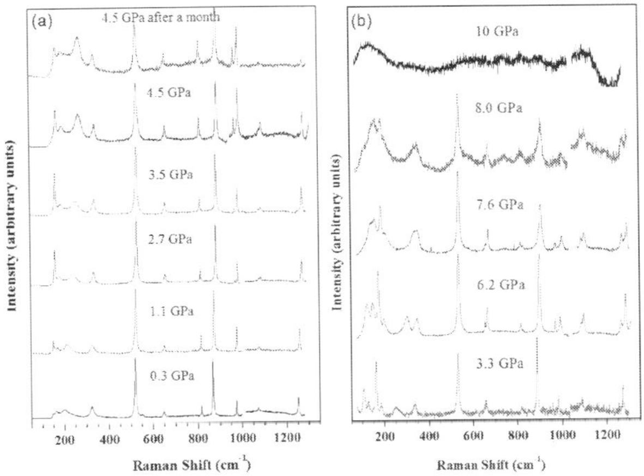

Figure 16. Raman spectra of acrylic acid at selected pressures upon compression in the pressure region of 0.3–4.5 GPa (a) and 3.3–10 GPa (b) in the spectral region of 100–1300 cm^{-1}. Reproduced with permission from reference [23].

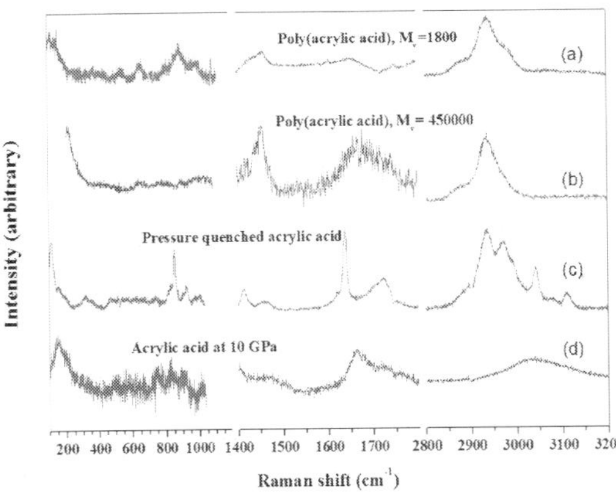

Figure 17. Raman spectrum of poly(acrylic acid) purchased from Aldrich with an average molecular weight of 1800 g/mol (a) and 450,000 g/mol (b) in comparison with that of recovered acrylic acid by decompression from 10 GPa (c) and that of acrylic acid at 10 GPa (d). Reproduced with permission from reference [23].

6.2. Ethylene glycol

Using combined high-pressure and photon excitations especially in the UV range has demonstrated strong potential to produce new molecular materials in a highly efficient way. Using multi-line UV radiation at ~350 nm, the photon-induced reactivity of liquid ethylene glycol (EG) at room temperature was investigated by FTIR spectroscopy [24]. Upon UV irradiation,

IR spectra of EG show two sets of distinctive profiles after specific reaction time, indicating multiple photon-induced chemical reactions, which can be designated as primary and secondary processes (**Figure 18**). Careful spectral analysis allows the identification of primary reaction products that include glycolaldehyde, acetaldehyde and methanol. Further photoreactions of these primary products led to the formation of the secondary products, which were identified as methane, formaldehyde, methoxymethanol, methylformate and carbon dioxide. Based on these reaction products, possible reaction mechanisms and production pathways were proposed. We also found that the initial loading pressure of EG plays an important role in influencing the reaction kinetics as well as in controlling the accessibilities for some reaction channels such as for CH_4 (**Figure 19a**). Quantitative analysis of the antisymmetric stretching mode of CO_2 formed at different loading pressures suggests the formation of CO_2 clathrate hydrates well as CO_2 clusters. The stabilities as well as relative abundance of these CO_2 species are found to be dependent on both pressure and radiation time (**Figure 19b**). These observations revealed interesting pressure-induced CO_2 sequestration behaviors as a result of photochemical reactions of ethylene glycol.

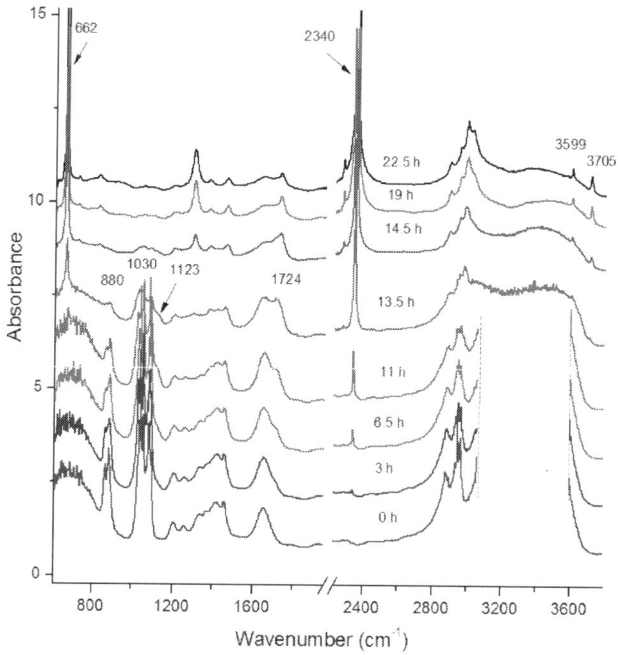

Figure 18. Selected FTIR spectra of EG with an initial loading pressure of 0.1 GPa upon UV irradiation (with λ of ~350 nm and power of ~700 mW) collected at different radiation time. The most characteristic new IR bands emerged at 13.5 h and observed at 22.5 h indicating sequential photochemical reactions are labeled. The spectral region in 3000–3500 cm^{-1} before 13.5 h is truncated due to the saturated IR absorption intensity. Reproduced with permission from reference [24].

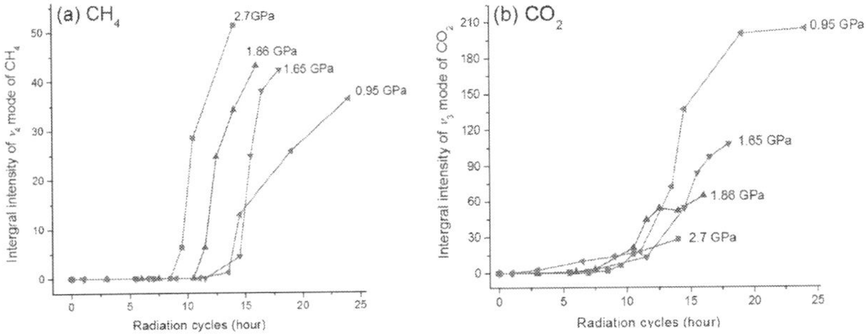

Figure 19. Relative photochemical reaction yields of CH_4 (a) and CO_2 (b) derived by integrating the intensity of the respective characteristic IR modes (ν_4 of CH_4 and ν_3 of CO_2) as a function of radiation time for EG samples with an initial loading pressure of 0.1, 0.5, 1.0 and 2.0 GPa. The pressures labeled for each sample indicate the final system pressure. Reproduced with permission from reference [24].

7. Porous materials and guest-host interactions

7.1. ZIF-8

ZIF-8 is a representative member of the zeolitic imidazolate framework (ZIF) family, an emerging class of porous materials with promising applications in gas storage and catalysis, etc. As a result, substantial interest has been focused on the investigation of its structure and properties under different conditions. Pressure tuning has proven an important and effective means to modify the structures and thus the associated properties of porous materials. Therefore, ZIF-8 was investigated under high pressures up to ~39 GPa using *in situ* IR spectroscopy [25]. Upon compression to 1.6 GPa followed by decompression, the structural modifications on ZIF-8 framework appear reversible (**Figure 20a**). However, further compression to higher pressures led to irreversible structural transitions to an amorphous phase characterized by the very broad IR profiles (**Figure 20b**). Nevertheless, the chemical structure of the framework was found to sustain extreme compression without permanent breaking down. Overall, the high-pressure behavior and especially the surprising chemical stability probed by *in situ* IR spectroscopy demonstrate strong promises storage applications of ZIF-8 under extreme conditions.

In a subsequent study, ZIF-8 framework was investigated when loaded with CO_2 in a diamond anvil cell at high pressures of 0.8 GPa, far beyond the conventional gas adsorption pressure also using *in situ* FTIR spectroscopy [26]. Upon loading, CO_2 molecules in two types of environment (i.e., outside as bulk medium and inside the framework) can be unambiguously differentiated by monitoring the combination IR bands of CO_2 (**Figure 21**). Furthermore, pressure was found to play a regulating role in the migration of CO_2 molecules with respect to the framework even at room temperature. The strong interactions between CO_2 and framework are evident from the IR features of the framework (e.g., C=C stretching region), providing valuable information about the possible interaction site. As guest molecules, CO_2 in turn can substantially enhance the structural stability of the ZIF-8 framework as compared to the empty framework (**Figure 22**). The enhanced CO_2 storage capacity of ZIF-8 at high pressure provides new insight into the gas capture and storage applications of ZIFs.

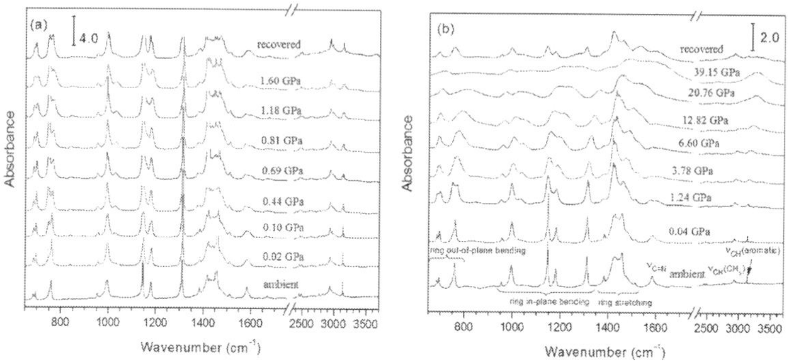

Figure 20. Selected IR spectra of ZIF-8 on compression to a highest pressure of 1.60 GPa and as recovered (a), and to another highest pressure of 39.15 GPa and as recovered (b). Reproduced with permission from reference [25].

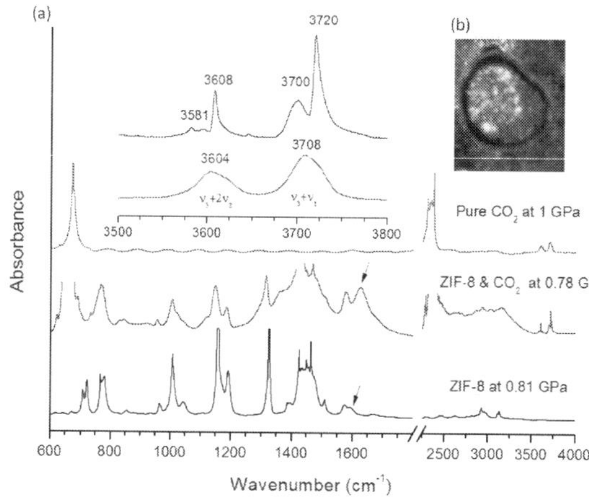

Figure 21. (a) The comparison of IR spectrum of pure CO_2 (top), ZIF-8 loaded with CO_2 (middle) and that of pure ZIF-8 (bottom) at similar pressures. The inset shows the spectral region for the combination modes of CO_2 loaded with ZIF-8 loaded (top) and pure CO_2 (bottom). (b) Photograph of ZIF-8 loaded with CO_2 obtained under an optical microscope. The arrows denote the positions of the C=C stretching mode of the imidazole ring. Reproduced with permission from reference [26].

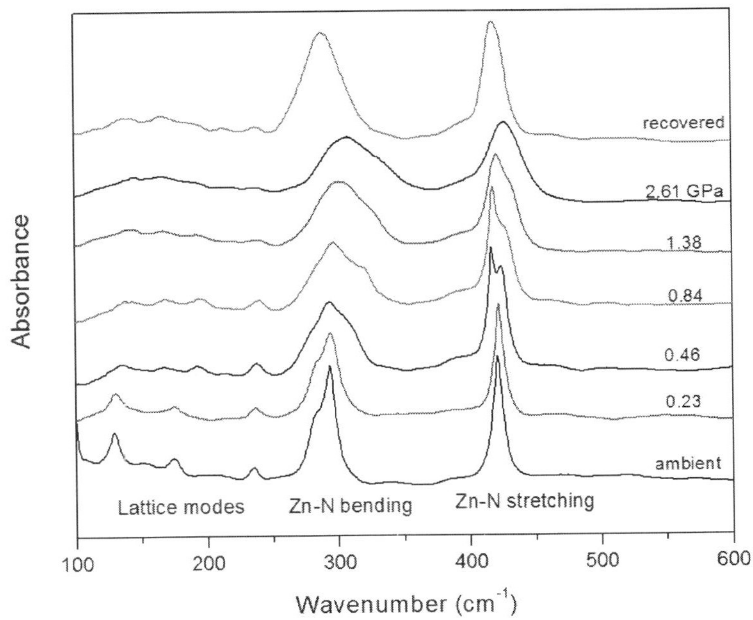

Figure 22. Far-IR spectra of empty ZIF-8 framework upon compression to 2.61 GPa and decompression to ambient pressure. These far-IR spectra suggest that pressure can significantly modify the crystal structures of empty ZIF-8 framework irreversibly. Reproduced with permission from reference [26].

7.2. MIL-68

As a promising candidate for the application of gas storage and separation, metal-organic framework (MOF) MIL-68 has unique structural topology that contains two types of channels with distinct pore sizes. Using *in situ* IR spectroscopy, the behavior of as-made and activated MIL-68 (In) and their structural reversibilities were investigated under high pressures [27]. Overall, the structures of both frameworks were found highly stable upon compression to 9 GPa. However, some modifications on the local structure especially the bridging O–H units, which are very sensitive to compression, can be clearly identified. The structural modifications are found to be completely reversible upon decompression for as-made MIL-68 (In) but irreversible for the activated framework. The different reversibility of framework is most likely associated with the solvent DMF molecules contained in the framework channels. Furthermore, the stability of the activated framework was investigated using PTM to achieve hydrostatic compression. The pressure-induced inclusion of PTM makes the framework more resilient to compression (18 GPa). As a result, structural modifications of the framework with PTM are completely reversible upon decompression (**Figure 23a**). Moreover, the performance of MIL-68 (In) for CO_2 adsorption under high pressure was investigated. Our results show that at relative low pressures such as below 0.35 GPa, the hexagonal pores are readily accessible for CO_2, while the triangular pores become accessible for CO_2 at higher pressures such as above 1.5 GPa (**Figure 23b**). Such pressure-regulated CO_2 occupation in different channels of the MIL-68 framework is completely reversible between compression and decompression (**Figure 23c**). The unique adsorption behavior of CO_2 in the MIL-68 is strongly correlated with the OH units contributing as the primary binding sites through hydrogen bonding with CO_2. Molecular dynamics simulations further support our analysis (**Figure 24**). The high framework stability and enhanced CO_2 adsorption of MIL-68 (In) under high pressure make it a promising candidate for greenhouse gas storage.

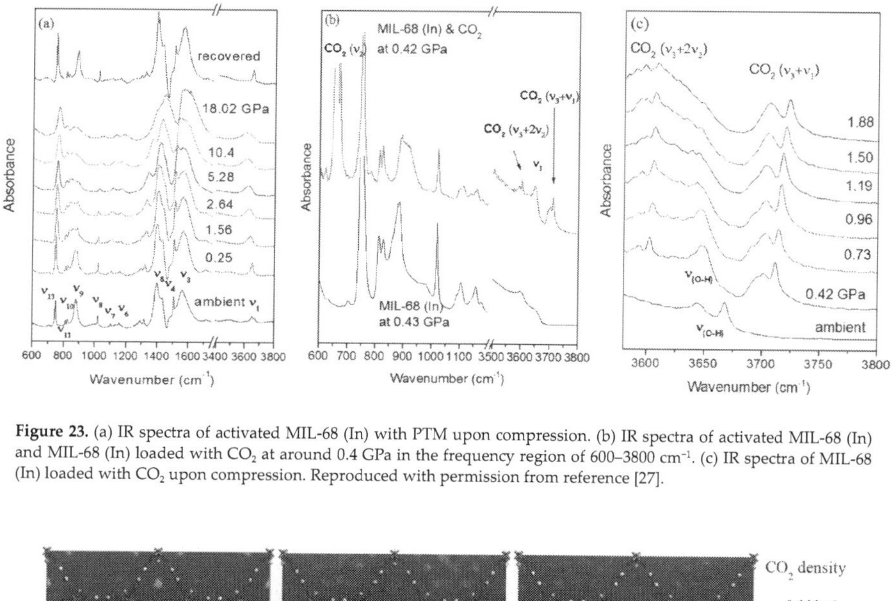

Figure 23. (a) IR spectra of activated MIL-68 (In) with PTM upon compression. (b) IR spectra of activated MIL-68 (In) and MIL-68 (In) loaded with CO_2 at around 0.4 GPa in the frequency region of 600–3800 cm^{-1}. (c) IR spectra of MIL-68 (In) loaded with CO_2 upon compression. Reproduced with permission from reference [27].

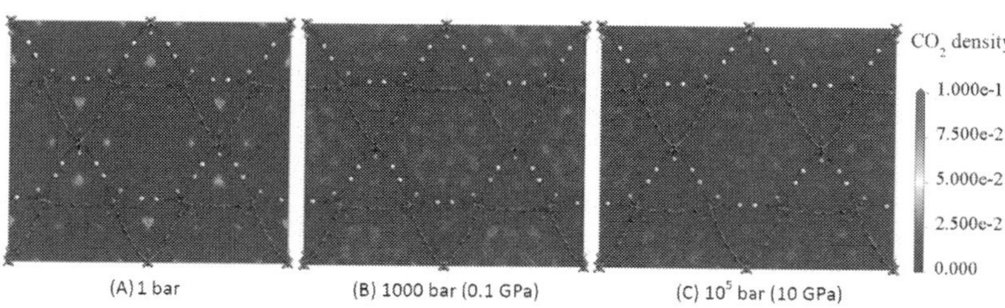

Figure 24. Simulated contour plots of the CO_2 probability density distributions along the hexagonal and triangular channels of MIL-68 (In) framework at (a) 1 bar, (b) 1000 bar or 0.1 GPa and (c) 10^5 bar (or 10 GPa). Reproduced with permission from reference [27].

http://dx.doi.org/10.5772/64617

8. Summary and future perspectives

In summary, this chapter demonstrated the application of *in situ* vibrational spectroscopy including Raman and FTIR spectroscopy in the elucidation of molecular structures and transformation mechanism for a wide variety of materials rendered under high-pressure conditions. Specifically, conformational changes, pressure-mediated hydrogen bonding interactions, molecular and crystal structural transitions, polymerizations and photon-assisted chemical reactions, as well as guest-host interactions of respective selected systems can be efficiently and accurately probed and characterized using *in situ* high-pressure Raman spectroscopy, FTIR spectroscopy or combination of both. These spectroscopic data provided enormously valuable information for us to understand the pressure-induced phenomena at microscopic level in-depth. Thus, vibrational micro-spectroscopy can be considered a routine but indispensable technique in any high-pressure materials research laboratories.

In addition to extreme pressure, extreme temperatures such as several Kelvin and several hundred degrees Celsius, and especially their combinations pose experimental challenges yet offer new and unexplored P-T domains for novel structures and properties of materials to be discovered. *In situ* vibrational micro-spectroscopy is expected to play an important role in structural characterization under these tough conditions. Although extremely convenient, one should realize that vibrational spectroscopy itself alone is seldom successful in solving totally unknown structures. Therefore, to realize the full potential of vibrational spectroscopy on materials under extreme conditions, other experimental techniques such as X-ray diffraction, synchrotron probes as well as theoretical modeling are essential to obtain the full structural information of novel materials.

Acknowledgements

This work was supported by a Discovery Grant, a Research Tools and Instruments Grant from the Natural Science and Engineering Research Council of Canada, a Leaders Opportunity Fund from the Canadian Foundation for Innovation, an Early Researcher Award from the Ontario Ministry of Research and Innovation, a Petro-Canada Young Innovator Award and by Defense Research and Development Canada under contract No. W7702-135601. The synchrotron IR measurements presented were performed at U2A beamline of National Synchrotron Light Source of US Brookhaven National Laboratory.

Author details

Yang Song

Address all correspondence to: yang.song@uwo.ca

Department of Chemistry, University of Western Ontario, London, Ontario, Canada

References

[1] Song Y, Manaa MR. New trends in chemistry and materials science in extremely tight space. J Phys Chem C. 2012;116:2059–60.

[2] Hemley RJ, Ashcroft NW. The revealing role of pressure in the condensed matter sciences. Phys Today. 1998;51:26–32.

[3] Hemley RJ. Effects of high pressure on molecules. Annu Rev Phys Chem. 2000;51:763–800.

[4] Schettino V, Bini R. Molecules under extreme conditions: Chemical reactions at high pressure. Phys Chem Chem Phys. 2003;5:1951–65.

[5] McMillan PF. Chemistry at high pressure. Chem Soc Rev. 2006;35:855–7.

[6] Grochala W, Hoffmann R, Feng J, Ashcroft NW. The chemical imagination at work in very tight places. Angew Chem-Int Edit. 2007;46:3620–42.

[7] Schettino V, Bini R. Constraining molecules at the closest approach: chemistry at high pressure. Chem Soc Rev. 2007;36:869–80.

[8] Song Y. New perspectives on potential hydrogen storage materials using high pressure. Phys Chem Chem Phys. 2013;15:14524–47.

[9] Mao HK, Bell PM, Shaner JW, Steinberg DJ. Specific volume measurements of copper, molybdenum, palladium, and silver and calibration of the ruby R1 fluorescence pressure gauge from 0.06 to 1 mbar. J Appl Phys. 1978;49:3276–83.

[10] Dong Z. High-pressure study of molecular solids and 1D nanostructures by vibrational spectroscopy and synchrotron X-ray diffraction [Thesis]: University of Western Ontario; 2012.

[11] Dong Z, Song Y. Transformations of cold-compressed multiwalled boron nitride nanotubes probed by infrared spectroscopy. J Phys Chem C. 2010;114:1782–8.

[12] Sabharwal RJ, Huang Y, Song Y. High-pressure induced conformational and phase transformations of 1,2-dichloroethane probed by Raman spectroscopy. J Phys Chem B. 2007;111:7267–73.

[13] Dong Z, Beilby NG, Huang Y, Song Y. Conformational and phase transformations of chlorocyclohexane at high pressures by Raman spectroscopy. J Chem Phys. 2008;128:074501.

[14] Murli C, Lu N, Dong Z, Song Y. Hydrogen bonds and conformations in ethylene glycol under pressure. J Phys Chem B. 2012;116:12574–80.

[15] Zhou L, Shinde N, Hu A, Cook C, Murugesu M, Song Y. Structural tuning of energetic material Bis(1H-tetrazol-5-yl)amine monohydrate under pressures probed by vibrational spectroscopy and X-ray diffraction. J Phys Chem C. 2014;118:26504–12.

[16] Dong Z, Seemann NM, Lu N, Song Y. Effects of high pressure on azobenzene and hydrazobenzene probed by Raman spectroscopy. J Phys Chem B. 2011;115:14912–8.

[17] Zhuravlev KK, Traikov K, Dong Z, Xie S, Song Y, Liu Z. Raman and infrared spectroscopy of pyridine under high pressure. Phys Rev B. 2010;82:064116.

[18] Xie S, Song Y, Liu Z. In situ high-pressure study of ammonia borane by Raman and IR spectroscopy. Can J Chem. 2009;87:1235–47.

[19] Liu A, Xie S, Dabiran-Zohoory S, Song Y. High-pressure structures and transformations of calcium borohydride probed by combined Raman and infrared spectroscopies. J Phys Chem C. 2010;114:11635–42.

[20] Liu A, Song Y. In situ high-pressure study of sodium amide by Raman and infrared spectroscopies. J Phys Chem B. 2011;115:7–13.

[21] Liu A, Song Y. In situ high-pressure and low-temperature study of ammonia borane by Raman spectroscopy. J Phys Chem C. 2012;116:2123–31.

[22] Chen JH, Couvy H, Liu HZ, Drozd V, Daemen LL, Zhao YS, et al. In situ X-ray study of ammonia borane at high pressures. Int J Hydrogen Energy. 2010;35:11064–70.

[23] Murli C, Song Y. Pressure-induced polymerization of acrylic acid: a Raman spectroscopic study. J Phys Chem B. 2010;114:9744–50.

[24] Guan J, Song Y. Pressure selected reactivity and kinetics deduced from photoinduced dissociation of ethylene glycol. J Phys Chem B. 2015;119:3535–45.

[25] Hu Y, Kazemian H, Rohani S, Huang Y, Song Y. In situ high pressure study of ZIF-8 by FTIR spectroscopy. Chem Commun. 2011;47:12694–6.

[26] Hu Y, Liu Z, Xu J, Huang Y, Song Y. Evidence of pressure enhanced CO_2 storage in ZIF-8 probed by FTIR Spectroscopy. JACS. 2013;135:9287–90.

[27] Hu Y, Lin B, He P, Li Y, Huang Y, Song Y. The structural stability of and enhanced CO_2 storage in MOF MIL-68 (In) under high pressures probed by FTIR spectroscopy. Chem Eur J. 2015;21:18739–48.

Chapter 9

Conformational Analysis of Molecules: Combined Vibrational Spectroscopy and Density Functional Theory Study

Partha P. Kundu and Chandrabhas Narayana

Additional information is available at the end of the chapter

http://dx.doi.org/10.5772/64452

Abstract

Vibrational spectroscopy can be broadly classified into Raman and infrared (IR). These two techniques are complementary to each other as the mechanisms behind these are different. Vibrational spectroscopy provides detail information about the structure of molecules. The advantage of this technique over X-ray diffraction is that it can be used to probe molecules in solid, liquid or gas phase. This is especially helpful for studying biomolecules as those molecules can be probed in their physiological environment. Over the last few decades, quantum mechanical calculation has become important tool to assign bands from vibrational spectra. Combination of these two techniques has been used widely in the field of chemistry and biochemistry. In this chapter, we review some of the works that combine both of these techniques. A brief theoretical background is given for understanding the principle of these two techniques.

Keywords: Raman, infrared (IR), density functional theory (DFT), conformation

1. Introduction

Spectroscopy is a subject that is related to the interaction of electromagnetic radiation with the atoms or molecules. It provides rich information about the structures, physical and chemical properties of the materials. Energy of a stationary molecule can be written as a sum of three parts: electronic, vibrational and rotational. In vibrational spectroscopy, the vibrational levels of a molecule are probed. Vibrational spectroscopy can be broadly classified into two: infrared (IR) and Raman. Even though both techniques probe the vibrational energy levels, physical processes leading to these spectra are different. Thus, these two techniques offer different information about the molecule and regarded as complementary to each other. To get the structural information from a vibrational spectrum, the first task is to assign the bands. Band assignments can be performed by comparing the modes of similar molecule already existing in literature. However, vibrational frequencies are sensitive to the small difference in the structure and also to the environment. Moreover, it is difficult to assign the large number of closely spaced bands arising from even a medium size molecule. Hence, for reliable assignments of these bands, it is essential to calculate the normal modes theoretically and compare those with the experimental spectra. Over the last few decades, calculation of molecular spectra using quantum mechanics has become a common practice. There are many quantum mechanical-based methods for calculation, of which density functional theory (DFT) is most popular as its computational cost is less without compromising significantly with the accuracy.

Here, we will introduce the basic theory behind these vibrational techniques. Also, the principle of DFT will be discussed in brief. We will provide references so that the interested readers can gain insights about these topics.

2. Theoretical background

2.1. Raman scattering

When a molecule is irradiated with a monochromatic light of wave number $\tilde{\nu}_0$, a small fraction of the incident light will be scattered. In the scattered radiation, major portion of the light will have the same wave number as the incident light; however, a tiny fraction of light will have the wave number $\tilde{\nu}' = \tilde{\nu}_0 \pm \nu_M$. The first kind of scattering is called Rayleigh scattering while the latter is called Raman scattering. Classically, the phenomenon of Raman scattering can be explained in the following paragraphs.

When a molecule is in an oscillating electric field $E(t)$ with angular frequency ω_L, an induced dipole will be created, Which in the linear approximation can be written (in complex notation) as,

$$\underline{\boldsymbol{p}}(\omega_L) = \hat{\alpha}_L(\omega_L).\underline{\boldsymbol{E}}(\omega_L) \tag{1}$$

where $\hat{\alpha}_L$ is the *polarizability tensor* and $\boldsymbol{E}(t) = \mathrm{Re}(\underline{\boldsymbol{E}} \times e^{-i\omega_L t})$. Here, it is to be mentioned that as the molecule vibrates, the polarizability tensor gets modulated. Polarizability thus not only depends on frequency but also on the atomic positions. To describe the vibrational pattern, *normal coordinates* (Q_k) are introduced. For a specific normal mode k (k = 1, 2, ..., 3N-6, where N is the total number of atoms in the molecule), all the atoms in the molecule will oscillate with the same frequency ω_k. If the polarizability tensor is expanded in Taylor series around the equilibrium position, it can be shown [1] that the induced dipole (real) can be expressed as

$$\boldsymbol{p}(t) = \boldsymbol{p}_L(t) + \boldsymbol{p}_S(t) + \boldsymbol{p}_{aS}(t), \quad (2)$$

where

$$\boldsymbol{p}_L(t) = \mathrm{Re}[\hat{\alpha}_L(\omega_L, 0)\underline{\boldsymbol{E}}(\omega_L)e^{-i\omega_L t}] \quad (3)$$

is the oscillating dipole with frequency ω_L representing Rayleigh scattering,

$$\boldsymbol{p}_S(t) = \frac{Q_k^0}{2}\mathrm{Re}[\hat{R}_k(\omega_L)\underline{\boldsymbol{E}}(\omega_L)e^{-i(\omega_L-\omega_k)t-i\phi}] \quad (4)$$

is oscillating with a frequency $\omega_S = \omega_L - \omega_k$ and produce Stokes Raman scattering,

$$\boldsymbol{p}_{aS}(t) = \frac{Q_k^0}{2}\mathrm{Re}[\hat{R}_k(\omega_L)\underline{\boldsymbol{E}}(\omega_L)e^{-i(\omega_L+\omega_k)t-i\phi}] \quad (5)$$

represents oscillating dipole with frequency $\omega_S = \omega_L + \omega_k$ and gives rise to anti-Stokes Raman scattering. Here, $\hat{R}_k(\omega_L)$ is called Raman tensor, Q_k^0 is the amplitude of oscillation of mode k and ϕ is an arbitrary phase.

Although classical approach could successfully explained the change in the frequency observed in the scattered radiation, it failed to account for the difference in the intensity observed in Stokes and anti-Stokes Raman. Also, it could not give reasons for the *resonant Raman scattering* phenomenon. Thus, quantum mechanics is required to understand the Raman scattering. According to the quantum picture, when a molecule makes a transition from one state to another with different discrete energies, radiation is absorbed or emitted. To describe the scattering process, it is necessary to treat both the molecule and radiation quantum mechanically. Such a rigorous treatment can be avoided by considering *semi-classical* approach, where the molecule is considered to be a quantum mechanical system, whereas the incident light can be considered as a perturbation to the energy level of the molecule. In this treatment, scattering is viewed as transition probabilities between initial state $|i\rangle$ of the molecule to the final state $|f\rangle$ in the presence of perturbing incident light. It can be shown that for such a transition, a Raman polatizability component is given by [1, 2],

$$\alpha_{kl} = \frac{1}{h}\sum_{r \neq i,f}\left\{\frac{\langle f|p_k|r\rangle\langle r|p_l|i\rangle}{\omega_r - \omega_i - \omega_L - i\Gamma_r} + \frac{\langle f|p_l|r\rangle\langle r|p_k|i\rangle}{\omega_r - \omega_f + \omega_L + i\Gamma_r}\right\} \tag{6}$$

where the sum is over all possible states $|r\rangle$ of the molecule (except for the initial and final states), p_k and p_l called quantum dipole moment operators, $\hbar(\omega_r - \omega_i)$ is the energy difference between state $|i\rangle$ and $|r\rangle$, and Γ_r is inversely proportional to the lifetime of the state $|r\rangle$. As it can be seen from the expression, polarizability depends on the excitation frequency ω_L. Thus by choosing ω_L close to the frequency corresponding to the transition between two states, transition polarizability can be enormously increased, which is essentially the resonance Raman scattering. Quantum mechanics also successfully explained the difference observed in the intensity of Stokes and anti-Stokes Raman.

Let us now consider the vibrational motion of a molecule consisting of *N* atoms. Since the motion of the electrons moves much faster than the nucleus, we can consider the motion of the nucleus separately. It can be shown [1] that potential energy of the nucleus can be approximately expressed by,

$$\begin{aligned} V^n &= \frac{1}{2}\sum_{i,j=1}^{3N} f_{i,j} q_i q_j \\ f_{i,j} &= \left(\frac{\partial^2 V^n}{\partial q_i \partial q_j}\right)_{q_i, q_j = 0} \end{aligned} \tag{7}$$

where q_is are called *reduced mass coordinates*. The scalar terms $f_{i,j}$ are called *force constants* represented as a real symmetric matrix called Hessian matrix $\hat{F}$. The dynamics of a molecule can be written as [1],

$$\hat{F}.\boldsymbol{A} = \omega^2 \boldsymbol{A} \tag{8}$$

where $\boldsymbol{A} = (A_i)_i$ [illegible]. For a particular mode, these quantities completely describe the dynamics of the system. These 3*N* normal modes form a complete system, any arbitrary pattern of the motion of the atoms can be expressed as a combination of these modes. In a molecule, there are 3*N*-6 (3*N*-5, for linear molecule) eigenvectors correspond to the *normal vibrational modes.* The atomic displacement can also be described by set of coordinates Q_k, called *normal coordinates*, defined as $Q_k = A_k . q$. For a single vibrational mode *k*, a single scalar Q_k can describe the atomic displacements.

2.2. Infrared (IR) spectroscopy

Like Raman spectroscopy, IR spectroscopy also involves the interaction of electromagnetic radiation with the molecule, but the nature of interaction is different. Here, the transition from a state n to m takes place as a result of absorption of photon. The process is mediated through the electric dipole moment operator μ_q, given by,

$$\hat{\mu}_q = \sum_{\alpha} e_{\alpha}.q_{\alpha} \tag{9}$$

where e_{α} is the effective charge on the atom α and q_{α} is the distance from the centre of gravity. It can be shown [3] that the transition probability is given by,

$$\left[\hat{\mu}_q\right] = \sum_{k=1}^{3N-6} \hat{\mu}_q^k \langle \psi_m^* | Q_k | \psi_n \rangle \tag{10}$$

where

$$\hat{\mu}_q^k = \left(\frac{\partial \mu_k}{\partial Q_k}\right)_0 \tag{11}$$

A particular mode k is IR active if both $\hat{\mu}_q^k$ and $\langle \psi_m^* | Q_k | \psi_n \rangle$ are non-zero.

2.3. Density functional theory

Over the past few decades, *density functional theory* has gain its popularity as computational tool in quantum chemistry because of its computational cost comparable to that of Hartree-Fock (HF) theory [4], yet accuracy similar to the computationally demanding post-Hartree-Fock methods. Earlier attempts were made to express energy in terms of electron density alone [5–7]. The most successful attempt, suggested by Kohn and Sham [8], was to divide the kinetic energy into two parts. The major contribution is analogous to the HF kinetic energy, which can be calculated precisely, with a small contribution due to correlation. The main idea of Kohn-Sham theory is to calculate the kinetic energy by assuming a non-interacting system. The missing kinetic energy term, existing in real system, is absorbed in a term called *exchange-correlation*. According to the Kohn-Sham approach, the DFT energy can be written as

$$E_{DFT}[\rho] = T_S[\rho] + E_{ne}[\rho] + J[\rho] + E_{xc}[\rho] \tag{12}$$

In the above equation, the first term corresponds to kinetic energy of non-interacting electrons, the second term denotes the nuclear-electron attraction, the third represents Coulomb electron-electron repulsion, and the last term is called the exchange-correlation. In Kohn-Sham theory, the only approximation to be made is for the exchange-correlation functional. Different DFT methods vary in the functional form of this exchange-correlation energy. Kohn-Sham approach is an independent particle model, similar to the HF theory, but simpler than many-particle (correlation) wave function methods. Once proper exchange-correlation functional has been chosen, the next job is to determine a set of orthogonal orbitals corresponding to the minimum energy. There are different types of exchange-correlation functionals. *Local density approximation* (LDA) [9–13] is the simplest exchange-correlation functional, where the density is treated as a uniform electron gas, or slowly varying function at a given point. A better approximation is the *generalized gradient approximation* (GGA) [14–16] that apart from the density itself includes the first derivative of the density as a variable. Another popular type is hybrid functional that is a combination of DFT correlation and DFT and HF exchange [17, 18].

Molecular orbitals (MOs) are generally expressed in terms of basis set. Although it requires infinite functions to represent a MO, the basis sets used are finite for practical purpose. Two common types of orbitals used to form basis set are Slater type orbitals (STOs) [19] and Gaussian type orbitals (GTOs) [20]. Another computationally less costly basis set is contracted basis set [4]. Some examples are 3-21G, 6-31G, 6-311G, etc. Polarization [21] and diffuse [22] functions can be added to each of these basis sets.

2.4. Geometry optimization

The first step of any quantum chemical calculation is the geometry optimization of the molecule. In general, optimization is performed on an isolated molecule, considering non-interacting system in the gas phase. Initial structure is either taken from the literature or obtained from *empirical force field model*. Geometry optimization starts with solving the Kohn-Sham equation self-consistently on the initial geometry. Energy and force on the molecule are calculated from the solution. If the force on the molecule is not zero, a different geometry is assumed. The process of finding a local minimum in the *potential energy surface* is achieved through the *conjugate gradient* [23] method.

2.5. Frequency calculations

Once the equilibrium atomic positions of the atoms are known, the electronic structure can be calculated on the optimized structure. The interaction of the atoms is now known, which enables to calculate force constants. Force constants can be calculated by displacing each atom from their equilibrium positions and recalculating the total energy of the deformed configuration. By numerical differentiation of the total energy, force constants on each atom can be calculated. This enables to construct Hessian matrix for the vibrational modes as described earlier. The frequency needs to be calculated at the same theoretical model and with same basis set as that used in the optimization procedure.

3. Applications of the combined study of vibrational spectroscopy and DFT for conformational analysis

Conformational analysis of molecules is very important in chemistry, biochemistry and structural biology as it determines their functions. Intra- and inter-molecular interactions play important role in determining structure of a molecule. Solvent has significant effect on the structure. The effect of solvent on structure of protein was studied by Zhu et al. [24] on a model tripeptide upon micro-solvation. Doubly terminated tripeptide Z-Aib-Pro-NHMe (Z = benzyloxycarbonyl), an important structure for many natural and synthetic peptides, was studied in solvent-free gas phase and complexed with one- and two-methanol molecules. There are two competing secondary structures present in the peptide as found in the literature. In the condensed phase, it prefers β-turn structure, whereas in gas phase, γ-turn structure is found. In this work, authors carried out IR study combined with the theoretical study of the peptide in the gas phase and in solvent. IR absorption of amide bands [25] and fingerprint region (<1500 cm^{-1}) was recorded. The DFT calculation was carried out at B3LYP/6-311++G(d, p) level for different conformers. Amide bands are sensitive to the secondary structure of the protein and can be used as marker bands. The three amide bands namely amide I, II and III were used to determine the secondary structures of the peptides. It was shown that the model tripeptide in the unsolvated form and one-methanol cluster prefers to form γ-turn structure, whereas in the two-methanol cluster, because of more H-bonding interaction with the methanol, γ-turn structure is favourable that is similar to the condensed phase. From the shift observed in the amide bands and the comparison of experimental spectrum with the theoretical one, it was also concluded that the methanol binding sites are the head and tail part of the tripeptide (see **Figure 1**).

Biological molecules remain in zwitterionic form naturally. Conformational analysis of these molecules is important as they are highly flexible, and their conformations determine H-bonding networks leading to the hydrophobic and hydrophilic interactions. Moreno et al. [26] have studied two amino acids, L-Phenylalanine (L-Phe) and L-Tyrosine (L-Tyr), by IR and Raman spectroscopy in the zwitterionic form to investigate their conformational preferences. For conformational analysis study, different conformers of zwitterionic forms were found by force field method. The DFT calculation was carried out with B3LYP and MO62X functionals together with 6-31+G(d) and 6-311++G(d,p) basis sets. The MO62X functional was found to be better for agreement with the experiment as it describes the non-covalent interactions present in the zwitterionic form well. It was found that the low-frequency region (30–500 cm^{-1} region) contained rich information about different conformers. Comparing the calculated spectra of different conformers and the experimental spectra, contribution on spectra from different conformers could be identified. This study might help in understanding the folding mechanism of proteins.

Conformational preferences of two tripeptides, *N*-acetyl-Phe-Pro-NH_2 and *N*-acetyl-Pro-Phe-NH_2, were studied by Chin et al. [27] in the gas phase with the help of IR spectroscopy and DFT. Modification of *N* and *C* termini helps in investigating the conformational preference of each residue isolated from the neighbouring residues. Initially, the lowest energy conformations were found out by exploring the potential energy surface. Further, geometry optimization and vibrational frequency were carried out by DFT calculation at B3LYP/6-31+G (d) level. The NH stretch region was found to be sensitive to the conformations and used as marker bands. Together with the DFT calculation, it was concluded that most of the conformation assumed repeated γ-turn structure; however, only one conformation of *N*-Ac-Phe-Pro-NH_2 took up β-turn structure. It was also observed that the conformation is dependent on the neighbouring residues of the Phe. In case of *N*-acetyl-Phe-NH_2 or *N*-acetyl-Phe-Pro-NH_2, Phe favours β conformation while in case of *N*-acetyl-Pro- Phe-NH_2, it prefers γ_L conformation.

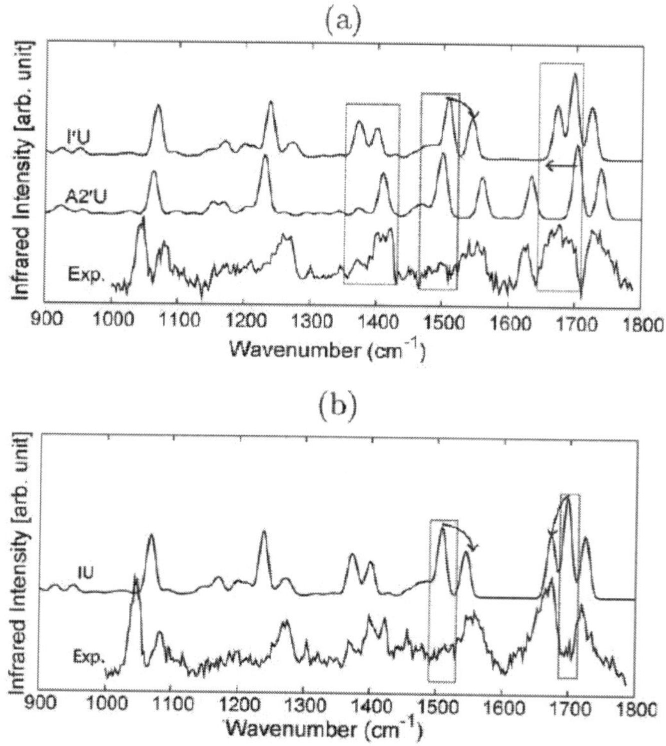

Figure 1. Comparison between theoretical and experimental mid-IR spectra for peptide with (a) one or (b) two methanol. Reproduced from Ref 24 with permission of The Royal Society of Chemistry.

Nicotinic acid and its derivatives are subjects of intense study because of their biological activity and versatile bonding capability. The conformational and vibrational studies (Fourier transform (FT) IR and FT-Raman) of two derivatives of nicotinic acids, 2-bromonicotinic acid and 6-bromonicotinic acids, were reported by Karabacak et al. [28]. The geometrical optimization and vibrational wave numbers were calculated by DFT method using B3LYP as a functional and the 6-311++G(d,p) as basis set. By varying the dihedral angle, energy of different conformations was calculated, and most stable conformation was found. Vibrational assignments were made based on the total energy distribution (TED) [29]. Raman activities calculated by Gaussian program [30] were converted to relative intensity [31]. The calculated and experimental vibrational spectra were compared. Disagreement between experimental and calculated spectra was ascribed to the neglect of inter-molecular interaction present in solid samples as well as the neglect of anharmonicity present in real system. After applying proper scaling factor, calculated spectra resembled well with the experimentally obtained spectra. Dimer structures of the two derivatives and the presence of inter-molecular H-bonding between the pyridine *N* atom and O-H group were also determined.

The room temperature ionic liquids (ILs) have gained interest over last decade due to their potential applications. They are environment friendly because of their non-volatility. Their high thermal stability and tuneable solvent property make them ideal candidates to use as reaction media for organic synthesis or as electrolytes in solar cell. To understand the relation between structure and function of these ILs, details structural and bonding data are required. In the absence of X-ray structural data for the liquids, vibrational spectroscopic data can be helpful in gaining insights into their structures. Several imidazolium-based ILs were studied by Katsyuba et al. [32] by vibrational spectroscopy along with the DFT calculations. The functional chosen for calculation was B3LYP with 6-31G* basis set. Multiple stable structures were found placing anions at different positions. It was shown that halide anions were able to occupy all the positions around the imidazolium ring, whereas perfluorinated anions prefer forward position. Vibrations of the cations depend upon the conformation as well as the interaction with the counterions. In the complex, only imidazolium C-H group vibrations (stretching and out-of-plane vibrations) and perfluoroanions stretching vibrations are affected. Thus, the study could shed some light on the relationship between the structure, vibrations and melting point of ILs.

In another study by the same group [33], combination of vibrational spectroscopy with theoretical calculation was used to quantitatively characterize the strength of H-bonding in ILs. DFT calculation was carried out by B3LYP functional with 6-31G** basis set. Since B3LYP functional does not take into account the van der Waals dispersion forces, dispersion-corrected energy was calculated using DFT-D3 [34] along with Becke-Johnson damping function [35]. It was found that various interactions affect the structure and vibrational spectra. These interactions led to the blue shift of the CH group stretching vibration. This study could help in understanding the role of H-bonding in ILs, which would help in synthesis of ILs of desired physical and chemical properties.

The spectroscopic distinction of bis(trifluoromethanesulphonyl)imide anion ($TFSI^-$) was carried out by Herstedt et al. [36]. The $TFSI^-$ anion can exist in two conformational states, a tronsoid form of C_2 symmetry and a cisoid form of C_1 symmetry. In this work, the effect on Raman and IR spectroscopy due to conformation was investigated with the help of theoretical calculation. DFT calculations were performed at B3LYP/6-31G** level. It was shown that even the effect of conformation on the vibrational spectra was subtle, still it was possible to distinguish TFSI⁻ conformational isomerism. In IR spectroscopy, the regions 130–1380 and 480–60 cm^{-1} were found to be sensitive to the conformational state. It was also possible to distinguish the change due to conformational state from that due to ionic interactions. The ratio of the two conformers in solution was found by measuring IR bands at 602 or 656 (for cisoid, C_1) and 618 cm^{-1} (for transoid, C_2). In Raman spectra, the marker bands were found at 629 (for transoid) and 653 cm^{-1} (for cisoid).

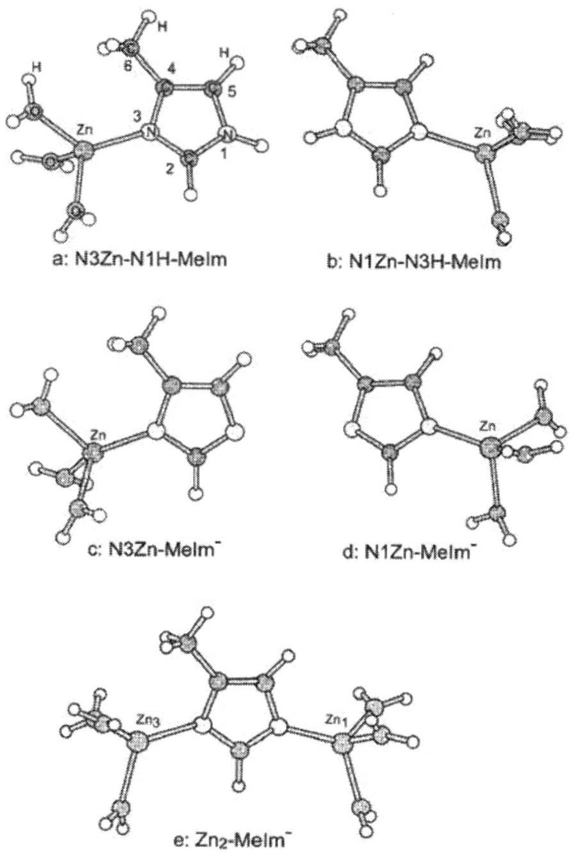

Figure 2. Optimized geometries of the zinc complexes of 4-MeIm in different protonation states. Reprinted with permission from The Journal of Physical Chemistry A, vol. 106, pp.3377–3390. Copyright 2012 American Chemical Society [39].

The alkaloids that are mainly found in plants, and to a lesser quantity in animals, have been studied by ATR-IR and FT-Raman spectroscopy for fast, reliable detection in pharmaceutical products. One such candidate in the group, morphine, was studied by Baranska and Kaczor [37]. For theoretical study, potential energy scanning (PES) was performed to find the minimum energy conformations. Subsequently, geometry optimizations and frequency calculations were carried out on these conformers. It was found that IR spectroscopy could help in monitoring the arrangement of hydroxyl group of the morphine molecule while Raman is not very sensitive to these changes. Chemical modifications of perfluoropolymers were studied using spectroscopic method and quantum calculation by Radice et al. [38]. In the IR absorption spectra, carbonyl stretching region was found to be sensitive to the different polymer degradation pathways.

Metal plays an important role in determining the structure and properties of molecules. Several authors investigated metal-molecule interactions using vibrational spectroscopy. When metal binds to the molecule, few peaks change considerably and those peaks can be used as marker bands to determine its coordination and protonation state. Hasegawa et al. [39] studied different protonated and metal-bound forms of 4-methylimidazole (4-MeIm) vibrational spectroscopy in conjunction with theoretical calculations. The N-H stretching frequencies were seen to be downshifted by ~50 cm^{-1} on complexation with Zn while CH stretching vibration showed upshifts. The CC and CN stretching showed complicated behaviour in the Zn-bound form as these vibrations are coupled with other vibrations, and upon metal binding, some of the coupling changed. From these changes, together with theoretical study, the different metal-bound forms of the histidine could be identified (see **Figure 2**).

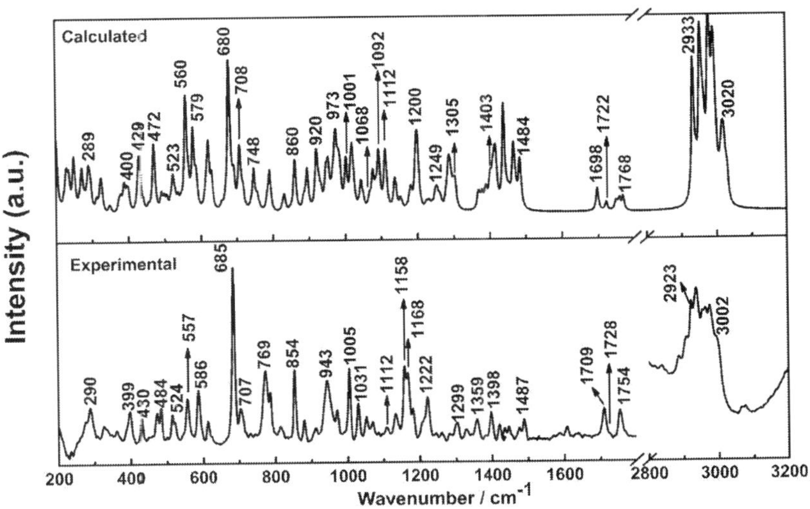

Figure 3. A comparison between experimental and calculated Raman spectrum. Reprinted from Journal of Molecular Structure, vol. 1102, Kundu *et al.*, Raman, IR and DFT studies of mechanism of sodium binding to urea catalyst, pp 267-274, Copyright (2015), with permission from Elsevier [41].

A bis-camphorsulphonyl was synthesized as a hydrogen-bonding catalyst [40]. It was found that the enantioselectivity capability of the catalyst is poor. However, when it was complexed with sodium ion, the selectivity of the catalyst increased significantly. The X-ray crystallographic structure showed that the native conformation of the catalyst is unfavourable for enantioselectivity. The X-ray data for the complex form could not be obtained as it was not soluble in most of the organic solvents, and thus crystal could not be formed. Since vibrational spectroscopy does not require crystalline samples, we were interested to probe the structure of the catalyst in its free and sodium-bound form by vibrational spectroscopy, both experimentally and theoretically [41]. For DFT calculations, we have chosen B3LYP/6-31G (d,p) level. To include the inter-molecular interaction present in the solid form, we considered dimer structure of the catalyst for frequency calculation. A comparison between calculated Raman spectrum and experimentally obtained one is shown in **Figure 3**. To study the effect of sodium

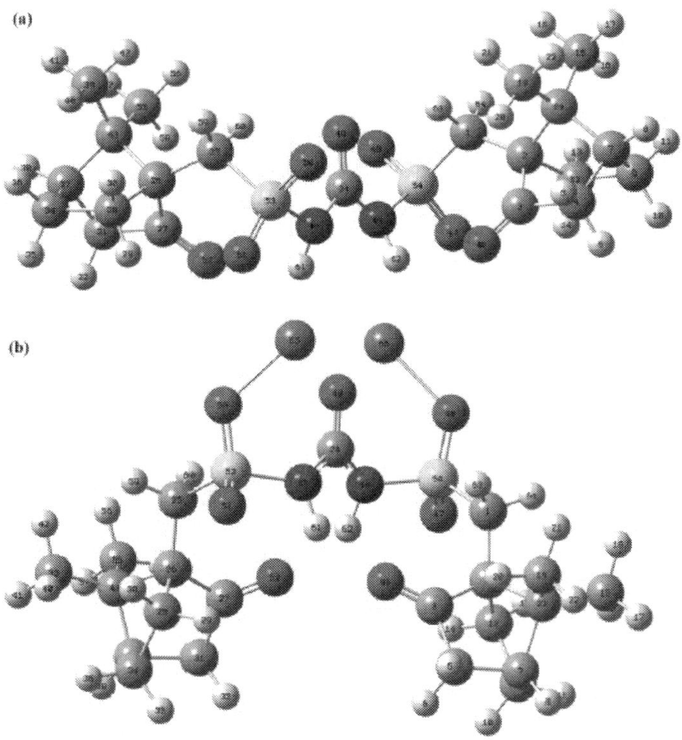

Figure 4. Optimized structure of urea catalyst in (a) free and (b) Na-bound form. Colour representation: white–hydrogen, red–oxygen, grey–carbon, blue–nitrogen, violet–sodium, yellow–sulphur. Reprinted from Journal of Molecular Structure, vol. 1102, Kundu *et al.*, Raman, IR and DFT studies of mechanism of sodium binding to urea catalyst, pp 267-274, Copyright (2015), with permission from Elsevier [41].

ion, we have compared the monomer form of the catalyst and its complex (see **Figure 4**). In our predicted structure, we considered two sodium forming bond with the oxygen atoms of the urea carbonyl and sulphonyl groups. The shift of stretching frequency of carbonyl, sulphonyl, C-N and N-H frequencies observed experimentally could be qualitatively reproduced in the theoretical calculation. The study showed how in the absence of any X-ray data, vibrational spectroscopy together with theoretical study was helpful in predicting the conformation of the catalyst in complex form (see **Figure 5**).

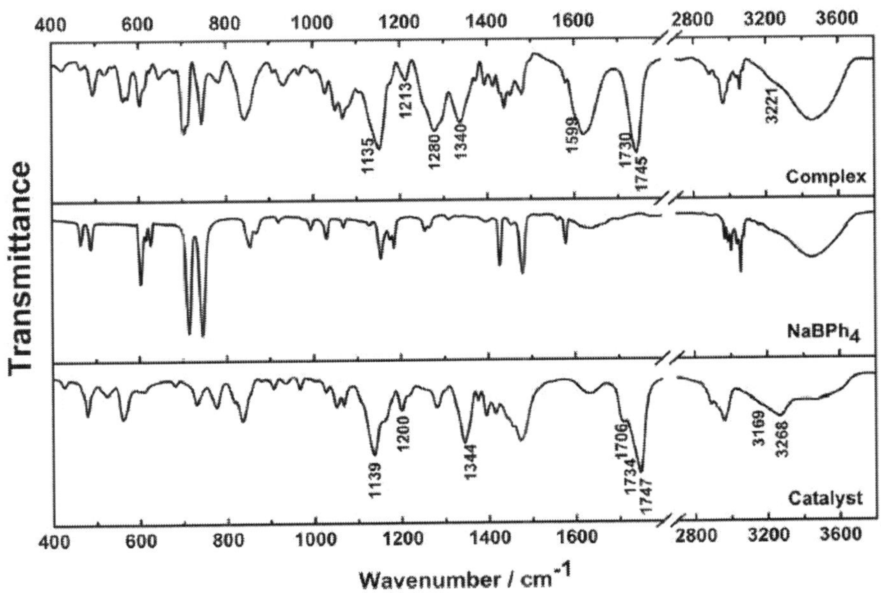

Figure 5. Experimental IR spectra of urea catalyst, NaBPh4 and complex showing the changes in the spectrum of catalyst upon sodium binding. Reprinted from Journal of Molecular Structure, vol. 1102, Kundu *et al.*, Raman, IR and DFT studies of mechanism of sodium binding to urea catalyst, pp 267–274, Copyright (2015), with permission from Elsevier [41].

4. Conclusions

In conclusion, in this chapter, we discuss two complementary vibrational techniques, Raman and IR. We give brief theoretical background for this technique. We also discuss about density functional theory, widely used to predict the spectrum of a molecule and help in assigning the bands. Then, we discuss the use of the combination of these two techniques in different molecular systems. We hope this will give the readers an idea of the potential of these techniques for various conformational analyses.

Author details

Partha P. Kundu[1] and Chandrabhas Narayana[2*]

*Address all correspondence to: cbhas@jncasr.ac.in

1 Department of Physics, M. S. Ramaiah University of Applied Sciences, Bangalore, Karnataka, India

2 Light Scattering Laboratory, Chemistry and Physics of Materials Unit, Jawaharlal Nehru Centre for Advanced Scientific Research, Bangalore, Karnataka, India

References

[1] Le Ru, E.; Etchegoin, P. *Principles of Surface-Enhanced Raman Spectroscopy: And Related Plasmonic Effects;* Elsevier Science: Amsterdam 2008.

[2] Long, D. A. *The Raman Effect: A Unified Treatment of the Theory of Raman Scattering by Molecules;* John Wiley and Sons Ltd: Chichester,West Sussex 2002.

[3] Sieber, F.; Hildebrandt, P. *Vibrational Spectroscopy in Life Science;* Wiley-VCH Verlag GmbH and Co. KGaA: Weinheim 2008.

[4] Jensen, F. *Introduction to Computational Chemistry;* Wiley: Chichester,West Sussex 2006.

[5] Thomas, L. H. *The Calculation of Atomic Fields*. Cambridge Univ Press: 1927; pp. 542–548.

[6] Fermi, E. A statistical method for the determination of some atomic properties and the application of this method to the theory of the periodic system of elements. *Zeitschrift für Physik* 1928, *48* (73), 29.

[7] Hohenberg, P.; Kohn, W. Inhomogeneous electron gas. *Physical Review* 1964, *136* (3B), B864.

[8] Kohn, W.; Sham, L. J. Self-consistent equations including exchange and correlation effects. *Physical Review* 1965, *140* (4A), A1133.

[9] Dirac, P. A. *Note on Exchange Phenomena in the Thomas Atom*. Cambridge Univ Press: 1930; pp. 376–385.

[10] Ceperley, D. Ground state of the fermion one-component plasma: A Monte Carlo study in two and three dimensions. *Physical Review B* 1978, *18* (7), 3126.

[11] Ceperley, D. M.; Alder, B. J. Ground state of the electron gas by a stochastic method. *Physical Review letters* 1980, *45* (7), 566–569.

[12] Vosko, S. H.; Wilk, L.; Nusair, M. Accurate spin-dependent electron liquid correlation energies for local spin density calculations: a critical analysis. *Canadian Journal of Physics* 1980, *58* (8), 1200–1211.

[13] Perdew, J. P.; Zunger, A. Self-interaction correction to density-functional approximations for many-electron systems. *Physical Review B* 1981, *23* (10), 5048–5079.

[14] Perdew, J. P.; Yue, W. Accurate and simple density functional for the electronic exchange energy: generalized gradient approximation. *Physical Review B* 1986, *33* (12), 8800.

[15] Becke, A. D. Density-functional exchange-energy approximation with correct asymptotic behavior. *Physical Review A* 1988, *38* (6), 3098.

[16] Lee, C.; Yang, W.; Parr, R. G. Development of the Colle-Salvetti correlation-energy formula into a functional of the electron density. *Physical Review B* 1988, *37* (2), 785–789.

[17] Becke, A. D. Density-functional thermochemistry. III. The role of exact exchange. *The Journal of Chemical Physics* 1993, *98* (7), 5648–5652.

[18] Stephens, P. J.; Devlin, F. J.; Chabalowski, C. F.; Frisch, M. J. Ab initio calculation of vibrational absorption and circular dichroism spectra using density functional force fields. *The Journal of Physical Chemistry* 1994, *98* (45), 11623–11627.

[19] Slater, J. C. Atomic shielding constants. *Physical Review* 1930, *36* (1), 57.

[20] Boys, S. F. Electronic wave functions. I. A general method of calculation for the stationary states of any molecular system. *Proceedings of the Royal Society of London. Series A. Mathematical and Physical Sciences* 1950, *200* (1063), 542–554.

[21] Francl, M. M.; Pietro, W. J.; Hehre, W. J.; Binkley, J. S.; Gordon, M. S.; DeFrees, D. J.; Pople, J. A. Self-consistent molecular orbital methods. XXIII. A polarization-type basis set for second-row elements. *The Journal of Chemical Physics* 1982, *77*, 3654.

[22] Frisch, M. J.; Pople, J. A.; Binkley, J. S. Self-consistent molecular orbital methods 25. Supplementary functions for Gaussian basis sets. *The Journal of Chemical Physics* 1984, *80* (7), 3265–3269.

[23] Dongarra, J. J.; Duff, I. S.; Sorensen, D. C.; Vorst, H. V. D. Solving linear systems on vector and shared memory computers. *Society for Industrial and Applied Mathematics* Philadelphia, PA, USA: 1990.

[24] Zhu, H.; Blom, M.; Compagnon, I.; Rijs, A. M.; Roy, S.; von Helden, G.; Schmidt, B. Conformations and vibrational spectra of a model tripeptide: change of secondary structure upon micro-solvation. *Physical Chemistry Chemical Physics* 2010, *12* (14), 3415–3425.

[25] Ferraro, J.; Nakamoto, K.; Brown, C. *Introductory Raman Spectroscopy;* Academic Press, San Diego: 2002.

[26] Moreno, J. R. A.; Moreno, M. d. M. Q.; Ureña, F. P.; González, J. J. L. Conformational preference of short aromatic amino acids from the FT-IR, FT-Raman and Far-IR

spectroscopies, and quantum chemical calculations: l-phenylalanine and l-tyrosine. *Tetrahedron: Asymmetry* 2012, *23* (14), 1084–1092.

[27] Chin, W.; Mons, M.; Dognon, J. P.; Piuzzi, F.; Tardivel, B.; Dimicoli, I. Competition between local conformational preferences and secondary structures in gas-phase model tripeptides as revealed by laser spectroscopy and theoretical chemistry. *Physical Chemistry Chemical Physics* 2004, *6* (10), 2700–2709.

[28] Karabacak, M.; Cinar, M.; Ermec, S.; Kurt, M. Experimental vibrational spectra (Raman, infrared) and DFT calculations on monomeric and dimeric structures of 2- and 6-bromonicotinic acid. *Journal of Raman Spectroscopy* 2010, *41* (1), 98–105.

[29] Baker, J.; Jarzecki, A. A.; Pulay, P. Direct scaling of primitive valence force constants: an alternative approach to scaled quantum mechanical force fields. *The Journal of Physical Chemistry A* 1998, *102* (8), 1412–1424.

[30] M.J. Frisch, G.W. Trucks, H.B. Schlegel, G.E. Scuseria, M.A. Robb, J.R. Cheeseman, J.A. Montgomery Jr., T. Vreven, K.N. Kudin, J.C. Burant, J.M. Millam, S.S. Iyengar, J. Tomasi, V. Barone, B. Mennucci, M. Cossi, G. Scalmani, N. Rega, G.A. Petersson, H. Nakatsuji, M. Hada, M. Ehara, K. Toyota, R. Fukuda, J.Hasegawa, M. Ishida, T. Nakajima, Y. Honda, O. Kitao, H. Nakai, M. Klene, X. Li, J.E. Knox, H.P. Hratchian, J.B.Cross, V. Bakken, C. Adamo, J. Jaramillo, R. Gomperts, R.E. Stratmann, O. Yazyev, A.J.Austin, R. Cammi, C. Pomelli, J.W. Ochterski, P.Y.Ayala, K. Morokuma, G.A. Voth, P. Salvador, J.J. Dannenberg, V.G. Zakrzewski, S. Dapprich, A.D.Daniels, M.C.Strain, O. Farkas, D.K. Malick, A.D. Rabuck, K. Raghavachari, J.B. Foresman, J.V.Ortiz, Q. Cui, A.G. Baboul, S. Clifford, J. Cioslowski, B.B. Stefanov, G. Liu, A. Liashenko, P. Piskorz, I. Komaromi, R.L.Martin, D.J. Fox, T. Keith, M.A. Al-Laham, C.Y. Peng, A. Nanayakkara, M. Challacombe, P.M.W. Gill, B. Johnson, W. Chen, M.W. Wong, C. Gonzalez, J.A. Pople, Gaussian 03, Revision C.02, Gaussian, Inc., Wallingford CT, 2004.

[31] Keresztury, G.; Holly, S.; Besenyei, G.; Varga, J.; Wang, A.; Durig, J. R. Vibrational spectra of monothiocarbamates-II. IR and Raman spectra, vibrational assignment, conformational analysis and ab initio calculations of S-methyl-N,N-dimethylthiocarbamate. *Spectrochimica Acta Part A: Molecular Spectroscopy* 1993, *49* (13–14), 2007–2026.

[32] Katsyuba, S. A.; Zvereva, E. E.; Vidiš, A.; Dyson, P. J. Application of density functional theory and vibrational spectroscopy toward the rational design of ionic liquids. *The Journal of Physical Chemistry A* 2007, *111* (2), 352–370.

[33] Katsyuba, S. A.; Vener, M. V.; Zvereva, E. E.; Fei, Z.; Scopelliti, R.; Laurenczy, G.; Yan, N.; Paunescu, E.; Dyson, P. J. How strong is hydrogen bonding in ionic liquids? Combined x-ray crystallographic, infrared/Raman spectroscopic, and density functional theory study. *The Journal of Physical Chemistry B* 2013, *117* (30), 9094–9105.

[34] Grimme, S.; Antony, J.; Ehrlich, S.; Krieg, H. A consistent and accurate ab initio parametrization of density functional dispersion correction (DFT-D) for the 94 elements H-Pu. *The Journal of Chemical Physics* 2010, *132* (15), 154104.

[35] Grimme, S.; Ehrlich, S.; Goerigk, L. Effect of the damping function in dispersion corrected density functional theory. *Journal of Computational Chemistry* 2011, *32* (7), 1456–1465.

[36] Herstedt, M.; Smirnov, M.; Johansson, P.; Chami, M.; Grondin, J.; Servant, L.; Lassegues, J. C. Spectroscopic characterization of the conformational states of the bis (trifluoromethanesulfonyl) imide anion (TFS⁻). *Journal of Raman Spectroscopy* 2005, *36* (8), 762–770.

[37] Baranska, M.; Kaczor, A. Morphine studied by vibrational spectroscopy and DFT calculations. *Journal of Raman Spectroscopy* 2012, *43* (1), 102–107.

[38] Radice, S.; Canil, G.; Millefanti, S.; Tortelli, V.; Milani, A.; Castiglioni, C. Chemical modification of Hyflon® AD copolymer end groups by means of physical and chemical treatments. A joint spectroscopic and quantum chemical investigation. *Journal of Molecular Structure* 2015, *1090*, 44–52.

[39] Hasegawa, K.; Ono, T. A.; Noguchi, T. Ab initio density functional theory calculations and vibrational analysis of zinc-bound 4-methylimidazole as a model of a histidine ligand in metalloenzymes. *The Journal of Physical Chemistry A* 2002, *106* (14), 3377–3390.

[40] Chittoory, A. K.; Kumari, G.; Mohapatra, S.; Kundu, P. P.; Maji, T. K.; Narayana, C.; Rajaram, S. Conformational change in a urea catalyst induced by sodium cation and its effect on enantioselectivity of a Friedel-Crafts reaction. *Tetrahedron* 2014, *70* (21), 3459–3465.

[41] Kundu, P. P.; Kumari, G.; Chittoory, A. K.; Rajaram, S.; Narayana, C. Raman, IR and DFT studies of mechanism of sodium binding to urea catalyst. *Journal of Molecular Structure* 2015, *1102*, 267–274.

Chapter 10

Geometric and Electronic Properties of Porphyrin and its Derivatives

Metin Aydin and Daniel L. Akins

Additional information is available at the end of the chapter

http://dx.doi.org/10.5772/64583

Abstract

In this chapter, we discuss protonation and substitution effects on the absorption spectra of porphyrin molecules based on density functional theory (DFT) and time-dependent DFT calculations. The results of the calculations are compared with experimental data. The calculations show that protonation of core nitrogen atoms of porphyrin and *meso*-substituted porphyrins produces a substantial shift in Soret and Q-absorption bands, relative to their positions in corresponding nonprotonated and nonsubstituted chromophores. A relaxed potential energy surface (RPES) scan has been utilized to calculate ground and excited state potential energy surface (PES) curves as functions of the rotation of one of the *meso*-substituted sulfonatophenyl groups about dihedral angles θ (corresponding to C_α—C_m—C_φ—C) ranging from 40 to 130°, using 10° increments. The ground state RPES curve indicates that when the molecule transitions from the lowest ground state to a local state, the calculated highest potential energy barrier at the dihedral angle of 90° is only 177 cm^{-1}. This finding suggests that the *meso*-sulfonatophenyl substitution groups are able to rotate around C_m—C_φ bond at room temperature because the thermal energy (k_BT) at 298 K is 207.2 cm^{-1}. Furthermore, the calculations show that the geometric structure of the porphyrin is strongly dependent on protonation and the nature of the *meso*-substituted functional groups.

Keywords: porphyrins, protonation, absorption, PES, DFT calculation

1. Introduction: overview of molecular spectroscopy and quantum calculations

Spectroscopy is the branch of science dealing with the interaction of electromagnetic and other forms of radiated energy with matter. The earliest prospect of making spectroscopic measurements came with the observation that visible light can be dispersed by an optical prism, and the concomitant recognition that matter could be intimately investigated through its response to optical radiative energy as a function of frequency, defining what is referred to as optical spectroscopy. As it turns out, optical spectroscopy is a useful approach for both qualitative and quantitative studies of physical and chemical processes involving matter in most of its states by measurement of absorption, emission, or scattering of electromagnetic radiation; moreover, optical spectroscopic measurements can be very sensitive, nondestructive, and typically require only small amounts of material for analysis.

Absorption spectra are usually acquired for analytes dissolved in nonabsorbing solvents. And, ideally the absorbance of a dissolved analyte depends linearly on concentration, thereby resulting in an absorption spectrum providing quantitative measurement of the analyte's concentration in solution, arrived at by applying the Beer-Lambert Law. In particular, since absorption spectra of molecules depend on their energy level structure, absorption spectra are not only useful for identifying isolated molecules, but also can be used to probe intermolecular interactions (e.g., effects of aggregation) that affect energy level structure.

It is to be noted that molecules that are excited to higher energy than the lowest excited state above the ground electronic state can relax to lower excited levels by a range of intrinsic processes. Included among such deactivation processes are emission of radiation, more popularly referred to as luminescence, as well as processes that are nonradiative in nature, where lower energy states can be directly populated without the emission of photons. Luminescence from such intermediate states can be defined as fluorescence or phosphorescence, where fluorescence is a process by which electronically excited molecules return to a lower electronic state of the same spin multiplicity (which is often the electronic ground state) by emitting a photon; phosphorescence, on the other hand, is the corresponding transition between states with different spin multiplicities. While fluorescence is a spin-allowed process and generally occurs rapidly, phosphorescence is spin forbidden and is typically a slower relaxation process.

Paths by which nonradiative relaxation can occur include, but are not limited to, such phenomena as collisional energy transfer, electron or proton transfer processes, change of molecular conformation, photochemistry, formation of excited state complexes (e.g., excimers or exciplexes), as well as the classic processes of internal conversion (IC) (e.g., vibrational relaxation) and intersystem crossing (ISC) (e.g., singlet-triplet conversion).

It is to be noted that transient intermediates are likely to form during IC and ISC radiationless processes, and detection of such species, if at all possible, often necessitates the use of highly sensitive ultrafast optical (or other) techniques.

The aforementioned phenomena are depicted more fully in **Figure 1** that shows a combined Perrin-Jablonski diagram illustrating the different processes involved in the interaction of a molecule with photons in the spectral region between 300 and 1500 nm. Photophysical processes for an isolated molecule would occur via transitions between the different internal energy states shown in **Figure 1**.

http://dx.doi.org/10.5772/64583

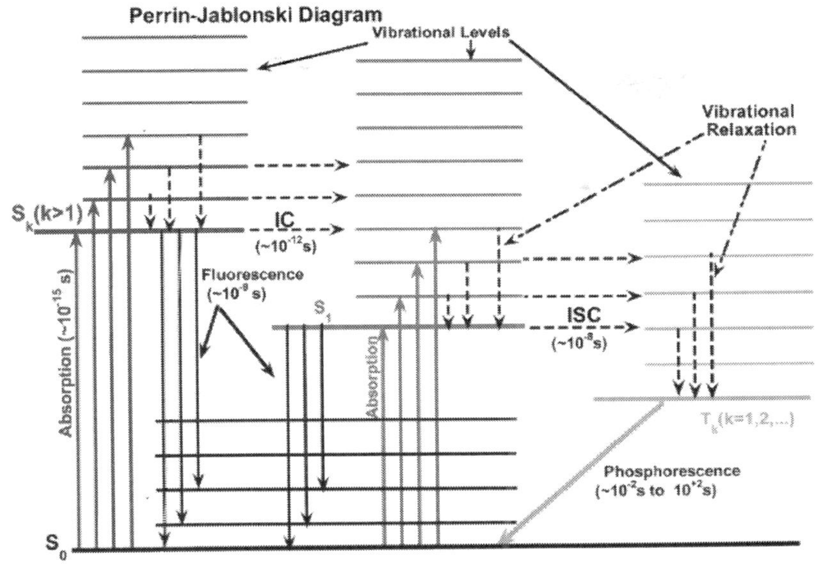

Figure 1. A general Perrin-Jablonski diagram, where S and T stand for singlet and triplet electronic states, respectively. IC and ISC represent "internal conversion" and "intersystem crossing," respectively.

Using **Figure 1** for discussion, in the gaseous or solution phase at room temperature a molecular system is generally in its ground state (S_0). The transition from the ground state to an upper vibroelectronic state by absorption of a photon would take place within ca. 10^{-15} s, which is much faster than the emission of the photon from an excited electronic state ($S_{k>1}$) to its ground state (ca. 10^{-8} s). As suggested earlier, all of the excited molecules might not directly return to their ground state by emission of radiation ($S_{k>0} \rightarrow S_0 + h\nu$), but some may return by internal conversion (IC). For example, when a molecule is excited to an upper vibroelectronic state ($S_k > 1$), it could undergo relaxation to the first excited singlet level S_1 (in 10^{-12} s) by way of vibrational coupling between these states before undergoing additional vibrational relaxation and returning to the lowest singlet electronic energy level (as per Kasha's Rule). As illustrated in **Figure 1**, another possible pathway is that a molecule in the first excited singlet level S_1 may undergo a transition to a triplet state by ISC, which can relax to the lowest triplet state (T_1) via vibrational relaxation and IC processes. The molecule can then return to its ground state through phosphorescence. There is also the possibility of a transition from T_1 to S_1 followed by the transition to S_0 by emitting a photon. This latter path is not shown in **Figure 1**.

During the past two decades, there has been very intense theoretical research on the physical and chemical properties of molecular structures. Computational chemistry is a powerful tool for investigation of molecules, surfaces, and interfaces at the electronic structure level. Various molecular properties directly comparable with experiment such as structural parameters, thermodynamic data, and vibrational spectra can be obtained by solving quantum mechanical equations. When the result of a theoretical prediction is consistent with an experimental measurement, one can more confidently interpret the experimental result. Computational studies are not only carried out in order to provide an understanding of experimental data, such as the position and source of spectroscopic peaks, but also can be used to predict the existence of unobserved molecules, intermediates, or to explore reaction mechanisms that are not readily studied experimentally.

The particular computational method chosen depends critically on the desired accuracy (qualitative vs. quantitative) sought, the size of the system, and available computational capacities. At a qualitative level, especially for large systems, molecules can be treated by classical mechanics using a class of methods called molecular mechanics. The structure of a protein containing hundreds of atoms might be calculated this way. Somewhat more quantitatively accurate are the semiempirical methods. These methods (such as PM3) use experimentally measured parameters that approximate parts of a quantum mechanical system. These latter methods can be fast, and give good results if the molecule of interest is very similar to those used to determine the parameters. However, many molecules of interest (i.e., transition metal complexes) do not have sufficiently good parameter sets to be accurately calculated using such methods. *Ab initio* methods, however, do not assume experimental parameters, but instead attempt to calculate the molecular wave functions directly using a variety of approximation techniques. These methods, such as Hartree-Fock (HF) and MP2, can be very accurate for some observable phenomena, but can also be computationally expensive. HF with a mid-level basis set will often be used as a good starting point for more accurate calculations, or as a relatively fast way of getting qualitative data.

The fourth type of computational methods are the density functional theory (DFT) methods. With a few exceptions, DFT is the most cost-effective method to achieve a given level of quantitative accuracy. It incorporates electron correlation and is computationally less expense.

In addition to choosing a method, one must also choose a basis set. A basis set is a set of functions that substitute for the "real" atomic orbitals (AOs) of a system and should approximate the real wave functions well enough to give chemically meaningful and close approximations to the correct values of measurable quantities being considered (e.g., geometry and energy). Using more complex basis sets improves results at the cost of increased computer time to make a calculation (i.e., increased computational expense). Basis sets, in order to allow electron-electron correlation to be taken into account, must incorporate polarization terms to allow distortions in orbital shapes; they must also incorporate diffuse functions (especially necessary when you have a molecule with weakly bound electrons) (as in the case of some anions and for some transition states); and they must account for relativistic effects (for heavier atoms).

In this chapter, we will discuss protonation and *meso*-substitution effects on geometric and electronic structures of the porphyrin macrocycle based on quantum chemical methods. This chapter is also a sister publication for an accompanying article in another chapter of the present book entitled "Infrared and Raman Spectroscopic Characterization of Porphyrin and its Derivatives." Details about the calculations that we have made are provided in Section 4.

2. Porphyrin macrocycle

Porphyrin and its derivatives have received extensive attention from both experimentalists and theoreticians since they have been found to have many important potential applications in a broad variety of high technology and biomedical fields. Indeed, in recent years analyses of geometric and spectroscopic properties of molecular systems incorporating porphyrins have produced a substantial body of information that has greatly expanded our knowledge of high efficiency utilization of solar energy [1–5] and the use of such synthetic molecular analogs as active agents in molecular electronic devices [6, 7]. Also, a great deal of interest has been shown for the use of porphyrin-like molecular systems as therapeutic drugs and photosensitizers in photodynamic therapy of cancer [8] and their possible use in the treatment of nonmalignant conditions such as psoriasis, treatment of blocked arteries, and for the treatment of pathological and bacterial viruses [9] and HIV [10]. The biological importance of porphyrins essentially derives from their physicochemical properties that basically determine their photophysical behavior. Additionally, aggregation and axial ligation lead to significant changes in absorption spectra as well as quantum yield, fluorescence lifetime, and triplet state lifetime [11–13]. More detailed information about porphyrins can be obtained in the *Handbook of Porphyrin Science* [14, 15].

Of particular note is the observation that optical properties of porphyrin can be altered by the protonation or metallation of nitrogen atoms in its core structure, with electronic changes as a result of structural alterations such as flattening and distortion from planarity of the macrocycle, interactions between porphyrins (aggregation), redox reactions, and solvent effects. A few porphyrins have been found to form aggregates; a requirement of being zwitterionic character upon protonation of macrocycle core nitrogen atoms. It has also been suggested that aggregation is facilitated by interaction with proteins [16, 17] and surfactants [18].

Indeed, aggregation of the anionic porphyrin *meso*-tetrakis(*p*-sulfonatophenyl)porphyrin (TSPP) has been discussed extensively [9, 19–21]. The electronic absorption spectrum of monomeric TSPP at neutral pH exhibits multiple electronic transition bands with an intense peak maximum at about 410 nm and several weak transitions in the region of 500 to 700 nm in aqueous solutions. The intense transition at 410 nm is known as Soret or B-band, and the weaker bands are termed Q-bands. In very acidic medium, the TSPP becomes protonated, and one finds that highly ordered molecular aggregates are formed. While the protonated TSPP (i.e., H_4TSPP) shows a strong absorbance peak at around 430 nm (Soret band) along with the weak bands in the Q-bands region, the aggregated-H_4TSPP spectrum displays a Soret band at about 490 nm [21] and difference Q-type bands at longer wavelengths.

The developments in computing facilities and the sophisticated computation programs, with increasingly efficient algorithms, especially the fundamental improvements in the treatment of electron correlation based on density-functional theory (DFT) [22], have combined to allow quantum chemical methods to routinely handle molecular systems containing hundreds of atoms. As a result, DFT has become one of the most important techniques used by theoreticians to provide deep insight into spectroscopic and structural properties, even for complex molecular systems, especially those of large sizes such as the porphyrinoids [23–28].

In this chapter, we discuss the effect of *meso*-substitution groups and protonation of the N atoms at the core (of parent-porphine or porphyrin macrocycle) of its geometric and electronic structures. The density functional theory (DFT) and time-dependent DFT (TD-DFT) have been employed to calculate the geometric structures and electronic transition energies of porphyrin and derivatives in water used as solvent. The compounds studied here are unsubstituted porphyrin (free-base porphin, FBP), *meso*-tetraphenylporphyrin (TPP), *meso*-tetrakis(*p*-sulfonatophenyl)porphyrin (TSPP), protonated-FBP (H_4FBP), deuterated-H_4FBP (D_4FBP), protonated-TPP (H_4TPP or dicationic TPP), deuterated-H_4TPP (D_4TPP), protonated-TSPP (H_4TSPP or dianionic-TSPP), deuterated-H_4TSPP (D_4TSPP),dicationic TSPP (H_8TSPP), and deuterated-H_8TSPP (D_8TSPP). The possible internal conversion (IC) and intersystem crossing (ISC) processes for the porphyrin and derivatives are also discussed based on the results of the TD-DFT calculations. Furthermore, the relaxed potential energy surface scans were employed to study the minimum potential energy pathways for the ground and excited states of the TSPP molecule as a function of rotation C_m—C_φ bond (or dihedral angle (C_α—C_m—C_φ—C(ph)). We would like to point out that the calculated and experimental data are taken from our prior work [29].

3. Structures of porphyrin and its derivatives

DFT theory at the B3LYP/6-311G(d,p) level was performed to predict the geometric parameters of the ground state of the parent porphyrin and its derivatives in water used as a solvent. The

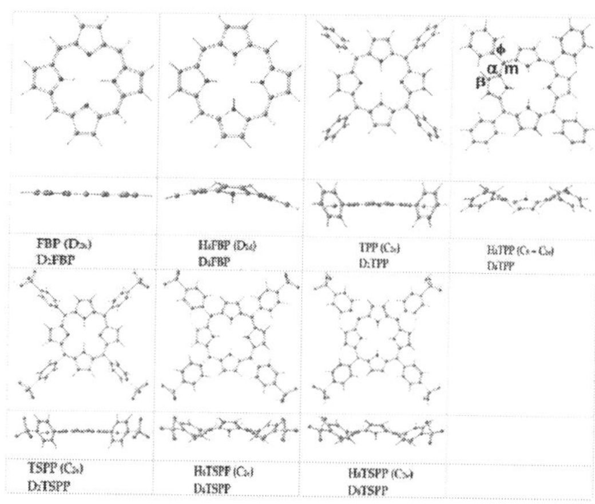

Figure 2. Optimized geometric structures of unsubstituted porphyrin (FBP), *meso*-tetraphenylporphyrin (TPP), anionic *meso*-tetrakis(*p*-sulfonatophenyl)porphyrin (TSPP), and their protonated derivatives (H_4FBP, H_4TPP, H_4TSPP, and H_8TSPP) in water at the B3LYP/6-311G(d,p) level of DFT.

optimized ground state geometry of these compounds is provided in **Figure 2**. The selected bond angles and dihedral angles are given in **Table 1**. Results of the calculations show that while the *meso* substitution of porphyrin with tetraphenyl or tetrasulfonatophenyl brings about slight out-of-plane distortion from the planar structure of the macrocycle within 3–5° for both TPP and TSPP, the protonation of the porphyrin core gives rise to a substantial distortion from planarity ranging from 10 to 20° for H_4FBP, H_4TPP, H_4TSPP, and H_8TSPP, principally due to repulsive interactions between the H atoms bonded to core N atoms. Moreover, with reference to the average plane of the macrocycle (**Figure 2** and **Table 1**), the peripheral phenyl and sulfonatophenyl substituents are tilted by an angle of about 72° for the nonprotonated structures TPP and TSPP, and about 48° for protonated H_4TPP, H_4TSPP, and H_8TSPP. Rotation of the *meso* substituents is attributed to repulsive interactions between H atoms on C_β and C in phenyl, as well electron correlation effect.

	FBP	TPP	TSPP	H_4FBP	H_4TPP	H_4TSPP	H_8TSPP
$D(C_\beta, C_\alpha, C_m, C_\varphi)$	N.A	3.6	4.0	N.A	19.4	19.6	19.0
$D(C_\beta, C_\alpha, C_m, C_{\alpha'})$	180.0	−176.5	−176.4	−169.5	−166.6	−160.5	−161.6
$D(C_\alpha, C_m, C_{\alpha'}, N)$	0.0	2.3	2.4	10.3	20.6	20.7	20.0
$D(C_\alpha, C_m, C_\varphi, C)$	N.A.	72.1	71.0	N.A	47.8	47.2	49.6
$D(C_\alpha, C_m, C_{\alpha'}, C_{\beta'})$	180.0	−177.2	−176.9	−169.5	−160.6	−160.5	−160.8
$D(C_m, C_{\alpha'}, C_{\beta'}, C_{\beta''})$	180.0	179.8	179.7	−178.8	−176.4	−174.4	−176.9
$D(C_{\alpha'}, C_{\beta'}, C_{\beta''}, C_{\alpha''})$	0.0	0.0	0.0	0.0	0.0	0.0	0.0
$D(C_{\beta'}, C_{\beta''}, C_{\alpha''}, C_{m'})$	180.0	−179.8	−179.7	178.8	176.4	176.4	176.9
$D(C_{\beta''}, C_{\alpha''}, C_{m'}, C_{\varphi'})$	N.A	−2.6	−2.6	N.A	−19.4	−19.4	−18.9
$D(C_{\alpha''}, C_{m'}, C_{\varphi'}, C)$	N.A	−72.1	−70.9	N.A	−47.8	−47.5	−49.9
$D(C_{\alpha'}, N, C_{\alpha''}, C_{\beta''})$	0.0	0.4	0.4	2.2	4.2	4.2	4.1
$A(C_\beta, C_\alpha, C_m)$	127.9	123.0	123.1	127.7	128.0	128.0	128.0
$A(C_\alpha, C_m, C_\varphi)$	N.A	118.2	118.2	N.A	118.3	118.3	118.2
$A(C_\varphi, C_m, C_{\alpha'})$	N.A	116.6	116.5	N.A	118.3	118.4	118.4
$A(C_\alpha, C_m, C_{\alpha'})$	127.0	125.2	125.3	127.4	123.4	123.4	123.9
$A(C_m, C_{\alpha'}, C_{\beta'})$	123.4	126.9	126.9	127.7	128.0	128.0	127.9
$A(N, C_\alpha, C_m)$	125.6	126.3	126.2	125.5	125.4	125.3	125.2
$A(C_m, C_{\alpha'}, N)$	125.7	126.6	126.6	125.5	125.4	125.3	125.3

Table 1. Comparison of selected dihedral angles (*D*) and bond angles (*A*) of the parent porphyrin and its derivatives: unsubstituted porphyrin (FBP), *meso*-tetraphenylporphyrin (TPP), and anionic *meso*-tetrakis(*p*-sulfonatophenyl) porphyrin (TSPP) with their protonated structures (H_4FBP, H_4TPP, H_4TSPP and H_8TSPP), calculated in water as solvent at the B3LYP/6-311G(d,p) level of DFT.

It is ascertained that the calculated bond lengths are consistent with X-ray data within ca. ± 0.01 Å. Hence, one can conclude that protonation of the porphyrin core, in addition to causing deviation from planarity of the macrocycle, also simply has an effect on the tilt angles of the phenyl and *p*-sulfonatophenyl substituent groups.

In Section 3.1, protonation and *meso* substitution of the porphyrin macrocycle is found to not only affect the geometric structure, but also, due to the change in molecular symmetry, lead to significant changes in electronic band positions as well as the number of bands.

3.1. Calculated electronic spectra of porphyrin and its derivatives

Porphyrin and its derivatives find use in a myriad of important natural and biomimetic processes, with the major focus in the latter case on processes such as conversion of solar energy into chemical energy, photodynamic therapy, and as active agents in optical sensors. The excited states of porphyrins play fundamental roles in essentially all processes involving porphyrin and its derivatives.

In this section, we discuss the *meso* substitution and protonation effects on the electronic energy levels of the porphyrin macrocycle. Up to 24 singlet and 24 triplet energy levels of porphyrins have been calculated in water used as solvent at the TD-B3LYP/6-31G(d,p) level; singlet-singlet absorption spectra have been calculated for these specific compounds and are provided in **Figure 3**.

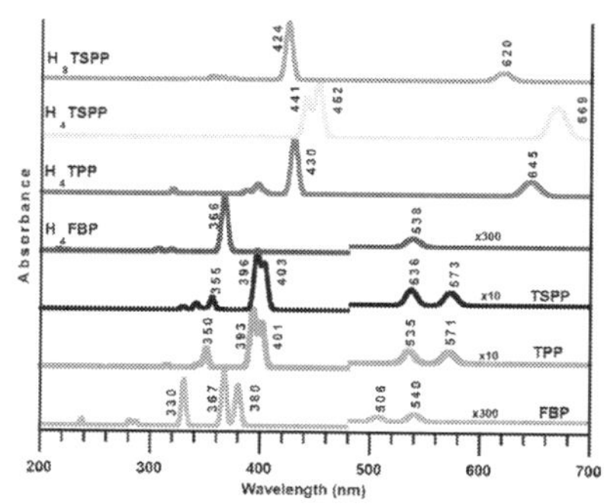

Figure 3. Comparison of calculated dipole-allowed electronic transitions of the parent porphyrin (FBP), *meso*-tetraphenylporphyrin (TPP), dianionic *meso*-tetrakis(*p*-sulfonatophenyl)porphyrin (TSPP), protonated-FBP (H_4FBP), protonated-TPP (H_4TPP), protonated TSPP (H_4TSPP), and dicationic-TSPP (H_8TSPP). The calculations were carried out in water used as a solvent at the TD-B3LYP/6-31G(d,p) level of TD-DFT.

The calculations mainly produce a strong electronic absorption band in the 360–450 nm range and a few weak or very weak electronic transitions below as well as above the strong bands (**Figure 3**). The strongest band is known as the Soret band (also referred to as the B-band), and weaker bands at longer wavelength, in the range 500–750 nm, are known as Q-bands that are usually quite weak. The results of calculations indicate that (1) the electronic bands in the parent porphyrin (FBP, neutral) are slightly blue-shifted in diprotonated-FBP (H_4FBP, dicationic) structure; (2) the bands in neutral TPP molecule (*meso*-phenyl substituted porphyrin) become significantly red-shifted in the dicationic or diprotonated-TPP (H_4TPP)—this observation indicates that the cationic macrocycle is stabilized by *meso*-substituted phenyl rings; (3) in the case of the *meso*-sulfonatophenyl substituted porphyrin ($TSPP^{-4}$, anionic), the electronic bands in the TSPP are significantly red-shifted in both of the diprotonated porphyrin cores (H_4TSPP, dianionic) and in the protonation of the N atoms and sulfonato (SO_3^-) groups (H_8TSPP, dicationic)—but the red shift in band positions for the H_4TSPP is greater than that in H_8TSPP; and (4) bands in the nonprotonated and diprotonated porphyrin are significantly more red-shifted in its corresponding *meso*-phenyl/sulfonatophenyl structures. The electronic spectra of porphyrin and its derivatives studied here are discussed in more details in the next sections.

3.2. The electronic spectra of FBP and protonated-FBP (H_4FBP)

The electronic spectrum of the FBP molecule exhibited two weak electronic bands at wavelength longer than that of the Soret band (B): one of the bands corresponds to $S_0 \rightarrow S_1$ (B_{1u}, at 540 nm with oscillator strength $f = 0.0005$) resulting from H − 1 → L + 1 (40%) and H → L (59%), where H and L represent HOMO and LUMO, respectively; the second band corresponds to $S_0 \rightarrow S_2$ (B_{2u}, at 506 nm and $f = 0.0003$) resulting from H − 1 → L + 1 (47%) and H → L + 1 (53%).

There are also two strong electronic bands in the spectral range of the Soret band (B): one of the bands corresponds to $S_0 \rightarrow S_4$ (B_{2u}, at 367 nm and $f = 1.1911$) resulting from H − 1 → L (50%) and H → L + 1 (47%); the second band corresponds to $S_0 \rightarrow S_3$ (B_{1u}, at 380 nm and $f = 0.8144$) resulting from H − 1 → L (23%), H − 1 → L + 1 (49%), and H → L (30%).

A relatively strong band at 330 nm is also calculated as existing. One of these is due to the $S_0 \rightarrow S_7$ transition (B_{1u}, at 330 nm and $f = 0.6934$), H − 3 → L (76%), H − 1 → L + 1 (12%), and H → L (12%), while the second band at ca. 330 nm is assignable to the transition $S_0 \rightarrow S_8$ (B_{2u} and $f = 0.2479$) and transition H − 3 → L + 1 (93%). It is to be noted that a few weak bands in range of 330–200 nm are also calculated as existing (see **Table 2**).

The experimental absorption spectrum of FBP [30] exhibits absorption bands at about 372 and 340 nm in the Soret-band region. In the Q-band region, bands at about 512 and 626 nm are observed. The measured bands in the FBP spectrum are in good agreement with calculated values for the B-bands at 380, 367, and 340 nm, and for the Q-bands at 540 and 506 nm, but not for the weak band at 626 nm. In these band regions, the calculation did not produce any dipole-allowed or forbidden singlet-singlet transition. Therefore, the free-base porphin (FBP) sample may contain the free-base aza-porphin(s) (as aza substitution at the *meso* position). Moreover, the calculations indicated the internal-conversion (IC) process, from the Soret band ($S_{3/4}$ at 376 nm) to Q-bands (S_2 at 506 nm and S_1 at 540 nm), which is experimentally observed from the fluorescence spectra of sulfite reductase porphin methyl ester in chloroform, at the exciting light of 380 nm, showed two peaks at 597 and 640 nm [31]. We also calculated 24 triplet states ($S_0 \rightarrow T_n$) in the range of 249–822 nm for the FBP molecule. There are two triplet states: one $T_8(B_{3g})$ at 373 nm and the other $T_9(B_{2u})$ at 366 nm. The later one, $T_9(B_{2u})$ at 366 nm, closely overlaps with the strongly dipole-allowed electronic energy level $S_4(B_{2u})$ at 367 nm. This finding implies that there is not only the possibility of the internal-conversion (IC) process from the $S_{3/4}$ (B_{2u} at 376 nm) to S_2 (B_{2u} at 506 nm) and S_1 (B_{1u} at 540 nm), but also the possibility of the intersystem-crossing (ISC) process by way of the strong vibrational coupling between the singlet and triplet electronic states at $S_{3/4}$ at 376 nm and $T_{8/9}$ at 367 nm. The lowest triplet state

($S_0 \rightarrow T_1(B_{2u})$) was predicted at 822 nm resulting from H − 1 → L (21%) and H → L + 1 (79%) transitions.

FBP: $S_0 \rightarrow S_n$						**$S_0 \rightarrow T_n$**				
S_n	**(eV)**	**(nm)**	**f**	**Sym**	**Major contrib's**	**T_n**	**(eV)**	**(nm)**	**Sym**	**Major contrib's**
1	2.30	540	0.0005	B_{1U}	H − 1 → L + 1 (40%), H → L (59%)	1	1.51	822	B_{2U}	H − 1 → L (21%), H → L + 1 (79%)
2	2.45	506	0.0003	B_{2U}	H − 1 → L (47%), H → L + 1 (53%)	2	1.82	682	B_{1U}	H → L (94%)
3	3.26	380	0.8144	B_{1U}	H − 3 → L (22%), H − 1 → L + 1 (48%), H → L (29%)	3	2.04	608	B_{2U}	H − 1 → L (78%), H → L + 1 (22%)
4	3.38	367	1.1911	B_{2U}	H − 1 → L (50%), H → L + 1 (47%)	4	2.07	598	B_{1U}	H − 1 → L + 1 (94%)
5	3.44	360		B_{3G}	H − 2 → L (98%)	7	2.90	428	B_{3G}	H − 2 → L (88%)
6	3.66	339		A_G	H − 2 → L + 1 (99%)	8	2.96	419	B_{1U}	H − 3 → L (86%)
7	3.76	330	0.6934	B_{1U}	H − 3 → L (76%), H − 1 → L + 1 (12%), H → L (12%)	9	3.15	393	A_G	H − 2 → L + 1 (93%)
8	3.76	330	0.2479	B_{2U}	H − 3 → L + 1 (93%)	11	3.33	373	B_{3G}	H − 8 → L + 1 (16%), H → L + 2 (72%)
16	4.33	287	0.0914	B_{2U}	H − 5 → L + 1 (97%)	13	3.39	366	B_{2U}	H − 3 → L + 1 (96%)
18	4.41	281	0.1037	B_{1U}	H − 5 → L (99%)	15	3.61	343	A_G	H − 8 → L (26%), H − 1 → L + 2 (68%)
23	5.22	237	0.1338	B_{1U}	H − 2 → L + 2 (98%)	16	3.64	340	B_{3G}	H − 4 → L + 1 (79%)
H_4FBP: $S_0 \rightarrow S_n$						**$S_0 \rightarrow T_n$**				
S_n	**(eV)**	**(nm)**	**f**	**Sym**	**Major contrib's**	**T_n**	**(eV)**	**(nm)**	**Sym**	**Major contrib's**
1	2.31	538	0.0007	E	H − 1 → L + 1 (48%), H → L (52%)	1	1.63	763	E	H − 1 → L + 1 (30%), H → L (70%)
2	2.31	538	0.0007	E	H − 1 → L (48%), H → L + 1 (52%)	2	1.63	763	E	H − 1 → L (30%), H → L + 1 (70%)
3	3.39	366	1.4554	E	H − 1 → L + 1 (52%), H → L (48%)	3	1.96	632	E	H − 1 → L + 1 (69%), H → L (31%)
4	3.39	366	1.4554	E	H − 1 → L (52%), H → L + 1 (48%)	4	1.96	632	E	H − 1 → L (69%), H → L + 1 (31%)
7	3.90	318	0.0597	E	H − 5 → L + 1 (45%), H − 4 → L + 1 (53%)	7	3.23	384	B1	H − 3 → L + 1 (28%), H − 2 → L (28%), H → L + 2 (31%)
8	3.90	318	0.0597	E	H − 5 → L (45%), H − 4 → L (53%)	8	3.32	374	E	H − 3 → L + 1 (44%), H − 2 → L (44%)
11	4.05	306	0.0695	E	H − 5 → L + 1 (54%), H − 4 → L + 1 (45%)	9	3.37	367	E	H − 5 → L + 1 (42%), H − 4 → L + 1 (48%)
12	4.05	306	0.0695	E	H − 5 → L (54%), H − 4 → L (45%)	10	3.37	367	E	H − 5 → L (42%), H − 4 → L (48%)

Table 2. The selected values of the calculated singlet-singlet (**$S_0 \rightarrow S_n$**) and singlet-triplet (**$S_0 \rightarrow T_n$**) vertical electronic transitions with their oscillator strengths (f) for the FBP and protonated-FBP (H4FBP). The calculations were carried out in water used as a solvent at the TD-B3LYP/6-31G(d,p) level of TD-DFT. The percentages in parenthesis are the contributions from the different HOMO(H) → LUMO(L) transitions to a desired electronic transitions. The minor contributions are not given here.

The predicted electronic spectrum of protonated-FBP (H_4FBP) exhibits Q- and B-bands for $S_0 \rightarrow S_{1/2}$ (E at 538 nm with $f = 0.0007$) owing to H − 1 → L (48%) and H → L + 1 (52%) transitions; $S_0 \rightarrow S_{3/4}$ (E at 366 nm and $f = 1.4554$) owing to H − 1/ → L + 1 (53%) and H → L (49%) transitions. In addition, a few weaker transitions exist up to 300 nm (**Table 2**).

Comparing the electronic spectrum of FBP with that of H_4FBP: in FBP three strong bands were predicted at 380, 360, and 330 nm, see **Figure 3** and **Table 2**, whereas the H_4FBP spectrum exhibited only one strong band at 366 nm in the Soret-band region; in the Q-band region, H_4FBP has a doubly degenerate band at 538 nm, while FBP has two very weak transitions at 540 and 506 nm. This reduction in the number of bands is due to the higher symmetry for H_4FBP.

Also for FBP, the calculated electronic spectrum of diprotonated-FBP (i.e., H_4FBP) molecule indicates two IC processes from the $S_{3/4}$ (the strongest bands or B-band) at 366 nm (with the symmetry E) to the $S_{1/2}$ (at 538 nm with symmetry E), as well as the ISC process between the $S_{3/4}$ (at 366 nm) and $T_{7/8}$ (at 367 nm), see **Table 2**.

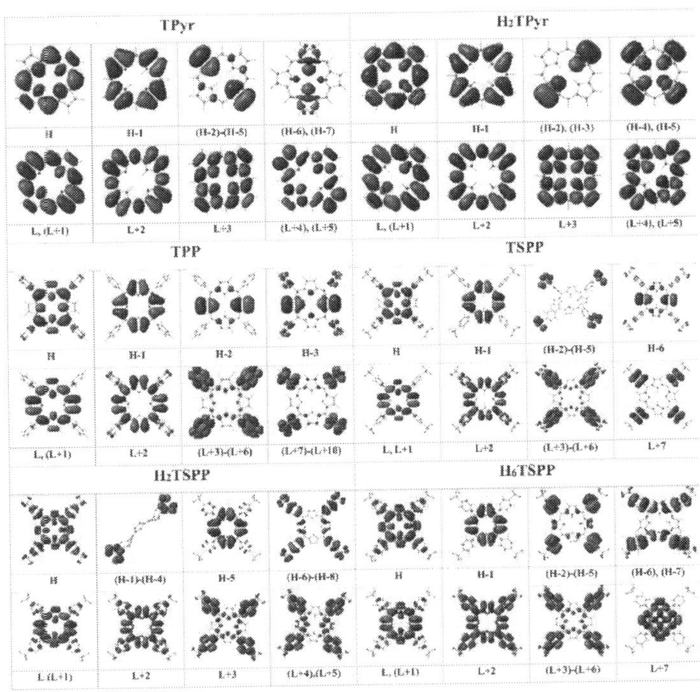

Figure 4. Plot of calculated electron densities in the desired HOMOs (H) and LUMOs (L) of parent porphyrin (FBP), *meso*-tetraphenylporphyrin (TPP), dianionic *meso*-tetrakis(*p*-sulfonatophenyl)porphyrin (TSPP), and protonated-FBP (H_4FBP), protonated-TPP (H_4TPP), protonated-TSPP (H_4TSPP), and dicationic-TSPP (H_8TSPP) molecules.

The electron density plots of the molecular orbitals (i.e., HOMOs (H) and LUMOs (L)), as seen in **Figure 4** and **Table 3**, show that the H − m and L + m (m = 0, 1, 2, …) are not just pure π and π^* molecular orbitals (MOs), in particular cases they also include nonbinding atomic orbitals (AOs).

FBP		**H_4FBP**	
H	$\pi(C_\beta{-}C_\beta/C_m{-}C_\alpha)$ + n(N)	H	$\pi(C_\beta{-}C_\beta/C_\alpha{-}C_m{-}C_\alpha)$ + n(N)
H − 1	$\pi(C_\beta{-}C_\alpha)$	H − 1	$\pi(C_\alpha{-}C_\beta)$
H − 2	$\pi(C_\alpha{-}N{-}C_\alpha/C_\beta{-}C_\beta)$	H − 2/H − 3	$\pi(C_\alpha{-}C_\beta)$ + n(N)
H − 3	$\pi(C_\alpha{-}N{-}C_\alpha/C_\beta{-}C_\beta)$ + n(minor, N/C_m)	H − 4/H − 5	$\pi(C_\beta{-}C_\beta)$ + n(N)
H − 4/H + 5	$\pi(C_\beta{-}C_\beta)$ + n(N)	H − 6/H − 7	$\pi(C_\beta{-}C_\alpha{-}C_m)$ + n(minor, N)
H − 6/H − 7	n(N) + σ(minor; $C_\beta{-}C_\alpha$)	H − 8	$\pi(C_\alpha{-}C_m{-}C_\alpha)$
L/L + 1	$\pi^*(C_\beta{-}C_\beta/C_\beta{-}C_\alpha/C_m{-}C_\alpha)$ + n(minor; N)	L/L + 1	$\pi(C_\beta{-}C_\alpha)$ + n(minor, N)
L + 2	$\pi^*(C_\beta{-}C_\alpha)$ + n(C_m)	L + 2	$\pi(C_\beta{-}C_\alpha)$ + n(C_m)
L + 3	$\pi^*(C_\beta{-}C_\beta/C_m{-}C_\alpha)$ + n(N/C_α)	L + 3	$\pi(C_\beta{-}C_\beta)$ + n(N/C_α/C_m)
L + 4/L + 5	$\pi^*(C_\beta{-}C_\beta/C_m{-}C_\alpha)$ + n(N/C_α/C_β)	L + 4	$\pi(C_\beta{-}C_\beta$ and $C_\alpha{-}C_m)$ + n(N/C_α/C_m)

Table 3. Bond type of the highest occupied molecular orbitals (H - m) and the lowest unoccupied molecular orbitals (L + m), m = 0, 1, 2, …

3.3. The electronic spectra of TPP and H_4TPP

While the calculated spectrum of the TPP molecule displayed two weak peaks at 571 ($S_0 \rightarrow S_1$, f= 0.00337) and 535 nm ($S_0 \rightarrow S_2$, f= 0.0359) in the region of the Q-absorption bands, the spectrum of diprotonated-TPP (H_4TPP) exhibited only a double degenerate peak that is also red-shifted to 645 nm ($S_0 \rightarrow S_{1/2}$, f = 0.3040). In the region of Soret band, two strong peaks were predicted at 401 and 393 nm ($S_0 \rightarrow S_{3/4}$, f = 1.2834/1.6972) in the TPP absorption spectrum, and the H_4TPP spectrum exhibited a doubly degenerate band at 430 nm ($S_0 \rightarrow S_{3/4}$, f = 1.209). Both spectra show a few weak and very weak-allowed electronic transitions in the high energy region as seen in **Table 4** and **Figure 5**. The observed spectrum of the **TPP** in **DMF** (Dimethylformamide) exhibited a strong band at about 412 nm with a shoulder at around 400 nm in Soret-band region, and four weak bands at about 513, 548, 590, and 645 nm in Q-band region (**Figure 5**). However, as shown in **Figure 5**, the calculated electronic spectrum of the TPP does not indicate any dipole-allowed or forbidden singlet-singlet transition with wavelength longer than 571 nm, rather, a weak singlet-singlet transition was predicated at 645 nm for the diprotonated-TPP (H_4TPP) molecule. This observation suggests that the TPP sample may contain a small percentage of aza-substituted TPP (at the beta or *meso* position) that produces weaker absorption peaks at around 590 and 646 nm. Another possibility might be that a small percentage of the TPP molecules in the sample might be at their local minima (instead of their ground state) due to rotation of the phenyl substitution at the *meso* position of TPP, resulting in weaker bands at around 590 and 646 nm.

TPP/$S_0 \rightarrow S_n$						$S_0 \rightarrow T_n$				
S_n	(eV)	(nm)	F	Sym	Major contrib's	T_n	(eV)	(nm)	Sym	Major contrib's
1	2.17	571	0.0337	B2	H − 1 → L + 1 (32%), H → L (67%)	1	1.40	884	B1	H − 1 → L (16%), H → L + 1 (84%)
2	2.32	535	0.0359	B1	H − 1 → L (37%), H → L + 1 (63%)	2	1.66	745	B2	H → L (98%)
3	3.09	401	1.2834	B2	H − 3 → L (10%), H − 1 → L + 1 (62%), H → L (27%)	3	1.99	623	B1	H − 1 → L (84%), H → L + 1 (15%)
4	3.16	393	1.6972	B1	H − 1 → L (62%), H → L + 1 (37%)	4	2.06	602	B2	H − 1 → L + 1 (97%)
6	3.54	350	0.5462	B2	H − 3 → L (87%)	5	2.84	436	A2	H − 2 → L (88%)
8	3.62	343	0.0909	B1	H − 3 → L + 1 (98%)	6	2.90	428	B2	H − 3 → L (82%)
19	3.95	314	0.0267	B1	H − 10 → L (39%), H − 8 → L + 1 (57%)	7	3.08	403	A1	H − 2 → L + 1 (91%)
20	3.95	314	0.0216	B2	H − 14 → L (15%), H − 11 → L (56%), H − 10 → L + 1 (14%), H − 8 → L (10%)	8	3.15	393	A2	H − 16 → L + 1 (10%), H → L + 2 (74%)
H_4TPP/$S_0 \rightarrow S_n$						$S_0 \rightarrow T_n$				
S_n	(eV)	(nm)	F	Sym	Major contrib's	T_n	(eV)	(nm)	Sym	Major contrib's
1	1.92	645	0.304	A′	H − 1 → L + 1 (16%), H → L (84%)	1	1.22	1020	A″	H → L + 1 (98%)
2	1.92	645	0.3039	A″	H − 1 → L (16%), H → L + 1 (84%)	2	1.22	1020	A′	H → L (98%)
3	2.89	430	1.2029	A′	H − 1 → L + 1 (74%), H → L (14%)	3	2.02	615	A″	H − 1 → L (95%)
4	2.89	430	1.2026	A″	H − 1 → L (74%), H → L + 1 (14%)	4	2.02	615	A′	H − 1 → L + 1 (95%)
10	3.13	396	0.2053	A″	H − 5 → L (89%)	5	2.67	464	A″	H − 3 → L (13%), H − 2 → L + 1 (13%), H → L + 2 (56%)
11	3.13	396	0.2056	A′	H − 5 → L + 1 (89%)	6	2.76	449	A′	H − 7 → L + 1 (16%), H − 6 → L (16%), H − 3 → L + 1 (29%), H − 2 → L (29%)
15	3.20	387	0.078	A″	H − 8 → L (80%)	7	2.88	431	A′	H − 3 → L + 1 (46%), H − 2 → L (46%)
16	3.20	387	0.0778	A′	H − 8 → L + 1 (80%)	8	2.88	430	A″	H − 3 → L (46%), H − 2 → L + 1 (46%)
21	3.45	360	0.0351	A′	H − 10 → L (12%), H − 9 → L (80%)	9	2.93	424	A′	H − 5 → L + 1 (82%)
22	3.66	339	0.039	A″	H − 10 → L + 1 (13%), H − 9 → L + 1 (80%)	10	2.93	424	A″	H − 5 → L (82%)

Table 4. Selected values of the calculated singlet-singlet (**$S_0 \rightarrow S_n$**) and singlet-triplet (**$S_0 \rightarrow T_n$**) vertical electronic transitions with their oscillator strength (f) for the TPP and H4TPP. The percentages in parenthesis are the contributions from the various HOMO (H) to LUMO (L) transitions to a desired electronic transitions. The minor contributions are not given here. The calculations were carried out in water as solvent at the TD-B3LYP/6-31G(d,p) level of TD-DFT.

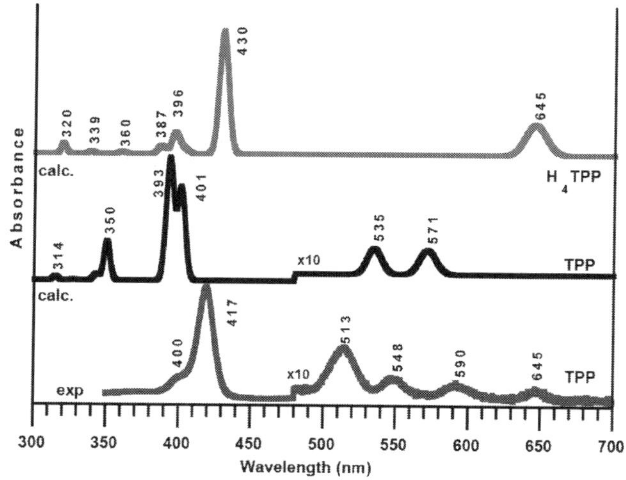

Figure 5. Calculated and measured absorption spectra of TPP and calculated absorption spectrum of protonated-TPP (H_4TPP).

TPP			
H	$\pi(C_\alpha - C_m - C_\alpha/C_\beta - C_\beta)$ + n(N and C_φ)	L/L + 1	$\pi^*(C_\beta - C_\beta/C_\beta - C_\alpha/C_m - C_\alpha)$ + n*(N(H))
H − 1	$\pi(C_\alpha - C_\beta)$	L + 2	$\pi^*(C_\alpha - C_m)$ + n*(C_φ)
H − m m = 2, 3	$\pi(C_\beta - C_\beta)/\ \pi(N - C_\alpha - C_m)$	L + m m = 3–6	π^*(C—C in phenyl)

Table 5. Bond type of the highest occupied molecular orbitals (H − m) and the lowest unoccupied molecular orbitals (L + m), m = 0, 1, 2, ….

Furthermore, Jiang *et al.* [32] measured absorption and EPR spectra of some porphyrins (TPP and derivatives) and metalloporphyrins compounds, and the measured absorption spectrum of TPP produced an intense electronic transition centered about 417 nm (B-band) and several Q-bands with weak intensities around 514, 550, 590, and 646 nm. The authors have also stated that both steric hindrance and electronic effects of the functional groups influenced the UV-vis absorption of the TPP, with the Soret bands of the *para*-substituted *meso*-tetraphenylporphine derivatives somewhat red-shifted (3–5 nm). This experimental observation is consistent with the result of our calculations (**Figure 5** and **Tables 2–5**).

Additionally, the results of the calculations for TPP (in solution/water) indicate the possibility of an IC process from S_6 (B_2 at 350 nm)/S_4 (B_1 at 393 nm)/S_3 (B_2 at 401 nm) to S_2 (B_1 at 535 nm) and S_1 (B_2 at 571 nm), which are verified by experimental measurements of the fluorescence spectrum of the TPP in different environments. Moreover, based on theoretical predictions, there are strong surface crossings between the singlet-triplet excited states of the TPP: S_4 (B_1 at 393 nm) and T_8 (A_2 at 393 nm), and S_3 (B_2 at 401 nm) and T_7 (A_1 at 403 nm), which may cause an ISC process in the excited state.

The results of the calculated electronic energy states of diprotonated-TPP molecule (H_4TPP) indicate the existence of an IC process from the $S_{3/4}$ (A′ and A″ at 430 nm) to $S_{1/2}$ (A′ and A″ at 645 nm), in addition to possibility of an ISC process between the $S_{3/4}$ (A′ and A″ at 430 nm) and $T_{7/8}$ (at 431 and 430 nm, with symmetry A′ and A″, respectively).

3.4. Calculated electronic spectra of TSPP, H_4TSPP, and H_8TSPP

In the Q-band region, while the calculations indicate the presence of two weak transitions at 573 nm ($S_0 \rightarrow S_1$ with symmetry B_2 and f = 0.0419) and at 536 nm ($S_0 \rightarrow S_2$ with symmetry B_1, f = 0.0506) in the TSPP spectrum, the calculated spectrum of H_4TSPP shows two doubly degenerate peaks at 669 nm ($S_0 \rightarrow S_{1/2}$ with symmetries B_2 and B_1 with f = 0.4223/0.4138) and at 528 nm ($S_0 \rightarrow S_{4/5}$ with symmetries A_1 and B_2 and f = 0.0001/0.0005), in addition to a band at 518 nm ($S_0 \rightarrow S_{10}$, B_2 symmetry and f = 0.0001). However, the spectrum of the dicationic-TSPP (H_8TSPP) indicates almost overlapping (or nearly degenerate) weak electronic transitions at 620 nm ($S_0 \rightarrow S_1$ with symmetries B_2, f = 0.2467) and at 619 nm ($S_0 \rightarrow S_2$ with symmetries B_1, f = 0.2377).

In the B-band (Soret band) region, while the calculated spectrum of the TSPP exhibited two strong bands at 403 nm ($S_0 \rightarrow S_3$ with B_2 symmetry and f = 1.4382) and at 396 nm ($S_0 \rightarrow S_4$ with symmetry B_1 and f = 1.8378), the H_4TSPP spectrum contains two strong bands at 452/451 nm ($S_0 \rightarrow S_{12/13}$, B_1 and B_2 symmetries, f = 0.7601/0.7369) and a medium intense band at 441 nm ($S_0 \rightarrow S_{16}$, B_2 symmetry and f = 0.5315). The calculated spectrum of the dicationic-TSPP (H_8TSPP) exhibits a double degenerate strong transitions at 424 nm ($S_0 \rightarrow S_{3/4}$, with B_1 and B_2 symmetries and f = 1.7153/1.7145, respectively). Additionally, many weak electronic transitions are predicted at longer wavelengths than these strong Soret bands (see **Table 6**). The predicted B-bands and Q-bands for the molecules studied here are compatible with the experimental data [21, 33].

Akins *et al.* [21] and Zhang *et al.* [33] have reported the UV-vis spectra of the free-base TSPP and the H_4TSPP (dianionic-TSPP). While the measured absorption spectrum of the TSPP displayed an intense band (S-band, also known as B-band) at ~412 nm, and several very weak broad bands (known as Q-bands) at about 517 (± 2), 555 (± 3), 581 (± 3), and 640 (± 3) nm, the H_4TSPP spectrum exhibited the B-band at 432 nm and very weak broad Q-bands at 589 (± 5) and 645 nm. However, in the Q-band region, the calculations produced only two very weak dipole-allowed electronic transitions at 536 and 573 nm for the TSPP and only one doubly degenerated band at 669 nm for the H_4TSPP. The calculated absorption spectra suggest that the two of four absorption bands in TSPP and one of two bands for the H_4TSPP, in the Q-region, must be due to vibrational progression such as from the lowest vibrational level in the electronic ground state to higher vibrational level in electronically excited state, $S_0(\nu'' = 0)$ to $S_Q(\nu' \geq 1)$.

TSPP:$S_0 \rightarrow S_n$						**$S_0 \rightarrow T_n$**				
S_n	**(eV)**	**(nm)**	**f**	**Sym**	**Major contrib's**	**T_n (eV)**		**(nm)**	**Sym**	**Major contrib's**
5	2.16	573	0.0419	B2	H − 1 → L + 1 (32%), H → L (67%)	1	1.40	884	B1	H − 1 → L (16%), H → L + 1 (86%)
6	2.31	536	0.0506	B1	H − 1 → L (36%), H → L + 1 (64%)	2	1.67	744	B2	H → L (97%)
10	3.07	403	1.4382	B2	H − 1 → L + 1 (62%), H → L (28%)	3	1.99	624	B1	H − 1 → L (84%), H → L + 1 (15%)
11	3.13	396	1.8378	B1	H − 1 → L (62%), H → L + 1 (36%)	4	2.05	604	B2	H − 1 → L + 1 (97%)
36	3.49	355	0.3924	B2	H − 10 → L (83%)	5	2.84	437	A2	H − 9 → L (49%), H − 7 → L (39%)
38	3.56	348	0.0392	B1	H − 10 → L + 1 (28%), H − 8 → L (69%)	6	2.89	429	B2	H − 11 → L (36%), H − 10 → L (48%)
47	3.64	341	0.2178	B2	H − 11 → L (83%)	7	3.07	404	A1	H − 9 → L + 1 (39%), H − 7 → L + 1 (51%)
48	3.77	329	0.0936	B1	H − 11 → L + 1 (82%), H − 10 → L + 1 (11%)	8	3.14	395	A2	H → L + 2 (70%)

H_4TSPP:$S_0 \rightarrow S_n$						**$S_0 \rightarrow T_n$**				
S_n	**eV**	**(nm)**	f	**Sym.**	**Major contrib's**	**T_n**	**eV**	**(nm)**	**Sym**	**Major contrib's**
1	1.85	669	**0.4223**	B2	H − 5 → L + 1 (13%), H → L (87%)	1	1.17	1056	B2	H → L (97%)
2	1.85	669	**0.4138**	B1	H − 5 → L (13%), H → L + 1 (87%)	2	2.01	618	B1	H − 5 → L (95%)
4	2.35	528	**0.0001**	A1	H − 3 → L + 1 (47%), H − 2 → L (53%)	3	2.34	530	A1	H − 3 → L + 1 (47%), H − 2 → L (52%)
5	2.35	528	**0.0005**	B2	H − 4 → L (53%), H − 1 → L + 1 (47%)	5	2.34	529	B1	H − 4 → L + 1 (47%), H − 1 → L (52%)
10	2.40	518	**0.0001**	B2	H − 4 → L (47%), H − 1 → L + 1 (53%)	9	2.39	518	A1	H − 3 → L + 1 (53%), H − 2 → L (47%)
12	2.74	452	**0.7601**	B1	H − 6 → L (79%), H − 5 → L (15%)	13	2.51	494	A2	H − 8 → L + 1 (34%), H − 7 → L (35%), H → L + 2 (23%)
13	2.75	451	**0.7369**	B2	H − 6 → L + 1 (83%), H − 5 → L + 1 (12%)	14	2.62	474	B1	H − 6 → L (90%)
16	2.81	441	**0.5315**	B2	H − 9 → L (17%), H − 6 → L + 1 (16%), H − 5 → L + 1 (57%)	17	2.65	467	A1	H − 8 → L (46%), H − 7 → L + 1 (46%)
18	2.81	441	**0.5073**	B1	H − 9 → L + 1 (16%), H − 6 → L (19%), H − 5 → L (54%)	18	2.69	462	A2	H − 8 → L + 1 (47%), H − 7 → L (47%)

19 2.97	417	**0.0013**	B1	H − 13 → L (14%), H − 10 → L + 1 (74%)	19 2.72	455	A2	H − 8 → L + 1 (13%), H − 7 → L (12%), H → L + 2 (63%)
22 3.10	400	**0.0081**	B2	H − 13 → L + 1 (12%), H − 10 → L (82%)	20 2.96	419	A2	H − 12 → L + 1 (43%), H − 11 → L (47%)

$H_8TSPP:S_0 \rightarrow S_n$					**$S_0 \rightarrow T_n$**			
S_n eV	**(nm)**	**f**	**Sym.**	**Major contrib's**	**T_n (eV)**	**(nm)**	**Sym**	**Major contrib's**
1 2.00	620	0.2467	B2	H − 1 → L + 1 (23%), H → L (77%)	1 1.31	946	B2	H → L (96%)
2 2.00	619	0.2377	B1	H − 1 → L (23%), H → L + 1 (77%)	2 1.31	945	B1	H → L + 1 (96%)
3 2.92	424	1.7153	B1	H − 1 → L (74%), H → L + 1 (23%)	3 1.94	639	B1	H − 1 → L (94%)
4 2.92	424	1.7145	B2	H − 1 → L + 1 (75%), H → L (22%)	4 1.94	638	B2	H − 1 → L + 1 (94%)
7 3.32	374	0.0199	B2	H − 4 → L (92%)	5 2.74	452	A2	H → L + 2 (72%)
8 3.32	373	0.0240	B1	H − 4 → L + 1 (92%)	6 3.00	413	A1	H − 7 → L (26%), H − 6 → L + 1 (26%), H − 3 → L (13%), H − 2 → L + 1 (18%)
11 3.41	363	0.0560	B1	H − 5 → L (88%)	7 3.08	403	A2	H − 3 → L + 1 (37%), H − 2 → L (40%)
12 3.42	363	0.0614	B2	H − 5 → L + 1 (88%)	8 3.09	401	A1	H − 3 → L (36%), H − 2 → L + 1 (34%), H − 1 → L + 2 (14%)
15 3.50	355	0.0822	B1	H − 8 → L (94%)	9 3.16	393	B2	H − 5 → L + 1 (16%), H − 4 → L (66%)
16 3.50	355	0.0918	B2	H − 8 → L + 1 (94%)	10 3.16	393	B1	H − 5 → L (17%), H − 4 → L + 1 (65%)
20 3.59	346	0.0365	B2	H − 10 → L (10%), H − 9 → L (84%)	11 3.20	387	A2	H − 12 → L + 1 (11%), H − 11 → L (12%), H − 7 → L + 1 (29%), H − 6 → L (30%)
21 3.69	336	0.0363	B1	H − 10 → L + 1 12%), H − 9 → L + 1 (81%)	12 3.23	384	B1	H − 8 → L (66%)

Table 6. The selected values of the calculated singlet-singlet (**$S_0 \rightarrow S_n$**) and singlet-triplet (**$S_0 \rightarrow T_n$**) vertical electronic transitions with oscillator strength (f) for the TSPP and protonated-TSPP (H4TSPP). The calculations were carried out in water used as a solvent at TD-B3LYP/6-31G(d,p) level of the TD-DFT. The percentages in the parenthesis indicate the contributions from the different HOMO(H) → LUMO(L) transitions to a desired electronic transitions. The minor contributions are not given here.

Also, Akins and coworkers have measured fluorescence spectra of free-base TSPP (pH = 12), monomeric H_4TSPP (pH = 4.5), and aggregate H_4TSPP in highly acidic situation. The authors reported that the fluorescence spectrum of the TSPP at 412 nm (B-band region) excitation displayed a peak at 642 nm with a red degraded shoulder at 702 nm. The spectrum of H_4TSPP upon excitation at 432 nm in the B-band region exhibited similar structure, for example, a strong emission peak at 665 nm with relatively weak shoulder at about 716 nm [21]. Both fluorescence spectra of the TSPP and deprotonated-TSPP (H_4TSPP) indicated that when excited in the Soret- or B-band region, initially internal conversion (IC) occurs from the B-band to the Q-bands, followed by a fluorescence from the lowest excited state(s) in Q-band region to the ground state S_0 a sequence of: $S_0 + h\nu_0 \rightarrow S$(B-band) $\rightarrow$ S(Q-band) $\rightarrow S_0 + h\nu$. These observations are consistent with our calculations. For instance, the predicted possible IC (internal-conversion) process may take place from the S_3 (at 403 nm) to S_1 (at 573 nm) and S_2 (at 536 nm) for the TSPP; and from the $S_{12/13}$ (at 452/451 nm)/S_{16} (at 441 nm) to S_{10} (at 518 nm)/$S_{4/5}$(at 528 nm)/$S_{1/2}$(at 669 nm) for the H_4TSPP (diprotonated- or dianionic-TSPP).

The calculations also indicate that there might be an ISC (intersystem crossing) process between the S_3 (at 403 nm)/S_4 (at 396 nm) and the T_7 (at 404 nm)/T_8 (at 395 nm), and between the S_1 (at 573 nm) and T_4 (at 604nm) for TSPP (where the energy difference between S_1 and T_4 states is about 0.11 eV or 896 cm^{-1}). For the H_4TSPP (or dianionic-TSPP), the ISC process may occur between the $S_{4/5}$ (at 528 nm) and $T_{3,4,5,6}$ (at 530 and 529 nm), and S_{10} (at 518 nm) and $T_{15,16,17,18}$ (at 518 nm), and between the S_1(669 nm) and T_3(618 nm) (where the energy difference between the S_1(669 nm) and T_3(618 nm) states is 0.15 eV or 1233 cm^{-1}) (**Figure 6**). The results of the calculations suggest that, depending on competition between the IC and ISC processes, there can be ISC through vibrational coupling or potential energy surface (PES) touching between singlet and triplet states.

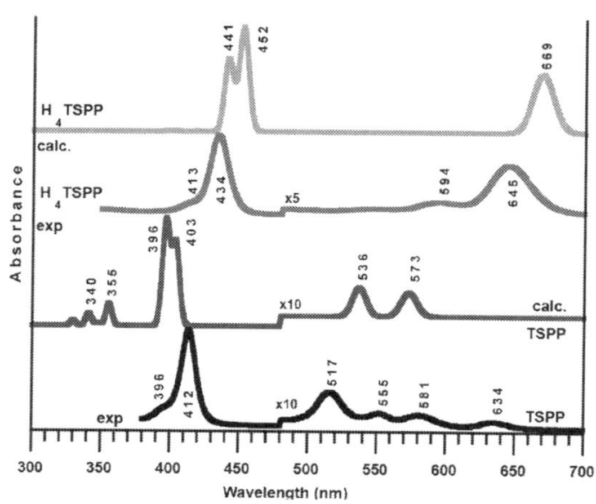

Figure 6. Calculated and measured absorption spectra of TSPP and protonated-TSPP (H_4TSPP).

Likewise, for the H_8TSPP (dicationic-TSPP molecule), the IC process may happen from the B-bands ($S_{3/4}$ at 424 nm) to the Q-bands ($S_{1/2}$ at 620/619 nm). Furthermore, the energy difference between the $S_{1/2}$ (at 424/424 nm) and $T_{3/4}$ (at 639/637 nm) is about 0.056 eV or 455 cm^{-1}, which may lead to a strong vibrational coupling in their excited vibroelectronic states. Owing to this small energy distance between singlet and triplet states of the H_8TSPP, there would likely occur ISC that may originate from B-bands ($S_{1/2}$) to triplet states ($T_{3/4}$).

TSPP		H_4TSPP		H_8TSPP	
H	$\pi(C_\alpha{-}C_m{-}C_\alpha/C_\beta{-}C_\beta)$ + n(N and C_φ/C(S))	H	$\pi(C_\alpha{-}C_m{-}C_\alpha/C_\beta{-}C_\beta$/C—C in phenyl) + n(N)	H	$\pi(C_\alpha{-}C_m{-}C_\alpha/C_\beta{-}C_\beta)$ + n(N and C_φ/C(S) in phenyl)
H − 1	$\pi(C_\alpha{-}C_\alpha)$	H − m m = 1–4	n(O)	H − 1	$\pi(C_\alpha{-}C_\beta)$
H − m m = 2–5	n(O)	H − 5	$\pi(C_\alpha{-}C_\beta)$	H − m m = 2–5	π(C—C—C in phenyl) + n(C_α and C_β, minor)
H − 6	$\pi(C_\beta{-}C_\beta$ and $C_\alpha{-}N{-}C_\alpha)$	H − m m = 6–8	π(C—C—C in phenyl) + minor n(O and C_β)	H − 6/H − 7	π(C(S)—C/C—C_φ in phenyl and $C_m{-}C_\alpha{-}C_\beta$)
L/L + 1	$\pi^*(C_\beta{-}C_\alpha/C_\beta{-}C_\beta$ and $C_\alpha{-}C_m)$ + n*(N(H))	L/L + 1	$\pi^*(C_\alpha{-}C_m/C_\beta{-}C_\beta)$ + n*(N)	L/L + 1	$\pi^*(C_\alpha{-}C_m/C_\beta{-}C_\beta)$ + n*(N);
L + 2	$\pi^*(C_\varphi{-}C_m/C_\alpha{-}C_m)$ + n*(C(S), minor)	L + 2	$\pi^*(C_\varphi{-}C_m/C_\beta{-}C_\alpha)$ and minor n*(C in phenyl)	L + 2	$\pi^*(C_\varphi{-}C_m/C_\beta{-}C_\beta$ and C—C/C—S in phenyl)
L + m m = 3–6	π*(C—C in phenyl)	L + m m = 3–5	π*(C—C in phenyl) + n*(C(S) and C_φ in phenyl)	L + m m = 3–6	π*(C—C/C—S in phenyl) and n*(C_φ)/n*(C_α and N, minor)

Table 7. Bond type of the highest occupied molecular orbitals (H − m) and the lowest unoccupied molecular orbitals (L + m), m = 0, 1, 2, ...

Consequently, the results of calculated absorption spectra for the porphyrin molecules studied here (**Tables 2–7**) reveal several important points: (1) protonation of the N atoms at the porphyrin core and the *meso* substitutions of the parent porphyrin with the phenyl or sulfonatophenyl groups lead to substantial red-shifts in the spectral position of the Soret bands and Q-bands; (2) the IC process takes place from the B-band(s) to the Q-band(s) for the all porphyrin derivatives; (3) an ISC process might be possible through the surface touching and/or strong vibrational coupling, but would be dependent on the competition with IC processes and the rate constant of fluorescence; and (4) deuteration of the N atoms at the core of the macrocycle and O atoms do not produce significant change in their corresponding spectra.

3.5. Relaxed potential energy surface (RPES) scan of TSPP molecule

The relaxed potential energy surface (RPES) scan was performed to calculate the ground state PES of the TSPP molecule in water by rotating one of four dihedral angles θ ($C_\alpha{-}C_m{-}C_\varphi{-}C$)

from 40 to 130° in 10° increments. The calculated ground state curve, S_0(RPES), shows two minima at dihedral angles of ~66 and 110° (see **Figure 7(B)** and **(C)**). These two minima on the S_0(RPES) curve represent the lowest ground state with C_{2v} symmetry and an energetically stable local state with C_2 symmetry, respectively. The local minima at 110° is about 106 cm^{-1} (0.0132 eV) above the lowest ground state at 66° as seen in **Figure 7(C)**. When the molecule goes from the lowest ground state to this local state, the predicted highest potential energy barrier at the dihedral angle of 90° is only 177 cm^{-1} (0.0219 eV). This finding suggests that the *meso*-substituted sulfonatophenyl groups are able to rotate around C_m—C_φ bond at room temperature because the thermal energy (k_BT) at 298 K is 207.2 cm^{-1}. Consequently, since the computed potential energy barrier is small as much as 106 cm^{-1} at the dihedral angle of 90°, the self-assembling of the TSPP molecules in any environment might be very easily formed. It should be point out that the calculated ground state RPES was carried out only for the rotation of one of four *meso*-sulfonatophenyl groups within the TSPP molecule. If the RPES scans were performed for the rotations of all four *meso*-substitutional groups, there would be more than a few different local minima with dissimilar potential energy barriers on the ground state RPES. Thereby, a slight change in the potential energy barrier distribution of the *meso*-substituted porphyrin molecules, such as the TSPP, can be used as a scanning nanocalorimetric measurement (or for other electronic purposes) of very small variations in energy.

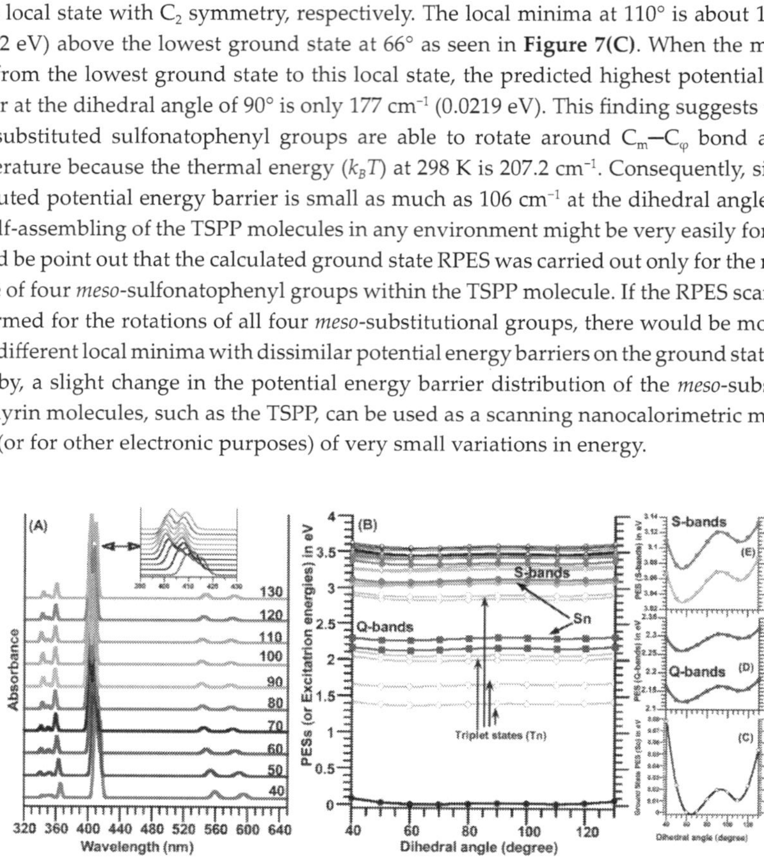

Figure 7. The calculated spectra of the TSPP as a function of the dihedral angle (C_α—C_m—C_φ—C(ph)) rotation varying from 40 to 130° with 10° increment: (A) plot of dipole-allowed singlet electronic transitions $S_0 \rightarrow S_n$, with n = 1–24; (B) the relaxed potential energy surfaces of the ground state (S_0) and upper singlet (S_n) and triplet (T_n) states, n = 1–24; (C)–(E) illustrate the RPES curves for the ground state S_0, Q-bands and Soret bands at a low scale for a better view. It is noteworthy that only one of the four *meso*-sulfonatophenyl groups is rotated about the C_m—C_φ bond and the uppercase letter S in (E) symbolizes the Soret band.

It is to be noted that the PES curves of the upper singlet (S_n) and triplet (T_n) energy states were calculated by the following Eqs. (1) and (2), respectively:

$$V_n(S_n, \theta) = E(\theta) - E_0 + E(S_0 \rightarrow S_n; \theta) \quad (1)$$

$$V_n(T_n, \theta) = E(\theta) - E_0 + E(S_0 \rightarrow T_n; \theta) \quad (2)$$

(1) In the aforementioned equations, E_0 and $E(\theta)$ symbolize the calculated global (total SCF) energies at the lowest ground state and the relaxed potential energy at the dihedral angle $\theta(C_\alpha{-}C_m{-}C_\varphi{-}C_1)$, respectively; $E(S_0 \rightarrow S_n/S_0 \rightarrow T_n; \theta)$ represents the vertical electronic transition energy from S_0 to excited electronic energy levels S_n/T_n at the dihedral angle θ. It is to be noted that **Figure 7(A)** displays the computed dipole-allowed electronic transitions at each rotated dihedral angle, $\theta(C_\alpha{-}C_m{-}C_\varphi{-}C_1)$, while **Figure 7(C)–(E)** shows the alteration in the calculated singlet and triplet electronic energy levels as a function of the $\theta(C_\alpha{-}C_m{-}C_\varphi{-}C_1)$. The PES curves of the excited states S_n and T_n are akin to the ground state RPES, S_0(RPES). Consequently, the results of the calculations indicated that the red-shift in spectral position of the Soret bands increases with increasing rotational dihedral angle in both right-handed and left-handed rotational directions around the equilibrium dihedral angle of ~66° in the ground state.

4. Calculation section

The calculations were carried out in water used as solvent at the B3LYP level of the density functional theory (DFT) [34, 35] with the 6-311G(d,p) basis set [36]. The solvent effects were considered by using the self-consistent reaction field (SCRF) calculations [37] with the conductor-like polarizable continuum model [38–40] and a dielectric constant of 78.39 for water; SCRF = (CPCM, solvent = water) as implemented within the Gaussian 09 software package [41]. All compounds studied here were optimized to minima on their ground state relaxed potential energy surfaces (RPESs) that were verified by revealing the absence of imaginary frequencies in calculated vibrational spectra. Time-dependent DFT (TD-DFT) was performed to calculate the first 24 singlet-singlet ($S_0 \rightarrow S_n$) and singlet-triplet ($S_0 \rightarrow T_n$; n = 1 to 24) vertical electronic transitions in water. Finally, to investigate the dependence of the potential energy of the ground state (S_0) and excited states (S_n and T_n) on the rotation of the $C_m{-}C_\varphi$ bond, we used the Gaussian keyword "Opt = ModRedundant." The calculated ground state (S_0) potential energies at each optimized structure were plotted as a function of the rotated dihedral angle θ in the region of 40–130° with 10° increment. The potential energy surfaces of the singlet and triplet excited states were conducted by calculating the singlet-singlet, $S_0 \rightarrow S_n$, and singlet-triplet, $S_0 \rightarrow T_n$, electronic transition energies for each optimized structure at the rotated dihedral angle θ, including the SCF energy correction to each calculated electronic transition energy, $\Delta ESCF = E(\theta) - E0$, where $E(0)$ and $E(\theta)$ represent the calculated global energies of the ground state and energetically most stable structure at the dihedral angle θ, respectively.

We would like to point out that the electron densities in HOMO and LUMO molecular orbitals, and electronic spectra of the molecules studied here were plotted using GaussSum software [42].

Acknowledgements

We would like to thank the following: the U.S. National Science Foundation (NSF) for support of research efforts under grant no. HRD-08-33180 and Ömer Andaç (of the Chemistry Department of Ondokuz Mayıs University) for kindly making available computing facilities and software setup. We also thank TUBITAK ULAKBIM, High Performance and Grid Computing Center (TR-Grid e-Infrastructure) for performing the calculations reported in this work.

Author details

Metin Aydin[1*] and Daniel L. Akins[2]

*Address all correspondence to: aydn123@netscape.net

1 Department of Chemistry, Faculty of Art and Sciences, Ondokuz Mayıs University, Samsun, Turkey

2 Department of Chemistry and Biochemistry, Center for Analysis of Structures and Interfaces (CASI), The City College of the City University of New York, New York, USA

References

[1] Dolphin, D. *The Porphyrins*; Dolphin, D., Ed.; Academic Press: New York, 1978; Volume 1–3.

[2] Shelnutt, J. A,; Alston, K.; Findsen, E. W.; Ondrias, M. R.; Rifkind, J. M. In *Polphyrins: Excited States and Dynamics*; Gouterman, M., Rentzepis, P. M., Straub, K. D., Eds.; ACS Symposium Series 321; American Chemical Society: Washington, DC, 1986; Chapter 16.

[3] Brunner, H.; Nishiyama, H.; Itoh, K. in *Catalytic Asymmetric Synthesis*; Ojima, I. (Ed.), VCH, New York, 1993, pp. 303–322.

[4] Lu, J., Li, H., Liu, S., Chang, Y-C., Wu, H-P., Cheng, Y., Diau, E. W-G., Wang, M. Novel porphyrin-preparation, characterization, and applications in solar energy conversion. *Phys.Chem.Chem.Phys*. 2016, 18, 6885.

[5] Bian, Y.; Zhang, Y.; Ou, Z.; Jiang, J. Chemistry of Sandwich Tetrapyrrole Rare Earth Complexes. Handbook of Porphyrin Science, World Scientific Publishing, Singapore, 2011, 14, 249–460.

[6] Holten, D.; Bocian, D.F. Probing electronic communication in covalently linked multiporphyrin arrays. A guide to the rational design of molecular photonic devices. *J.S. Acc. Chem. Res*. 2002, *35*, 57–69.

[7] Hopfield, J.J.; Onuchic, J.N.; Beratan, D.N. A molecular shift register based on electron transfer. *Science* 1988, *241*, 817–820.

[8] Bonnett, R. Photosensitizers of the porphyrin and phthalocyanine series for photodynamic theory. *Chem. Soc. Rev.* 1995, *24*, 19–33.

[9] Andrade, S.M.; Costa, S.M.B. Spectroscopic studies on the interaction of a water soluble porphyrin and two drug carrier proteins. *Biophys. J.* 2002, *82*, 1607–1619.

[10] Ben-Hur, E.; Horowitz, B. Advances in photochemical approaches for blood sterilization. *Photochem. Photobiol.* 1995, *62*, 383–388.

[11] Uhehara, K.; Hioki, Y.; Mimuro, M. The chlorophyll *a* aggregate absorbing near 685 nm is selectively formed in aqueous tetrahydrofuran. *Photochem. Photobiol.* 1993, *58*, 127–132.

[12] Tominaga, T.T.; Yushmanov, V.E.; Borissevitch, I.E.; Imasato, H.; Tabak, M. Aggregation phenomena in the complexes of iron tetraphenylporphine sulfonate with bovine serum albumin. *J. Inorg. Biochem.* 1997, 65, 235–244.

[13] Togashi, D.M.; Costa, S.M.B. Absorption, fluorescence and transient triplet-triplet absorption spectra of zinc tetramethylpyridylporphyrin in reverse micelles and microemulsions of aerosol OT-(AOT). *Phys. Chem. Chem. Phys.* 2000, *2*, 5437–5444.

[14] Ferreira, G. C.; Kadish, K. M.; Smith, K. M.; Guilard, R., Eds. In *The Handbook of Porphyrin Science*; World Scientific Publishers: Singapore, 2013, Vol. 27.

[15] Gillam, M. E.; Hunter, G. A.; Ferreira, G. C. The ultimate step of heme biosynthesis: orchestration between iron trafficking and porphyrin synthesis. In *Heme Biochemistry*; Ferreira, G.C., Ed., of *The Handbook of Porphyrin Science* series, Kadish, K. M.; Smith, K. M.; Guilard, R. series Eds., World Scientific Publishing Co., New Jersey, USA, 2013, Vol. 26, pp. 129–189.

[16] Borissevitch, I.E.; Tominaga, T.T.; Imasato, H.; Tabak M. Fluorescence and optical absorption study of interaction of two water soluble porphyrins with bovine serum albumin: the role of albumin and porphyrin aggregation. *J. Luminesc.* 1996, *69*, 65–76.

[17] Huang, C.Z.; Li, Y.F.; Li, K.A.; Tong, S.Y. Spectral characteristics of the aggregation of α, β, γ, δ-tetrakis(*p*-sulfophenyl)porphyrin in the presence of proteins. *Bull. Chem. Soc. Jpn.* 1998, *71*, 1791–1797.

[18] Maiti, N.C.; Ravikanth, M.; Mazumdar, S.; Periasamy N. Fluorescence dynamics of noncovalent linked porphyrin dimers and aggregates. *J. Phys. Chem.* 1995, *99*, 17192–17197.

[19] Ohno, O.; Kaizu, Y.; Kobayashi, H. J-aggregate formation of a water-soluble porphyrin in acidic aqueous media. *J. Chem. Phys.* 1993, *99*, 4128–4139.

[20] Akins, D.L.; Zhu, H.-R.; Guo, C. Absorption and Raman scattering by aggregated meso-tetrakis(p-sulfonatophenyl)porphine. *J. Phys. Chem.* 1994, *98*, 3612–3618.

[21] Akins, D.L.; Özçelik, S.; Zhu, H.R.; Guo, C. Fluorescence decay kinetics and structure of aggregated tetrakis(*p*-sulfonatophenyl)porphyrin. *J. Phys. Chem.* 1996, *100*, 14390–14396.

[22] Labanowski, J.K.; Andzelm, J.W. *Density Functional Theory Methods in Chemistry*; Springer-Verlag: New York, NY, USA, 1991.

[23] Aydin, M. DFT and Raman spectroscopy of porphyrin derivatives: tetraphenylporphine (TPP). *Vib. Spectrosc.* 2013, *68*, 141–152.

[24] Słota, R.; Broda, M.A.; Dyrda, G.; Ejsmont, K.; Mele, G. Structural and molecular characterization of meso-substituted zinc porphyrins: a DFT supported study. *Molecules* 2011, *16*, 9957–9971.

[25] Wang, C.; Yang, G.; Li, J.; Mele, G.; Słota, R.; Broda, M.A.; Duan, M.; Vasapollo, G.; Zhang, X.; Zhang, F.X. Novel meso-substituted porphyrins: synthesis, characterization and photocatalytic activity of their TiO2-based composites. *Dyes Pigments* 2009, *80*, 321–328.

[26] Zhang, Y.H.; Zhao, W.; Jiang, P.; Zhang, L.J.; Zhang, T.; Wang, J. Structural parameters and vibrational spectra of a series of zinc meso-phenylporphyrins: a DFT and experimental study. *Spectrochim. Acta A* 2010, *75*, 880–890.

[27] Yao, P.; Han, S.; Zhang, Y.; Zhang, X.; Jiang, J. Structures and spectroscopic properties of mesotetrasubstituted porphyrin complexes: Meso-Substitutional and central metallic effect study based on density functional theory calculations. *Vib. Spectrosc.* 2009, *50*, 169–177.

[28] De Oliveira, V.E.; Cimini Correa, C.; Pinheiro, C.B.; Diniz, R.; de Oliveira, L.F.C. Structural and spectroscopy studies of the zinc complex of p-hydroxyphenylporphyrin. *J. Mol. Struct.* 2011, *995*, 125–129.

[29] Aydin, M. Comparative study of the structural and vibroelectronic properties of porphyrin and its derivatives. *Molecules* 2014, *19*, 20988–21021.

[30] Edwards, L.; Dolphin, D. H.; Gouterman, M.; Adler, A. D. Porphyrins XVII. Vapor absorption spectra and redox reactions: Tetraphenylporphins and porphin. J Mol Spectrosc 1971, 38, 16.

[31] Murphy, M. J.; Siegel, L. M.; Kamin, H.; Rosenthal, D. Reduced nicotinamide adenine dinucleotide phosphate-sulfite reductase of enterobacteria. II. Identification of a new class of heme prosthetic group: an iron-tetrahydroporphyrin (isobacteriochlorin type) with eight carboxylic acid groups. J Biol Chem. 1973, 248(8), 2801–2814.

[32] Lan, M.; Zhao, H.; Yuan, H.; Jiang, C.; Zuo, S.; Jiang, Y. Absorption and EPR spectra of some porphyrins and metalloporphyrins. *Dyes Pigments* 2007, *74*, 357–362.

[33] Zhang, Y.H.; Chen, D.M.; He, T.; Liu, F.C. Raman and infrared spectral study of meso-sulfonatophenyl substituted porphyrins (TPPSn, n_/1, 2A, 2O, 3, 4). *Spectrochim. Acta Part A* 2003, *59*, 87–101.

[34] Becke, A.D. Density-functional thermochemistry. III. The role of exact exchange. *J. Chem. Phys.* 1993, *98*, 5648–5652.

[35] Lee, C.; Yang, W.; Parr, R.G. Development of the Colle-Salvetti correlation-energy formula into a functional of the electron density. *Phys. Rev. B* 1988, *37*, 785–789.

[36] Krishnan, R.; Binkley, J.S.; Seeger, R.; Pople, J.A. Self-consistent molecular orbital methods. XX. A basis set for correlated wave functions.. *J. Chem. Phys.* 1980, *72*, 650–654.

[37] Tomasi, J.; Mennucci, B.; Cammi, R. Quantum mechanical continuum solvation Models. *Chem. Rev.* 2005, *105*, 2999–3093.

[38] Barone, V.; Cossi, M. Quantum calculation of molecular energies and energy gradients in solution by a conductor solvent model. *J. Phys. Chem. A* 1998, *102*, 1995–2001.

[39] Miertus, S.; Scrocco, E.; Tomasi, J. Electrostatic interaction of a solute with a continuum. A direct utilization of ab initio molecular potentials for the prevision of solvent effects. *Chem. Phys.* 1981, *55*, 117–129.

[40] Cossi, M.; Rega, N.; Scalmani, G.; Barone, V. Energies, structures, and electronic properties of molecules in solution with the C-PCM solvation model. *J. Comput. Chem.* 2003, *24*, 669–681.

[41] Frisch, M.J.; Trucks, G.W.; Schlegel, H.B.; Scuseria, G.E.; Robb, M.A.; Cheeseman, J.R.; Scalmani, G.; Barone, V.; Mennucci, B.; Petersson, G.A.; *et al. Gaussian 09, Revision A. 02.* Gaussian. Inc.: Wallingford, CT, 2009, 200.

[42] O'Boyle, N. M.; Tenderholt, A. L.; Langner, K. M. A library for package-independent computational chemistry algorithms. *J. Comp. Chem.* 2008, *29*, 839–845.

Section 3

Biological and Biomedical Applications of Molecular Spectroscopy

Chapter 11

Applications of Molecular Spectroscopic Methods to the Elucidation of Lignin Structure

Tingting You and Feng Xu

Additional information is available at the end of the chapter

http://dx.doi.org/10.5772/64581

Abstract

Lignin in plant cell wall is a complex amorphous polymer and is biosynthesized mainly from three aromatic alcohols, namely, *p*-coumaryl, coniferyl, and sinapyl alcohols. This biosynthesis process consists of mainly radical coupling reactions and creates a unique lignin polymer in each plant species. Generally, lignin mainly consists of *p*-hydroxyphenyl (H), guaiacyl (G), and syringyl (S) units and is linked by several types of carbon-carbon (β-β, β-5, β-1, and 5–5) and ether bonds. Due to the structural complexity, various molecular spectroscopic methods have been applied to unravel the aromatic units and different interunit linkages in lignin from different plant species. This chapter is focused on the application of ultraviolet (UV) spectroscopy, Fourier transform infrared (FT-IR) spectroscopy, Fourier transform Raman (FT-Raman) spectroscopy, fluorescence spectroscopy, and nuclear magnetic resonance (NMR) spectroscopy to lignin structural elucidation.

Keywords: lignin, structure elucidation, 2D HSQC NMR, composition, linkages

1. Introduction

Plant cell walls in higher plants are mainly consisted of cellulose, hemicelluloses, and lignin. As a major cell wall component, lignin in plants provides rigidity, internal transport of water and nutrients, and protection against attack by microorganisms [1]. It has been reported that lignin in lignified plants accounts for 16–36% by weight [2]. Due to the high content and complex structure, lignin plays a key role in pulping and other chemical conversion process of plants. Most importantly, lignin currently attracts widespread attention as a feedstock for biofuels and biochemical production [3–5]. Broadening the knowledge of structural features of lignin is therefore necessary to help to effectively improve the economics of these processes.

1.1. Lignin structure

Lignin is the most abundant renewable aromatic biopolymer present in nature. To the best of current knowledge, lignin shows a heterogeneous composition and lacks a defined primary structure due to the special biosynthesis processes [6]. Lignin is generally considered to be synthesized mainly from three *p*-hydroxycinnamyl alcohols precursors, namely, *p*-coumaryl, coniferyl, and sinapyl alcohols. During the lignification process, each of these precursors gives rise to a different type of lignin unit called *p*-hydroxyphenyl (H), guaiacyl (G), and syringyl (S) units, respectively. This biosynthesis process consists of mainly radical coupling, and creates a unique lignin polymer in each plant species, even in different tissues of the same individual [7]. In general, softwood (spruce, pine, etc.) lignin consists almost entirely of G units,

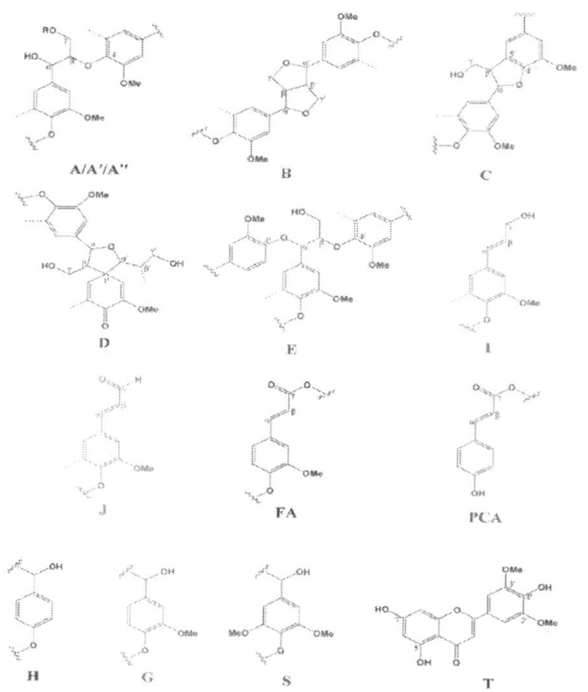

Figure 1. Main structures of lignin, involving different side-chain linkages and aromatic units: (A) β-O-4 linkages; (A') β-O-4 linkages with acetylated γ-carbon; (A'') β-O-4 linkages with *p*-coumaroylated γ-carbon; (B) resinol structures formed by β-β, α-O-γ and γ-O-α linkages; (C) phenylcoumarance structures formed by β-5 and α-O-4 linkages; (D) spirodienone structures formed by β-1 and α-O-α linkages; (E) α, β-diaryl ether substructures; (H) *p*-hydroxyphenyl unit; (G) guaiacyl unit; (S) syringyl unit; (I) cinnamyl alcohol end-groups; (J) cinnamyl aldehyde end-groups; (FA) ferulate; (PCA) *p*-coumarate; (T) tricin.

hardwood (beech, poplar, etc.) lignin is a mixture of G and S units, and herbaceous lignin (bamboo, reed, etc.) is composed of all the three units. These units are linked by several types of carbon-carbon (β-β, β-5, β-1, and 5–5) and ether bonds (β-O-4 and α-O-4) with various percentage. The β-O-4 linkages are the main lignin interunit linkages, accounting for more than 60% among various linkages. However, the less C—C linkages constitute some of the most difficult bonds to break [8]. In addition to lignin, the local noncovalent interaction and oxidative reactions among carbohydrates, phenolic components, and lignin render and control the formation of lignin-carbohydrate complex (LCC) during lignin biosynthesis processes [9, 10]. Other components, such as hydroxycynnamic acids and tricin, have been demonstrated to be incorporated into herbaceous lignin, apparently making the structure of lignin more complex [7]. **Figure 1** shows the structure of main components of lignin.

1.2. Methods for structural elucidation of lignin

Lignin chemists have devoted their efforts to reveal the molecular details of lignin structure over the past few decades. Various wet-chemistry methods have been developed. Permanganate oxidation, nitrobenzene oxidation, GC-MS pyrolysis, thioacidolysis, and derivatization followed by reductive cleavage (DFRC) methods partially degrade lignin polymer and release diagnostic monomers, which reveals H/G/S composition in lignin. The presence of acylating groups in some herbaceous lignins is well known for a long time. A modified DFRC method (DFRC′) was developed for naturally acetylated lignin moieties determination [11]. Though significant strides have been made in elucidating the chemical structure of lignin by wet-chemistry methods, they have not yet been completely elucidated. Only a fraction of lignin was analyzed and significantly different results were obtained even for the same sample by different wet-chemistry methods. Moreover, all protocols provide relative comparison in an array of samples rather than giving the absolute quantitative values.

Attempts through the years have been made to develop nondestructive and quantitative methods for w using molecular spectroscopic methods. Molecular spectra see differences in chemical structure of lignin that it is invisible for other analytical methods. Acid-soluble lignin in biomass is determined to be less than 4% by ultraviolet (UV) spectroscopy based on Beer′ Law. On the other hand, the level of purity originated from different resources was easily determined by comparison of the extinction coefficients [12]. Despite some constituents can also be found from the UV spectra, further evidence is needed to confirm that. To date, natural lignin is found to contain several functional groups and chromophores which can be qualitatively determined by Fourier transform infrared (FT-IR) spectroscopy, FT-Raman spectroscopy, and fluorescence spectroscopy [13–15]. FT-IR spectroscopy is the most widely distributed as a modern and powerful analytical technique. The structural changes during physical and chemical pretreatment were monitored, which revealed the mechanism [16, 17]. However, no quantitative information about both lignin interunit linkages and S/G/H composition was achieved until the application of nuclear magnetic resonance (NMR). Development of 1D, 2D, and 3D NMR spectroscopic methods provides a powerful tool for lignin analysis. Evidence of the presence of lignin substructure dibenzodioxocine and spirodienone in lignin was first observed by the NMR spectroscopic methods [18, 19]. It is noted that more detailed structure of the whole macromolecule can be obtained by NMR spectroscopic methods. In addition to the structure details, the absolute amount of the side chain moieties and functional groups can be determined by combination of ^{13}C NMR and 2D HSQC NMR spectroscopic methods [20].

This chapter describes the application of UV spectroscopy, fluorescence spectroscopy, FT-IR spectroscopy, FT-Raman spectroscopy, and various one-dimensional and multidimensional NMR spectroscopic methods in lignin structure determination.

2. UV spectroscopy

Ultraviolet spectroscopy has been widely applied to quantitatively measure the acid-soluble lignin content, semiquantitatively determining the lignin purity, and predicting the possible lignin constituents [21–23]. The location of maximum absorption and the extinction coefficient of it are the two main factors that determine the properties of lignin.

Capitalizing on the stronger absorbance of lignin compare with carbohydrates in the UV region, the amount of acid-soluble lignin can be determined by applying Beer' Law. In 1985, a standard protocol was developed by the Technical Association of the Pulp and Paper Industry (TAPPI) to determine the amount of acid-soluble lignin and was applied widely across the world [24–26]. However, the accuracy is affected by measuring the absorbance at 200–205 nm, where carbohydrate monomers may also absorb light. Moreover, the extinction coefficient that is used varied with the type of lignin. In 2008, more accurate method of laboratory analytical procedure (LAP) of biomass was provided by the National Renewable Energy Laboratory (NREL) [27]. For the measurement, the absorbance of ASL was recorded at the recommended wavelength, which varied among plant species. Nowadays, many scientists use LAP for the determination of ASL in untreated and pretreated biomass. Our previous research applied LAP procedure to determine the content of ASL before and after ionic liquid-acid pretreatment, and an increased ASL was found [17]. Likewise, Rajan et al. used NREL methods to reveal the effect of dilute acid pretreatment on ASL of wheat straw [28].

Generally, the UV spectrum of lignin exhibits several absorption maxima at around wavelengths of 200, 240, 280, and 320 nm, which originate from the intrinsic structure [29]. Occurrence of the maximum absorption at around 200 nm in UV spectra is corresponding to the $\pi \rightarrow \pi^*$ electronic transition in the aromatic ring of lignin structure. Maxima at around 240 and 282 nm probably originate from the free and etherified hydroxyl groups. Phenolic structure and phenylpropane units (S, G) in lignin also can be detected from the UV spectra [22, 30]. It is reported that a pure syringyl lignin has an absorption maximum at 270–273 nm, whereas a red shift to 280–282 nm and three times stronger extinction coefficient are found in a pure guaiacyl lignin [22]. As for the maximum absorption at around 320 nm, it is attributed to $\pi \rightarrow \pi^*$ transitions in lignin units with $C_\alpha=C_\beta$ linkages conjugated with aromatic ring and $n \rightarrow \pi^*$ transition in lignin units containing $C_\alpha=O$ groups. In herbaceous plants, bound hydroxycinnamic acid especially a predominance of esterified *p*-coumaric acid or etherified ferulic acid may contribute to the appearance of it [7, 31].

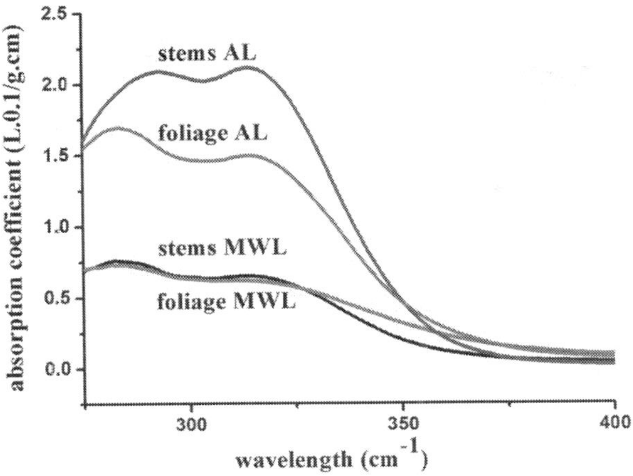

Figure 2. UV spectra of stem milled wood lignin (MWL), foliage MWL, stem alkaline lignin (AL), and foliage AL from *A. donax*. The lignin fractions were dissolved in DMSO, and scanned from 500 to 190 nm on a UV 2300 spectrophotometer (reprinted with permission from [7]. Copyright 2013 American Chemical Society).

The UV spectroscopic patterns of lignin cary among plant species. It has been demonstrated that the maximum of the absorption curve of milled wood lignin (MWL) from spruce fibers is at about 280 nm, whereas the UV spectra of acid insoluble lignin fractions from shrubs *Caligonum monogoliacum* and *Tamarix* spp. exhibits two absorption maxima at around 245 and 280 nm [22, 32]. It is noted the UV spectra of grasses are somewhat different from wood species. Apart from the first maximum absorption at 284 nm, a shoulder peak at around 320 nm is always appeared in the UV spectrum of lignin from grasses [7, 33]. For instance, esterified *p*-coumaric acid is the main component if the wavelength of maximum absorption is shorter than 320 nm. In contrast, the wavelength of maximum absorption at 325 nm is indicative of rich etherified ferulic acid in lignin. **Figure 2** shows UV absorption spectra of stem MWL, foliage MWL, stem alkaline lignin (AL), and foliage AL from energy crops *Arundo donax* Linn. that reprinted from our previous article [7]. As illustrated, two maximum absorptions were found at around 284 and 310 nm in the spectra of lignin fractions. It can be inferred that the four lignin fractions may contain free and etherified hydroxyl groups, and bound *p*-coumaric acid.

UV spectroscopy has been used to semiquantitatively determine the purity of lignin with respect to the concentration. According to Beer's Law ($A = \varepsilon cd$, where A = absorbance, ε = extinction coefficient, d = path length, c = concentration), the value of extinction coefficient reveals the concentration of lignin. The low extinction coefficient of lignin is due to the high amount of nonlignin materials. As aforementioned, lignin in plant tissue does not exist as an independent polymer but bonded with polysaccharides, which can be coextracted. The presence of a variety of variable abundances of lignin-carbohydrate bonds among different plant species makes it difficult to isolate lignin purely and completely. Sugar analysis of isolated cellulolytic enzyme lignin from Douglas fir, redwood, white fir, *Eucalyptus globulus*

Labill., *A. donax*, and poplar wood revealed that lignin contained relatively noticeable amount (7.74–20%) of associated carbohydrates [34–36], which reduces the level of purity. Moreover, lignin fractions isolated from lignocellulosic biomass still contain other nonlignin contaminants, for instance, ash. Specifically, the ash content in technical lignins such as lignosulfonate (LS) and kraft lignin (KL) is up to 9.4 and 27.1%, respectively [37]. A certain amount of ash in lignin was demonstrated to decrease the extinction coefficient. Sun et al. found that the relatively lower absorption of lignin fractions was probable due to the higher amounts of coextracted nonlignin materials such as ash and salts [21].

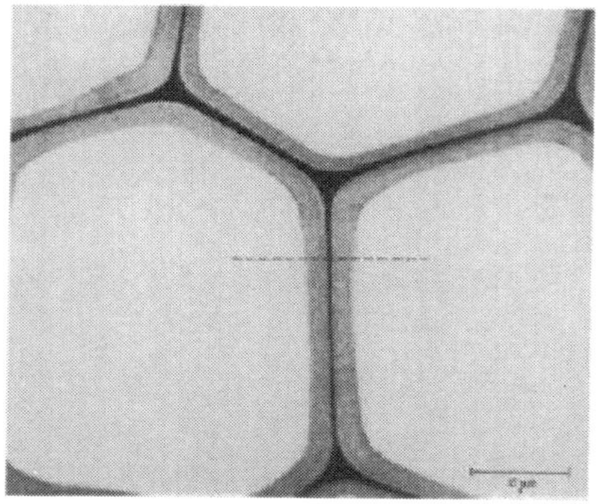

Figure 3. Cross section of Epon embedded tracheids of black spruce earlywood photographed in ultraviolet light of wavelength 240 nm. The densitometer tracing was taken across the tracheid wall on the dotted line (reprinted with permission from [38]. Copyright 1969 Springer).

Apart from UV spectroscopy, UV microscopy has been used in a number of studies to monitor the lignin distribution among various tissues of gymnosperm and dicotyledonous angiosperm in respect to the concentration by comparison of UV absorbance. UV light transmits through ultrathin sections (0.5 μm) of wood and measure within the wavelength range 240–320 nm of lignin absorption. From the absorbance and the cell wall dimensions, the concentrations of lignin in cell wall layering structure of earlywood of black spruce (**Figure 3**) are determined to decrease in the order: the cell comers > the compound middle lamella > the secondary wall [22, 38].

3. Fluorescence spectroscopy

Fluorescence spectroscopy has been used for the analysis of lignin constituents in wastewaters from pulp mills in the 1970s [14]. Subsequently, scientists focus a lot on the fluorescence properties of lignin. Lundquist et al. have investigated the fluorescence spectra of a variety of

http://dx.doi.org/10.5772/64581

model compounds, lignin, and lignin-related products to establish a basis for the interpretation of the fluorescence results [39]. By comparing the fluorescence spectra (emission spectra and excitation spectra) of lignin with the structural elements, the possible chromophores can be determined.

Dioxane-water or water is a good solvent for the dissolution of lignin as the absorbance of these solutions is less than 0.05. This implied that the intensity of the emitted light (*Q*) can be expressed by the following equation [39, 40]: $Q = I_0\ (2.3\varepsilon cd)\ \Phi_f$, where ($I_0$ = intensity of the incident light, ε = molar absorptivity, c = concentration in moles per liter, b = sample path length, Φ_f = quantum efficiency for fluorescence). Albinsson et al. dissolved the untreated and borohydride-reduced MWL from spruce in dioxane-water 9:1 for the fluorescence spectra collecting [14]. Lundquist et al. used either water or dioxane-water 1:1 as the solvents to explore the fluorescence properties of lignin sulfonate, MWL, and kraft lignin [39].

Sample*	Excitation spectrum λ_{max} (nm)	Emission spectrum λ_{max} (nm)
MWL from spruce [39]	285, 240	358
MWL from spruce [14]	280	364
Borohydride-reduced MWL from spruce [39]	284, 240	358
MWL from birch [39]	282, 240	350
Kraft lignin [39]	334	398
Lignin sulfonate [39]	287, 240, 315	393

Note: *Reference.

Table 1. Fluorescence properties of lignin materials.

Nonradiative energy transfer from lignin chromophores is excited to an acceptor and then emits the fluorescent light. Excitation spectra and emission spectra were collected on a spectrofluorimeter. Generally, the emission spectra are recorded at the wavelengths of the excitation maxima and the excitation spectra are recorded at the wavelengths of the emission maxima. **Table 1** shows the fluorescence properties of lignin materials [14, 39]. As can be seen, emission spectra of MWL from spruce exhibit a maximum at about 360 nm on excitation at different wavelengths in the range 240–320 nm. For the MWL from birch, emission spectra exhibit a maximum at 350 nm. A great difference is found for the fluorescence properties of technical lignin, such as kraft lignin or lignin sulfonate. The emission spectra of these lignins exhibit a maximum at about 400 nm on excitation at different wavelengths in the range 240–350 nm.

Examination of the fluorescence spectra of lignin samples and model compounds suggested the possible chromophores. If the structural elements spectra closely match the emission from the lignins, it points to the possibility that the lignin fluorescence is mainly emitted from that structure of lignin. Based on these, small amounts of phenylcoumarone structures are found in lignin from pretreated acid or balling materials [14]. Reduction of carbonyl in lignin by borohydride does not change the position of emission maximum but increase the fluorescence intensity due to some "energy sink" structure in lignin. It has been determined that the "energy sink" structure could be arylconjugated carbonyl groups such as cinnamyl alcohol or phenylcoumarone type and stilbene structure.

Lignocellulosic biomass is known to be autofluorescent. Compared with holocellulose, the autofluorescent of lignin is generally much brighter [41, 42]. Laser scanning confocal fluorescence microscopy (LSCFM) allowed direct visualization of the relative amounts of lignin in different cell types on a semiquantitative basis. Based on the brightness of fluorescence images, the relative amounts of lignin in different regions of the cell wall in different cell types can be measured [43]. Donaldson et al. used confocal fluorescence microscopy (FM) to provide semiquantitative information in different regions based on lignin autofluorescence, and by staining with acriflavine [44]. The level of lignification in different plant species was then determined. FM was also used to investigate the cell wall structure changes during chemical pretreatment of biomass [45].

4. FT-IR and FR-Raman spectroscopy

Fourier transform infrared and FT-Raman spectroscopic methods have been described as an efficient measurement of valuable plant components, such as lipids, fatty acids, carbohydrates, phenolic substances, and so forth [46]. Results from both FT-IR and FT-Raman spectroscopy are in general agreement and provide complimentary information. Both techniques are nondestructive, rapid, and accurate and use only microscale samples. Differed from FT-IR spectroscopy, FT-Raman spectroscopy is insensitive to water. Hence, it is more suitable to perform in situ studies of fresh plant materials that contained some moisture by FT-Raman spectroscopy. The application of these two techniques to numerous research areas has already provided useful information on lignin.

4.1. FT-IR spectroscopy

FT-IR spectroscopy is a nondestructive, noninvasive, high sensitivity, and rapid method for lignin structure investigation or wood constituent determination widely use by lignin and wood chemists as a molecular probe [13, 47, 48]. This method opens perspective to quantify lignin in samples and semiquantitative and qualitative analyses of lignin structure characteristic.

4.1.1. FT-IR spectra assignment

Much work has been published on the characterization of lignin, and a lignin FT-IR-spectrum library has been established over the past few decades. Transmission or diffuse reflectance spectra in the midinfrared (4000–200 cm^{-1}) have been shown to provide reliable information on the chemical properties of lignin fractions or lignin in wood. Before the analysis, all of the spectra should be baseline corrected and normalized. In the region 3800–2750 cm^{-1}, several bands are observed which are caused by the presence of alcoholic and phenolic hydroxyl groups and the methyl and methylene groups in lignin [47]. More bands are clearly discernible with deconvolution. In more detail, a wide absorption band appearing at 3580–3550 cm^{-1} is derived from free hydroxyl group in phenolic and alcoholic structures. Additionally, signals in the region 3000–2750 cm^{-1} are predominantly arising from C-H stretching in aromatic methoxyl groups and in methyl and methylene groups of the side chain. Fatty acid present in lignin undoubtedly increases the intensity of C-H stretching [49]. Demethylation or methylation affects the intensity of these bands sharply. It has been reported that the intensity of O-H stretching peak reduces dramatically upon methylation, whereas the intensity of peaks corresponding to C-H stretching increased simultaneously [50].

The investigation of more complex fingerprint region is necessary to facilitate understanding of the intact lignin characteristics. Bands found at around 1735 and 1714 cm^{-1} are originated from unconjugated carbonyl-carboxyl stretching in ketones, carbonyls, and ester groups [13]. Esterified phenolic acids and acetyls from associated hemicelluloses are the contributors to these absorption bands. The intensity of these bands also increased when a ketone or an aldehyde structure is produced [51]. However, the occurrence of a serial of absorption peaks at range 1675–1655 cm^{-1} is corresponding to conjugated carbonyl-carboxyl stretching. Hergert et al. concluded that the peak at 1660 cm^{-1} was originated from a ketone group located at α position, whereas the peak at 1712 cm^{-1} was assigned to a ketone group located at β position [52]. It is noteworthy that sharp bands at 1653 cm^{-1} from the spectrum of oven-dried samples are probably arising from the tricin associated with lignin especially from nonwood biomass [7, 25, 53].

Every lignin FT-IR spectrum shows prominent absorptions at around 1600, 1510, and 1420 cm^{-1} and the C—H deformation combined with aromatic ring vibration at 1460 cm^{-1}. The first three bands are assigned to the aromatic skeleton vibrations in lignin, which is the "core" structure of lignin. According to the classification of lignin proposed by Faix, the FT-IR spectra of lignin are divided into three categories, i.e., G type, GS type, and GSH type [13]. The spectra of type G lignins show typical feature at 1140 cm^{-1}, which originates from aromatic C—H in-plane deformation. Structure features of G type lignin are primary found in the spectrum of softwood lignin. Generally, the spectra of lignin samples from softwood, such as pine and spruce, show absorptions at 1269 cm^{-1} (G ring and C═O stretch), 1140, 854, and 817 cm^{-1} (C—H out-of-plane vibrations at positions 2, 5, and 6 of G units). Lignin from hardwood such as poplar, birch, and beech belongs to GS type according to the IR classification criteria [13]. GS type lignin exhibites typical features at a wavenumber at around 1128 (aromatic C—H in-plane deformation), 1328 (S ring plus G ring condensed), and 834 cm^{-1} (C—H out-of-plane in position 2 and 6 of S). Furthermore, the spectra within the GS category are subdivided into four groups based on different intensities of each band. In the samples from *A. donax* and bamboo, the absorption band at 1167 cm^{-1} which is attributed to C═O in ester groups (conjugated) is additionally present compared to the spectra of GS and G type lignin [7, 54]. Hence, maxima absorption at 1167 cm^{-1} is typical only for GSH type lignin. Signals from lignin functional groups such as phenolic hydroxyl group can be found at 1370–1375 cm^{-1}. As small amount of carbohydrate is apt to associate with lignin, aromatic C—H deformation at 1035 cm^{-1} appears as a complex vibration associated with the C—O, C—C stretching and C—OH bending in polysaccharides. For more details, see **Table 2** [7, 13, 47, 55].

Frequency (cm^{-1})	Assignment	Comments
3000–2840	ν (C—H)	C—H stretching vibrations in methyl and methylene of side chains
1738–1709	ν (C=O)	C=O stretching in unconjugated ketone, carbonyl and ester; C=O stretching in conjugated aldehydes and carboxylic acids absorb around and below 1700 cm^{-1}
1699–1633	ν (C=O)	C=O stretching in conjugated aldehydes and carboxylic acids absorb around and below 1700 cm^{-1}
1684	ν (C=O)	β-enone carbonyl stretching modes
1655–1675	ν (C=O)	C=O stretching in conjugated para-substitute aryl ketones
1593–1605	ν (Ar), ν(C=O)	Aromatic skeletal vibrations of S and G (S > G) plus C=O Stretching; S > G and G Condensed > G etherified
1510	ν (Ar)	aromatic skeletal vibrations of S and G (G > S)
1460	ν (C—H)	Asymmetric C—H deformations in $-CH_3$ and $-CH_2$
1420	ν (Ar)	Aromatic skeletal vibrations combined with C—H in-plane deform
1365–1370	ν (C—H), ν (O—H)	Aliphatic C—H stretching in CH_3 and phenolic OH
1267	ν (Ar), ν(C=O)	G ring and C=O stretching
1221–1230	ν (C=O), ν (C—O), ν (C—C)	C—C, C—O and C=O stretching (G condensed > G etherified)
1167	ν (C=O)	Typical for HGS type lignin; C=O stretching inconjugated ketone, ester groups
1124	δ (C—H)	Aromatic C—H bending in-plane (typical for S units)
1030–1035	ν (C—H)	The aromatic C—H deformation acting as a complex vibration associated with the C—O, C—C stretching and C—OH bending
966–990	ν (—HC=CH—)	—HC=CH- out-of-plane deformation (trans)
853–858	ν (C—H)	C—H out of plane in position 2, 5, and 6 of G units
834	ν (C—H)	C—H out of plane in position 2 and 6 of S units, and in all positions of H units

Table 2. Main assignments of lignin in FT-IR bands.

4.1.2. Qualitative and semiquantitative analysis

FT-IR spectroscopy reflects the chemical structure of lignin. As a result, the native characteristics are uncovered and the structural changes taking place in samples are monitored. Huang et al. found that some tannin was possibly condensed with bark lignin by comparing with the FT-IR spectra discrepancy of MWLs from loblolly pine stem wood, residue, and bark [26]. Moreover, all of the three MWLs belonged to G type lignin. Differed from the softwood lignin, MWL from energy crops *A. donax* showed features of GSH type lignin. A treatment is involved in efficient utilization of lignocellulosic biomass. It is demonstrated that the "core" structure of lignin samples isolated by alkaline, ionic liquid, organic solvents, acid, and thermal treatment does not change significantly. However, the absorption frequencies correspond to the vibrational motions of the nuclei of a functional group show distinct changes when the chemical environment of the functional group is modified. Jia et al. found that a ketone structure was produced during acidic ionic liquid treatment of lignin model compound by comparing the FT-IR spectra that resulted from the cleavage of β-O-4 linkage [51]. Chen et al. have reported that the shift of 1505–1510 cm^{-1} accompanied with the increased intensity of the band at 1321 cm^{-1} identified by attenuated total reflectance (ATR)-FTIR spectra of lignin indicated the occurrence of condensed reaction during the heat treatment [56]. It should be noted that the ATR-FTIR only quantitatively determined the chemical changes in the surface of the samples. Further evidence was needed to confirm that.

Apart from the qualitative analysis, FT-IR spectroscopy has been utilized as a means of relative quantifying lignin in samples. Raiskila et al. have developed a method based on the FT-IR spectra to determine the relative amount of lignin in a large amount of samples [57]. In addition to the lignin content determination, the condensation indices (i.e., the cross-linking indexes) of lignin reflect the condensation degree of lignin, which can be expressed by the following formula [58, 59]:

$$CI = \frac{\text{Sum of all minima between 1500 and 1050 cm}^{-1}}{\text{Sum of all mixima between 1600 and 1030 cm}^{-1}}$$

Lignin condensation always occurred during the acid treatment or the severe ball milling. The quantitatively analysis of the CI by using FT-IR spectroscopy is extremely convenient and time saving. Furthermore, the Abs. 1742/Abs. 1768 ratio is positive in connection with the phenolic hydroxyl group in lignin [60].

4.2. FT-Raman spectroscopy

1064 nm excited FT-Raman spectroscopy overcomes the obstacle of lignin autofluorescence, and the Raman spectra of acceptable quality for lignin or lignin-related materials can be obtained. Compared with the FT-IR spectra, more bands can be detected in the FT-Raman spectra. It has been reported that in neat state only about 60% of the bands of the total detected FT-Raman bands are detected in the FT-Raman spectra, implying that 40% of the bands could only be detected by FT-Raman [62]. In addition, fresh plants even with some extract present in them do not affect the FT-Raman spectral data. It is therefore significant to use both Raman and IR analyses to obtain the most detailed chemical information of lignin.

4.2.1. FT-Raman spectra assignment

The band assignment information in a lignin FT-Raman spectrum has been achieved primarily by Agarwal and his group [15, 61–64]. In general, the spectra are divided into three regions: 3200–2700, 1850–1350, and 1450–250 cm^{-1} region. In the region 3200–2700 cm^{-1}, several bands derived from the aliphatic and aromatic C—H stretches are detected. By studying the spectra of benzene derivatives and lignin models, it is determined that band at around 3070 cm^{-1} is likely to be due to the aromatic C—H stretch. The bands at approximately

3007 and 2938 cm^{-1} are originated from the asymmetric aliphatic C—H stretch like methoxy and acetoxy groups, whereas the band at 2843 cm^{-1} can be assigned to the symmetric aliphatic C—H stretch. A high amount of S units and acetylation or methylation treatments give particularly more intense 2938 cm^{-1} band [15]. Another band at 2890 cm^{-1} is assigned to the C—H stretch in R_3C—H structures.

The 1850–1350 cm^{-1} region is the most informative region for lignin. This region contains bands due to aromatic rings, ring-conjugated ethylenic C=C, α- and γ-C=O, and the *o*- and *p*-quinones. Every FT-Raman spectrum of lignin exhibited a strong band at about 1600 cm^{-1}, and can be used to normalize the spectra. This band is due to the aromatic ring stretch, and the band intensity is enhanced by the conjugation and resonance Raman effect [62]. Nonsymmetrical shape of the band suggested that two more components are included. After curve-fit treatment of the band of lignin from hardwood samples, S-marker and G-maker bands occurred. The calculation of these two bands would be a good probe to quantify the S/G ratio [65]. The band at 1660 cm^{-1} is assigned to ethylenic C=C bond in coniferyl alcohol units and the γ-C=O in coniferaldehyde units in lignin, whereas the band at 1630 cm^{-1} is found to be associated with the ring-conjugated C=C bond in coniferaldehyde units.

The 1450–250 cm^{-1} region reveals more details of lignin substructures. FT-Raman study of a large number of dehydrogenation polymer (DHP) lignin and hydroxycinnamic acid standard compounds is essential to establish a basis for the interpretation of FT-Raman spectra for lignin. Agarwal et al. collected and compared the spectra of G-DHP, S-DHP, and H-DHP lignin [63]. It was demonstrated that the bands at 371, 1041, 1333, and 1456 cm^{-1} were mainly resulted from the S type lignin, whereas the band at 1271 cm^{-1} was originated from G type lignin. Further investigation of softwood and hardwood MWLs confirmed these assignments. After a preliminary study of the spectra of hydroxycinnamic acid standard compounds and lignin, we recently showed that the band at 1173 cm^{-1} was assigned to C=O vibration form esterified or free hydroxycinnamic acid from grass *A. donax* [66]. A similar band at around 1173 cm^{-1} was also found in the spectra of switchgrass. Another band at about 1202 cm^{-1} is attributed to ring deformation and aryl-OCH_3 and aryl-OH in-plane bending from H type lignin [67].

4.2.2. *Qualitative and semiquantitative analysis*

Similar to FT-IR spectroscopy, FT-Raman spectroscopy is useful to the rapid characterization of lignin. The basic lignin units and functional groups in different lignocellulosic biomass are easy to identify from the FT-Raman spectra according to the bands assignments aforementioned. Raman spectral changes show the modification of lignin aroused by chemical, mechanical, or biological treatment [15, 64]. The spectral data changes of lignin treated by acetylation, methylation, diimide treatment, and alkaline hydrogen peroxide bleaching are achieved [15].

Some of the bands in the Raman spectra are applied to quantify lignin content or lignin structure. It is interesting to found that the intensity (peak area) of band at 1600 cm^{-1} of lignin is linearly related to the kappa number of pulp (R^2 = 0.98). The residual lignin content in bleaching pulp is therefore obtained by calculating the peak area of band at 1600 cm^{-1}. It is well known that significant variation in the S/G ratio exists among different plant species. A spectral

http://dx.doi.org/10.5772/64581

deconvolution method based on FT-Raman spectroscopy holds significant promise in the rapid and accurate determination of S/G ratio quantitatively [65].

5. NMR spectroscopy

The structure of lignin macromolecule in plants is extreme complex. To investigate the structure of the whole lignin macromolecule, various nuclear magnetic resonance spectroscopic methods (one-dimensional and multidimensional NMR) in both solid and solution states are frequently utilized. Compare with other spectroscopic methods mentioned above, NMR spectroscopic methods have much higher resolution and enable a larger amount of information to be obtained. Solid-state ^{13}C cross-polarization magic angle spinning (CPMAS) NMR spectroscopy allows the investigation of lignin structure in the native state and simultaneously avoids the chemical modification for sample preparation [68]. It is suitable for the analysis of lignin samples that have restricted solubility. However, due to the low resolution, only some structural features of lignin can be observed. Despite some improvements for the solid-state NMR have been made, it is still not routinely used. Solution-state NMR spectroscopy is more powerful in lignin structural elucidation. In solution, a much better resolution is obtained and a more detailed characterization of lignin aromatic and side chain is possible. Moreover, absolute quantification or relative quantification of each substructures and linkages can be unambiguously achieved [69–71]. One of the major factors that impede the application of solution-state NMR is the difficulty in dissolving lignin. In order to enhance the solubility, lignin is subjected to acetylation by anhydride/pyridine solution before the solution-state NMR spectra collection [72]. Briefly, 100 mg of lignin is dissolved in 4 mL of a solution of acetic anhydride: pyridine (1:1). After stirring for 24 h at room temperature under the exclusion of sunlight, the mixture is concentrated under reduced pressure. Then the mixture is dropped slowly into 200 mL of ice water (pH = 2.0) to induce precipitation, and the precipitate is washed with deionized water for several time. After centrifugation and freeze-drying, acetylated lignin is obtained. It is important to note that the chemical shift of lignin moieties will somewhat shift to a higher field after acetylation [69]. Recent advances in characterization of lignin polymer by solution-state NMR methodology have been reviewed [73].

5.1. One-dimensional NMR spectroscopy

1D NMR spectroscopic methods including the solution-state ^{1}H NMR, solution- or solid-state ^{13}C NMR, and solution-state ^{31}P NMR have been utilized to routinely determine the amount of hydroxyl groups (aliphatic, phenolic, and carboxylic acid), interunit linkages, S units, G units, and H units in lignin. The databases of chemical shifts of these spectroscopies have been well established based on comparison with synthetic model compound data [74–78].

5.1.1. ^{1}H NMR spectroscopy

The ^{1}H NMR spectra of lignin can be obtained within a few minutes. **Table 3** lists the signal assignment of ^{1}H NMR data [79–82]. The integral of all signals between 6.0 and 8.0 ppm belongs to aromatic protons in G, S, and H units, whereas those between 1.60 and 2.40 ppm and from 5.76 to 6.18 ppm are due to the hydroxyl groups in lignin. Signals derived from linkages such as β-O-4 and β-1 substructures in lignin are found in the range of 2.98–3.14 and 5.76–6.18 ppm. Moreover, signals between 4.93 and 5.09 ppm are appeared in the spectrum of lignin that contained some carbohydrates. Integration of the spectra allows the quantification of some specific moieties, such as phenolic hydroxyl groups [83].

Signal (ppm)	Assignment
9.9–9.7	Aldehyde protons from cinnamaldehyde and benzaldehyde
7.15–6.75	Aromatic protons from G
6.75–6.25	Aromatic protons from S
6.18–5.76	Benzylic OH from β-O-4 and β-1
5.09–4.93	Protons from carbohydrates
4.00–3.33	Protons from methoxyl
3.14–2.98	H_β from β-1
2.40–2.20	Protons from acetylated phenolic OH groups
2.20–1.60	Protons from acetylated aliphatic OH groups and 5-5

Note: [*]$CDCl_3$ was used as the solvent.

Table 3. Signal assignment for ^1H NMR spectroscopy of lignin[*].

5.1.2. Quantitative ^{13}C NMR spectroscopy

Owing to the complex structure of lignin, there are many overlapping resonances on ^1H NMR spectra and only some of structure features are detected. The accuracy of calculation based on ^1H NMR spectra of lignin is relative low. To obtain the detailed molecular structures, both solid-state and solution-state ^{13}C NMR spectroscopic methods have been applied to investigate the structural difference between lignin fractions from different plant species since 1981 [68, 84–86]. In addition to ^{13}C NMR spectra, the collection of distortionless enhancement by polarization transfer (DEPT) CH ($\theta = 135°$) spectra of lignin has been found to have a synergetic effect [33]. ^{13}C NMR spectroscopy allows the classification and quantification of lignin nondestructively. However, more than 24 h is needed to collect the quantitative ^{13}C NMR spectra. To decrease the experiment time without affecting the quality of spectra, 0.01M chromium (III) triacetylacetonate ($Cr(acac)_3$) is considered a relaxant which allows a 4-fold decrease in the experiment time. One has to take into account that the use of tetramethylsilane (TMS) as an internal reference (0.00 ppm) is significant. Additionally, it is noteworthy that differences in the location of some structures of lignin may happen due to their strong solvent dependency [19].

G units are identified by signals at 149.3 and 149.1 (C-3, G etherified), 147.8 and 147.4 (C-4, G etherified), 134.2 (C-1, G etherified), 119.0 (C-6, G), 114.8 and 114.6 (C-5, G), and 110.9 ppm

(C-2, G). The S units are verified by signals at 152.0 (C-3/C-5, S), 147.8 and 147.4 (C-3/C-5, S nonetherified), 134.2 (C-1, S etherified), and 104.1 ppm (C-2/C-6 S). The H units are detected at 127.6 (C-2/C-6, H), 115.5 (C-3/C-5, H), and 159.8 ppm (C-4, H). Signals related to lignin linkages are also present. The resonance of C-β, C-α, and C-γ in β-O-4 linkages produces signals at 85.9, 72.1, and 59.5 ppm. Signals from β-β were detected at 71.3 (C-γ) and 54.1 ppm (C-β). Other C—C linkages such as signals from β-5 and 5–5 linkages are found to be 66.2 (C-γ) and 125.9 ppm (C-5), respectively. Additionally, signals from different functional (carbonyl, carboxyl, and hydroxyl) groups are also detected. Apart from signals from lignin, the carbohydrates impurity also produce signals at 100, 72.3, and 63.4 ppm. It is noted that the ^{13}C NMR spectra of lignin is very complex and some signals can be overlapped by the impurity such as residual solvent and carbohydrates. Therefore, it is recommended that only a relative pure lignin fraction is suggested to analysis by this technique.

Estimation of lignin moieties and functional group is significant permitting more comprehensive information about the architecture and reactivity of lignin. The amount of side-chain moieties and functional groups can be estimated by integral at corresponding chemical shift in the spectra of lignin. The integral at 160–102 ppm is always set as the reference, assuming that it includes six aromatic carbons and 0.12 vinylic carbons; therefore, all moieties can be based on equivalences per aromatic ring [69]. However, signals belonging to carbon atoms of the same groups may be derived from different moieties. To calculate the amount of one of the moieties, the content of other moieties should be calculated first by other methods. For instance, signals at 50–48 ppm in the spectrum of both lignin belong to carbon atoms of phenylcoumaran and β-1 moieties [69]. Firstly, the amount of phenylcoumaran structures (0.03/Ar) was estimated from the resonance at about 87 ppm. Then, the integral at 50–48 ppm in the spectrum is calculated to be 0.05/Ar. Finally, the content of β-1 moieties can be calculated to be 0.02/Ar.

5.1.3. ^{31}P NMR spectroscopy

Phosphitylation of hydroxyl groups in lignin followed by quantitative ^{31}P NMR provides a valuable characterization tool for determination of the content of aliphatic hydroxyl groups, phenolic hydroxyl groups, and carboxyl group. Application of this method in lignin characterization has been reviewed by Pu et al [70]. ^{31}P NMR spectra of MWL derivatized with 2-chloro-4,4,5,5-tetramethyl-1,3,2-dioxaphospholane (TMDP) reproduced from our previous study are illustrated in **Figure 4**, and the signal assignments are labeled [17]. The quantitative results of these hydroxyl groups were obtained by peak integration with cyclohexanol (signals at 133.8–133.3 ppm) as internal standard (IS). Typical phosphitylating procedures are shown as follows.

Lignin sample (20 mg) was dissolved in anhydrous pyridine and deuterated chloroform (1.6:1, v/v, 500 μL) under stirring. Cyclohexanol (10.85 mg/mL, 100 μL) was added as an internal standard, followed by addition of chromium (III) acetylacetonate solution (5 mg/mL in anhydrous pyridine and deuterated chloroform 1.6:1, v/v, 100 μL) as a relaxation reagent. The mixture was reacted with TMDP (phosphitylating reagent, 100 μL) for about 10 min and placed into the NMR tube for ^{31}P NMR analysis [7, 17].

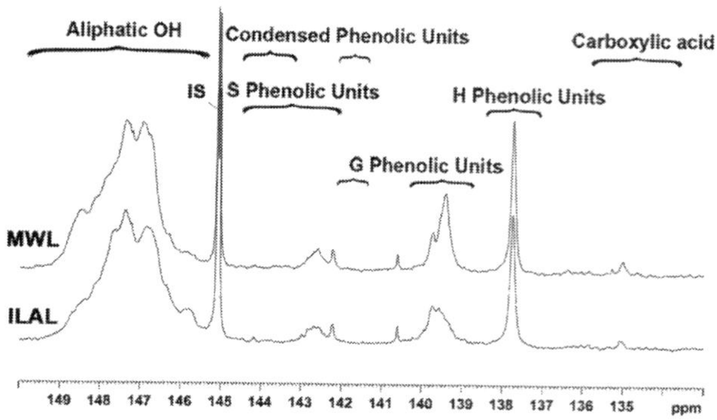

Figure 4. Quantitative ^{31}P NMR spectrum of a *A. donax* ball-milled lignin derivatized with TMDP using cyclohexanol as internal standard (reprinted with permission from [17]. Copyright 2015 American Chemical Society).

5.2. Multidimensional NMR spectroscopy

In traditional one-dimensional ^{1}H and ^{13}C NMR spectra, the signals are heavily overlapped due to the very complex and heterogeneous structure of lignin and instrumental limitations [87]. In the course of time, modern solution-state two- and three-dimensional methods are developed as efficient tools to investigate the structure of lignin. Besides better resolution, the multidimensional methods provide more reliability to the assignments [88].

5.2.1. Two-dimensional NMR spectroscopy

Two-dimensional NMR spectroscopic methods such as heteronuclear multiple quantum coherence (HMQC) spectroscopy, homonuclear Hartmann-Hahn (HOHAHA) spectroscopy, total correlation (TOCSY) spectroscopy, rotating-frame Overhauser experiment (ROESY) heteronuclear single quantum coherence NMR (HSQC) spectroscopy, and heteronuclear multiple bond coherence (HMBC) have been employed in lignin structure characterization [25, 89–91]. Among these, advanced 2D HSQC NMR is the most extensively used due to its versatility in illustrating structural features and structural transformations of isolated lignin fractions. The interpretation of 2D HSQC NMR spectra of lignin has been facilitated by the application of HOHAHA, HMQC, TOCSY, and ROESY techniques. For instance, del Río et al. performed HMBC experiment to give important information about the connectivity of the ester moiety to the lignin skeleton, and ether linkages between lignin and tricin were proposed [25]. ^{1}H-^{1}H TOCSY correlation NMR analysis can further confirm the doubted assignment of cross-peak in the spectra of HMQC or HSQC [69]. The cellulolytic enzyme lignin of *A. donax* was shown in **Figure 5**, and the main substructures are depicted in **Figure 1**.

The basic composition (S, G, and H units) and various substructures linked by ether and carbon–carbon bonds (β-O-4, β–β, β-5, etc.) can be observed in the 2D HSQC spectra (**Figure 5**). In the aromatic region (δ_C/δ_H 150–90/8.0–6.0 ppm), S, G, and H units show prominent correlations at δ_C/δ_H 103.7/6.71 ($S_{2,6}$), δ_C/δ_H 106.7/7.28 (C_α-oxidized S units S′), 110.7/6.98 (G_2),

114.9/6.72 and 6.94 (G_5), 118.7/6.77 (G_6), 127.8/7.22 ($H_{2,6}$), and 115.4/6.63 ($H_{2,6}$), respectively. It is noted that some signals may be overlapped by the others. It has been reported that signals in H units that are assigned to $C_{3,5}$-$H_{3,5}$ at δ_C/δ_H 115.4/6.63 overlapped with those from G 5-position [92]. After the alkaline treatment, intensity of signals correlated to S units is sharply increased [33]. Typically, as in spectra from grasses, prominent signals corresponding to *p*-coumarate (PCA) and ferulate (FA) structures are observed at δ_C/δ_H 115.5/6.77 ($PCA_{3,5}$), 129.9/7.46 ($PCA_{2,6}$), 111.0/7.32 (FA_2), and 111.0/7.32 (FA_6), respectively. The olefinic correlations of the cinnamyl aldehyde end-group structures (J) are observed at δ_C/δ_H 112.24/7.25 and 122.3/7.10. However, the aromatic cross-signals of the cinnamyl alcohol end-groups (I) are overlapped with the same signals in S and G units. In the side-chain region (δ_C/δ_H 90–50/6.0–2.7 ppm), cross-signals of methoxy groups (δ_C/δ_H 55.9/3.73) and β-O-4 linkages are the most predominant. The C_α—H_α correlations in the β-O-4 linkages are observed at δ_C/δ_H 72/4.7 to 4.9, while the C_β—H_β correlations are observed at δ_C/δ_H 84/4.3 and 86/4.1 for the substructures linked to the G/acylated S and S units, respectively. The C_γ-H_γ correlations in the β-O-4 substructures are found at δ_C/δ_H 60.1/3.40 and 3.72. The presence of acylating groups in some lignin produced additional intense signals. After acylation, the intense signals corresponding to acylated γ-carbon in β-O-4 substructure (A′/A″) are found in the range between δ_C/δ_H 62.7/3.83 and δ_C/δ_H 62.7/4.30. C_α—H_α correlation of carbon-carbon linkages such as resinol β-βsubstructures (B), phenylcoumaran β-5 substructures (C), spirodienone β-1 substructures (D), and α,β-diaryl ether substructures (E) are identified by the cross-peaks at δ_C/δ_H 84.8/4.66, 86.6/5.47, 81/5.01, and 79.2/5.52, respectively. In our previous research, the intensity of signals derived from β-O-4 substructures decreased during ionic liquid pretreatment suggesting the partial cleavage of this substructure [17]. With respect to signals from lignin, signals derived from carbohydrates are also found in this region and somewhat overlap with signals from C_γ-H_γ correlations in the β-O-4 substructures. For lignin-carbohydrate complex linkages investigation, the regions of δ_C/δ_H 90–105/3.9–5.4, 81.5–80/5.3–4.3, and 65–62/4.5–4.0 are of significance. Accordingly, benzyl ether LCC linkage could be detected in the region δ_C/δ_H 81.5–80/5.3–4.3; intense cross-peaks of phenol glycoside LCC linkages can be observed in the area of δ_C/δ_H 102.6–101.4/5.17–4.94, whereas cross-peaks of γ-ester linkages can be observed at δ_C/δ_H 65–62/4.5–4.0 [35, 93].

A quantitative evaluation of the lignin structure moieties has been performed successfully, and some of the quantitative methods frequently used are described. All C9 units in lignin are used as an internal standard. Integrate the G_2, $0.5S_{2,6} + G_2$, and $0.5S_{2,6} + G_2 + 0.5H_{2,6}$ signals as ISs are for softwood [91], hardwood lignin [91], and grass lignin [17, 33, 35], respectively. Based on the internal standard (all C9 units), the amount of S, G, H, and different interunit linkages could be obtained. In that way, the amount of β-O-4 linkages (% of C_9 units) was determined to be 47.0–49.4 in softwood lignin, and 60.3 in beech wood lignin, and more than 50 in grass MWL [91]. Another semiquantitative strategy of interunit linkages based on the total side chains is also acceptable for directly comparison [7, 25]. The actual extent of lignin acylation can also be estimated in the similar way particularly in grass samples [94]. In addition, Zhang and Gellerstedt proposed a quantitative method by combining ^{13}C NMR and 2D HSQC spectra, resulting in significant progress in the characterization of lignin moieties by NMR [71]. According to the method, the quantitative ^{13}C NMR spectrum was used as a reference.

Consequently, the absolute amounts of lignin substructures and even LCC moieties were quantitative calculated and expressed per 100 Ar.

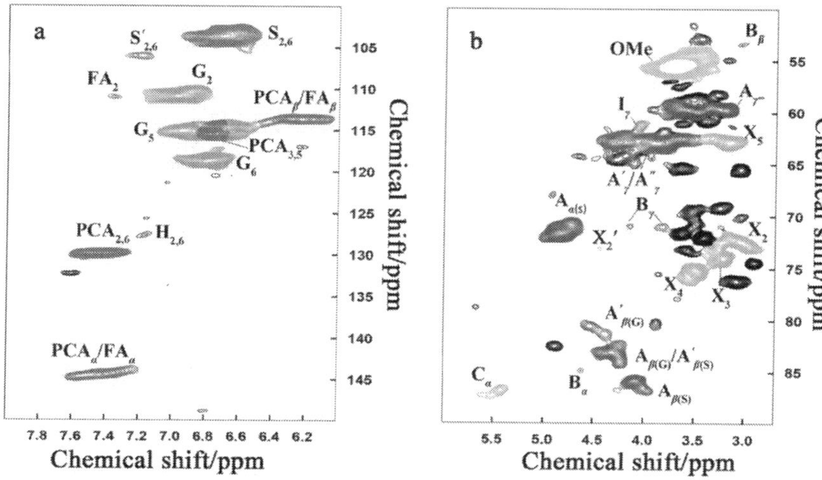

Figure 5. HSQC spectra of cellulolytic enzyme lignin from *A. donax* a aromatic region (δ_C/δ_H 150–90/8.0–6.0 ppm); b side-chain region (δ_C/δ_H 90–50/6.0–2.7 ppm). See **Figure 1** for the main lignin structures identified (reprinted with permission from [35]. Copyright 2015 Elsevier).

5.2.2. Three-dimensional NMR spectroscopy

As discussed above, the overlapping of the lignin signals cannot be fully avoided even by 2D experiments. Thanks to the rapid advances in NMR technology, the three-dimensional HSQC-TOCSY and HMQC-HOHAHA techniques are utilized to elucidate the ^1H-^1H and ^1H-^{13}C correlations of individual spin systems and thus indicate a certain lignin side chain structure [87, 88, 95]. The 3D spectra provide more reliability to the assignments, as the connectivity can be cross-checked from different planes of the 3D spectrum. However, long measurement time is required for the 3D experiments. As most of the structural information of lignin can be obtained by 1D and 2D NMR spectroscopic methods, the applications of 3D NMR to lignin structure characterization are still limited.

6. Conclusion and outlook

Lignin is an aromatic polymer essential for defense, water and nutrient transport, and mechanical support in vascular terrestrial plants. To reveal the molecular details of lignin structure nondestructively, various molecular spectroscopic methods have been routinely utilized. It can be inferred that the extinction coefficients of UV spectra demonstrate the purity of lignin. Moreover, the functional groups and possible lignin composition can be obtained by the FT-IR, FT-Raman, and fluorescence spectroscopy spectral features, whereas more accurate composition and contents can be calculated using one-dimensional and multidimensional NMR spectroscopic methods. The combination of these molecular spectroscopic methods provides a comprehensive and systematic evaluation of lignin from different plant species. It was demonstrated that herbaceous plants and wood species displayed different structural characteristics of lignin, and structural modification was occurred during various treatment. Overall, these nondestructive techniques provide alternative safe, rapid, accurate, and nondestructive technology for lignin structure determination. The information of lignin presented by these molecular spectroscopic methods contributes to the understanding of native recalcitrance and facilitates the design of more effective strategies to produce lignin-based value-added materials, biochemicals, and biofules.

Acknowledgements

The authors gratefully acknowledge the financial support from National Science Foundation for Distinguished Young Scholars of China (31225005) and Chinese Ministry of Education (113014A).

Author details

Tingting You[1] and Feng Xu[1,2*]

*Address all correspondence to: xfx315@bjfu.edu.cn

1 Beijing Key Laboratory of Lignocellulosic Chemistry, Beijing Forestry University, Beijing, China

2 Key Laboratory of Pulp and Paper Science and Technology, Ministry of Education, Qilu University of Technology, Shandong, China

References

[1] Buranov AU, Mazza G. Lignin in straw of herbaceous crops. Ind Crop Prod. 2008;28(3): 237–59.

[2] Bujanovic B, Ralph S, Reiner R, Hirth K, Atalla R. Polyoxometalates in oxidative delignification of chemical pulps: effect on lignin. Materials. 2010;3(3):1888–903.

[3] Bouxin FP, McVeigh A, Tran F, Westwood NJ, Jarvis MC, Jackson SD. Catalytic depolymerisation of isolated lignins to fine chemicals using a Pt/alumina catalyst: part 1-impact of the lignin structure. Green Chem. 2015;17(2):1235–42.

[4] Xu CP, Arancon RAD, Labidi J, Luque R. Lignin depolymerisation strategies: towards valuable chemicals and fuels. Chem Soc Rev. 2014;43(22):7485–500.

[5] Laskar DD, Tucker MP, Chen XW, Helms GL, Yang B. Noble-metal catalyzed hydrodeoxygenation of biomass-derived lignin to aromatic hydrocarbons. Green Chem. 2014;16(2):897–910.

[6] Ralph J, Lundquist K, Brunow G, Lu F, Kim H, Schatz PF, et al. Lignins: natural polymers from oxidative coupling of 4-hydroxyphenyl-propanoids. Phytochem Rev. 2004;3(1–2): 29–60.

[7] You TT, Mao JZ, Yuan TQ, Wen JL, Xu F. Structural elucidation of the lignins from stems and foliage of *Arundo donax* Linn. J Agric Food Chem. 2013;61(22):5361–70.

[8] Chakar FS, Ragauskas AJ. Review of current and future softwood kraft lignin process chemistry. Ind Crop Prod. 2004;20(2):131–41.

[9] Balakshin MY, Capanema EA, Chang HM. MWL fraction with a high concentration of lignin-carbohydrate linkages: isolation and 2D NMR spectroscopic analysis. Holzforschung. 2007;61(1):1–7.

[10] Grabber JH, Ralph J, Hatfield RD. Cross-linking of maize walls by ferulate dimerization and incorporation into lignin. J Agric Food Chem. 2000;48(12): 6106–13.

[11] Ralph J, Lu F. The DFRC method for lignin analysis. 6. A simple modification for identifying natural acetates on lignins. J Agric Food Chem. 1998;46(11): 4616–9.

[12] Xu F, Sun RC, Zhai MZ, Sun JX, Jiang JX, Zhao GJ. Comparative study of three lignin fractions isolated from mild ball-milled Tamarix austromogoliac and Caragana sepium. J Appl Polym Sci. 2008;108(2):1158–68.

[13] Faix O. Classification of lignins from different botanical origins by FT-IR spectroscopy. Holzforschung-Int J Biol Chem Phys Technol Wood. 1991;45(s1):21–8.

[14] Albinsson B, Li S, Lundquist K, Stomberg R. The origin of lignin fluorescence. J Mol Struct. 1999;508(1):19–27.

[15] Agarwal UP, McSweeny JD, Ralph SA. FT-Raman investigation of milled-wood lignins: softwood, hardwood, and chemically modified black spruce lignins. J Wood Chem Technol. 2011;31(4):324–44.

[16] Sannigrahi P, Kim DH, Jung S, Ragauskas A. Pseudo-lignin and pretreatment chemistry. Energ Environ Sci. 2011;4(4):1306–10.

[17] You TT, Zhang LM, Guo SQ, Shao LP, Xu F. Unraveling the structural modifications in lignin of *Arundo donax* Linn. during acid-enhanced ionic liquid pretreatment. J Agric Food Chem. 2015;63(50):10747–56.

[18] Ralph J, Lapierre C, Lu FC, Marita JM, Pilate G, Van Doorsselaere J, et al. NMR evidence for benzodioxane structures resulting from incorporation of 5-hydroxyconiferyl alcohol into lignins of O-methyltransferase-deficient poplars. J Agric Food Chem. 2001;49(1):86–91.

[19] Zhang LM, Gellerstedt G, Ralph J, Lu FC. NMR studies on the occurrence of spirodienone structures in lignins. J Wood Chem Technol. 2006;26(1):65–79.

[20] Capanema EA, Balakshin MY, Kadla JF. A comprehensive approach for quantitative lignin characterization by NMR spectroscopy. J Agric Food Chem. 2004;52(7):1850–60.

[21] Sun R, Lu Q, Sun X. Physico-chemical and thermal characterization of lignins from Caligonum monogoliacum and Tamarix spp. Polym Degrad Stab. 2001;72(2):229–38.

[22] Adler E. Lignin chemistry—past, present and future. Wood Sci Technol. 1977;11(3):169–218.

[23] Hatfield R, Fukushima RS. Can lignin be accurately measured? Crop Sci. 2005;45(3): 832–9.

[24] TAPPI method T 222 om-83. Acid-insoluble lignin in wood and pulp. In: Test Methods 1998–1999. Atlanta: TAPPI Press; 1999.

[25] Del Río JC, Rencoret J, Prinsen P, Martínez AT, Ralph J, Gutiérrez A.. Structural characterization of wheat straw lignin as revealed by analytical pyrolysis, 2D-NMR, and reductive cleavage methods. J Agric Food Chem. 2012;60(23):5922–35.

[26] Huang F, Singh PM, Ragauskas AJ. Characterization of milled wood lignin (MWL) in Loblolly pine stem wood, residue, and bark. J Agric Food Chem. 2011;59(24):12910–6.

[27] Sluiter A, Hames B, Ruiz R, Scarlata C, Sluiter J, Templeton D, et al. Determination of structural carbohydrates and lignin in biomass. National Renewable Energy Laboratory (NREL) Laboratory Analytical Procedures (LAP) for standard biomass analysis. 2007. NREL/TP-510-42618.

[28] Rajan K, Carrier DJ. Effect of dilute acid pretreatment conditions and washing on the production of inhibitors and on recovery of sugars during wheat straw enzymatic hydrolysis. Biomass Bioenerg. 2014;62:222–7.

[29] Pinto PC, Oliveira C, Costa CA, Gaspar A, Faria T, Ataíde J, et al. Kraft delignification of energy crops in view of pulp production and lignin valorization. Ind Crop Prod. 2015;71:153–62.

[30] Lai YZ, Funaoka M. The distribution of phenolic hydroxyl groups in hardwood lignins. J Wood Chem Technol. 1993;13(1):43–57.

[31] Schwarz P, Youngs V, Shelton D. Isolation and characterization of lignin from hard red spring wheat bran. Cereal Chem. 1989;66(4):289–95.

[32] Lange PW. The distribution of the components in the plant cell wall. Sweden: Stockholm University; 1954.

[33] Seca AM, Cavaleiro JA, Domingues FM, Silvestre AJ, Evtuguin D, Neto CP. Structural characterization of the lignin from the nodes and internodes of *Arundo donax* reed. J Agric Food Chem. 2000;48(3):817–24.

[34] Guerra A, Filpponen I, Lucia LA, Argyropoulos DS. Comparative evaluation of three lignin isolation protocols for various wood species. J Agric Food Chem. 2006;54(26): 9696–705.

[35] You TT, Zhang LM, Zhou SK, Xu F. Structural elucidation of lignin-carbohydrate complex (LCC) preparations and lignin from *Arundo donax* Linn. Ind Crop Prod. 2015;71:65–74.

[36] Yuan TQ, Sun SN, Xu F, Sun RC. Structural characterization of lignin from triploid of Populus tomentosa Carr. J Agric Food Chem. 2011;59(12):6605–15.

[37] El Mansouri NE, Salvado J. Structural characterization of technical lignins for the production of adhesives: application to lignosulfonate, kraft, soda-anthraquinone, organosolv and ethanol process lignins. Ind Crop Prod. 2006;24(1):8–16.

[38] Fergus B, Procter A, Scott J, Goring D. The distribution of lignin in sprucewood as determined by ultraviolet microscopy. Wood Sci Technol. 1969;3(2):117–38.

[39] Lundquist K, Josefsson B, Nyquist G. Analysis of lignin products by fluorescence spectroscopy. Holzforschung-Int J Biol Chem Phys Technol Wood. 1978;32(1):27–32.

[40] Parker C. Photoluminescence of Solutions. Amsterdam: American Elsevier; 1968.

[41] Olmstead J, Gray D. Fluorescence spectroscopy of cellulose, lignin and mechanical pulps: a review. J Pulp Pap Sci. 1997;23(12):J571-J81.

[42] Donaldson LA, Singh AP, Yoshinaga A, Takabe K. Lignin distribution in mild compression wood of Pinus radiata. Can J Botany. 1999;77(1):41–50.

[43] Kutscha N, McOrmond R. The suitability fo using fluorescence microscopy for studying lignification in Balsam Fir. Tech. Bull Life Sci, agr Exp Sta, Univ Maine. 1972;(62):15.

[44] Donaldson L, Hague J, Snell R. Lignin distribution in coppice poplar, linseed and wheat straw. Holzforschung. 2001;55(4):379–85.

[45] Ji Z, Zhang X, Ling Z, Zhou X, Ramaswamy S, Xu F. Visualization of Miscanthus× giganteus cell wall deconstruction subjected to dilute acid pretreatment for enhanced enzymatic digestibility. Biotechnol Biofuels. 2015;8(1):1.

[46] Schulz H, Baranska M. Identification and quantification of valuable plant substances by IR and Raman spectroscopy. Vib Spectrosc. 2007;43(1):13–25.

[47] Popescu CM, Popescu MC, Singurel G, Vasile C, Argyropoulos DS, Willfor S. Spectral characterization of eucalyptus wood. Appl Spectrosc. 2007;61(11):1168–77.

[48] Pandey K. A study of chemical structure of soft and hardwood and wood polymers by FTIR spectroscopy. J Appl Polym Sci. 1999;71(12):1969–75.

[49] Boeriu CG, Bravo D, Gosselink RJA, van Dam JEG. Characterisation of structure-dependent functional properties of lignin with infrared spectroscopy. Ind Crop Prod. 2004;20(2):205–18.

[50] Sen S, Patil S, Argyropoulos DS. Methylation of softwood kraft lignin with dimethyl carbonate. Green Chem. 2015;17(2):1077–87.

[51] Jia S, Cox BJ, Guo X, Zhang ZC, Ekerdt JG. Cleaving the β□ O□ 4 bonds of lignin model compounds in an acidic ionic liquid, 1-H-3-methylimidazolium chloride: an optional strategy for the degradation of lignin. ChemSusChem. 2010;3(9):1078–84.

[52] Hergert HL, Kurth E. The isolation and properties of catechol from white fir bark. J Organic Chem. 1953;18(5):521–9.

[53] Jiao JJ, Zhang Y, Liu CM, Liu J, Wu XQ, Zhang Y. Separation and purification of tricin from an antioxidant product derived from bamboo leaves. J Agric Food Chem. 2007;55(25):10086–92.

[54] Wen JL, Sun ZJ, Sun YC, Sun SN, Xu F, Sun RC. Structural characterization of alkali-extractable lignin fractions from bamboo. J Biobased Mater Bio. 2010;4(4): 408–25.

[55] Sipponen MH, Lapierre C, Mechin V, Baumberger S. Isolation of structurally distinct lignin-carbohydrate fractions from maize stem by sequential alkaline extractions and endoglucanase treatment. Bioresour Technol. 2013;133:522–8.

[56] Chen Y, Fan YM, Gao JM, Li HK. Coloring characteristics of in situ lignin during heat treatment. Wood Sci Technol. 2012;46(1–3):33–40.

[57] Raiskila S, Pulkkinen M, Laakso T, Fagerstedt K, Loija M, Mahlberg R, et al. FTIR spectroscopic prediction of Klason and acid soluble lignin variation in Norway spruce cutting clones. Silva Fennica. 2007;41(2):351.

[58] Faix O. Condensation indices of lignins determined by FTIR-spectroscopy. Eur J Wood Wood Products. 1991;49(9):356.

[59] Seca AM, Cavaleiro JA, Domingues FM, Silvestre AJ, Evtuguin D, Pascoal Neto C. Structural characterization of the bark and core lignins from kenaf (Hibiscus cannabinus). J Agric Food Chem. 1998;46(8):3100–8.

[60] Faix O, Stevanovic-Janezic T, Lundquist K. The lignin of the diffuse porous angiosperm tree Triplochyton scleroxylon K. Schum with low syringyl content. J Wood Chem Technol. 1994;14(2):263–78.

[61] Agarwal UP. An overview of Raman spectroscopy as applied to lignocellulosic materials. In: Advances in lignocellulosics characterization. Argyropoulos, D. S. (Ed.); Atlanta: TAPPI Press; 1999. p. 201–225.

[62] Agarwal UP, Reiner RS, Pandey AK, Ralph SA, Hirth KC, Atalla RH, editors. Raman spectra of lignin model compounds. 59th Appita Annual Conference and Exhibition: Incorporating the 13th ISWFPC (International Symposium on Wood, Fibre and Pulping Chemistry), Auckland, New Zealand, 16–19 May 2005: Proceedings; 2005: Appita Inc.

[63] Agarwal U P, Terashima N. FT-Raman study of dehydrogenation polymer (DHP) lignins. Processing of 12th International Symposium of Wood and Pulping Chemistry. Madison, WI: Department of Forest Ecology and Management, University of Wisconsin, 2003, 3: 123–126.

[64] Agarwal UP, Ralph SA. FT-Raman spectroscopy of wood: identifying contributions of lignin and carbohydrate polymers in the spectrum of black spruce (Picea mariana). Appl Spectrosc. 1997;51(11):1648–55.

[65] Sun L, Varanasi P, Yang F, Loque D, Simmons BA, Singh S. Rapid determination of syringyl: guaiacyl ratios using FT-Raman spectroscopy. Biotechnol Bioeng. 2012;109(3): 647–56.

[66] You TT, Ma JF, Guo SQ, Xu F. Study of alkaline lignin from *Arundo donax* Linn based on FT Raman spectroscopy. Spectroscopy Spectral Anal 2014;8:023.

[67] Larsen KL, Barsberg S. Theoretical and Raman spectroscopic studies of phenolic lignin model monomers. J Phys Chem B. 2010;114(23):8009–21.

[68] Liitiä T, Maunu S, Sipilä J, Hortling B. Application of solid-state 13 C NMR spectroscopy and dipolar dephasing technique to determine the extent of condensation in technical lignins. Solid State Nucl Magn Reson. 2002;21(3):171–86.

[69] Capanema EA, Balakshin MY, Kadla JF. Quantitative characterization of a hardwood milled wood lignin by nuclear magnetic resonance spectroscopy. J Agric Food Chem. 2005;53(25):9639–49.

[70] Pu Y, Cao S, Ragauskas AJ. Application of quantitative 31P NMR in biomass lignin and biofuel precursors characterization. Energ Environ Sci. 2011;4(9):3154–66.

[71] Zhang L, Gellerstedt G. Quantitative 2D HSQC NMR determination of polymer structures by selecting suitable internal standard references. Magn Reson Chem. 2007;45(1):37–45.

[72] Pan XJ, Kadla JF, Ehara K, Gilkes N, Saddler JN. Organosolv ethanol lignin from hybrid poplar as a radical scavenger: relationship between lignin structure, extraction conditions, and antioxidant activity. J Agric Food Chem. 2006;54(16): 5806–13.

[73] Wen JL, Sun SL, Xue BL, Sun RC. Recent advances in characterization of lignin polymer by solution-state nuclear magnetic resonance (NMR) methodology. Materials. 2013;6(1):359–91.

[74] Ralph S, Ralph J, Landucci L, Landucci L. NMR Database of Land Cell Wall Model Compounds. US Forest Prod Lab; Madison, WI (http://ars usda gov/Services/docs htm); 2004.

[75] Lundquist K. NMR studies of lignins. 4. Investigation of spruce lignin by 1H NMR spectroscopy. Acta Chem Scand B. 1980;34:21–6.

[76] Robert D. Carbon-13 nuclear magnetic resonance spectrometry. In: Methods in lignin chemistry. Lin, S. Y., & Dence, C. W. (Eds.); Berlin Heidelberg: Springer; 1992. p. 250–273.

[77] Kringstad KP, Mörck R. 13C-NMR spectra of kraft lignins. Holzforschung-Int J Biol Chem Phys Technol Wood. 1983;37(5):237–44.

[78] Saake B, Argyropoulos DS, Beinhoff O, Faix O. A comparison of lignin polymer models (DHPs) and lignins by 31 P NMR spectroscopy. Phytochemistry. 1996;43(2):499–507.

[79] Faix O, Argyropoulos DS, Robert D, Neirinck V. Determination of hydroxyl groups in lignins evaluation of 1H-, 13C-, 31P-NMR, FTIR and wet chemical methods. Holzforschung-Int J Biol Chem Phys Technol Wood. 1994;48(5):387–94.

[80] Fasching M, Schroder P, Wollboldt RP, Weber HK, Sixta H. A new and facile method for isolation of lignin from wood based on complete wood dissolution. Holzforschung. 2008;62(1):15–23.

[81] Lundquist K. Proton (1H) NMR spectroscopy. In: Methods in lignin chemistry. Lin, S. Y., & Dence, C. W. (Eds.); Berlin Heidelberg: Springer; 1992. 1992. p. 242–249.

[82] Tejado A, Pena C, Labidi J, Echeverria J, Mondragon I. Physico-chemical characterization of lignins from different sources for use in phenol–formaldehyde resin synthesis. Bioresour Technol. 2007;98(8):1655–63.

[83] Rokhin A, Kanitskaya L, Kushnarev D, Kalabin G, Smirnova L, Abduazimov KA, et al. Quantitative1H and13C NMR spectroscopies of cottonplant dioxane lignin. Chem Nat Compd. 1994;30(6):745–53.

[84] Nimz H, Robert D, Faix O, Nemr M. Carbon-13 NMR spectra of lignins, 8. Structural differences between lignins of hardwoods, softwoods, grasses and compression wood. Holzforschung-Int J Biol, Chem, Phys Technol Wood. 1981;35(1):16–26.

[85] Holtman KM, Chang HM, Jameel H, Kadla JF. Quantitative 13C NMR characterization of milled wood lignins isolated by different milling techniques. J Wood Chem Technol. 2006;26(1):21–34.

[86] Akim L, Shevchenko S, Zarubin MY. 13C NMR studies on lignins depolymerized with dry hydrogen iodide. Wood Sci Technol. 1993;27(4):241–8.

[87] Liitiä T. Application of modern NMR spectroscopic techniques to structural studies of wood and pulp components. Finland: University of Helsinki; 2002.

[88] Liitia TM, Maunu SL, Hortling B, Toikka M, Kilpelainen I. Analysis of technical lignins by two- and three-dimensional NMR spectroscopy. J Agric Food Chem. 2003;51(8): 2136–43.

[89] Ede RM, Brunow G. Application of two-dimensional homo-and heteronuclear correlation NMR spectroscopy to wood lignin structure determination. J Organic Chem. 1992;57(5):1477–80.

[90] Kilpelaeinen I, Sipilae J, Brunow G, Lundquist K, Ede RM. Application of two-dimensional NMR spectroscopy to wood lignin structure determination and identification of some minor structural units of hard-and softwood lignins. J Agric Food Chem. 1994;42(12):2790–4.

[91] Sette M, Wechselberger R, Crestini C. Elucidation of lignin structure by quantitative 2D NMR. Chem—Eur J. 2011;17(34):9529–35.

[92] Martínez ÁT, Rencoret J, Marques G, Gutiérrez A, Ibarra D, Jiménez-Barbero J, et al. Monolignol acylation and lignin structure in some nonwoody plants: a 2D NMR study. Phytochemistry. 2008;69(16):2831–43.

[93] Balakshin M, Capanema E, Gracz H, Chang H-m, Jameel H. Quantification of lignin–carbohydrate linkages with high-resolution NMR spectroscopy. Planta. 2011;233(6): 1097–110.

[94] del Río JC, Rencoret J, Marques G, Gutiérrez A, Ibarra D, Santos JI, et al. Highly acylated (acetylated and/or p-coumaroylated) native lignins from diverse herbaceous plants. J Agric Food Chem. 2008;56(20):9525–34.

[95] Ämmälahti E, Brunow G, Bardet M, Robert D, Kilpeläinen I. Identification of side-chain structures in a poplar lignin using three-dimensional HMQC-HOHAHA NMR spectroscopy. J Agric Food Chem. 1998;46(12):5113–7.

Chapter 12

Using Fluorescence Spectroscopy to Diagnose Breast Cancer

Tatjana Dramićanin and Miroslav Dramićanin

Additional information is available at the end of the chapter

http://dx.doi.org/10.5772/63534

Abstract

Optical spectroscopy methods have had considerable impact in the field of biomedical diagnostics, providing novel methods for the early or noninvasive diagnosis of various medical conditions. Among them, fluorescence spectroscopy has been the most widely explored mainly because fluorescence is highly sensitive to the biochemical makeup of tissues. It has been shown that tumors were easily detected on account of altered fluorescence properties with respect to fluorescence of ordinary tissue. Breast cancer is one of the most commonly diagnosed cancers among women in the world and also it is one of the leading causes of deaths from cancer for the female population. However, when detected in early stage, it is one of the most treatable forms of cancer. Therefore, fluorescence technologies could be highly beneficial in early detection and timely treatment of cancer. This chapter presents main results and conclusions that have been reported on the use of fluorescence spectroscopy for the investigation of breast cancer. It also gives an overview on the instruments and methodology of measurements, on the main endogenous fluorophores present in tissues, on the tissue fluorescence, and on the statistical methods that aid interpretations of fluorescence spectra. Finally, examples of using various fluorescence techniques, such as excitation, emission and synchronous spectroscopy, excitation-emission matrices, and lifetimes, for the breast cancer diagnosis are presented.

Keywords: fluorescence, breast cancer, fluorophores, tissue fluorescence, cancer diagnosis

1. Introduction

According to World Cancer Research Fund International (WCRFI), 1.7 million women have been diagnosed with breast cancer in 2012 [1]. With 25% share of all diagnosed cancers, breast cancer is a leading malignant disease in female population throughout the world. It is also present in male population, but with about 100 times less incidence that is fairly stable over the last 30 years. Breast cancer incidence is highest in North America, West Europe, and Oceania (in highly developed countries); in 2012, record-high values of age-standardized rate per 100,000 were recorded in Belgium (111.9) and Denmark (105.0). Breast cancer is age- and hormone-status–related, being highest for woman 50–70 years old. Fortunately, prognosis is very good for early diagnosed disease. Five-year survival rate (see **Table 1**) is 100% for 0 and I stage conditions [2]. Therefore, the early diagnosis of breast cancer (and also all cancers) is a key for the successful treatment. On the other hand, in the absence of any early detection or screening and treatment intervention, patients are diagnosed at very late stages when curative treatment is no longer an option [3].

Stage	0	I	II	III	IV
5-year survival rate	100%	100%	93%	72%	22%

Table 1. Five-year survival rate of patients diagnosed with breast cancer of different stage.

Generally, tumor-observing methods can be categorized to screening, diagnostic, and monitoring tests. Clinical breast examination, mammography, and ultrasonography are the most common methods used for screening. Clinical examination by a specialist is usually the first step in breast cancer detection. However, this test is quite subjective and the results depend on the experience and skill of the examiner and most importantly, it does not detect small-size tumors. Mammography is the most cost-effective test, but with moderate sensitivity of 67.8% and specificity of 75%. Mammography combined with clinical breast examination slightly improves sensitivity (77.4%). For screening, ultrasonography is usually recommended to younger persons (premenopause), but it fails to detect microcalcifications and shows poor specificity (34%). Modern imaging techniques, such as positron emission tomography (PET), magnetic resonance imaging (MRI), and computed tomography (CT) are important tools for cancer detection and treatment. Some important features of the most common breast cancer tests are listed in **Table 2**.

Medical diagnostics of breast cancer based on optical spectroscopy and optical imaging are in the early stage of development in comparison with the traditional methods. But the need for sensitive and early cancer detection along with advances in technology played, and still plays, an essential role for the extensive research in this field. For example, lasers have provided a new technology for excitation, and microchips and miniaturized sensors have eased signal detection, while optical fibers have transformed the ways of access to the object of examination. The most attention among a variety of optical techniques is given to fluorescence, Raman spectroscopy, diffuse reflectance, elastic scattering spectroscopy, Fourier transform infrared microspectroscopy, near-infrared imaging, and optoacoustic tomography. These techniques offer several principal advantages over the traditional methods, including (a) noninvasiveness through the use of safe, nonionizing radiation, (b) display of contrast between soft tissues based on optical properties, and (c) a facility for continuous bedside monitoring. Detailed description

http://dx.doi.org/10.5772/63534

of optical methods in cancer research is beyond the scope of this chapter, which is devoted to fluorescence techniques. Those interested in more information on this subject can read some of topical books (e.g., Biomedical Photonics Handbook edited by T. Vo-Dinh [4]).

Method	Benefit	Deficiency
Clinical Breast Examination	no equipment needed	highly subjective, large FP and FN values
Mammography	suitable for screening	large FP and FN values, use of X-ray radiation
Ultrasonography	use of non-ionizing radiation, noninvasive, able to analyze lesions in dense breasts	large FP and FN results, not sensitive enough
Magnetic Resonance Imaging (MRI)	able to detect small DCIS, use of non-ionizing radiations	expensive, can not be used with patients with metal implants
Computed tomography (CT)	painless, noninvasive, accurate	expensive, use of X-radiation
Positron Emission Tomography (PET)	can indicate cancer from disordered metabolism,accurate	expensive, allergic reactions to RP, use of injection RP

Note: FP – false positive, FN – false negative, RP – radioactive pharmaceutic, DCIS – ductal carcinoma in situ.

Table 2. Some important features of the most common breast cancer tests.

In the past several decades, fluorescence spectroscopy has been applied to many different types of samples, ranging from individual biochemical species to organs of living people. It has been applied so far for almost every type of cancer, both *in*- and *ex vivo*, and it has demonstrated advantages over other light-based methods in terms of sensitivity, speed, and safety. Since the early twentieth century, it is known that tissues fluoresce when exposed to light of a suitable wavelength and that infiltrating tumors can be detected on account of altered fluorescence signals. These alternations are the result of the exceptionally high sensitivity of fluorescence on the biochemical makeups of tissues, and are premise for the diagnoses of tissue pathologies by fluorescence. Tissue fluorescence comprises emissions of a number of natural fluorophores (endogenous fluorophores) that have unique spectral characteristics when excited with ultraviolet or visible light. Among them, regarding fluorescence, the most important are tryptophan, reduced nicotinamide adenine dinucleotide (phosphate) (NAD(P)H), elastin, collagen, flavins, and porphyrins. The changes in the concentrations and microenvironments of fluorophores alter tissue fluorescence to a sufficient extent to detect metabolic and pathological changes related to precancerous and cancerous growth, even though fluorescence measurements are not capable of detailing structural changes in tissues. The fluorescence of tissue and tissue fluorophores are discussed later in detail in a separate section.

It is impossible to present beauty, complexity, versatility, and usefulness of fluorescence in a single chapter. Many good treatises could be used for this purpose [5–7]. Thus, with this chapter we intended to provide an overview of applications of fluorescence spectroscopy for the breast cancer diagnosis. This application is a small part of the extremely large and growing field of biomedical fluorescence. To do so, brief descriptions of basic principles of fluorescence, instruments, and techniques are given in Section 2. Tissue fluorescence, management, and interpretation of fluorescence data are explained in Sections 3 and 4. Fluorescence measurements ought to be subjected to mathematical quantification and interpretation for obtaining appropriate data for cancer diagnosis. Useful mathematical tools, mainly modern statistical tools, are described in Section 5. Examples of applications are given in Section 6, and they are selected to cover most of the fluorescence techniques listed in Section 2. These examples are mainly drawn from our own work, simply because we are most familiar with them.

2. Fluorescence spectroscopy methods and instrumentation

Photoluminescence is the physical process of light emission from any substance that has not been heated and takes place from the electronically excited states. Depending upon the nature of the excited state, luminescence can be strictly divided into two types, fluorescence and phosphorescence. Fluorescence is a rapid and spin-allowed emission of light from singlet excited states, while in phosphorescence emission of light comes due to spin-forbidden transitions from triplet excited states to the ground state. The fluorescence process, depicted in **Figure 1**, is ruled by three important events: (1) excitation of a molecule by an incoming photon from the ground state S_0, (2) vibrational relaxation of S_1' excited state electrons to the lowest excited energy state S_1, and, finally, (3) emission of a lower-energy photon and return of the molecule to the ground state.

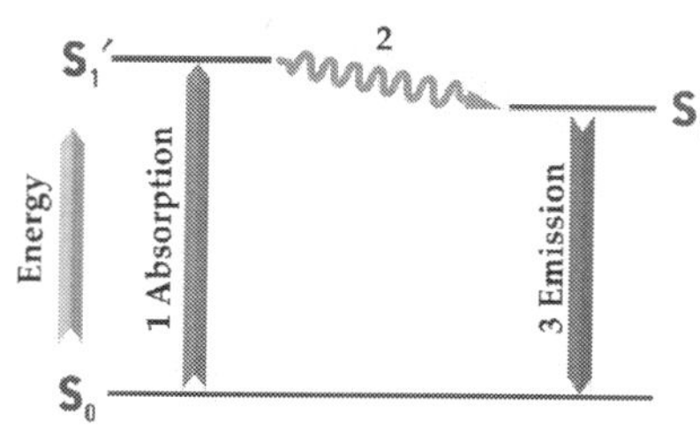

Figure 1. One form of a Jablonski diagram explaining the process of fluorescence.

Fluorescence measurements are the core of any fluorescence technology, and they are presently widely utilized by scientists from many disciplines. Depending on the temporal nature of excitation and detection, fluorescence measurements can be classified as steady-state (implemented with constant illumination and observation) and time-resolved (performed with pulsed excitation). Generally, steady-state measurements are simpler than time-resolved since they require less complex instrumentation and are easier for interpretation. Also, they are sufficient for many applications, so they are more common in practice. However, being simply an average of time-resolved phenomena, steady-state measurements do not account all of the molecular information imparted to fluorescence. Characteristic examples are information on the distribution of emission decays or the nature of fluorescence quenching.

Fluorescence measurements can also be classified according to the type of observed data. Emission spectra are obtained by recording emission intensity over the spectral range of interest for the fixed energy of excitation. Excitation spectra are created by measuring the variation of emission intensity at the fixed energy (wavelength) while changing excitation energy in the desired spectral region. Generally, these spectra are symmetric (mirror image) because the same transitions are involved in both absorption and emission, and similar vibrational levels are present in both ground and excited state. Certainly, many exceptions may occur, but their explanation is beyond the scope of this chapter. Synchronous fluorescence method involves simultaneously scanning both emission and excitation wavelengths while maintaining the interval constant between emission and excitation (constant-wavelength mode) or maintaining the frequency gap constant (constant-energy mode). When selected interval corresponds to the difference between the excitation and emission maxima of a specific molecule, the emission of that molecule is maximally intensified in the measured spectrum. Consequently, synchronous luminescence spectra have sharper spectral characteristics than conventional, characteristics that are particularly valuable for discrimination between different biological tissues.

Biological materials are complex and comprise many different fluorophores, chromophores (molecules that absorb light but do not emit one), and light scatters. Because of that, the measurement of a single spectra, either emission or excitation or synchronous, is sometimes insufficient for the analyses and diagnostics. In this case, more comprehensive measurements are required to thoroughly combine excitation and emission features of tissue; commonly, this type of measurement is referred as multidimensional fluorescence measurement. Excitation-emission matrices (EEMs), also termed excitation-emission landscapes, are the most widely used type of multidimensional measurements. They combine in a three dimensional (3D) space the set of emission spectra excited by the light of different wavelengths. One can say the EEM of a sample is its characteristic fluorescent fingerprint, taking that the majority of the fluorescence characteristics of the sample are included in EEM. Alternatively, three-dimensional total synchronous fluorescence spectroscopy (3D-TSFS) can be used for the same purpose. During 3D-TSFS measurements, a series of SFS spectra are recorded for a range of synchronous intervals; therefore, 3D-TSFS is a multidimensional extension of SFS. The term "three-dimensional" refers to the space defined by the excitation or emission wavelength, synchronous interval, and the fluorescence emission intensity.

For medical diagnostics, measurements of quantum yield, polarization, and excited state lifetime may also be valuable. Quantum yield is the ratio of the number of photons emitted from fluorophore to the number of photons absorbed. Polarization gives information on the movement of fluorophore, if there is any, during the time between the absorption and emission of light, namely during the excited state lifetime.

To stimulate and measure fluorescence from a sample, one needs to use instrument with five basic components: (1) light source, (2) wavelength selector elements on the excitation and emission paths to/from sample, (3) sample holder/positioner, (4) polarizers, and (5) detector (**Figure 2**).

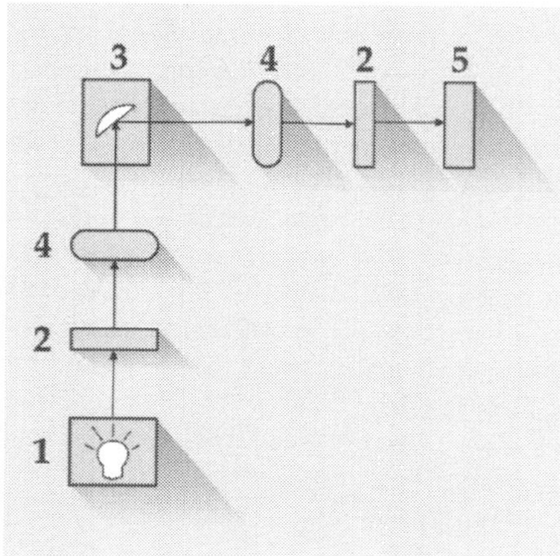

Figure 2. Five basic components of spectrofluorometer.

Polychromatic light from light source is dispersed on a dispersing element from which light beam of selected wavelength is directed on a sample to excite fluorescence. Most frequently, Xe arc lamps are used. These days, deuterium, tungsten, or halogen lamps are rarely utilized for excitation, since they are relatively weak sources for fluorescence. Lasers, laser diodes, and light-emitting diodes (LEDs) are also frequently used. They produce intense monochromatic light, so there is no need for the wavelength-selecting element on the excitation path. The drawback is that excitation spectra cannot be measured with these light sources. Tuneable and supercontinuum lasers are an exception, which can produce emissions over the given spectral interval, but these are rather expensive devices. Light dispersion can be accomplished with prisms and diffraction gratings; the latter is dominantly used in the modern spectrofluorometers. The detector can be either single-channeled or multi-channeled. The single-channeled detector, usually photomultiplier tube (PMT) or semiconductor, detects the intensity of one wavelength at a time. Multi-channeled detectors, such as charge-coupled device cameras (CCDs), record the intensity of emission over the range of wavelengths simultaneously. In addition, modern instruments include some other important components, such as polarizers, filters, and optical fiber connectors. Laboratory spectrometers are complex and robust devices capable of versatile and sensitive measurements. However, for many applications, and for the convenience and mobility of measurements fluorimeters can be designed as miniaturized devices. These devices are particularly suitable for clinical investigations of tissue fluorescence [8].

http://dx.doi.org/10.5772/63534

3. Breast tissue fluorescence

Native fluorescence (autofluorescence) of breast tissue comes from a number of fluorescent biochemical species (fluorophores) whose excitations and emissions are strongly influenced by the absorption and scattering of tissue components. Since the latter processes are both depth- and wavelength-dependent, the observed tissue fluorescence is subject not only to the concentration and microenvironment of fluorophores, absorbers (chromophores), and scatters but also it substantially depends on the geometry of excitation and observation.

The breast, or mammary gland, is an organ composed of four major structures: skin, subcutaneous tissue, breast tissue, and the nipple centered on the round pigmented skin area (areola). In the breast tissue, distinct optical characteristics show glandular, adipose (fat), and fibrous tissue. The differences in autofluorescence of these tissues come from the structural and compositional differences of which the concentration of fluorophores and their distribution has the largest influence. Most fluorophores are associated with the structural matrix of tissues, such as collagen and elastin, or are involved in cellular metabolic processes such as reduced nicotinamide adenine dinucleotide (NADH) and flavins. Other fluorophores include the aromatic amino acids (e.g., tryptophan, tyrosine, phenylalanine), various porphyrins, and lipopigments (e.g., ceroids, lipofuscin) that are the end products of lipid metabolism [9]. Absorption and fluorescence emission spectra of the most important tissue fluorophores are shown in **Figure 3**. The excitation (λ_{ex}), emission (λ_{em}), and absorption (λ_{ab}) maxima for some important tissue fluorophores are given in **Table 3**.

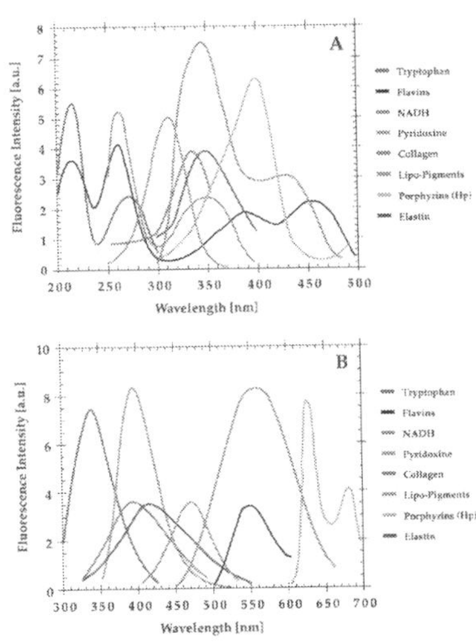

Figure 3. Absorption (A) and fluorescence emission (B) of the most important fluorophores present in tissue. (Adapted from Wagnières et al. [9] with permission of John Wiley and Sons).

Endogenous fluorophore	λ_{ex} (nm)	λ_{em}(nm)	λ_{ab}(nm)
Amino acids			
Tryptophan	280	350	280
Tyrosine	275	300	275
Phenylalanine	260	280	257
Structure proteins			
Collagen	325	400,405	325
Elastin	290,325	340,400	325
Coenzymes			
FAD, flavin	450	535	450
NADH	290,351	440,460	260
NADPH	336	464	340
Vitamins			
Vitamin A	327	510	300 - 350
Vitamin K	335	480	249
Vitamin D	390	480	280
Vitamin B6 complex			
Pyridoxine	332,340	400	291
Pyridoxamine	335	400	-
Pyridoxal	330	385	-
Pyridoxal 5'-phosphate	330	400	-
4-Pyridoxic acid	315	425	307
Vitamin B12	275	305	361
Lipids			
Phospholipid	436	540,560	234
Lipofuscin	340-395	540,430-460	335, 435
Ceroid	340-395	430-460,540	-
Porphyrin	400-450	630,690	408

Table 3. Tissue fluorophores and approximate values of their excitation (λ_{ex}), emission (λ_{em}), and absorption (λ_{ab}) maxima.

Regarding breast cancer, it has been shown that distinct fluorescence response of tumors compared to one ordinary tissue is a result of notable differences in concentrations of collagen, elastin, NADH, and flavin adenine dinucleotide (FAD) [10]. Generally, fluorescence measurements have indicated lower concentrations of collagen and FAD and increased concentrations of NAD(P)H in malignant tissues compared to normal breast tissue [11]. Transformation from normal to malignant tissue leads to degradation and changes in the cross-links of collagen; breaking the cross-links in collagen is a consequence of the increased presence of collagenase in the tumor cells [12]. Modulation of the extracellular matrix is a common characteristic of invading tumor cells and usually involves increased production of collagenases by the tumor cells or stromal fibroblasts [13]. It has also been shown that the changes in the metabolic status of tissue have influence on the variation in concentrations of both NADPH and NADH. A shift from aerobic to anaerobic metabolism accompanied with damaged mitochondrial metabolism caused by malignant alterations leads to an increased concentrations of electron carriers such as NADH. Increased levels of these coenzymes have been observed in a high-grade malignant tissue [12, 14].

4. Interpretation of tissue fluorescence data

As in all fluorescence measurements, acquiring experimental data on tissue fluorescence is associated with procedural and instrumental/technical difficulties. One should note that unlike some other types of spectroscopy measurements, for example, absorption measurements, the intensity of recorded fluorescence signal and the shape of spectra strongly depend on the characteristics and settings of the instrument. One can see later in the text that establishing diagnosis from fluorescence measurement can be successfully done only if data were acquired from a statistically significant number of samples. With this in mind, each study related to tissue fluorescence should be done on a single instrument and with exactly the same instrument settings (such as PMT voltage and monochromator slit widths) for each sample. Even so, if measurements are done over the longer times some of instrument characteristics may change, such as the strength of the excitation lamp. Samples most frequently have different surface morphologies and can be differently or partly illuminated when they are small sized. All of this may produce unwanted artificial differences in fluorescence spectra, commonly in the intensity of observed signal. Normalization of measured spectra may resolve this problem; however, there is no universal method or procedure for this task. For *ex vivo* measurements, it is important to note that all excised tissues change over the time. The fluorescence measurements on different tissue samples should be done, in principle, at the similar time interval after excision.

It is also important to keep in mind distinction between technical (uncorrected for instrument characteristics) and molecular (corrected) spectrum. Modern instruments commonly produce both types of spectra. In most cases, technical emission spectra are sufficient for the analysis and presentation. However, when it is important to compare emission spectra obtained on different instruments, corrected spectra must be used. Excitation spectra, and conversely SFS, EEMs, and 3D-TSFS, are more instrument-dependent, being largely influenced with spectral characteristic of excitation sources. These spectra are, therefore, analyzed and presented after corrections. In addition, spectra may contain features that originate from physical processes apart from fluorescence, such as Rayleigh and Raman scatters, which should be excluded from subsequent analyses.

Fluorescence spectra of tissues are generally poorly resolved since they are composed from a broad emission of many fluorophores and affected by strong re-absorption processes. When needed, better-resolved spectra can be achieved with synchronous scanning.

5. Mathematical tools for fluorescence-based diagnosis of breast cancer

To draw conclusions from experimental data of tissue fluorescence measurements, these data ought to be interpreted using adequate mathematical methods, most frequently statistical methods. Even if one analyze fluorescence from a single type of tissue, one must take into account that tissue characteristics considerably differ between individuals, statistically speaking they exhibit "a large in-group variances". For this reason, proper analysis is achieved only if it is done on a statistically significant number of samples and measurements. The differences in fluorescence, for example, differences in wavelengths of emission maxima, integrated emission intensities over some spectral range, excited state lifetimes, etc., between groups of different-type tissues can be asserted by methods of exploratory data analysis, one of which is hypothesis testing. One-tailed and two-tailed t-tests are commonly used to find if there is a statistically significant difference between the mean values of analyzed parameters in two data groups. Analysis of variance (ANOVA) and Tukey's test are used when more than two groups of data should be compared and studied. ANOVA is considered the most used and most useful statistical technique in biomedical research, even though this method is not easy to learn and should be implemented with caution.

Frequently, data from measurements of tissue fluorescence are vast, especially in cases of EEM and 3D-TSFS measurements. The size of data can be significantly reduced without loss of variance using some of statistical tools available for reducing the dimensionality. Principal component analysis (PCA) is a common tool for this purpose, and it is also capable of revealing hidden structures in a data set. Generally, PCA is used to recognize the main variations in a data set in an unsupervised way. It is also able to study these variations and to aid in visualizing them. PCA transforms and compresses input variables, which may be correlated, into uncorrelated variables of fewer numbers in such a way that they preserve the majority of variance of the input data. These new variables are called principal components (PCs) and they are obtained by a linear orthogonal transformation of input variables, so that each PC is, in fact, the linear combination of inputs. The largest portion of the input data variance is accumulated in the first PC, then less variance in the second PC, and so on. Then, for further analyses, one can use just few PCs, which help to alleviate numerical problems associated with large data sets and enable efficient visualization and interpretation of data.

Many features of fluorescence differ between healthy and cancer tissues; the selected examples are shown and discussed in the following section. Even though fluorescence measurements contain information needed for the cancer diagnosis, the observed data are subtly related in ways that are often difficult to express in the form of diagnostic rules just by observing spectra and must be processed for tissue classification purposes. To do so, mathematical algorithms ought to be developed and optimized to classify tissues into their respective histological categories. Several methods have been successfully used for this purpose.

Linear discriminant analysis (LDA) can be used for the discrimination of two or more groups from one or more linear functions (latent variables) of input data. As a consequence of singularity problems (caused by fewer samples in group than variables or highly correlated variables), LDA is incapable of solving high-dimensional data problems. In such cases, reduction of data dimensionality is required for the successful application of LDA.

Partial least squares (PLSs) discriminant analysis (DA) is a method based on the PLS regression, which constructs linear discriminant models with no restriction for processing high-dimensional data. Input data are transformed by the PLS-DA into a set of linear components (latent variables) further used to predict the dependent variable. The dependent variable is in fact a dummy variable that only serves to show whether a particular sample belongs to the specific class. With the LDA model, one is able to perform classification of data, that is, to predict the class to which new, unknown samples belong.

Artificial neural network (ANN) is a model used to approximate unspecified relations between a large number of input data in a robust manner. ANN learns from data, so it can be regarded as a machine-learning method and is able to handle a variety of problems including complex ones such as those involved in medical diagnostics. For example, ANNs can perform data classifications in both a supervised and unsupervised way, that is, both with prior knowledge of the data class and without. Unlike human decisions, those made by ANNs are invariably consistent since they are not liable to suffer from fatigue or bias. Kohonen's self-organizing map (SOM) and feed-forward neural network (FFNN) are among the most popular neural network architectures. SOM converts high-dimensional nonlinear statistical relationships into simple geometric relationships in an unsupervised way. On the other hand, FFNN uses a supervised training method which requires data on sample's group membership entered along with other data at the network inputs.

Support vector machine (SVM) is a statistical technology developed by the machine-learning community that can be used for both classification and regression. Compared with other machine-learning methods, SVM has such advantages as it does not require a large number of training samples for developing model and is not affected by the presence of outliers. Having high generalization procedure and feasibility to extract higher order statistics, the SVM has become extremely popular in terms of classification and prediction. In the context of classification, SVM is transforming original data space into a much higher dimension space in which classification groups would be linearly separable. For the data that belong to one of two classes (binary classification), SVM aims to derive the hyperplane in the transformed space so that the data of one class are on one side of the plane, and the data of other class are lying on the other side of the plane. The position of the hyperplane should be such that the greatest possible fraction of data is correctly placed and that the distance of both classes from the hyperplane is maximal. Such conditions minimize the risk of misclassifying data.

The performance of binary classification, the most probable classification case in breast cancer diagnoses, can be evaluated from a receiver operating characteristic (ROC) curves. The ROC curve is in fact graphic that shows the performance of a classification at different discrimination threshold values. It helps in finding how good classification model discriminates samples belonging to the particular group from all other samples. To construct the ROC, sensitivity and specificity of the model are calculated for different threshold values and plotted. Generally, an excellent model has an area under the ROC curve between 0.9 and 1.

Some other mathematical tools may also be useful for the study of breast cancer fluorescence. We believe that it is important to mention parallel factor analysis (PARAFAC) since this method is capable of modeling florescence response of complex fluorescence systems, the category in which all biological systems fall. Nowadays, PARAFAC has commonly used with EEMs in the analysis of fluorescence in many fields since it is a multi-way decomposition method capable of analyzing complex high-dimensional data and perform second-order calibration. Its main advantages are the uniqueness and simplicity of its solutions. When the correct number of components and multilinear data are used to build the PARAFAC model, the true underlying phenomenon is revealed. An additional advantage of this form of analysis is that it can predict the concentrations of different chemical compounds in complex systems, a phenomenon known as second-order advantage. For these reasons, PARAFAC is used to detect fluorophores in multi-component systems and precisely calculate their concentrations.

6. Examples of applications of fluorescence spectroscopy in breast cancer research

So far, different types of fluorescence measurements have been employed for the breast cancer diagnosis and they are listed in **Table 4** along with comments on the used instrumentation and obtained sensitivity and specificity of detection.

Measurement type	Instrumentation specifics	Measurement setup (nm)	SE, SP (%)	Comment	Ref.
Emission spectra	Argon laser	Ex 457.9, 488 Em 460-700	-	*Ex vivo*	[15]
Emission spectra	N_2 laser	Ex 337 Em 360-560	99.6, 99.6	*Ex vivo*	[16]
Excitation spectra	lamp-based spectrophotometer	Ex 250-320 Em 340	> 90, -	*Ex vivo*	[17]
Emission spectra	N_2 laser	Ex 458 Ex 480-700	100, 100	*Ex vivo*	[18]
EEM	lamp-based spectrophotometer with fiber optics probe	Ex 300-460 Em 310-600	70, 91.7	*Ex vivo*	[19]
EEM	lamp-based spectrophotometer	Ex 260-540 Em 275-700	-	MDA231 MCF10 T47D cell line	[20]

Measurement type	Instrumentation specifics	Measurement setup (nm)	SE, SP (%)	Comment	Ref.
EEM	lamp-based spectrophotometer	Ex 335-470 Em 430-640	83.9, 88.9	*Ex vivo*	[11, 21]
3D-TSFS	lamp-based spectrophotometer	330-650, $\Delta\lambda$ 30-120	100, 100 87.1, 91.7	*Ex vivo*	[10, 26] [27]
Synchronous spectra	lamp-based spectrophotometer	Ex 340-650 $\Delta\lambda$ 30, 80	93.3, 91.9	*Ex vivo*	[22]
Polarized fluorescence spectra	Argon laser	Ex 488 Em 540-700	92.6, 91.2	*Ex vivo*	[23]
Emission spectra	laser diode Argon laser Ex	Ex 405, 458 Em 428-700, Em 470-750	-	*In vivo*	[8]
Synchronous spectra	lamp-based spectrophotometer	Ex 250-650 $\Delta\lambda$ 40	90, 79	*Ex vivo*	[24]
Lifetime	Supercontinuum pulsed laser	Ex 447 Em 532, 562, 632, 684	89.8, 93.8	*Ex vivo*	[25]

Note: Ex – excitation, Em – emission, $\Delta\lambda$ – synchronous interval, SE – sensitivity, SP – specificity.

Table 4. Fluorescence methods used for breast cancer diagnosis.

6.1. Breast cancer diagnosis using emission and excitation spectra

The first report on the characterization of human breast tissues by fluorescence emission spectroscopy has been published by Alfano et al. [15] They have obtained normal and cancerous breast tissues from two different individuals and have measured emission spectra (460–700 nm) after excitation with an argon ion laser (at 488 and 457.9 nm). However, the assessment of method's capability for diagnosis has been impossible because of small number of investigated samples. Gupta et al. [16] have measured N_2 laser-excited (337 nm) emission spectra (360–560 nm) from different sites on benign (fibroadenomas, 35 patients), cancerous (ductal carcinomas, 28 patients), and normal specimens (the uninvolved areas of the resected cancerous specimens). Intensities of emission were much higher from cancerous sites compared to those of benign tumor and normal breast tissue sites (**Figure 4**). In this *ex vivo* study, cancer tissues have been discriminated from benign and normal ones with a sensitivity and specificity of up to 99.6% using absolute intensity of emission (with aid of stepwise multivariate linear regression statistical method).

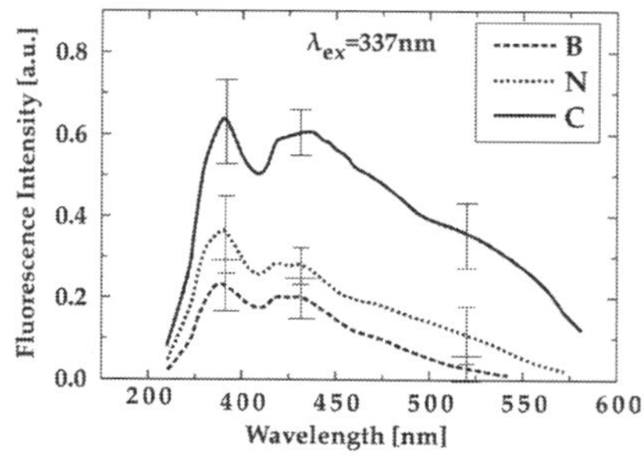

Figure 4. Mean emission spectra from 245 cancerous (C), 230 normal (N), and 436 benign (B) breast tissue sites. (Adapted from Gupta et al. [16] with permission of John Wiley and Sons.)

Yang et al. [17] have reported the use of excitation spectra for breast cancer detection. They have measured excitation spectra in the 250–320-nm spectral region recording emission at 340 nm from 103 malignant and 63 benign breast tissues. The averaged spectra and their difference (**Figure 5**) distinct spectral features can be clearly evidenced. Authors have estimated above 90% sensitivity and specificity of cancer diagnosis for this method.

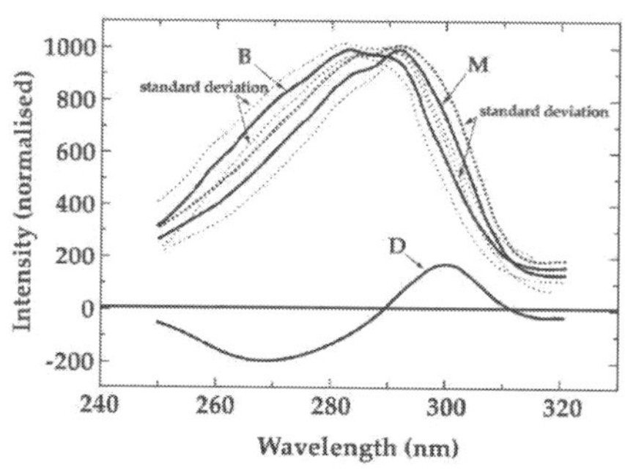

Figure 5. Averaged excitation spectra for emission at 340 nm from 103 malignant and 63 breast tissues. D is the difference spectrum (M – B). B is benign; M is malignant. (Adapted from Yang et al. [17] with permission of John Wiley and Sons.)

6.2. Breast cancer diagnosis using EEMs

Considering that EEMs incorporate more information pertinent for the breast cancer diagnosis than the single emission or excitation spectra, several studies have been conducted using this method (see **Table 4**). Palmer et al. [19] have measured EEMs (excitation range: 300–460 nm; emission range: 310–600 nm) of 20 malignant, 15 normal/benign fibrous, and 21 adipose tissue samples. They have found four peaks in the spectra of malignant and normal/benign fibrous tissues which occurs in similar locations (excitation/emission pairs of (300, 340), (340, 390), (360,

460), and (440, 520) nm). The spectra of adipose tissue were distinctively different; particularly, the peak at (340, 390) nm was weakly present and the peak at (360, 460) nm has been shifted to approximately (360, 520) nm. Using PCA as a data reduction technique and SVM for classification, authors have showed that EEM was successful in discriminating malignant and nonmalignant tissues with a sensitivity and specificity of 70% and 92%, respectively.

6.3. Breast cancer diagnosis using SFS and 3D-TSFS

Synchronous spectrum, in majority of cases, has more features and provides more information than the conventional emission or excitation spectra. This comes as a result of simultaneous acquisition of both excitation and emission spectral characteristics of the sample, which brings more information to a single spectrum. It has been shown that SFS measurements have better selectivity and decreased bandwidths than emission or excitation spectra. For these reasons, they can be used to detect fine differences in the fluorescence of complex systems. Dramićanin et al. [22] have measured SFS with different synchronous intervals on 21 normal and 21 malignant breast tissue samples. The largest differences between spectra have been found for synchronous intervals of 30 and 80 nm, **Figure 6**. Diagnostic criteria have been established from the areas below different spectral peaks and from the ratios of these areas. The first

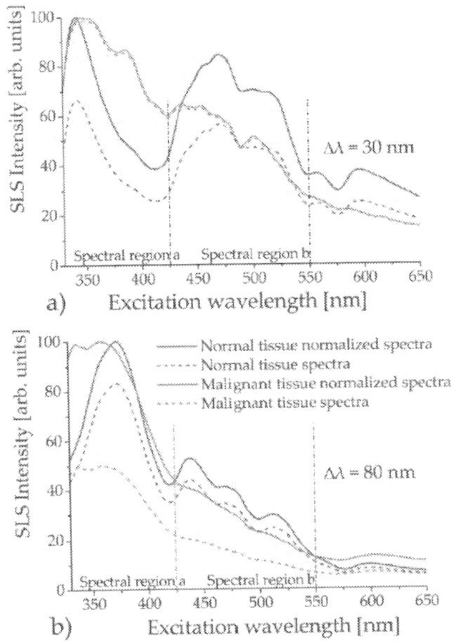

Figure 6. Synchronous fluorescence spectra taken with (a) 30 nm and (b) 80 nm synchronous intervals; the average spectra of 21 normal breast samples and 21 malignant are represented with the red and blue lines, respectively; dashed lines stand for the raw spectra and full lines for those additionally normalized. (Adapted from Dramićanin et al. [22].)

criterion yielded a sensitivity of detection of 77.4% and a specificity of 86.1%, while the second one presented a sensitivity of 90.3% and a specificity of 94.4%.

3D-TSFS and first derivate 3D-TSFS spectra have been measured on the same samples [10], and showed significant differences between normal and malignant tissues. Based on these differences, an artificial neural network (SOM) diagnosis method has been established [26], which provided a sensitivity of 87.1% and a specificity of 91.7% when using 3D-TSFS data, and 100% sensitivity and 94.4% specificity when using first derivate 3D-TSFS data. Diagnosis based on the use of support vector machine algorithm provided 100% specificity and sensitivity for both 3D-TSFS and first derivate 3D-TSFS [27].

6.4. Breast cancer diagnosis using lifetime and polarization measurements

In contrast to emission or excitation measurements, excited state lifetime measurements are not affected by the morphology of specimen and geometry of the measurements. Therefore, no normalization procedures are needed to compare results obtained with different instruments, which is quite important for the applicability of derived diagnostic methods. Sharma et al. [25] have measured fluorescence lifetimes at multiple emission wavelengths (532, 562, 632, and 644 nm) under excitation at 447 nm on 93 sites from 6 specimens (34 infiltrated ductal carcinoma, 31 benign fibrous, and 28 adipose sites). They have found that lifetime values measured at 532 and 562 nm significantly differ between normal and malignant sites, and achieved 92.3% accuracy for classifying infiltrated ductal carcinoma.

Choi et al. [23] have performed analysis of parallel and perpendicularly polarized fluorescence from 36 normal and 36 cancerous tissue samples. First, they have performed random forest algorithm to the four data sets: the fluorescence intensity and the curvature features for parallel and perpendicular components. Then, SVM classifiers were designed using the significant features from the random forest algorithm, which provided >90% specificity and sensitivity of breast cancer detection.

7. Conclusion

To conclude, fluorescence is sensitive to the biochemical makeup of tissue. There are several natural fluorophores that exist in tissues and cells that, when excited with ultraviolet and visible light, fluoresce over well-defined spectral regions. Among them, the fluorescence of collagen, elastin, NADH, and FAD contribute the most to the different fluorescence of cancerous tissue in respect to the normal because of altered concentrations and local environment of fluorophores. Distinct differences in the fluorescence of cancerous and normal tissues can be observed with almost all conventional fluorescence techniques. For this purpose, up to now, emission spectroscopy and EEMs have been used most extensively. However, even though fluorescence differences between tissues are obvious, the development of a sensitive diagnostic method based on these differences is not an easy task. First, a large number of specimens must be measured under identical conditions. The specimens should be taken from different individuals and some gold diagnostic standard must be provided (e.g., histopathology). Fluorescence data ought to be processed with statistical tools that include analysis of variance, data reduction, and regression to obtain diagnostic criteria, and further with statistical tools to validate diagnostic results. On the other hand, once established, fluorescence methods are a great aid for early cancer detection, since methods can be noninvasive and do not require expensive and sophisticated equipment.

Acknowledgements

The authors acknowledge the financial support of the Ministry of Education, Science and Technological Development of the Republic of Serbia (Project no 45020).

Author details

Tatjana Dramićanin and Miroslav Dramićanin*

*Address all correspondence to: dramican@vinca.rs

Vinča Institute of Nuclear Sciences, University of Belgrade, Belgrade, Serbia

References

[1] World Cancer Research Fund International. 2015. Available from: http://www.wcrf.org/ [Accessed: 2016-02-01]

[2] American Cancer Society. 2016. Available from: http://www.cancer.org/cancer/breast-cancer/detailedguide/breast-cancer-survival-by-stage [Accessed: 2016-01-15]

[3] Richards MA, Westcombe AM, Love SB, Littlejohns P, Ramirez AJ. Influence of delay on survival in patients with breast cancer: a systematic review. The Lancet. 1999;353:1119–1126. DOI: 10.1016/S0140-6736(99)02143-1

[4] Vo-Dinh T, editor. Biomedical Photonics Handbook. Boca Raton: CRC; 2003. DOI: 10.1201/9780203008997

[5] Jameson DM. Introduction to Fluorescence. Boca Raton: Tayor & Francis; 2014.

[6] Valuer B, Berberan-Santos MN. Molecular Fluorescence: Principles and Applications, 2nd ed. Weinheim: Wiley-VCH; 2012. DOI: 10.1002/9783527650002

[7] Lakowicz JR. Principles of Fluorescence Spectroscopy. 3rd ed. New York: Springer-Verlag US; 2006. DOI: 10.1007/978-1-4615-7658-7

[8] Alchab L, Dupuis G, Balleyguier C, Mathieu MC, Fontaine-Aupart MP, Farcy R. Towards an optical biopsy for the diagnosis of breast cancer in vivo by endogenous

fluorescence spectroscopy. Journal of Biophotonics. 2009;1–12. DOI: 10.1002/jbio.200900070

[9] Wagnières GA, Star WM, Wilson BC. In Vivo fluorescence spectroscopy and imaging for oncological applications. Photochemistry and Photobiology. 1998;68:603–632. DOI: 10.1111/j.1751-1097.1998.tb02521.x

[10] Dramićanin T, Dramićanin MD, Jokanović V, Nikolić-Vukosavljević D, Dimitrijević B. 3-dimensonal total synchronous luminescence spectroscopy criteria for discrimination between normal and malignant breast tissues. Photochemistry and Photobiology. 2005;81:1554–1558. DOI: 10.1562/2005-02-15-RA-442

[11] Dramićanin T, Lenhardt L, Zeković I, Dramićanin MD. Biophysical characterization of human breast tissues by photoluminescence excitation-emission spectroscopy. Journal of Research in Physics. 2012;36:53–62. DOI: 10.2478/v10242-012-0013-z

[12] Georgakoudi I, Jacobson BC, Muller MG, Sheets EE, Badizadegan K, Carr-Locke DL. NAD(P)H and collagen as in vivo quantitative fluorescent biomarkers of epithelial precancerous changes. Cancer Research. 2002;62:682–687. DOI: NEMA

[13] Fenhalls G, Geyp M, Dent MD, Parker MI. Breast tumour cell-induced down-regulation of type I collagen mRNA in fibroblasts. British Journal of Cancer. 1999;81:1142–1149. DOI: 10.1038/sj.bjc.6690821

[14] Uppal A, Gupta PK. Measurement of NADH concentration in normal and malignant human tissues from breast and oral cavity. Biotechnology and Applied Biochemistry. 2003;37:45–50. DOI: 10.1042/BA20020052

[15] Alfano RR, Tang GC, Pradhan A, Lam W, Choy DSJ, Orpher E. Fluorescence spectra from cancerous and normal human breast and lung tissues. IEEE Journal of Quantum Electronics. 1987;23:1806–1811. DOI: 10.1109/JQE.1987.1073234

[16] Gupta PK, Majumder SK, Uppal A. Breast cancer diagnosis using nitrogen laser excited autofluorescence spectroscopy. Lasers in Surgery and Medicine. 1997;21:417–422. DOI: 10.1002/(SICI)1096-9101(1997)21:5<417::AID-LSM2>3.0.CO;2-T

[17] Yang Y, Katz A, Celmer EJ, Zurawska-Szezepaniak M, Alfano RR. Fundamental differences of excitation spectrum between malignant and benign breast tissues. Photochemistry and Photobiology. 1997;66:518–522. DOI: 10.1111/j.1751-1097.1997.tb03183.x

[18] Hage R, Galhanone PR, Zângaro RA, Rodrigues KC, Pacheco MTT, Martin AA, Neto MM, Soares FA, da Cunha IW. Using the laser-induced fluorescence spectroscopy in the differentiation between normal and neoplastic human breast tissue. Lasers in Medical Science. 2003;18:171–176. DOI 10.1007/s10103-003-0271-9

[19] Palmer GM, Zhu C, Breslin TM, Xu F, Gilchrist KW, Ramanujam N. Comparasion of multiexcitation fluorescence and diffuse reflectance spectroscopy for the diagnosis of

breast cancer. IEEE Transaction on Biomedical Engineering. 2003;50:1233–1242. DOI: 10.1109/TBME.2003.818488

[20] Palmer GM, Keely PJ, Breslin TM, Ramanujam N. Autofluorescence spectroscopy of normal and malignant human breast cell lines. Photochemistry and Photobiology. 2003;78:462–469. DOI: 10.1562/0031-8655(2003)0780462ASONAM2.0.CO2

[21] Dramićanin T, Zeković I, Dimitrijević B, Ribar S, Dramićanin MD, Optical biopsy method for breast cancer diagnosis based on artificial neural network classification of fluorescence landscape data. Acta Physica Polonica A. 2009;116:690–692.

[22] Dramićanin T, Dramićanin MD, Dimitrijević B, Jokanović V, Lukić S. Discrimination between normal and malignant breast tissues by synchronous luminescence spectroscopy. Acta Chimica Slovenica. 2006;53:444–449.

[23] Choi J, Gupta S, Park I, Lee D, Ye JC. Breast cancer diagnosis from fluorescence spectroscopy using support vector machine. In: Proceedings of the (SPIE '07); Optical tomography and spectroscopy of tissue VII, edited by Britton Chance, Robert R. Alfano, Bruce Jason Tromberg, Mamoru Tamura, Eva Marie Sevick-Muraca, 2007; 6434: 64340P-1-10. DOI: 10.1117/12.700800

[24] Pu Y, Wang W, Yang Y, Alfano RR. Stokes shift spectroscopic analysis of multifluorophores for human cancer detection in breast and prostate tissues. Journal of Biomedical Optics. 2013;18:017005. DOI: 10.1117/1.JBO.18.1.017005

[25] Sharma V, Shivalingaiah S, Peng Y, Euhus D, Gryczynski Z, Liu H. Auto-fluorescence lifetime and light reflectance spectroscopy for breast cancer diagnosis: potential tools for intraoperative margin detection. Biomedical Optics Express. 2012;3:1825–1840. DOI: 10.1364/BOE.3.001825

[26] Dramićanin T, Dimitrijević B, Dramićanin MD. Application of supervised self-organizing maps in breast cancer diagnosis by total synchronous fluorescence spectroscopy. Applied Spectroscopy. 2011;65:293–297. DOI: 10.1366/10-05928

[27] Dramićanin T, Lenhardt L, Zeković I, Dramićanin MD. Support vector machine on fluorescence landscapes for breast cancer diagnostics. Journal of Fluorescence. 2012;22:1281–1289. DOI: 10.1007/s10895-012-1070-0

Chapter 13

Applications of ^{1}H Nuclear Magnetic Resonance Spectroscopy in Clinical Microbiology

Lara García-Álvarez, Jesús H. Busto,
Jesús M. Peregrina, Alberto Avenoza and
José Antonio Oteo

Additional information is available at the end of the chapter

http://dx.doi.org/10.5772/64450

Abstract

Proton nuclear magnetic resonance (^{1}H NMR) is a spectroscopic technique usually used for structural determination of molecules. In recent years, this technique has been employed for easy and quick recognition of microorganisms, in antimicrobial susceptibility tests and even for the diagnosis of different infectious conditions. Though ^{1}H NMR shows great potential for expanded applications in microbiological studies, to date applications of proton NMR to microbiological research are not totally standardized. In this chapter, we summarize the state of knowledge about ^{1}H NMR and its current and potential applications in this field.

Keywords: nuclear magnetic resonance, ^{1}H NMR, applications, clinical microbiology, microorganisms

1. Introduction

Scientific progress made in the recent years has enabled the development of new techniques that facilitate and improve microbiological study. In this way, nuclear magnetic resonance (NMR) is a spectroscopic technique easy to use and quick to recognize microorganisms and provides sensitivity to antimicrobials. Anyway, to date we have not consensus about the usefulness of these techniques that are not totally standardized. In this chapter, we summarize the state of knowledge about NMR in microbiological studies.

2. Nuclear magnetic resonance (NMR) spectroscopy

NMR is a spectroscopic technique initially developed by Felix Bloch and Edward M. Purcell that relies on the magnetic properties of the atomic nucleus. Since 1946, it has become a powerful and extremely valuable tool for chemists, physicists, biochemists and more recently for the medical practitioners [1–5].

Although the most widespread application of NMR is the structural determination of molecules, the technique offers the advantage of direct mixture analysis, and therefore, NMR has demonstrated a unique potential to be used for metabolic mixture analysis, fastidious bacteria included [6]. In comparison with other techniques employed for mixture analysis, NMR can be used to directly investigate biological samples and cell cultures without requiring significant sample preparation. Moreover, the technique allows the determination of compound ratios in a highly reproducible manner. For these reasons, metabolomics and metabonomics are driving new technological advances in the NMR field. Thus, the combination of sophisticated NMR-based methods for mixture analysis with the power of statistical and chemometric methods makes NMR spectroscopy the technique of choice for complex biological mixture analysis, especially in clinical and biomedical researches [3, 4].

In the biological field, the NMR technique [7] is employed to determine the structure and function of macromolecules. Additionally, NMR allows the determination of metabolic changes in organisms in response to external stimuli, through the identification and quantification of metabolites (metabolomics/metabonomics). Metabolomics is the study of global metabolite profiles of a cell under a given set of conditions; however, the terms metabolomics and metabonomics are used in the literature interchangeably [8]. Jeremy Nicholson was the pioneer of the approach, 'the distinction between the two terms metabolomics/metabonomics is mainly philosophical rather than technical' [9].

Since NMR is the only technique that allows to carry out *in vivo* analysis, it has applications in medical diagnosis, for example, magnetic resonance tomography (MRT).

The use of NMR in metabolic studies was reported in 1977, by Brown et al., who determined the presence of lactate, pyruvate, creatinine, and alanine in a blood cell suspension by proton NMR (^{1}H NMR) [10]. Since then, the analysis of biological fluids, tissues, and cell extracts has been carried out successfully by NMR, especially in the context of research on diseases and evaluation of toxicological processes [11, 12].

2.1. Basics of NMR spectroscopy

The NMR phenomenon can only be carefully described and fully understood using quantum mechanics. Therefore, a complete understanding of the technique would require an exhaustive knowledge of the properties of the angular momentum in the quantum mechanics field, along with statistical thermodynamics knowledge to describe the floating processes needed in the liquid state.

However, since the theory and fundamentals of NMR have been fully developed over the last few years [13, 14] and its detailed description would get away from the objective of this chapter, we describe below some aspects of the technique, which could help to understand the equipment and methodologies used in metabolic research.

The basis of the technique is to use the magnetic properties of atomic nuclei that are defined by their angular momentum and their associated magnetic moment. Both moments are vector quantities, and they are related by a constant called gyromagnetic constant (γ), which is specific to each type of atomic nucleus. According to quantum mechanics, both moments are quantized and their value depends on a quantum number called spin (I). Not all nuclei are valid to get NMR signals, only those whose spins are greater than zero are valid, but ^{1}H or proton is the most used nucleus in NMR studies. Moreover, higher the value of constant γ, more sensitive will be the active nucleus in NMR. So, ^{1}H is the nucleus used in NMR studies because of its great abundance (100%) and high value of γ. However, other magnetically active nuclei with lower sensitivity may be used, such as ^{2}H (or D), ^{13}C, ^{19}F, ^{15}N, ^{31}P, and ^{23}Na. Focusing on the proton, in the presence of an external magnetic field (B0), two different energy levels appear. The magnetic moments of these nuclei try to align with B0, resulting in two possible orientations at that time. Each of these orientations corresponds to a different energy level. That is, in the presence of B0, the cores can be arranged in two new states with different energies. This phenomenon, from a vectorial viewpoint is known as magnetization.

To obtain a NMR signal, the sample is irradiated with a radiofrequency wave, perpendicular to B0, which compels it to reach the state of resonance, where the nuclei gyrate with a resonance frequency (f0), specific to each atomic nucleus and called Larmor frequency. For this reason, NMR spectrometers are designated by the ^{1}H resonance frequency instead of the magnetic field (for example, on a 14.1 T field, ^{1}H resonates at 600 MHz). After the pulse, the excited nuclei return to the initial equilibrium state emitting a radiofrequency signal, which decays with the time, a phenomenon known as relaxation. The resonance of the excited nuclear magnets is detected as an oscillating current in a coil placed around the sample. This signal is the FID (free induction decay), which arrives at the receiver and provides a spectrum formed by lines defining frequencies and widths by a mathematical operation known as Fourier transform. The widths are formed from the contributions of all nuclei of the sample, so that this quality allows quantitative measurements. A line in the NMR spectrum obtained at a certain frequency (or chemical shift) corresponds to an atomic nucleus with a given chemical environment, which allows structural information about the molecule it belongs.

2.2. Equipment

The NMR spectrometer involves several parts such as a superconducting magnet, a radio transmitter, a probe, a radio receiver, an analog-to-digital converter (ADC), and finally, a computer. The magnet is the main element and consists of a solenoid of superconducting Nb/Ti alloy wire immersed in liquid helium (4 K) that is charged to generate the essential field strength. The helium is protected with a vacuum jacket and further cooled by an outer dewar of liquid nitrogen (77 K).

The probe head is a coil of wire positioned around the sample (NMR tube) that alternately transmits and receives radio frequency signals. The probe head is usually hosted into the magnet from the bottom and is connected to at least three radiofrequency channels providing the ^{2}H lock, ^{1}H frequency, and one X-nucleus frequency. In general, devices to control temperature (heater, thermoelement, and air) are needed. New developments include the digital transmission of the probe-head parameters to the console.

The computer addresses the transmitter to send a high-power and very short duration pulse of radio frequency to the probe-head. Instantly, after the pulse of radio frequency, a weak signal (FID) from the sample received by the probe-head is amplified and sampled at intervals by the ADC to produce a digital FID signal, which is just a list of numbers. The computer automatically determines the timing and power of pulses output by the transmitter and receives and processes the digitalization. After the computer performs the mathematical processes of Fourier transform in order to convert time domain into frequency domain, the resulting spectrum can be displayed on the computer monitor, transferred to other computers or plotted on a paper.

2.3. Sample preparation

The sample volume should be about 0.6 mL, which gives a 4.0 cm depth in a standard 5 mm NMR tube. Volume of the samples is less important than the concentration of targets. Very small volumes could not be studied by NMR equipment with low magnetic field but once the volume is established, it is more important to ensure that the metabolites to be studied are in

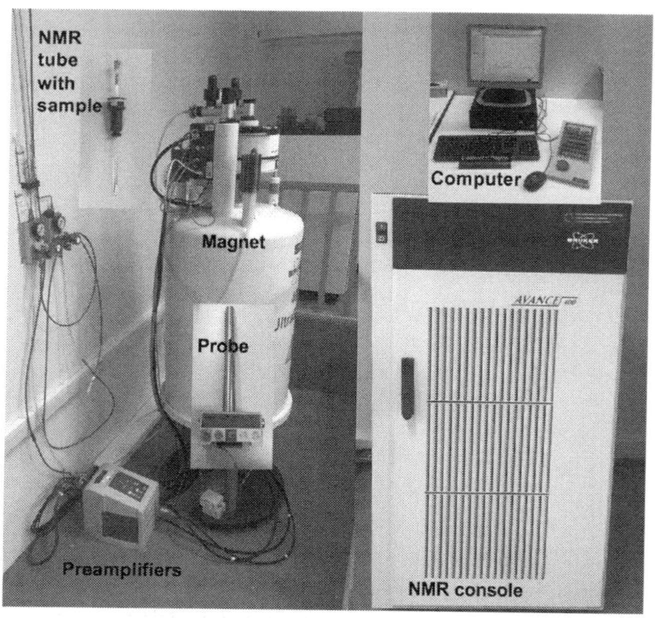

Figure 1. Components of an NMR equipment.

http://dx.doi.org/10.5772/64450

an appropriate concentration. For biological samples, the ideal solvent is D_2O or a mixture of H_2O/D_2O. In this latter case, the ^{1}H NMR spectra will be recorded using the pulse sequence for presaturation or the equivalent in order to avoid the signals from water. This is the typical and most simple method used to record NMR spectra of biological samples.

Usually, the protocol of preparation for biological and aqueous samples is simple, quick and involves two steps. Samples are prepared from inocula of the microorganisms and the bacterial concentration is adjusted using 0.5–2 McFarland standards [15, 16]. Cultures are incubated at the optimum temperature and time to preserve the same growth conditions than the reference method, if possible. After incubation, the suspensions are removed by centrifugation and the supernatants are decanted and used for NMR experiments. Because the measurements are carried out on supernatants, there is no need to quench cell growth rapidly. Furthermore, pH is measured and fixed by the slow addition of aqueous 1 M HCl and 1 M NaOH solutions or with a buffer solution in order to fix the chemical shifts. Next, a biological sample is added to a 5 mm NMR tube together with D_2O with the addition of the sodium salt (trimethyl)propanoic-2,2,3,3-d4 acid (TSP) for the chemical shift calibration. **Figure 1** shows an overview of the components of NMR equipment.

2.4. Sample treatment

The quality of the obtained NMR spectra depends on several variables that influence the process from sample collection to final data collection. The sample collection involves the sample, containers used, additives (preservatives, stabilizers), and time (collection, transport, storage) [17].

Depending of the source of the biological sample, two different methodologies can be used for the experiment acquisition:

(A) Sample concentration by lyophilization and subsequent reconstitution with a deuterated solvent: using this methodology, NMR experiments can be performed without solvent suppression, allowing an increase in sensitivity and stopping enzyme activity. The risks in using this method include the possibility of introducing contaminants into the sample and more importantly, the loss of volatile compounds.

(B) The addition of a small amount of D_2O to the aqueous sample: the corresponding NMR experiment is performed with a pulse program that can remove the water signal, which would otherwise mask the signals from the rest of the sample. This method involves minimal sample handling and the ability to detect volatile compounds, making it a more suitable method for metabolite analysis of biological samples.

In addition to these processes, due to the infectious potential of microorganisms contained in the sample, all steps of sample processing must be performed safely, through protocols and in laboratories appropriate to the biosafety level for the organism—until the organism is inactivated.

2.5. Data processing

Interpretation of the NMR data is essential to complete metabonomics studies to draw conclusions and trends. The first thing is to perform a pre-processing of NMR data in which NMR spectra are cleaned up in standard ways. After treatment of the spectra, it is possible to get information of metabolites, either through direct quantification by parameters [18, 19] or by applying the methods of data analysis and modelling. In this case, chemometric techniques and multivariate analysis are used to identify and quantify the different metabolites present in the sample. **Figure 2** shows an example of NMR spectrum with the main metabolites obtained. The existence of NMR databases of metabolites can greatly facilitate the latter processes.

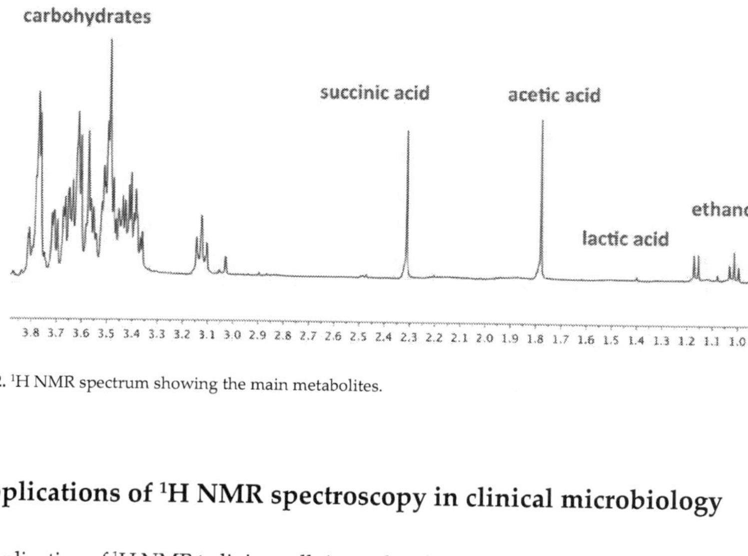

Figure 2. ^{1}H NMR spectrum showing the main metabolites.

3. Applications of ^{1}H NMR spectroscopy in clinical microbiology

The application of ^{1}H NMR to living cells is used to determine metabolites in complex mixtures and has been widely used for identification and quantification of the bacterial species [15, 20]. This technique has also been applied for antimicrobial drug susceptibility studies on different species of yeast, and in the last few years, it has also been developed for bacterial studies. Furthermore, other determinations directly in body fluids have emerged to help in the diagnosis of different diseases and conditions.

3.1. Bacterial identification and metabolic studies

^{1}H NMR spectroscopy has been used for bacterial identification and quantification and for metabolic pathways studies. Several studies have been conducted for the diagnosis of the bacteria that cause urinary tract infections (UTI). These focus on the use of ^{1}H NMR spectroscopy for the identification and quantification of common uropathogens such as *Pseudomonas aeruginosa, Klebsiella pneumoniae, Escherichia coli,* and *Proteus mirabilis* in urine samples. These studies are based on specific properties of the metabolism of the studied bacteria, and the results showed that ¹H NMR is a simple and fast tool compared with the traditional methods [16, 21, 22].

The qualitative and quantitative determination of *P. aeruginosa* using NMR spectroscopy is based on the specific property of the bacteria to metabolize nicotinic acid (NA) to 6-hydroxy-nicotinic acid (6-OHNA). Only this bacterium can produce this reaction. The addition of NA to urine samples after incubation and the subsequent analysis by ¹H NMR spectroscopy showed that NA signals disappeared from the medium after some time, while the appearance of new signals of the metabolite 6-OHNA indicated the presence of *P. aeruginosa*. The increase in the intensity of the metabolite signals, together with the decrease in the NA signals, involved a proportional increase in the number of bacteria. This shows the potential offered by this technique for quantitative and qualitative identification, simultaneously, on the bacteria [21].

A similar process occurs in the determination of *K. pneumoniae*. In this case, the specific metabolic reaction is the transformation of glycerol into 1,3-propanediol, so the substance that is added into the medium is glycerol. Despite *Citrobacter freundii* also being capable of carrying out this reaction, both bacteria can be easily differentiated using microscopic examination by observing their motility. *K. pneumoniae* is not mobile while *C. freundii* is. In addition, *C. freundii* is not a common nosocomial urinary tract infection agent. The combination of both methods showed very good sensitivity and specificity (90 and 100%, respectively), suggesting the potential usefulness of NMR for bacterial diagnosis [22].

The same experiment carried out on *E. coli* and *P. mirabilis* revealed that the specific metabolites of the bacteria are lactate and 2-hydroxy-4-(methylthio)butyric acid (MOBA) after incubation with lactose and methionine, respectively [16].

The results obtained using this alternative technique provided the warrant for the development of this method in bacterial identification and quantification and the technical development with other microorganisms. With experience, spectral analysis and data interpretation could be quick and reliable [16].

3.2. Antimicrobial susceptibility assays

The use of ¹H NMR spectroscopy for antimicrobial susceptibly tests has been not highly studied despite its powerful utility in this area of study [5]. Application of ¹H NMR spectroscopy to antimicrobial susceptibility studies was first carried out on different species of yeast. The standardized methods currently available for fungal susceptibility studies are unreliable and relatively slow, so, ¹H NMR spectroscopy can be a simple indicator, an objective and fast method (metabolic changes detected by this method are more easily observed than growth inhibition in broth). ¹H NMR spectroscopy is potentially valuable in determining the metabolic composition of yeast suspensions incubated with a drug. In addition, it is a high performance automated method with low operating costs, so that both operator time and reagent cost are greatly reduced. Therefore, it has great potential to emerge as an alternative method for the antifungal drug susceptibility determination of different yeast species [15, 20].

One of the few studies in which the ^{1}H NMR technique has been applied to observe the effect on microorganisms upon exposure to several drugs, has been carried out with medically relevant fungi. The fungal species analysed were *Cryptococcus neoformans*, different species of *Candida* and *Aspergillus* spp. These studies are based on the identification of the fungal metabolites produced, on their comparative profile implementation and on the monitoring of the nutrient utilization of the incubation medium in the presence of certain drug concentrations (caspofungin, amphotericin B, and voriconazole). The spectra obtained after subjecting the sample to the ^{1}H NMR were interpreted based on the metabolites produced (fumarate, malate, ethanol, etc.) and/or the metabolites consumed (tyrosine, phenylalanine, valine, etc.). This interpretation established a measurable parameter, the metabolic end point (MEP), from the spectral peak area. The MEP is defined as the lowest drug concentration at which nutrient utilization from the medium or the production of fungal metabolites is inhibited ≥ 50% and compared with minimum inhibitory concentrations (MIC) used in the reference method. The results of MEP generally showed a good correlation with MICs, which were determined by a modification of the reference method in broth microdilution M27-A of the CLSI. Discrepancies, which may arise between MEPs and MICs, could be due to differences in the culture medium and incubation time. In addition, ^{1}H NMR spectroscopy is a potentially valuable method for determining the metabolism of microorganisms incubated with the drug because it is a reproducible and relatively quick method (it takes 16 h versus 48 h required by the reference method) suggesting its consolidation as a platform for rapid determination of antifungal susceptibility [15, 20].

In reference to bacteria, there are very few studies based on their antimicrobial susceptibility according to metabolic profiling by ^{1}H NMR. One of these studies focused on a bacterial disease called 'Withering Syndrome' of abalone (a type of mollusc belonging to *Haliotis* spp. important in aquaculture). It is caused by a pathogen of the Rickettsiaceae family, '*Candidatus Xenohaliotis californiensis*' that infects digestive epithelial cells [23, 24]. On this basis, the effects of the antibiotic oxytetracycline in the metabolic profiles were observed by ^{1}H NMR. This drug, used to treat bacterial infections in aquatic species, reduces the severity and mortality of the Withering Syndrome. The aim of this study was to observe whether the recovery of the metabolism during treatment with oxytetracycline coincided with the disappearance of the disease caused by the bacteria. To this end, they examined the metabolic constituents present in the foot muscle of the mollusc after oxytetracycline treatment during several established days at two different seawater temperatures (13.4 and 17.3°C) [24]. Metabolic changes were observed at both temperatures: levels of taurine, glycine-betaine, and homarine increased and the amino acid and carbohydrate levels decreased. The detection of metabolic differences between animals treated and untreated with antibiotics was observed only at the highest temperature. Therefore, oxytetracycline eradicated the infection and at the highest temperature it reduced the metabolic changes due to the syndrome. The conclusions drawn from these experiments drive the development of ^{1}H NMR based on metabolic studies and its complementarity with other techniques, such as the histology for the identification of pathological processes in the aquatic species and for the optimization of drug therapy. This tool displays the performance by analysing quickly and cheaply the functional status of an organism [23, 24].

We have analysed the metabolism and antimicrobial susceptibility of *Escherichia coli* ATCC 25922 in the presence of gentamicin using ^{1}H NMR and compared with a reference method [25]. The MIC, determined by the reference method used in this study, would correspond to the termination of the bacterial metabolism observed using NMR. To carry out these experiments, serial dilutions of gentamicin were tested. Furthermore, two controls were also analysed (one was the medium with an inoculum of bacterium (control I), and the other was only the medium (control II)). The comparison of the two control ^{1}H NMR spectra showed different signals. Succinic acid, acetic acid, and ethanol were only detected in the control I spectra and threonine was only detected in the control II medium. According to the results obtained by visual turbidity, the lowest concentration of drug that completely inhibited visible growth (MIC) was 0.5 μg/ml (MIC: 0.25–1 μg/ml) [26]. When we registered antibiotic spectra at different concentrations, we detected the presence of succinic acid, acetic acid, and ethanol only in samples with concentrations of gentamicin lower than the MIC. Moreover, when the concentration of gentamicin was greater than the MIC, we detected the presence of threonine. These data suggested that the results obtained by ^{1}H NMR spectroscopy were in agreement with those obtained by visual turbidity. These results confirm that *E. coli* is able to metabolize components of the medium to produce succinic acid, acetic acid, and ethanol. Furthermore, threonine only appeared in the spectra of those samples with gentamicin concentrations of ≥0.5 μg/ml. Differences in peak intensities for the metabolites observed in spectra allowed the determination of the MIC for gentamicin using ^{1}H NMR spectroscopy. Consumption of the amino acid threonine, present in the culture medium, was interrupted when MIC was performed. Therefore, we assume that succinic acid, acetic acid, and ethanol are metabolites produced by the bacteria and threonine is the amino acid consumed by *E. coli*.

Furthermore, to evaluate the potential of this tool, we also performed the same biological experiments but using an NMR tube as the incubation reactor. The bacterial activity occurred effectively within the NMR tube, and the metabolic process started around 3 h 20 min and ended at 6 h. Moreover, when samples containing gentamicin were analysed in the same way the ethanol signal appeared later using the lowest concentration of gentamicin (4 h 40 min) compared with the experiments performed in the absence of the antibiotic (3 h 20 min) and much later (8 h 40 min) when the gentamicin concentration used was close to MIC. Similarly, threonine consumption by bacteria was delayed when the concentration of antibiotic in the medium was higher [25].

3.3. Applications in biofluids

In the last few years, ^{1}H NMR has been used to directly analyse biofluids and to diagnose different diseases directly from body fluids. In this sense, it has been applied to analyse human microbiota from faeces and urine samples, to study the metabolic implications that take place in sepsis, or even to diagnose hepatitis C virus infection, distinguish HIV-1 positive patients from negative individuals or to diagnose pneumonia from urine [27–33].

As mentioned above, ^{1}H NMR has been employed to study gut microbiome focusing on the metabolite profiling obtained by the analysis of different human samples. In this sense, Jacobs et al. [27] studied faeces from healthy subjects after consuming placebo, grape juice, or a mix of grape juice and wine during four weeks by ^{1}H NMR. The comparison of the NMR profiling of the samples indicated that only the mixture of wine and grape juice was able to modulate gut microbiota assessed by the reduction in isobutyrate observed only in the samples from subjects who had consumed the mixture. These results confirm that ^{1}H NMR could determine the impact of the nutrition in human microbiome [27]. Furthermore, the use of ^{1}H NMR to understand gut microbiota has been also extended to the study its impact on obesity by analysing metabolite profiling obtained from urine samples [28, 34].

^{1}H NMR has been also used to study different conditions as sepsis. Stringer et al. used ^{1}H NMR to compare the metabolic profile of whole blood from serum. This study revealed that the use of ^{1}H NMR in whole blood allowed obtaining more metabolic details that serum, and, in some cases, the metabolites detected were in higher concentrations in whole blood than in serum. Furthermore, whole blood allowed the determination of metabolites associated with red blood cell metabolism and observed that alterations in their metabolism could be in relationship with sepsis due to the haemolysis they cause [29]. The same authors have carried out several experiments in the same way. In these experiments, they were able to observe that blood samples of sepsis-induced acute lung injury patients were measurable and distinguishable from healthy blood due to differences in metabolites using ^{1}H NMR [35]. Other studies have been carried out to evaluate sepsis by ^{1}H NMR, but in rats [30]. Metabolic profiles obtain from the ^{1}H NMR analysis reveal changes in the metabolites involved in energy metabolism, especially in the serum of septic rats. From these results, authors concluded that according to the metabonomic approach, ^{1}H NMR has the potential for the early prognostic evaluation of sepsis [30].

As discussed above, ^{1}H NMR has been applied to the diagnosis of hepatitis C virus infection [31]. This study performed in urine samples was able to identify infected patients and negative individuals with good sensitivity and specificity using a metabonomics model based on the spectra obtained from the urine samples analysis. In this study, although the differences observed in the spectra allowed comparison of both groups, the chemical structures showed in the spectra are still being analysed [31]. In a similar way, ^{1}H NMR has been also used to distinguish between HIV-1 positive/AIDS patients on antiretroviral treatment and HIV-1 negative individuals [32]. These experiments were carried out in serum samples and differences in the metabolite profiling showed the distinction between the two groups. The authors also suggested that ARV-associated side effects could be monitored using ^{1}H NMR [32].

Pneumococcal pneumonia is another condition that has been diagnosed using ^{1}H NMR [33]. The use of the technique applied to urine samples of patients diagnosed with pneumonia has enable to distinguish pneumonia caused by *Streptococcus pneumoniae* from that caused by other microbes such as viruses or other bacteria. This distinction is observed due to the differences in the metabolomic profiles. So the use of ^{1}H NMR-metabolite profiling could result in a rapid, specific and sensitive tool for the diagnosis of pneumococcal pneumonia. In this study, it is also observed that the metabolic profile shown in the samples of patients with pneumonia caused by *S. pneumoniae* changed to a more normal metabolite profiling when specific treatment is administrated, suggesting that the urinary profiles were specific to the infection [33].

http://dx.doi.org/10.5772/64450

3.4. Other types of analyses

The combination of NMR spectroscopy, with the use of isotopically substituted molecules as tracers is a well-established protocol in microbiology. These NMR analyses appear to be the most appropriate for such studies because of their analytical power (provided that the labelling of products can be easily monitored non-invasively), their non-destructive features, and the large number of compounds that can be analysed simultaneously [6, 36–40]. However, despite the great potential of this combination in clinical practice, these analyses are out of the aim of this chapter.

4. Conclusion

In conclusion, ^{1}H NMR is an emerging technique in the microbiological field that promises to be a useful tool for the diagnosis of a broad range of conditions, including rapid identification of microorganisms, antimicrobial susceptibility and infectious-related syndromes. It can be also employed for knowing the metabolic pathways used by microorganisms, allowing the performance strategies for fighting against the infection.

Author details

Lara García-Álvarez[1], Jesús H. Busto[2], Jesús M. Peregrina[2], Alberto Avenoza[2] and José Antonio Oteo[1*]

*Address all correspondence to: jaoteo@riojasalud.es

1 Infectious Diseases Department, San Pedro Hospital-Center for Biomedical Research of La Rioja (CIBIR), Logroño, Spain

2 Department of Chemistry, Center for Research in Chemical Synthesis, University of la Rioja, Logroño, Spain

References

[1] Friebolin H. Basic one- and two-dimensional NMR spectroscopy. Wiley-VCH Verlag. Weinheim, 2005.

[2] Jacobsen NE. NMR spectroscopy explained: simplified theory, applications and examples for organic chemistry and structural biology. Wiley-Interscience: Cambridge, UK, 2007.

[3] Lindon JC, Nicholson JK, Holmes E. The Handbook of Metabonomics and Metabolomics. Elsevier. Amsterdam, 2007.

[4] Powers R. NMR metabolomics and drug discovery. Magn Reson Chem. 2009; 47: S2–S11. DOI: 10.1002/mrc.2461

[5] García-Álvarez L, Busto JH, Peregrina JM, Fernández Recio MA, Avenoza A, Oteo JA. Nuclear magnetic resonance applied to antimicrobial drug susceptibility. Future Microbiol. 2013; 8: 537–547. DOI: 10.2217/fmb.13.8.

[6] García-Álvarez L, Busto JH, Peregrina JM, Santibáñez S, Portillo A, Avenoza A, et al. Nuclear Magnetic Resonance (NMR) as a tool for the study of the metabolism of Rickettsia slovaca. Microbes Infect. 2015; 17: 850–855. DOI: 10.1016/j.micinf.2015.09.022.

[7] Gadian DG. NMR and its applications to living systems. Oxford University Press. Oxford, 1995.

[8] Tang J. Microbial metabolomics. Curr Genomics, 2011; 12: 391–403. DOI: 10.2174/138920211797248619

[9] Nicholson JK, Lindon JC. Systems biology: metabonomics. Nature. 2008; 455: 1054–1056. DOI: 10.1038/4551054a.

[10] Brown FF, Campbell ID, Kuchel PW, Rabenstein DC. Human erythrocyte metabolism studies by H-1 spin-echo NMR. FEBS Lett. 1977; 82: 12–16.

[11] Nicholson JK, Connelly J, Lindon JC, Holmes H. Metabonomics: a platform for studying drug toxicity and gene function. Nat Rev Drug Discov. 2002; 1: 153–161.

[12] Lindon JC, Nicholson JK, Holmes E, Antti H, Bollard ME, Keun H, et al. Contemporary issues in toxicology—the role of metabonomics in toxicology and its evaluation by the COMET project. Toxicol Appl Pharm. 2003; 187: 137–146.

[13] Ernst R, Bodenhausen G, Wokaun A. Principles of nuclear magnetic resonance in one and two dimensions. Oxford University Press. Oxford, 1990.

[14] Goldman M. Quantum description of high-resolution NMR in liquids. Oxford University Press. Oxford, 1991.

[15] Coen M, Bodkin J, Power D, Bubb WA, Himmelreich U, Kuchel PW, et al. Antifungal effects on metabolite profiles of medically important yeast species measured by nuclear magnetic resonance spectroscopy. Antimicrob Agents Chemother. 2006; 50: 4018–4026.

[16] Gupta A, Dwivedi M, Mahdi AA, Gowda GA, Khetrapal CL, Bhandari M. 1H nuclear magnetic resonance spectroscopy for identifying and quantifying common uropathogens: a metabolic approach to the urinary tract infection. BJU Int. 2009; 104: 236–244. DOI: 10.1111/j.1464-410X.2009.08448.x.

[17] Price KE, Vandaveer SS, Lunte CE, Larive CK. Tissue targeted metabonomics: metabolic profiling by microdialysis sampling and microcoil NMR. J Pharm Biomed Anal. 2005; 38: 904–909.

[18] López-Rituerto E, Cabredo S, López M, Avenoza A, Busto JH, Peregrina JM. A thorough study on the use of quantitative 1H NMR in Rioja red wine fermentation processes. J Agric Food Chem. 2009; 57: 2112–2118. DOI: 10.1021/jf803245r.

[19] Avenoza A, Busto JH, Canal N, Peregrina JM. Time course of the evolution of malic and lactic acids in the alcoholic and malolactic fermentation of grape must by quantitative 1H NMR (qHNMR) spectroscopy. J Agric Food Chem. 2006: 54: 4715–4720.

[20] Plummer R, Bodkin J, Power D, Pantarat N, Bubb WA, Kuchel PW, et al. Effect of caspofungin on metabolite profiles of Aspergillus species determined by nuclear magnetic resonance spectroscopy. Antimicrob Agents Chemother. 2007; 51: 4077–4084.

[21] Gupta A, Dwivedi M, Nagana Gowda GA, Ayyagari A, Mahdi AA, Bhandari M, et al. 1H NMR spectroscopy in the diagnosis of Pseudomonas aeruginosa-induced urinary tract infection. NMR Biomed. 2005; 18: 293–299.

[22] Gupta A, Dwivedi M, Gowda GA, Mahdi AA, Jain A, Ayyagari A, et al. 1H NMR spectroscopy in the diagnosis of Klebsiella pneumoniae-induced urinary tract infection. NMR Biomed. 2006; 19: 1055–1061.

[23] Viant MR, Rosemblum ES, Tjeerdema RS. NMR-based metabolomics: a powerful approach for characterizing the effects of environmental stressors on organism health. Environ Sci Technol. 2003; 37: 4982–4989.

[24] Rosemblum ES, Tjeerdema RS, Viant MR. Effects of temperature on host- pathogen-drug interactions in red abalone, *Haliotis rufescens*, determined by 1H NMR metabolomics. Environ Sci Technol. 2006; 40: 7077–7084.

[25] García-Álvarez L, Busto JH, Avenoza A, Sáenz Y, Peregrina JM, Oteo JA. Proton nuclear magnetic resonance spectroscopy as a technique for gentamicin drug susceptibility studies with *Escherichia coli* ATCC 25922. J Clin Microbiol. 2015; 53: 2433–2438. DOI: 10.1128/JCM.00604-15

[26] Clinical and Laboratory Standards Institute. Performance standards for antimicrobial susceptibility testing; 24th informational supplement. CLSI M100-S124. Clinical and Laboratory Standards Institute, Wayne, PA, 2014.

[27] Jacobs DM, Deltimple N, van Velzen E, van Dorsten FA, Bingham M, Vaughan EE, et al. (1)H NMR metabolite profiling of feces as a tool to assess the impact of nutrition on the human microbiome. NMR Biomed. 2008; 21(6): 615–626.

[28] Calvani R, Brasili E, Praticò G, Sciubba F, Roselli M, Finamore A, Marini F, et al. Application of NMR-based metabolomics to the study of gut microbiota in obesity. J Clin Gastroenterol. 2014; 48: S5–S7. DOI: 10.1097/MCG.0000000000000236.

[29] Stringer KA, Younger JG, McHugh C, Yeomans L, Finkel MA, Puskarich MA, et al. Whole blood reveals more metabolic detail of the human metabolome than serum as measured by 1H-NMR spectroscopy: implications for sepsis metabolomics. Shock. 2015; 44: 200–208. DOI: 10.1097/SHK.0000000000000406.

[30] Lin ZY, Xu PB, Yan SK, Meng HB, Yang GJ, Dai WX, et al. A metabonomic approach to early prognostic evaluation of experimental sepsis by (1)H NMR and pattern recognition. NMR Biomed. 2009; 22: 601–608. DOI: 10.1002/nbm.1373.

[31] Godoy MM, Lopes EP, Silva RO, Hallwass F, Koury LC, Moura IM, et al. Hepatitis C virus infection diagnosis using metabonomics. J Viral Hepat. 2010; 17: 854–858. DOI: 10.1111/j.1365-2893.2009.01252.x.

[32] Hewer R, Vorster J, Steffens FE, Meyer D. Applying biofluid 1H NMR-based metabonomic techniques to distinguish between HIV-1 positive/AIDS patients on antiretroviral treatment and HIV-1 negative individuals. J Pharm Biomed Anal. 2006; 41: 1442–1446.

[33] Slupsky CM, Rankin KN, Fu H, Chang D, Rowe BH, Charles PG, et al. Pneumococcal pneumonia: potential for diagnosis through a urinary metabolic profile. J Proteome Res. 2009; 8: 5550–5558. DOI: 10.1021/pr9006427.

[34] Calvani R, Miccheli A, Capuani G, Tomassini Miccheli A, Puccetti C, Delfini M, et al. Gut microbiome-derived metabolites characterize a peculiar obese urinary metabotype. Int J Obes (Lond). 2010; 34: 1095–1098. DOI: 10.1038/ijo.2010.44.

[35] Stringer KA, Serkova NJ, Karnovsky A, Guire K, Paine R 3rd, Standiford TJ. Metabolic consequences of sepsis-induced acute lung injury revealed by plasma ^{1}H-nuclear magnetic resonance quantitative metabolomics and computational analysis. Am J Physiol Lung Cell Mol Physiol. 2011; 300: L4–L11. DOI: 0.1152/ajplung.00231.2010.

[36] Mobashery S, Lerner SA, Johnston M. Monitoring beta-lactamase activity in vivo by 13C nuclear magnetic resonance spectroscopy. Antimicrob Agents Chemother. 1988: 32: 1196–1203.

[37] Aguayo JB, Gamcsik MP, Dick JD. High resolution deuterium NMR studies of bacterial metabolism. J Biol Chem. 1988; 263: 19552–19557.

[38] Ramos A, Jordan KN, Cogan TM, Santos H. C Nuclear Magnetic Resonance Studies of Citrate and Glucose Cometabolism by *Lactococcus lactis*. Appl Environ Microbiol. 1994; 60: 1739–1748.

[39] Neves AR, Ramos A, Shearman C, Gasson MJ, Almeida JS, Santos H. Metabolic characterization of Lactococcus lactis deficient in lactate dehydrogenase using in vivo 13C-NMR. Eur J Biochem. 2000; 267: 3859–3868.

[40] Chalk PA, Roberts AD, Blows WM. Metabolism of pyruvate and glucose by intact cells of *Helicobacter pylori* studied by 13C NMR spectroscopy. Microbiology. 1994; 140: 2085–2092.

Section 4

Analysis and Characterization of Foods and Beverages Using Molecular Spectroscopy

Chapter 14

Improving Food Safety by Using One- and Two-Photon-Induced Fluorescence Spectroscopy for the Detection of Mycotoxins

Lien Smeesters, Wendy Meulebroeck and Hugo Thienpont

Additional information is available at the end of the chapter

http://dx.doi.org/10.5772/103856

Abstract

The presence of mycotoxins in food products is a major worldwide problem. Nowadays, mycotoxins can only be detected by the use of sample-based chemical analyses. Therefore, we demonstrate the use of one- and two-photon-induced fluorescence spectroscopy for the non-destructive detection of mycotoxins in unprocessed food products. We first explain our optical set-up, which is able to measure the localized one- and two-photon-induced fluorescence spectra. Following, as a case study, the detection of aflatoxin in maize kernels is discussed. We present our research methodology, from the characterization of the fluorescence of pure aflatoxin, to the study of the one- and two- photon-induced fluorescence spectra of maize kernels and the development of an optical detection criterion. During both one- and two-photon-induced fluorescence processes, the fluorescence of the aflatoxin influences the intrinsic fluorescence of the maize. Based on the fluorescence spectrum between 400 and 550 nm, a detection criterion to sense the contaminated kernels is defined. Furthermore, we successfully monitored the localized contamination level on the kernel's surface, showing both contaminated kernels with a high contamination in a limited surface area (a few square millimetres) and kernels with a low contamination spread over a large surface area (up to 20 mm^2). Finally, the extensibility of our research methodology to other fluorescent mycotoxins is discussed.

Keywords: spectroscopy, fluorescence, two-photon-induced fluorescence, optical sensing, aflatoxin, multiphoton processes, spectrum analyses

INTECH
open science | open minds

1. Introduction

Fluorescence can be defined as the emission of electromagnetic radiation during the relaxation process of an excited electron that was excited from the ground state to a higher energetic state by the absorption of light. Fluorescence spectroscopy studies the fluorescence light emission as function of the wavelength. To date, fluorescence spectroscopy is already integrated in an extensive range of applications in, amongst others, the biochemical, medical and petrochemical industry [1–5]. In biology and chemistry, it facilitates the study of proteins and the tracking of biochemical reactions. In the medical industry, it is used to follow smart drug delivery systems that map the drug interaction with diseased tissue, while in the petroleum industry, it enables to characterize crude oils during refining processes.

Considering food applications, fluorescence spectroscopy has currently been demonstrated for the screening of products on basis of their chlorophyll concentration [6–8]. Particularly, the detection of foreign objects in food streams has been presented, by comparing the fluorescence spectra of the food products and the foreign objects. For example, when considering a mixture of green glasses and peas, a distinction can be made on basis of the chlorophyll fluorescence intensity, since the peas emit strong fluorescence signals while the glasses emit no fluorescence. In addition, also a distinction between peas, garden beans and sprouts has been demonstrated by evaluating the shape of the chlorophyll fluorescence spectrum [6]. However, toxic contaminants in food and feed products can hardly be optically detected due to the presence of large background fluorescent signals emitted by natural food constituents, like proteins, and the localized presence of the toxins. The current published fluorescence measurements only allow the identification of mycotoxins in homogeneous liquids, like beer or wine, in which no or very low background fluorescent elements are present [9, 10]. Therefore, we investigate the use of fluorescence spectroscopy for the detection of mycotoxins in unprocessed, solid food products.

Mycotoxins are secondary metabolites of toxic fungi, produced by various moulds on a wide range of food products, like nuts, maize, pistachios and peanuts [11]. Six types of mycotoxins are identified as major threats to food safety, of which we focus on the fluorescent mycotoxins, being aflatoxin, ochratoxin and zearalenone. Mycotoxins are observed under a diverse range of environments, both before and after the harvest [11]. Moreover, they appear both in the raw as processed products, since they cannot be destroyed during food processing, like cooking, freezing and roasting. The Food and Agriculture Organization (FAO) estimates that 25% of the world's food crops are affected by mycotoxin-producing fungi. The accumulation of mycotoxins in food and feed products represents a major threat to human and animal health, because they can induce cancer, liver diseases, immune-system suppression, mutagenicity and nervous disorders [9]. Therefore, to decrease the exposure to mycotoxins, international regulations were established [12, 13]. For example, for aflatoxins, the European Commission stated maximum allowed concentrations in the range between 2 and 15 ppb, depending on the commodity, while the USA food safety regulations included a maximum contamination level of 20 ppb for all food products. To fulfil these limitations, mycotoxins are nowadays detected by using time-consuming, sample-based chemical analyses, like liquid chromatography-dual mass spectrometry (LC-MS/MS). However, due to the localized presence of the toxin in the food products and crops, these analyses often give a limited view on the degree of contamination, inducing a large amount of food waste, without entirely preventing the toxins to enter the food chain [11]. Consequently, to increase food safety, the development of a non-destructive detection methodology, which is able to identify the localized contamination on the food products, is indispensable.

We investigate the use of fluorescence spectroscopy, including one- and two-photon-induced fluorescence (OPIF and TPIF) spectroscopy, as a non-destructive, optical detection technique for the sensing of fluorescent mycotoxins. OPIF and TPIF are two types of fluorescent processes, both resulting in the emission of a fluorescent photon during the relaxation process of an excited electron, but with a different excitation process [14]. To generate an OPIF photon, only a single excitation photon needs to be absorbed, while during the generation of a TPIF photon, two excitation photons need to be absorbed simultaneously to excite the electron and generate the fluorescence signal (**Figure 1**). In the case of OPIF, the energy of the incident photon equals the energy difference between the electronic states, such that the electron obtains sufficient energy to bridge the energy gap between the ground state and the excited state (**Figure 1(a)**). Considering TPIF, the sum of the energies of the two incident photons must equal the bandgap energy (**Figure 1(b)**). The first incident photon excites the electron to a virtual state, which does not need to correspond to any electronic or vibrational energy eigenstate. The second incident photon excites the electron from the virtual state to its higher-energy excited state. Because for both OPIF and TPIF the electron is excited to the same energetic state, they give rise to the same fluorescence wavelengths, but with the use of another excitation wavelength. Generally, OPIF requires ultraviolet (UV) excitation wavelengths, while TPIF uses near-infrared (NIR) light. If the energy of the incident photons does not match with the bandgap, no fluorescence emission will occur. As a result, because the bandgap is a molecular specific property, fluorescence is known to be a selective process. In addition, because TPIF is a non-linear process that requires the simultaneous absorption of two photons instead of one, it is considered to be more selective than OPIF [10].

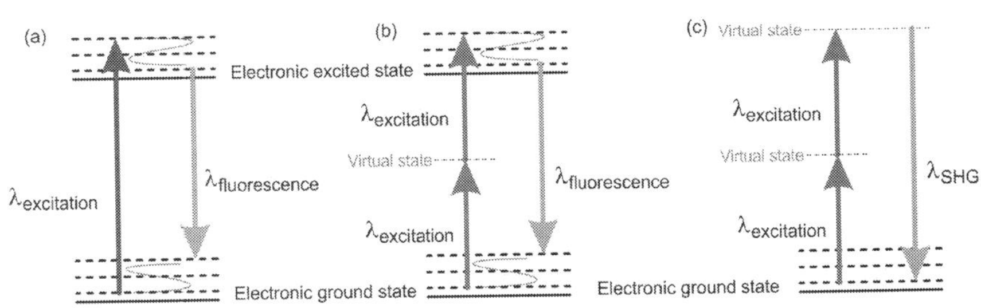

Figure 1. Jabłoński diagram for (a) OPIF, in which the electron is excited by the absorption of a single incident photon; (b) TPIF, in which the electron is excited by the simultaneous absorption of two incident photons; (c) second harmonic generation, in which two incident photons recombine to a new photon with the double energy.

Two-photon excitation was first theoretically described by Mario Göppert-Mayer in 1931. However, two-photon fluorescence has only been experimentally demonstrated in the 1960s by Kaiser and Garret when exciting an europium-doped crystal [15]. To date, two-photon fluorescence is mostly used in biomedical engineering, when studying living cells and tissue by using two-photon fluorescence microscopy [16–18]. Intuitively, we would expect that all fluorescent molecules, able to emit OPIF, generate a TPIF signal if they are excited with laser light with the double wavelength and a sufficiently high excitation power density. However, according to the quantum physics selection rules that describe the electron transitions in molecules, one- and two- photon absorption follow different selection rules [19]. More specifically, the occurrence of both OPIF and TPIF electronic transitions is only allowed in non-centrosymmetric molecules. When considering centrosymmetric molecules, the electron transition between the ground state and the excited state is allowed for either OPIF or TPIF. For these molecular structures, no TPIF signals can be observed if OPIF signals can be emitted.

Two-photon-induced fluorescence should not be confused with second-harmonic generation (SHG) (**Figure 1(c)**). SHG is generally defined as a non-linear optical process in which photons with the same energy interacting with a non-linear material are combined to generate new photons with the double energy and thus half the wavelength of the initial photons [20]. Considering the Jabłoński diagram for SHG, the two incident photons do not excite the electron but recombine to a new photon with an energy equal to the total energy of the two incident photons. Since no electron is excited, the process does not contain any information about the molecular structure of the products, making it unusable for the detection of toxins.

OPIF and TPIF spectroscopy are both considered as promising optical detection techniques. OPIF shows stronger fluorescence signals than TPIF, giving rise to a strong mycotoxin signal of the solid food products that enables the detection of low contamination levels. Particularly, the OPIF intensity increases linearly with the excitation power, while the TPIF intensity increases with the square of the excitation power. However, the TPIF process features a more selective excitation of the mycotoxins, which can reduce the influence of the background fluorescent signals emitted by the natural fluorescent substituents of the food products. Moreover, TPIF is obtained after excitation with NIR laser light that is more widely commercially available than the required UV laserlines used during OPIF. In addition, compact NIR lasers generally feature higher output powers than UV lasers, enhancing the fluorescence signal. When using UV light in an optical set-up, the optical mirrors and lenses need to be made from or coated with fused silica, resulting in a more expensive set-up than for NIR wavelengths.

In this chapter, we demonstrate the use of OPIF and TPIF as valuable tools for the non-destructive detection of mycotoxins. We first present our measurement configuration, able to study both OPIF and TPIF. Second, we discuss our measurement methodology, including the selection of the excitation wavelengths. Following, as a case study, we investigate the detection of aflatoxin B1 in individual maize kernels. This is because aflatoxin is considered as one of the most dominant mycotoxins and maize cultivates in climates that show an extensive presence of the fungi, giving rise to permanent high aflatoxin contamination levels. We first characterize the fluorescence of pure aflatoxin B1. Subsequently, we study the OPIF and TPIF spectra of healthy and contaminated maize kernels, containing an aflatoxin B1 contamination lower than 1 ppb and higher than 70 ppb, respectively. The emission wavelengths and intensities of the obtained spectra are investigated and a comparison between the performance of OPIF and TPIF is given. Afterwards, we examine the development of an optical detection criterion and study the localized contamination on the samples surfaces. Finally, we discuss the extensibility of our research methodology to other fluorescent mycotoxins.

2. Methodology and measurement set-up

We pursue a fluorescence measurement set-up suitable for both OPIF and TPIF fluorescence measurements. However, to obtain reliable measurement data, four challenges need to be tackled. (1) The selection of an optimal excitation wavelength is of major importance, to maximize the mycotoxin fluorescence and to minimize the influence of the intrinsic fluorescence signal. (2) The excitation spot size and excitation laser power should be optimized to enable the measurement of the weak mycotoxin fluorescence without damaging the food samples. (3) The natural variation within food products should be monitored, to assure we are measuring the influence of the toxin instead of the optical contrast between different product batches. (4) A careful selection of the products and the illumination positions on their surfaces is indispensable to deal with the localized presence of the mycotoxins. Due to the inhomogeneous presence of the toxin, measurements on individual products are much harder than optical measurements on homogeneous solutions and powders.

Keeping these challenges in mind, we first present our fluorescence measurement configuration, after which we explain the selection of the optimal excitation wavelengths.

2.1. OPIF and TPIF measurement set-up

To measure the OPIF and TPIF spectra of the localized contaminants, we require the generation of both UV and NIR illumination wavelengths, in combination with an accurate detection of the fluorescence signal while scanning the sample surface.

Generally, our measurement set-up can be divided into different building blocks, comprising an illumination laser system, the sample objective and the detecting optical spectrum analyser (**Figure 2**). A frequency-doubled Nd:YAG pump laser (Spectra-Physics Millennia Prime 10sJSPG) pumps a tunable titanium-sapphire laser (Spectra-Physics Tsunami laser). The wavelength of the titanium-sapphire laser can be tuned from 710 to 835 nm by the use of an internal slit, which selects the preferred wavelength after the dispersion of the generated laser light by internal prisms. The maximum output power ranges between 1.20 and 1.50 W, depending on the selected wavelength. The laser light of the titanium-sapphire laser can follow one out of two optical paths, depending on whether OPIF or TPIF measurements are performed (**Figure 2(a)**). For TPIF, the selected laser light is immediately directed towards the sample. In the case of OPIF, the fundamental laser light of the titanium-sapphire laser is directed towards a harmonic generating unit, comprising a second- and third-harmonic-generating crystal to generate the UV excitation wavelengths (**Figure 2(b)**). The second-harmonic-generating crystal is able to generate wavelengths between 355 and 417 nm, with a maximal output power between 200 and 450 mW. The third-harmonic-generating crystal enables us to use wavelengths between 240 and 275 nm, with a maximal output power between 60 and 200 mW. After the generation of the UV light, by either the second- or third-harmonic-generating crystal, the laser light is directed towards the sample. During both OPIF and TPIF measurements, the sample is illuminated with a circular beam, with a spot diameter of 951 and 231 μm, respectively. To maximize the TPIF irradiance, the spot size of the illumination laser beam was minimized by the use of an additional focusing lens positioned in front of the sample (**Figure 2(d)**).

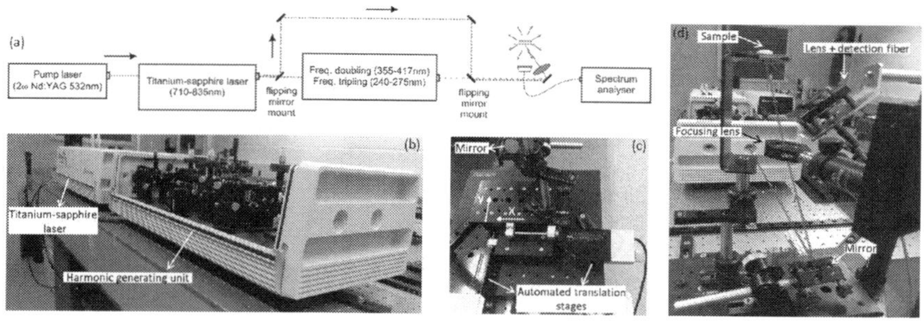

Figure 2. Measurement set-up that allows the investigation of both OPIF and TPIF: (a) schematic representation of the set-up; (b) tunable titanium-sapphire laser (710–835 nm) and harmonic generating unit with the frequency doubling (355–417 nm) and tripling crystals (240–275 nm); (c) automated translation stages on which the sample is mounted, enabling an accurate scanning in the X and Y direction; (d) optical path for the excitation of the sample, after which the fluorescence spectrum is captured by the detecting fibre. The focusing lens minimizes the spot size during the TPIF measurements.

The sample is positioned on a sample holder, containing a circular aperture with a diameter of 7 mm, enabling to position the kernels directly on this aperture. The sample holder is mounted on two automated translation stages (Newport 850G linear actuators) to accurately scan the product in both the X and Y direction (**Figure 2(c)**). For each X and Y position, the incident laser beam illuminates different parts of the product, allowing to study the localized contamination of the sample. Both automated translation stages have a travel range of 5 cm, scanning a maximum surface area of 25 cm^2. The translation stages are driven by a motion controller (Newport ESP300), enabling a high movement speed while assuring a high movement accuracy of 1 μm. During the scanning measurements, we use movement steps between 0.5 and 1.0 mm, as a trade-off between the measurement resolution and the illumination time of the sample. In order to avoid damage to the samples, the illumination time for each scanning measurement was restricted to 3 min, limiting the scanning resolution of the surface measurements.

After the excitation of the sample, the fluorescent signals are captured by a collimating lens, coupled into a broadband optical fibre (UVIR600 fibre of Avantes, transmitting light between 250 and 2500 nm) and guided towards the spectrum analyser (**Figure 2(d)**). We use a collimating lens in combination with a large fibre core diameter (600 μm) to obtain a total acceptance angle of 4.1°, allowing to capture the weak fluorescence signals of a surface area of 39 mm^2 (corresponding to the area within the circular aperture of the sample holder). In front of the detecting fibre, we additionally mounted an interference-based filter to eliminate the excitation light. During the OPIF measurements, we implement a long-wave pass filter transmitting from 405 nm onwards (Semrock 405 nm EdgeBasic filter BLP01-405R-25), while during the TPIF measurements, a short-wave pass filter with a cut-on wavelength of 650 ± 5 nm is used (Newport 10SWF-650-B). Without the use of an optical filter, the excitation signals would saturate the measured fluorescence spectrum, leading to incorrect measurement data. Furthermore, we make use of the AvaSpec2048 spectrum analyser, which is able to measure the spectrum between 300 and 1100 nm with a resolution of 8 nm. This instrument has a wide entrance slit of 200 μm that allows capturing the weak fluorescence signals.

To obtain the absolute fluorescence spectrum of a sample, the measured spectra are corrected by a transfer function, taking the wavelength-dependent transmittance of the optical fibre and the sensitivity of the detector inside the spectrum analyser into account. This transfer function was obtained after the measurement of a calibrated light source with an a priori known absolute spectrum. We first connected the calibration light source to the spectrum analyser by using the UVIR600 fibre. Following, we measured the spectrum of the calibration light source and compared the measured photon counts for each wavelength with the specified output power of the manufacturer, in μW/cm^2. Using this measurement data, the Avasoft 8 software is able to calculate the absolute fluorescence spectrum (in μW/cm^2/nm) of a sample (I_{sample}), by using the following equation [21]:

$$I_{sample}(\lambda) = \text{Caldata}(\lambda)\frac{\text{sample}(\lambda) - \text{dark}(\lambda)}{\text{refcal}(\lambda) - \text{darkcal}(\lambda)}\frac{\text{Int}_{cal}}{\text{Int}_{sample}} \quad (1)$$

With Caldata(λ) the intensity of the calibrated light source in μW/cm^2, obtained from the manufacturer; sample(λ) the measured spectrum of the sample, in A/D counts; dark(λ) the dark signal during the sample measurements, in A/D counts; refcal(λ) the measured spectrum of the calibrated light source, in A/D counts; darkcal(λ) the dark signal during the measurements with the calibrated light source, in A/D counts; Int_{cal} the integration time during the measurements of the calibrated light source; and Int_{sample} the integration time during the measurements of the sample.

During the fluorescence characterization measurements, a study of the fluorescence intensity as function of the excitation power is essential. Therefore, the excitation laser power is tuned by mounting attenuation filters behind the laser output, in front of the tilted mirror (mirror indicated in **Figure 2(d)**; the attenuation filters are not presented in this picture).

2.2. Selection of the excitation wavelengths

To maximize the mycotoxin fluorescence intensity, we select an excitation wavelength that is strongly absorbed by the mycotoxins, since an increasing amount of absorbed photons gives rise to an increasing amount of excited electrons and thus also to an enlarged number of fluorescent photons. Furthermore, by using an excitation wavelength with a strong mycotoxin absorbance, the influence of the mycotoxin fluorescence emission onto the natural fluorescence spectrum of the food products can be maximized. When considering the absorption spectrum of aflatoxin B1, ochratoxin A and zearalenone, the strongest absorbances are observed within the 200–400-nm spectral region (**Figure 3(a)**) [9]. Specifically, aflatoxin B1 shows the strongest absorbance in the range between 200 and 275 nm and around 365 nm. Ochratoxin A has a strong absorbance between 200 and 250 nm and around 335 nm, while zearalenone shows a strong absorbance in the 200–350 nm spectral range.

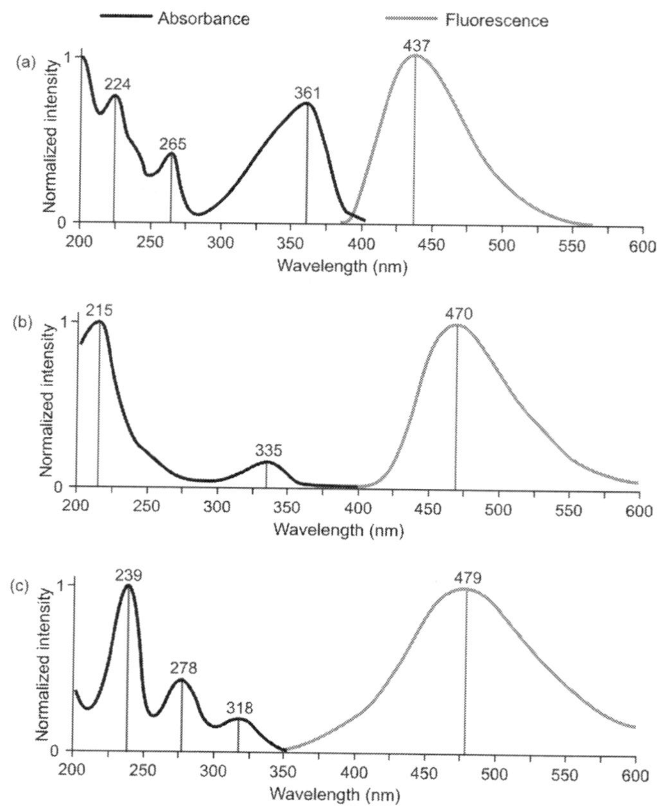

Figure 3. Absorbance and fluorescence spectrum of fluorescent mycotoxins: (a) aflatoxin B1, (b) ochratoxin A, (c) zearalenone.

In addition to the absorbance of the mycotoxins, the intrinsic fluorescence of the food products needs to be taken into account. Food products often contain several proteins that show fluorescence emission after excitation with certain wavelengths [22]. For example, maize contains the fluorescent proteins tryptophan (Trp), tyrosine (Tyr) and phenylalanine (Phe), which all show a strong absorbance in the 200–300 nm wavelength range (**Figure 4**). If we would excite maize kernels with a wavelength between 240 and 275 nm, the fluorescence of these proteins would disturb our measurements. The proteins show no fluorescence after excitation with 355–417 nm light, since they only show a weak absorbance from 300 nm onwards.

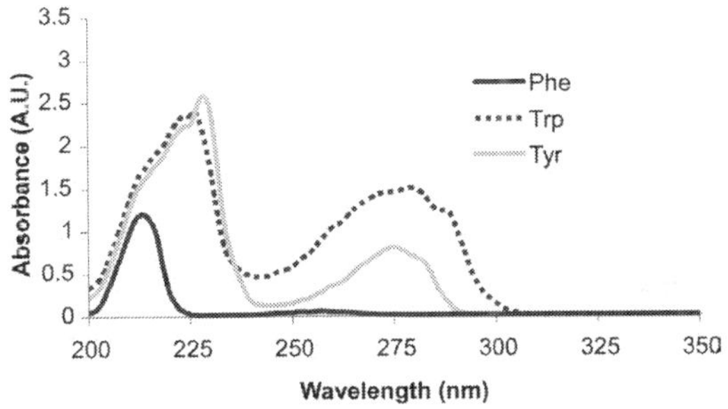

Figure 4. Absorbance spectrum of different fluorescent proteins in maize: phenylalenine (Phe), tryptophan (Trp) and tyrosine (Tyr) [22].

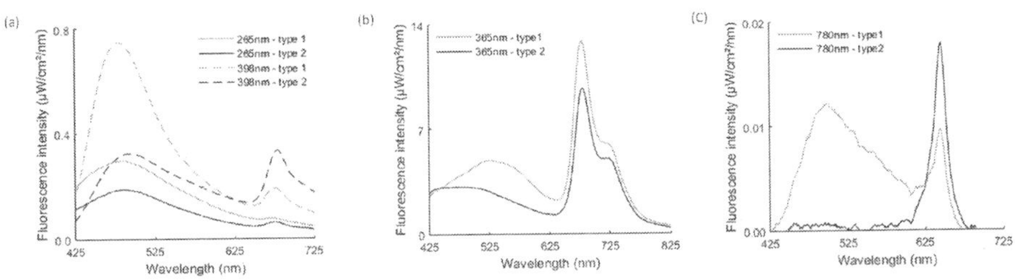

Figure 5. Intrinsic fluorescence of healthy food products: (a) OPIF of coffee beans after excitation with 265 and 398 nm, (b) OPIF of pistachio nuts after excitation with 365 nm, (c) TPIF of pistachio shells after excitation with 780 nm.

Furthermore, also the natural variation within food products should be examined during the fluorescence measurements. Due to the large internal variation in density, texture and substituents concentration, various batches of a product type can show different fluorescence spectra or intensities. As an illustration, we present the intrinsic fluorescence of healthy coffee beans and pistachio nuts, showing significant differences within the fluorescence spectra of different batches (**Figure 5**). When illuminating two batches of coffee beans with 265 and 398 nm laser light, different ratios of the local fluorescence maxima at 480 and 680 nm can be observed for each batch (**Figure 5(a)**). Two varieties of pistachio nuts, after illumination with 365 nm, show a different fluorescence spectrum between 425 and 625 nm, with a maximal difference at 530 nm (**Figure 5(b)**). Both types of pistachio nuts do show chlorophyll fluorescence, present between 625 and 825 nm. Considering the outer pistachio shells, after illumination with 780 nm, a clear contrast between the fluorescence of both pistachio types can be identified between 425 and 625 nm (**Figure 5(c)**). The fluorescence spectrum of the first type shows a clear two-photon-induced fluorescence signal between 425 and 625 nm and a chlorophyll fluorescence maximum at 645 nm, while the second type only shows chlorophyll fluorescence. As a result, to account for the natural variation, different independent sets of products should always be compared.

Finally, when selecting the excitation wavelengths, we keep in mind the commercial available laserlines. We prefer using commercially available wavelengths to enable the integration of our developed optical detection criterion into industrial scanning-based sorting devices.

When considering our case study, the sensing of the aflatoxin contamination in maize kernels, we use an excitation wavelength of 365 nm during the OPIF measurements, since at this wavelength, the absorbance of the aflatoxin B1 is maximized, while the absorbance of the proteins is minimized. Consequently, to study the TPIF spectrum, excitation wavelengths close to 730 nm are required. In addition, when studying the maize kernels, we also investigate the fluorescence spectrum after excitation with 750 and 780 nm, to study the influence of the matrix constituents onto the fluorescence spectrum.

3. Case-study: localized detection of the aflatoxin contamination

As a case study, we focus on the detection of the localized aflatoxin contamination in individual maize kernels. Specifically, within this section, we present an in-depth explanation of our research methodology, giving rise to the development of an accurate optical sensing criterion. We first characterize the OPIF and TPIF spectra of pure aflatoxin, after which we investigate the fluorescence properties of the healthy and contaminated maize kernels.

3.1. Characterization of the aflatoxin fluorescence spectrum

Aflatoxin is one of the most dominant and toxic mycotoxins. It is produced by the fungi *Asperigullus flavus* and *Asperigullus parasiticus*. To perform a basic characterization of the fluorescence properties of aflatoxin, we measured the OPIF and TPIF spectrum of pure aflatoxin B1 powder (98% or better purity, produced by the *A. flavus* fungi) that we purchased from Sigma-Aldrich. It is a white to yellow crystalline powder that we measured in its solid state.

We present the mean OPIF and TPIF spectra of pure aflatoxin B1 powder (**Figure 6**). The OPIF spectrum is obtained after excitation with 365 nm, with an excitation power density of 42 mW/mm^2, while the TPIF spectrum is measured after excitation with 730 nm, with an excitation power density of 14.3 W/mm^2. Both fluorescence spectra show their maximal fluorescence intensity at 428 nm, which corresponds with the expected aflatoxin B1 fluorescence maximum (**Figure 3(a)**). The OPIF spectrum shows higher fluorescence intensities than the TPIF spectrum. The maximal OPIF intensity (at 428 nm) is equal to 1.68 ± 0.93 $\mu W/cm^2$, while we observe a maximal TPIF intensity of 0.20 ± 0.09 $\mu W/cm^2$. The measured TPIF intensity is hence 10 times weaker than the OPIF intensity. The shape of the TPIF spectrum shows a narrower peak than the OPIF spectrum due to the more selective excitation during two-photon absorption than during one-photon absorption. Furthermore, in the TPIF spectrum at 365 nm, we observe a second-harmonic generation signal. This peak is created by the recombination of two illumination photons to a new photon with the double energy. Around 400 nm, the OPIF spectrum shows a steeper slope than the TPIF spectrum, because of the presence of a long-wave pass filter suppressing light below 405 nm during the OPIF measurements.

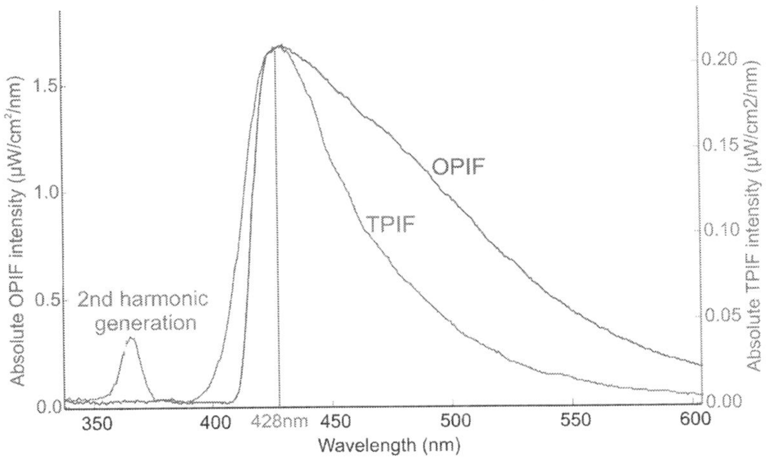

Figure 6. OPIF and TPIF spectra of the pure aflatoxin B1 powder, after excitation with 365 and 730 nm, respectively.

To characterize the aflatoxin B1 fluorescence spectrum, we studied the integrated fluorescence intensity as function of the excitation power, while maintaining a constant spot diameter of 951 μm during the OPIF measurements and 231 μm during the TPIF measurements (**Figure 7**). For each excitation power, we integrated the measured mean fluorescence spectrum between 400 and 600 nm to obtain the depicted integrated fluorescence intensity. Studying the integrated OPIF intensity as function of the excitation laser power, we observe a linear relationship, representing the linear one-photon absorption. However, at higher excitation powers, starting from 35 mW onwards, the integrated OPIF intensity deviates from the linear relationship and starts saturating. In the saturated region, the maximum fluorescence intensity is reached since then all electrons are excited to the higher energy state. The TPIF intensity shows a quadratic dependence on the excitation laser power, confirming the occurrence of non-linear two-photon absorption. Moreover, we can observe that TPIF requires much higher excitation powers, starting from 100 mW onwards. The variation of the measured data is less than 10 and 12% during the OPIF and TPIF measurements, respectively. Both the linear and exponential fits show an adjusted r-square value of 0.98 and 0.97, respectively, ensuring that the fitted function is a good representation of the measured data.

TPIF requires a higher excitation power density than OPIF [23]. The minimal excitation power density for OPIF is 5 mW/mm^2, while TPIF requires a minimal power density of 2 W/mm^2. As a result, the measurement constraints are more severe for TPIF than for OPIF. A high excitation power, a small spot size and a sensitive detector are indispensable to measure TPIF signals. However, considering the implementation in practical applications, TPIF uses NIR excitation wavelengths which are more widely commercially available than the required UV excitation wavelengths for OPIF.

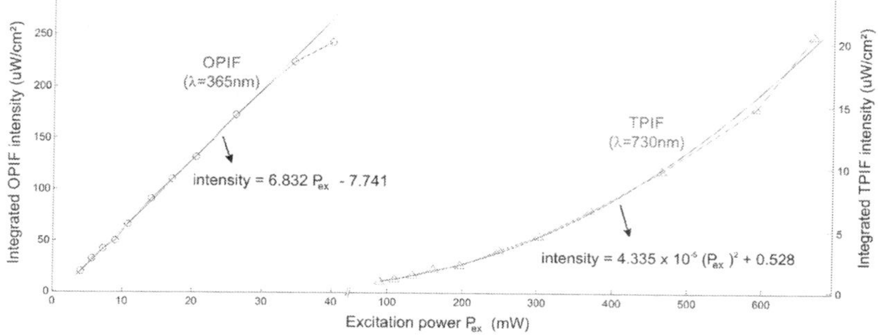

Figure 7. Intensity of the integrated OPIF spectrum increases linearly with the excitation power, while the intensity of the integrated TPIF spectrum shows a quadratic dependence on the excitation power.

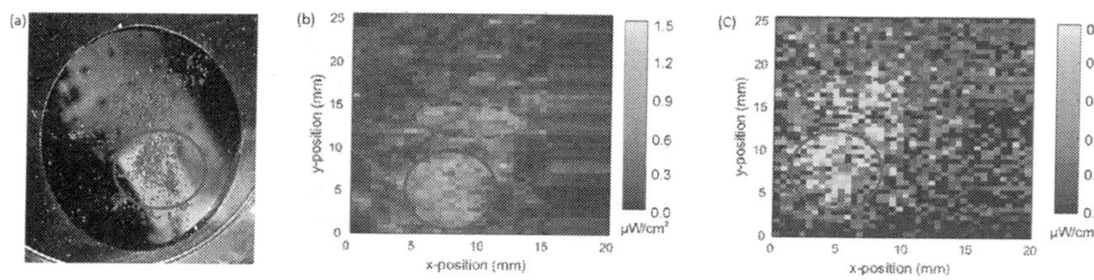

Figure 8. Scanning of the aflatoxin B1 powder, visualizing the location of the emitted fluorescence signals: (a) aflatoxin B1 power, (b) OPIF surface plot and (c) TPIF surface plot, indicating the localized fluorescence intensity at 428 nm.

To validate the automated screening of the samples, a scanning of the pure aflatoxin B1 powder is performed (**Figure 8**). Particularly, we consider a Petri dish on which the aflatoxin B1 powder is disposed in two separate areas, of which we screen an area of 20.0 mm by 25.0 mm, with a resolution of 0.5 mm. By adjusting the X and Y position of the automated translation stages, while measuring the fluorescence signal for every position, the aflatoxin powder can be identified (indicated by the solid and dotted circles in **Figure 8**). During both the OPIF and TPIF measurements, the aflatoxin B1 fluorescence is detected. The encircled regions in the surface plots indicate the areas from which a minimum fluorescence intensity of 0.8 and 0.1 μW/cm^2 at 428 nm is detected during the OPIF and TPIF measurements, respectively. The aflatoxin powder consists of granules positioned at two areas on the Petri dish (**Figure 8(a)**). If a small aflatoxin granule is illuminated, we detect a fluorescence signal. However, we obtain a smoother surface plot for OPIF than for TPIF, due to the larger spot diameter of the illumination beam (951 μm) during the OPIF measurements. During the OPIF measurements, most illumination positions around the powder excite one or multiple aflatoxin granules, since the surface area in-between the granules is smaller than the spot diameter. When using a smaller spot diameter, like during the TPIF measurements (231 μm), no fluorescence signal can be captured when the illuminating beam is positioned in-between the aflatoxin granules. Consequently, a more fragmented surface plot is obtained (**Figure 8(c)**).

Generally, our fluorescence measurements of the pure aflatoxin B1 powder correspond with the theoretical characteristics, validating a correct measurement of the fluorescence signals. In addition, the powder could be successfully visualized, indicating a good operation of the automated scanning of the samples.

3.2. Sensing of the localized aflatoxin contamination in maize kernels

We investigate the sensing of aflatoxin-contaminated maize kernels by using fluorescence spectroscopy. We first give an overview of the investigated samples. Following, we study the OPIF and TPIF spectra of healthy and contaminated maize kernels, including the investigation of the intrinsic fluorescence spectra and its dependency on the excitation wavelength. Based on this study, we develop a detection criterion enabling to sense the contaminated kernels. Finally, the detection of the localized contamination areas onto the kernel's surfaces are studied and visualized with surface plots.

3.2.1. Overview of the investigated samples

We consider two different independent maize batches, each with a healthy and contaminated subsample. One healthy and one contaminated maize batch were harvested in 2012 and provided by an Italian company. The second set of healthy and contaminated maize samples was collected from Croatian farmers, after the harvest in 2013. From each maize batch, a subsample of 25 g was drawn for the analytical determination of the aflatoxin contamination level. Each sample was chemically analysed using the ToxiQuant mycotoxin testing system of ToxiMet [24]. Considering the Italian maize kernels, the contaminated sample shows 72.1 ppb aflatoxin B1 and the healthy one 0.0 ppb. The maize samples from Croatia show approximately the same aflatoxin B1 contamination level, being 78.9 ppb for the contaminated sample and 0.8 ppb for the healthy sample. After our measurements were performed, the contamination level of the maize samples was confirmed by the CODA-CERVA, the Belgian Reference Laboratory for Mycotoxins. Comparing the contamination levels with the international regulations, both the Italian and Croatian contaminated batches can be considered as highly contaminated. The European Commission states the maximum allowed total aflatoxin concentration in maize to be 10 ppb, while the USA food safety regulations included a limit of 20 ppb of total aflatoxins in all food products [12].

During our fluorescence measurements, we investigated the fluorescence spectra of 45 healthy and contaminated Croatian maize kernels to obtain a statistic relevant distribution of our measurement data. To observe the influence of the sample type and the harvest environments on the fluorescence spectra, we also measure the fluorescence spectra of 15 healthy and contaminated Italian maize kernels. The Italian company provided only small maize batches, which limits the number of studied maize kernels, but is sufficient to monitor the environmental influences on the spectra. Because aflatoxin is sensitive to light, the samples are permanently stored in a dark enclosure to reduce the environmental influences onto the measurements. In-between the measurements, the samples are stored in a fridge to avoid cross-contamination.

3.2.2. *OPIF and TPIF spectra of the maize kernels*

To investigate the optical detection of aflatoxin B1, we study the OPIF and TPIF spectra of healthy and contaminated maize kernels of both the Croatian and Italian maize batches. We measured the OPIF spectrum after excitation with 365 nm, with an excitation power density of 317 mW/mm^2 (**Figure 9(a)**). The TPIF spectra are measured after excitation with 730, 750

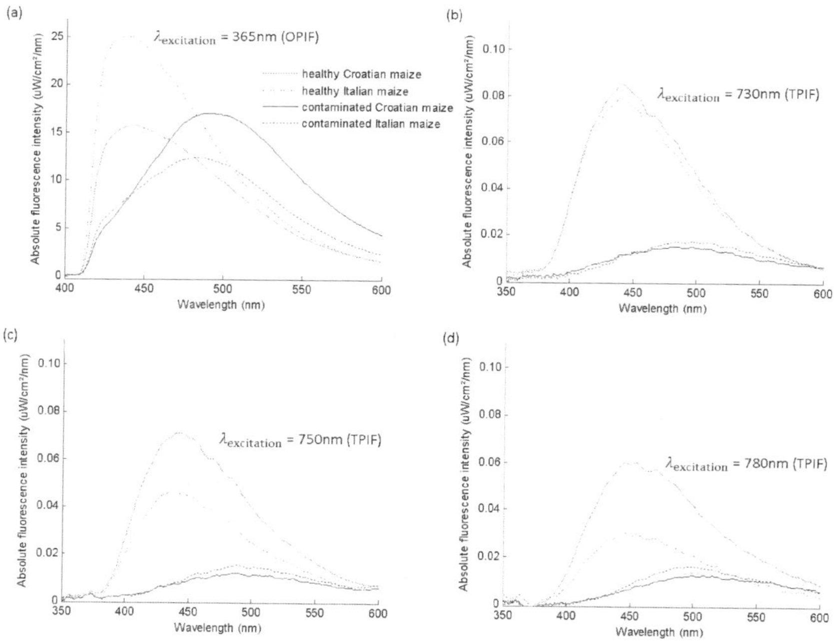

Figure 9. Fluorescence spectra of the healthy and contaminated maize kernels of the Croatian and Italian maize batches: (a) OPIF spectrum after excitation with 365 nm; TPIF spectrum after excitation with (b) 730 nm, (c) 750 nm and (d) 780 nm.

and 780 nm, with an excitation power density of 26.2, 29.1 and 36.0 W/mm^2, respectively (**Figure 9(b)**, **(c)** and **(d)**). During the TPIF measurements, we investigate the fluorescence spectrum after excitation with multiple wavelengths to monitor the influence of the illumination wavelength onto the fluorescence emission of the maize kernels. To maximize the signal to noise ratio, we illuminate the maize kernels with the maximal output power of the titanium-sapphire laser and harmonic generating unit. However, we limited the measurement time to 20 and 200 ms for the OPIF and TPIF measurements, respectively, to avoid damage to the sample surface. For every maize batch, the mean fluorescence spectrum is depicted. Both the healthy and the contaminated samples show a fluorescence signal due to the intrinsic fluorescence of the maize kernels (**Figure 9**). On the basis of these measurements, a comprehensive evaluation can be made by comparing the fluorescence spectra of the Italian and Croatian maize batches, by evaluating the performance of OPIF and TPIF and by defining the spectral differences between the healthy and contaminated samples.

Excitation wavelength	Maximum fluorescence intensity			
	Croatian maize (μW/cm²/nm)		Italian maize (μW/cm²/nm)	
	Healthy	Contaminated	Healthy	Contaminated
365 nm	15 ± 8	17 ± 10	25 ± 17	13 ± 9
730 nm	0.09 ± 0.05	0.02 ± 0.01	0.08 ± 0.05	0.02 ± 0.01
750 nm	0.07 ± 0.03	0.02 ± 0.01	0.05 ± 0.02	0.02 ± 0.01
780 nm	0.06 ± 0.04	0.02 ± 0.01	0.03 ± 0.01	0.02 ± 0.01

Table 1. Measured maximum fluorescence intensity and its variation for the different excitation wavelengths.

Considering the different maize batches, the fluorescence spectra of the Croatian and Italian maize correspond well within the variances of the measurements (**Table 1**). For all excitation wavelengths, both maize types emit corresponding fluorescence signals, showing a similar spectral shape in the same wavelength regions. The present intensity variances are caused by the differences within the molecular structure of the maize kernels. Because maize is a natural product, the maize kernels show a large internal variation in surface shape, density and natural composition. In correspondence with the measurements on the pure aflatoxin B1 powder, the fluorescence intensity induced by one-photon excitation is much stronger than the intensity induced by two-photon excitation. Specifically, the OPIF signal is approximately 500 times stronger than the TPIF signal.

Comparing the fluorescence spectra of the healthy and contaminated maize kernels, we observe significant differences in intensity and emission wavelength. We do not observe the aflatoxin fluorescence directly, but we measure its influence on the intrinsic fluorescence of the maize. In the contaminated samples, the aflatoxin is bonded to the different natural constituents of the healthy maize, changing the molecular structure of the constituents and therefore influencing the intensity and wavelength of the emitted fluorescence signals. During both the OPIF and TPIF measurements, we observe higher fluorescence intensities for the healthy maize kernels than for the contaminated ones (**Table 1**). The intensity contrast is the largest for the

TPIF spectra, since the different bonds inside the maize are more selectively excited during the TPIF process than during the OPIF one. The largest intensity differences between the healthy and contaminated samples are observed after excitation with 730 nm, where the mean fluorescence intensity of the healthy maize is four times stronger than the mean fluorescence intensity of the contaminated maize. In contrast, the minimum intensity differences are observed after excitation with 365 nm. The observed large intensity variations within the healthy and contaminated samples are caused by a combination of the curved surface of the maize kernels and the non-homogeneous presence of the toxins inside the maize.

In addition to the fluorescence intensity differences, we observe a wavelength shift between the fluorescence maxima of the healthy and contaminated maize samples (**Table 2**). Generally, the OPIF and TPIF spectra show a wavelength shift of approximately 50 nm. To compare the obtained wavelength shift for the different excitation wavelengths, we calculate the class difference for both the Croatian and Italian maize. The class difference (D) is a measure for the difference between the average values (μ) of two product types, taking the standard deviation (σ) and the amount of measured samples (N) into account [25]:

$$D = \frac{\left|\mu_{\text{contaminated}} - \mu_{\text{healthy}}\right|}{\sqrt{\frac{\sigma^2_{\text{contaminated}}}{N_{\text{contaminated}}} + \frac{\sigma^2_{\text{healthy}}}{N_{\text{healthy}}}}} \tag{2}$$

Excitation wavelength	Dominant emission wavelength Croatian maize (nm)		Class difference Croatian maize	Dominant emission wavelength Italian maize (nm)		Class difference Italian maize
	Healthy	Contaminated		Healthy	Contaminated	
365 nm	443 ± 5	497 ± 15	9.7	442 ± 5	486 ± 6	10.8
730 nm	441 ± 3	496 ± 20	6.1	441 ± 3	496 ± 16	4.1
750 nm	444 ± 7	497 ± 24	3.8	440 ± 5	490 ± 17	2.7
780 nm	450 ± 10	503 ± 19	5.2	450 ± 7	498 ± 10	4.8

Table 2. Emission wavelength at maximum fluorescence intensity and its variation for the different excitation wavelengths.

The OPIF measurements show the largest class differences, indicating the largest optical contrast between the healthy and contaminated maize kernels. The variation of the class differences between the OPIF and TPIF measurements are mainly caused by their different variances. The TPIF spectra show a larger variance than the OPIF spectra, which decreases their class difference. TPIF signals show a weaker intensity, resulting in a stronger relative noise signal and therefore a lower signal to noise ratio.

To validate the influence of the aflatoxin onto the intrinsic fluorescence spectrum of the maize, we tune the excitation wavelength towards the wavelengths for which aflatoxin B1 shows a weaker absorbance. We measured the fluorescence spectrum of healthy and contaminated maize kernels after excitation with 405 and 810 nm laser light, during the OPIF and TPIF measurements, respectively (**Figure 10**). Both the OPIF and TPIF measurements show a small difference between the fluorescence spectrum of the healthy and contaminated samples. We measure a wavelength shift of 8 ± 6 nm and 26 ± 12 nm, after excitation with 405 and 810 nm, respectively. Moreover, we observe only small differences between the fluorescence intensities. Because aflatoxin B1 shows a weak absorbance at 405 nm (**Figure 3(a)**), it only minimally influences the intrinsic fluorescence spectrum of the maize kernels.

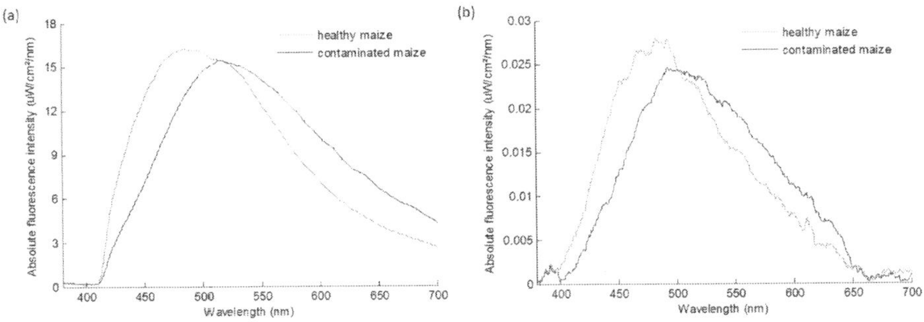

Figure 10. Fluorescence spectra of the healthy and contaminated maize kernels: (a) OPIF spectrum after excitation with 405 nm and (b) TPIF spectrum after excitation with 810 nm.

Summarized, both OPIF and TPIF show a spectroscopic contrast between the healthy and contaminated maize kernels. In the next step, we use the above quantitative evaluation to define and test our optical detection criterion.

3.2.3. *Development of an optical detection criterion*

To obtain an accurate optical detection criterion, taking the shape, intensity and maximum emission wavelength of the fluorescence spectra into account, we examine the integral of the fluorescence spectra in various wavelength intervals. Keeping in mind the wavelength ranges of the fluorescence spectra and the maximum fluorescence intensities, both depicted in **Figure 9** and presented in **Tables 1** and **2**, we obtain an optimal contrast between the healthy and contaminated samples by considering the ratio of the integrated fluorescence intensity from 475 to 550 nm and the integrated fluorescence intensity from 400 to 475 nm. When visualizing the ratio of these integrals for each maize kernel, a clear distinction between the healthy and contaminated samples is visible (**Figure 11**). Particularly, the ratios show a minimum contrast of 0.24, 0.13, 0.19, 0.25 and a maximum contrast of 2.15, 1.93, 2.87, 3.39 for excitation with 365, 730, 750 and 780 nm, respectively. The largest difference between the mean ratio of the healthy and contaminated samples is observed after excitation with 780 nm. Considering the class differences of the integral ratio between the healthy and contaminated samples, we obtain a value of 116.2, 111.1, 48.8 and 54.9 for excitation with

365, 730, 750 and 780 nm, respectively. The largest class difference can thus be found after excitation with 365 nm. Regarding the TPIF measurements, the largest class difference is obtained after excitation with 780 nm. The TPIF measurements show generally a smaller class difference than the OPIF ones, due to their larger measurement variation. However, the contaminated samples can still be properly identified with TPIF, allowing using NIR excitation wavelengths, instead of the UV wavelengths for OPIF.

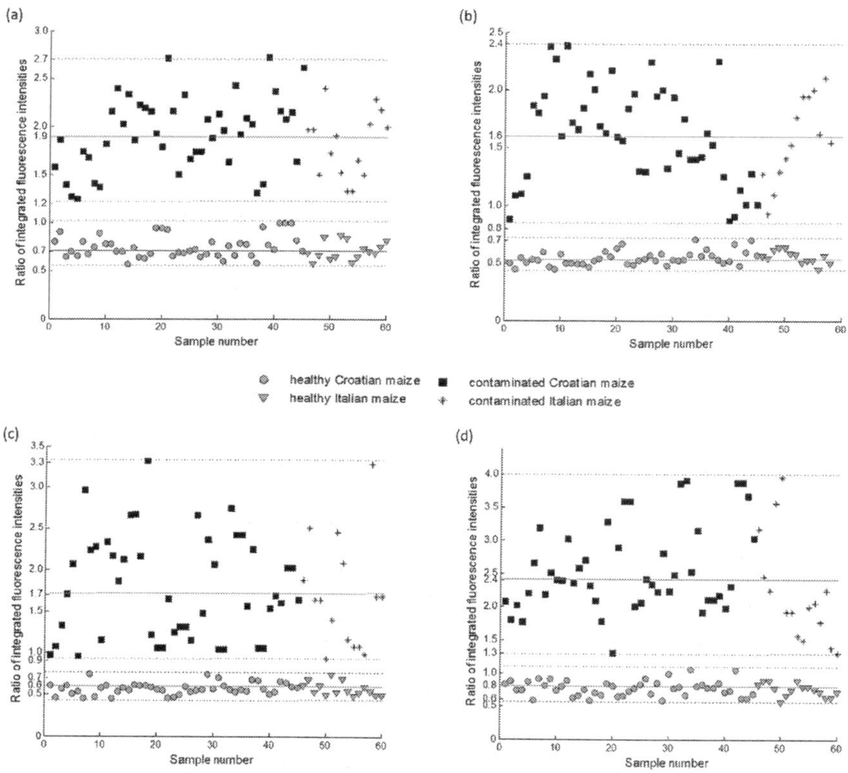

Figure 11. Contrast between the fluorescence spectra of the healthy and contaminated maize samples, visualized by the ratio of the integrated fluorescence spectrum from 475 nm until 550 nm to the integrated fluorescence spectrum from 400 nm until 475 nm, after excitation with (a) 365 nm, (b) 730 nm, (c) 750 nm and (d) 780 nm.

The ratios of the integrated fluorescence intensities of the contaminated samples generally show a larger variation than the ratios of the healthy samples, because of the localized presence of the aflatoxin and the variable aflatoxin concentration in the maize kernels. Depending on the illumination position on the maize kernel's surface, a different contamination level may be present, influencing the wavelength and intensity of the fluorescence spectrum. To visualize the localized aflatoxin contamination on the maize kernel's surfaces, scanning measurements needed to be performed.

3.2.4. Monitoring of the localized presence of the aflatoxin

To monitor the localized presence of the aflatoxin, we scanned the maize kernels by using the automated translation stages. We investigated the localized aflatoxin contamination when illuminating the samples with 365 and 780 nm laser light, since these wavelengths showed the largest spectroscopic differences during the OPIF and TPIF measurements, respectively [26].

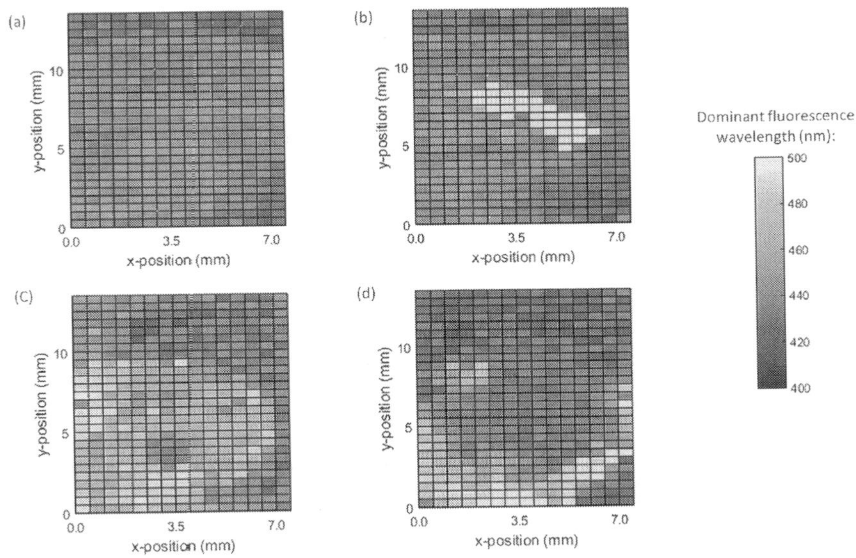

Figure 12. Visualization of the localized aflatoxin contamination, measured during OPIF, by mapping the dominant fluorescent emission wavelength along the kernel's surface: (a) healthy kernel, (b) kernel with a small localized contamination area, (c) kernel with homogeneous contamination and (d) kernel with a medium to large contamination at the edge.

During the OPIF measurements, we scanned the maize kernels with a resolution of 0.5 mm. For every sample, we recorded 405 spectra, each of which corresponding with a different illumination position of the laser beam. Considering the fluorescence wavelength showing the maximum fluorescence intensity, different regions in the kernels can be identified (**Figure 12**). In correspondence with the previous measurements, the fluorescence spectra of the healthy maize kernels show their maximum fluorescence intensity between 400 and 450 nm (**Figure 12(a)**). The contaminated areas show their maximum fluorescence intensity at wavelengths longer than 450 nm. By interpreting the surface plots, various areas containing different contamination levels can be observed on different positions of the kernel's surface (**Figure 12(b)**, **(c)** and **(d)**). As an example, the contaminated kernels can show a small localized contaminated area, a homogeneous contamination along the maize kernel or a medium to large contamination at the edge of the kernel. When considering the ratio of the integrated fluorescence spectra, specifically the ratio of the integrated fluorescence intensity between 475 and 550 nm and the integrated fluorescence intensity between 400 and 475 nm, a similar observation can be made (**Figure 13**). In correspondence with **Figure 11**, the high-contaminated areas show the largest ratios. Furthermore, the gradient of the contamination level is clearly visible. When considering **Figure 13(b)**, two highly contaminated areas surrounded by a medium-contaminated area can be identified, while **Figure 13(d)** shows a highly contaminated area at the edge that flows into a healthy area in the centre of the kernel. Concluding, localized contaminated areas can be accurately identified and characterized by mapping the integral ratio on the screened kernel's surfaces.

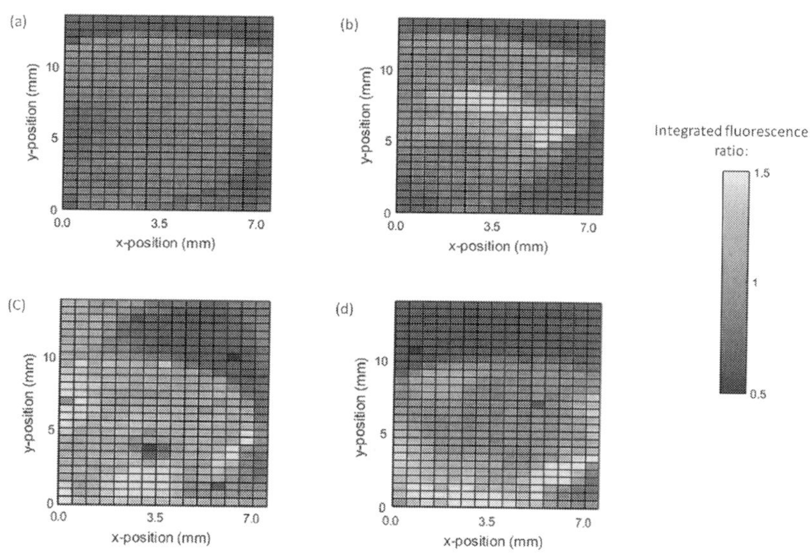

Figure 13. Visualization of the localized aflatoxin contamination during OPIF, by mapping fluorescence ratios (integrated fluorescence intensity between 475 and 550 nm divided by the integrated fluorescence intensity between 400 and 475 nm) along the kernel's surface: (a) healthy kernel, (b) kernel with small localized contamination area, (c) kernel with homogeneous contamination and (d) kernel with a medium to large contamination at the edge.

Considering the TPIF measurements, the maize kernels could only be partially scanned with a lower resolution of 1 mm (instead of the 0.5 mm resolution used during the OPIF scanning measurements). When accurately scanning a complete maize kernel's surface during the TPIF measurements, the kernel's surface would be damaged, due to the required long illumination time and high illumination power density. Specifically, the illumination time is significantly longer during the TPIF measurements than during the OPIF measurements (approximately 10 times longer) due to the longer measurement time of the spectrum analyser required to capture the weak TPIF signals. Consequently, to avoid measuring the influence of damaging effects, the kernels are measured with a lower accuracy (**Figure 14**). Studying the corresponding surface plots, regions with a high and low contamination level can still be observed. Specifically, the regions with a dominant fluorescence wavelength longer than 450 nm can be considered as contaminated. Furthermore, the longer the fluorescence wavelength, the higher the contamination level on the kernel's surface. The maize kernel depicted in **Figure 14(a)** shows generally a lower contamination than the one depicted in **Figure 14(b)**.

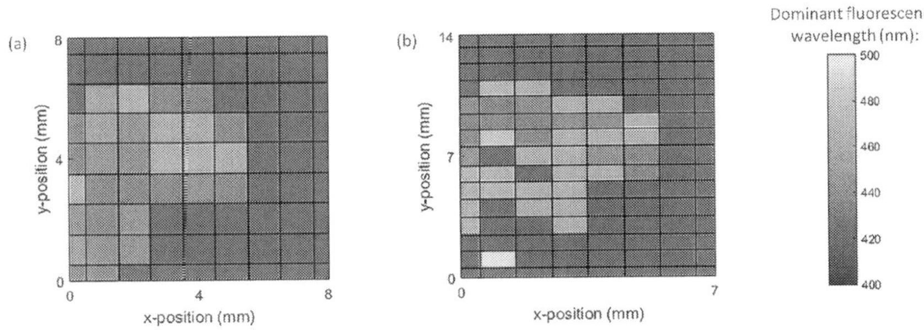

Figure 14. Visualization of the localized aflatoxin contamination during TPIF, by mapping the dominant fluorescent emission wavelength along the kernel's surface: (a) low-contaminated kernel and (b) high-contaminated kernel.

The scanning plots demonstrate a successful monitoring of the localized aflatoxin contamination. In contrast to the destructive chemical analyses that measure the mean contamination of a certain number of maize kernels, we are able to measure the localized contamination in individual unprocessed food products. By monitoring the localized contamination level of the kernels, we allow to sort-out the maize kernels that contain a low mean contamination level but feature a small area with a high contamination level, which would not be detected by any type of chemical analyses. As a result, our developed optical detection methodology contributes to an improved food safety.

4. Extensibility of our research methodology to the detection of ochratoxin and zearalenone

Next to aflatoxins, also ochratoxin and zearalenone may be present in contaminated food products. Ochratoxins are mainly produced by the fungi *Penicillium verrucosum* and *Aspergillus ochraceus*, and occur most frequently in cereals, fruits and coffee beans. Cereal and cereal products are considered as the main contributors to the ochratoxin intake in Europe [27]. There exist five types of ochratoxins, of which only ochratoxin A and ochratoxin B occur naturally in food products. Particularly, ochratoxin A is considered as being the most common and most toxic ochratoxin [28]. The European regulations limit the maximum allowed ochratoxin A concentration from 2 to 15 ppb, depending on the commodity [12]. Zearalenone, on the other hand, is a mycotoxin that can be produced by several fungi of the genus *Fusarium*, like *Fusarium graminearum*, *Fusarium culmorum* and *Fusarium cerealis* [29]. It mostly occurs in cereals as barley, oats, wheat, rice and maize. The European limitation on the zearalenone concentration ranges between 20 and 400 ppb [12].

In this section, we investigate the fluorescence characteristics of ochratoxin and zearalenone and discuss the applicability of our developed measurement methodology onto these toxins. Specifically, we performed measurements on pure ochratoxin A and zearalenone powder, purchased from Sigma-Aldrich.

4.1. Ochratoxin A detection

Ochratoxin A shows theoretically a strong absorbance in-between 200 and 275 nm and around 335 nm, while its theoretical fluorescence emission wavelength is situated around 475 nm (**Figure 3(b)**). The chemical analyses of ochratoxin A by using liquid chromatography in combination with fluorescence detection is widely discussed in literature [30–33]. During the execution of the fluorescence analysis, excitation wavelengths of 330, 332, 333 and 390 nm are commonly used, while the fluorescent detection occurs at 440, 460 and 476 nm.

Based on the absorbance spectrum of ochratoxin, while keeping in mind the wavelengths that are available with our fluorescence measurement configuration, we excite the ochratoxin A powder with 244, 250, 260, 356, 365, 375 and 395 nm during the OPIF measurements (**Figure 15**). The 244, 250 and 260 nm wavelengths are generated with the third-harmonic-generating crystal, while the 356, 365, 375 and 395 nm wavelengths are generated by the second-harmonic-generating crystal. To be able to compare the captured fluorescence intensities, we rescaled the fluorescence spectra to an excitation power density of 1 mW/mm^2 for all excitation wavelengths (illumination powers used during the measurements varied between 60 and 450 mW, depending on the excitation wavelength). For all excitation wavelengths, the measured fluorescence spectra indicate a maximum around 480 nm. Furthermore, the highest fluorescence intensity is obtained after excitation with 356 nm. In correspondence with the theoretical absorbance spectrum, the fluorescence intensity decreases when tuning the excitation wavelength from 244 to 260 nm and from 356 to 395 nm.

To measure the TPIF spectrum of the ochratoxin A powder, we used an excitation wavelength of 710 nm (with an excitation power of 562 mW and an excitation spot diameter of 375 μm), since the OPIF measurements showed the strongest fluorescence intensities after excitation with 356 nm. The obtained TPIF spectrum occurs in the same wavelength region as the OPIF spectrum, with a maximum fluorescence intensity at 480 nm (**Figure 16(a)**). However, the TPIF intensity is 1000 times weaker than the OPIF intensity and shows a larger noise contribution. To experimentally validate that we are measuring the TPIF spectrum of ochratoxin A, the integrated fluorescence intensity, obtained by integrating the fluorescence spectrum between 400 and 600 nm, is studied as a function of the excitation power (**Figure 16(b)**). Similar as for aflatoxin B1, the integrated fluorescence intensity shows a quadratic dependence on the excitation power, corresponding with the non-linear behaviour of two-photon absorption.

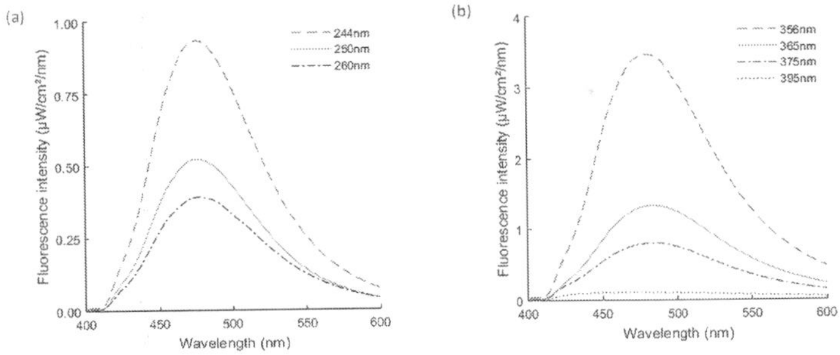

Figure 15. OPIF spectra of ochratoxin A: (a) spectra after excitation with 244, 250, 260 nm and (b) spectra after excitation with 356, 365, 375 and 395 nm.

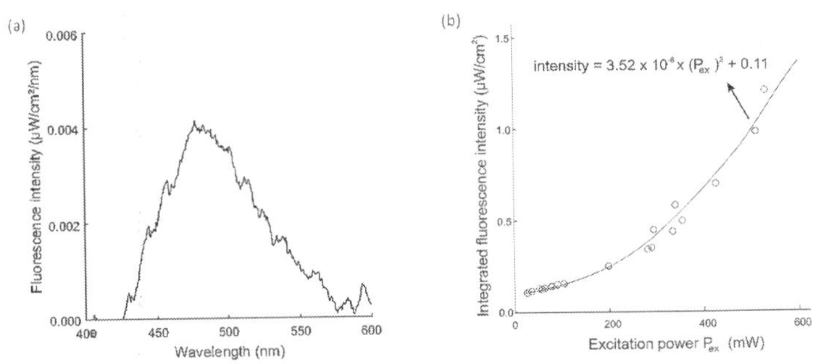

Figure 16. Study of TPIF of ochratoxin A: (a) TPIF spectrum after excitation with 710 nm and (b) quadratic relationship between the integrated fluorescence intensity and the excitation power, validating the two-photon absorption.

The fluorescence signal of the ochratoxin A powder shows a similar performance as the fluorescence of the aflatoxin B1 powder, but with excitation wavelengths of 356 and 710 nm during the OPIF and TPIF processes, respectively, instead of the 365 and 730 nm excitation wavelengths used during the aflatoxin characterization. As a result, the same research methodology as used during the aflatoxin detection on maize kernels can be applied to ochratoxin-contaminated food products.

4.2. Zearalenone detection

Zearalenone shows theoretically a strong absorbance between 200 and 350 nm and a maximum fluorescence intensity around 475 nm (**Figure 3(c)**). Similar to ochratoxin, the analysis of zearalenone by using high-performance liquid chromatography in combination with fluorescence detection is widely discussed in literature [34–37]. During the execution of the fluorescence step, an excitation wavelength of 254, 274, 275 or 356 nm is often used, while the detection of the fluorescence signal occurs at 440, 446 or 470 nm. Applying the same measurement procedure to the zearalenone powder as to the ochratoxin A, we studied the zearalenone OPIF signal after excitation with 244, 250, 260, 365, 366, 375 and 395 nm (**Figure 17**). Particularly, to compare the fluorescence intensity for all excitation wavelengths, we present the fluorescence spectrum rescaled to an excitation power density of 1 mW/mm^2. The zearalenone fluorescence spectrum reaches its maximum between 425 and 500 nm. Its fluorescence spectrum shows a broader peak than the aflatoxin and ochratoxin fluorescence spectra. Furthermore, the zearalenone fluorescence emission shows approximately a 10 times weaker intensity than for ochratoxin A. However, similar to the ochratoxin A fluorescence, the largest fluorescence intensities are obtained after excitation with 244 and 356 nm.

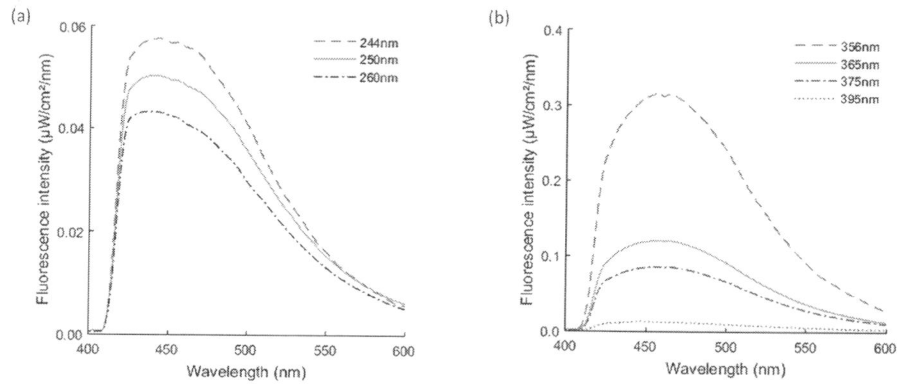

Figure 17. OPIF spectra of zearalenone: (a) spectra after excitation with 244, 250, 260 nm and (b) spectra after excitation with 356, 365, 375 and 395 nm.

To study the TPIF signal of zearalenone, we illuminated the zearalenone powder with 730, 750 and 780 nm, with a spot diameter of 231 μm and an excitation laser power of 1.10, 1.22 and 1.51 W, respectively (**Figure 18**). For all three excitation wavelengths, the measured spectra show two light signals: one at the excitation wavelength and one at half the excitation wavelength. The signal at the excitation wavelength represents the illumination laser light that reflects on the sample and reaches the spectrum analyser. Due to imperfections of the optical short-wave pass filter, small fractions of the high-power incident light are still able to reach the spectrum analyser. The signal at half the excitation wavelength, at 365, 375 and 385 nm for excitation with 730, 750 and 780 nm, respectively, represents SHG instead of TPIF. The absence of the TPIF signal can be explained by the study of the molecular structure of the zearalenone (**Figure 19**). Comparing the different molecular structures, zearalenone shows a significant different structure than aflatoxin B1 and ochratoxin A. Generally, molecules show a strong, two-photon absorption if they feature a long conjugation system in combination with strong donor and acceptor groups, because this induces non-linearity in the system and increases the potential for charge transfer [19]. Consequently, the molecular structure of zearalenone favours SHG instead of TPIF. In comparison with TPIF, the SHG shows a smaller linewidth. During SHG, only one wavelength will be emitted for each illumination wavelength, because two incident photons will always recombine to the double energy. In contrast, during the TPIF process, different emission wavelengths are captured, since the relaxation of the exited electrons can start from different excited energy levels, dependant on their vibrational energy loss. Furthermore, the stronger the captured illumination signal, the weaker the generated SHG signal. When illuminating the sample, part of the excitation photons recombine to the SHG signal, while another part of the excitation signal reflects on the samples surface. The higher the amount of photons that recombine, the stronger the SHG signal and the lower the reflection on the sample.

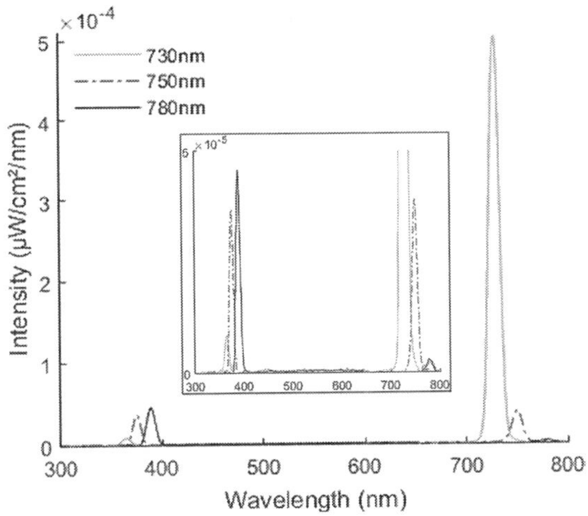

Figure 18. Second harmonic generation signal at 365, 375 and 385 nm, observed after illumination of the zearalenone powder with 730, 750 and 780 nm laser light, respectively. The inner graph presents a close-up on the weakest captured signals.

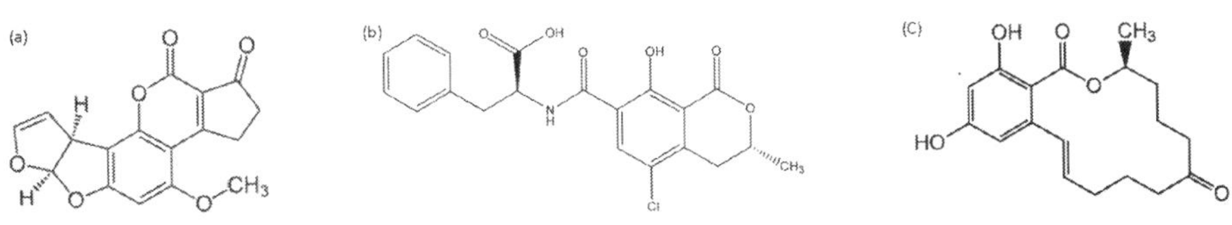

Figure 19. Chemical structure of (a) aflatoxin B1, (b) ochratoxin A and (c) zearalenone.

As a result, OPIF seems the most promising technique for the detection of zearalenone. In contrast to aflatoxin and ochratoxin, zearalenone does not show TPIF. Consequently, we consider TPIF as a less promising technique for the zearalenone detection. However, the influence of the zearalenone onto the intrinsic fluorescence of food products should be further investigated.

5. Conclusion

We successfully demonstrated the use of one- and two-photon-induced fluorescence spectroscopy for the detection of fluorescent mycotoxins in solid, unprocessed food products. First, we developed a sensitive measurement configuration able to study the localized one- and two-photon-induced fluorescence spectra of food products. Afterwards, as a case study, we investigated the detection of aflatoxin in individual maize kernels. We presented our research methodology, starting from the characterization of the aflatoxin fluorescence, to the measurement of the one- and two-photon-induced fluorescence spectra of the maize kernels and the development of a spectroscopic detection criterion. During both one- and two-photon-induced fluorescence processes, the fluorescence of the aflatoxin influenced the intensity and wavelength of the intrinsic fluorescence of the maize kernels. Based on the fluorescence spectrum between 400 and 550 nm, we defined an optical detection criterion, indicating a maximal contrast between the healthy and contaminated maize kernels for excitation with 365 and 780 nm, during one- and two-photon-induced fluorescence, respectively. Both one- and two-photon-induced fluorescence processes show a similar contrast. However, two-photon-induced fluorescence requires higher excitation power densities and a more sensitive detector than one-photon-induced fluorescence, but uses NIR laser wavelengths that are widely commercially available. Besides, in contrast to the chemical analyses that investigate the mean contamination of a batch of maize kernels, we successfully monitored the localized contamination level on the maize kernel surfaces. Both kernels containing a small area with a high contamination level and kernels containing a large region with a medium contamination level could be identified. Finally, we discussed the extensibility of our research methodology to the detection of ochratoxin A and zearalenone. Ochratoxin A showed a characteristic one- and two-photon-induced fluorescence signal, while for zearalenone only OPIF could be observed. Generally, we can conclude that we demonstrated the use of fluorescence spectroscopy as a valuable tool for the sensitive optical detection of fluorescent mycotoxins, paving the way for a non-destructive, real-time and high-sensitive industrial scanning-based detection.

Acknowledgements

This work was supported in part by FWO (G008413N), IWT (IWT120528), the COST Action (MP1205), the Methusalem and Hercules Foundations and the OZR of the Vrije Universiteit Brussel (VUB). The authors would also like to thank Dr Ir Alfons Callebaut and Dr Ir Bart Huybrechts of the CODA-CERVA, the National Reference Laboratory for Mycotoxins, for their valuable feedback and know-how about mycotoxins. We are very grateful to them for the execution of the liquid chromatography-tandem mass spectroscopy analytical technique. Parts of this chapter are reproduced from authors' recent conference publication, cited in reference number [26].

Author details

Lien Smeesters*, Wendy Meulebroeck and Hugo Thienpont

*Address all correspondence to: lsmeeste@b-phot.org

Department of Applied Physics and Photonics (TONA), Brussels Photonics Team (B-PHOT), Faculty of Engineering, Vrije Universiteit Brussel, Brussels, Belgium

References

[1] Jiskoot W, Hlady V, Naleway JJ, Herron JN. Physical Methods to Characterize Pharmaceutical Proteins [Internet]. Herron JN, Jiskoot W, Crommelin DJA, editors. Pharmaceutical Biotechnology. Boston, MA: Springer US; 1995. 1-63 p. (Pharmaceutical Biotechnology; vol. 7). Available from: http://link.springer.com/10.1007/978-1-4899-1079-0

[2] Hof M, Hutterer R, Fidler V. Fluorescence Spectroscopy in Biology. Vol. 3. Berlin Heidelberg: Springer; 2005. 131–286 p.

[3] Falla Sotelo F, Araujo Pantoja P, López-Gejo J, Le Roux GAC, Quina FH, Nascimento CAO. Application of Fluorescence Spectroscopy for Spectral Discrimination of Crude Oil Samples. Brazilian J Pet Gas. 2008;2(2):63–70.

[4] Wagnières GA, Star WM, Wilson BC. In Vivo Fluorescence Spectroscopy and Imaging for Oncological Applications. Photochem Photobiol. 1998;68(5):603–32.

[5] Shahzad A, Köhler G, Knapp M, Gaubitzer E, Puchinger M, Edetsberger M. Emerging applications of fluorescence spectroscopy in medical microbiology field. J Transl Med. 2009;7:99.

[6] Meulebroeck W, Thienpont H. Optical detection techniques for laser sorting machines. In: SPIE Optical Sensing II. Strasbourg: SPIE; 2006. p. 1–12.

[7] Grzesiak MT, Rzepka A, Hura T, Grzesiak S, Hura K, Filek W, et al. Fluorescence excitation spectra of drought resistant and sensitive genotypes of triticale and maize. Photosynthetica. 2007;45(4):606–11.

[8] Nawrocka A, Lamorska J. Determination of Food Quality by Using Spectroscopic Methods. In: Grundas S, Stepniewski A, editors. Advances in Agrophysical Research. Rijeka: InTech; 2013. p. 347–68.

[9] Rasch C, Kumke M, Löhmannsröben HGG. Sensing of Mycotoxin Producing Fungi in the Processing of Grains. Food Bioprocess Technol [Internet]. Springer New York; 2010 Apr 29 [cited 2013 Jan 31];3(6):908–16. Available from: http://dx.doi.org/10.1007/s11947-010-0364-y

[10] Rasch C, Böttcher M, Kumke M. Determination of aflatoxin B1 in alcoholic beverages: Comparison of one- and two-photon-induced fluorescence. Anal Bioanal Chem. 2010;397(1):87–92.

[11] Hruska Z, Yao H, Kincaid R, Darlington D, Brown RL, Bhatnagar D, et al. Fluorescence imaging spectroscopy (FIS) for comparing spectra from corn ears naturally and artificially infected with aflatoxin producing fungus. J Food Sci. 2013;78(8).

[12] Romer Labs. Mycotoxin regulations [Internet]. 2012. p. 1–5. Available from: http://www.romerlabs.com/en/knowledge/mycotoxin-regulations/regulations-usa/

[13] European Commission. Commission regulation (EC) No 1881/2006 setting maximum levels for certain contaminants in foodstuffs. Official Journal of the European Union. Brussels; 2006 Dec.

[14] Lakowicz JR. Principles of Fluorescence Spectroscopy. Third Edit. New York: Springer Science + Business Media; 2006. 607-613 p.

[15] Kaiser W, Garrett CGB. Two-Photon Excitation in Ca:F2 Eu2+. Phys Rev Lett. 1961;7(6): 229–31.

[16] Benninger RKP, Piston DW. Two-Photon Excitation Microscopy for the Study of Living Cells and Tissues. In: Current protocols in cell biology. Canada: John Wiley & Sons, Inc.; 2013. p. 4.11.1–4.11.24.

[17] Mulligan S, MacVicar B. Two-photon fluorescence microscopy: basic principles, advantages and risks. Mod Res Educ Top Microsc [Internet]. 2007;881–8. Available from: http://formatex.org/microscopy3/pdf/pp881-889.pdf

[18] Svoboda K, Yasuda R. Principles of Two-Photon Excitation Microscopy and Its Applications to Neuroscience. Neuron. 2006;50(6):823–39.

[19] Pawlicki M, Collins HA, Denning RG, Anderson HL. Two-photon absorption and the design of two-photon dyes. Angew Chemie–Int Ed. 2009;48(18):3244–66.

[20] Krueger A. MULTIPHOTON MICROSCOPY?: Turnkey femtosecond lasers fuel growth of multiphoton imaging. LaserFocusWorld. 2010;46(10):1–7.

[21] Avantes. Irradiance. In: AvaSoft for AvaSpec-USB2 User's Manual. Avantes; 2012. p. 63–4.

[22] Held P. Quantitation of Peptides and Amino Acids with a Synergy™ HT using UV Fluorescence. BioTek Instruments. 2006. p. 1–8.

[23] Smeesters L, Meulebroeck W, Raeymaekers S, Thienpont H. The use of one- and two-photon induced fluorescence spectroscopy for the optical characterization of carcinogenic aflatoxins. In: SPIE Ultrafast Nonlinear Imaging and Spectroscopy II [Internet]. San Diego: SPIE; 2014. p. 1–11. Available from: http://proceedings.spiedigitallibrary.org/proceeding.aspx?doi=10.1117/12.2061314

[24] ToxiMet. ToxiMet Advanced mycotoxin testing system [Internet]. 2014. Available from: http://www.toximet.com/technology/

[25] Downie NM, Health RW. Basic statistical methods. New York Harper; 1970. chapter 12.

[26] Smeesters L, Meulebroeck W, Raeymaekers S, Thienpont H. One- and two-photon induced fluorescence spectroscopy enabling the detection of localized aflatoxin contamination in individual maize kernels. In: Proc SPIE Optical Sensing and Detection IV. Brussels: SPIE; 2016. p. 98990X – 1–98990X – 15.

[27] Veterinary and Agrochemical Research Centre. Toxins and natural compounds: Ochratoxine (OTA) [Internet]. 2016 [cited 2016 Feb 27]. p. 2–4. Available from: http://coda-cerva.be/index.php?option=com_content&view=article&id=134&Itemid=293&lang=en

[28] Lawley R. Ochratoxins [Internet]. Food Safety Watch. Food Safety Watch; 2013 [cited 2016 Feb 21]. Available from: http://www.foodsafetywatch.org/factsheets/ochratoxins/

[29] Veterinary and Agrochemical Reseach Centre. Toxins and natural compounds: Zearalenone (ZEA) [Internet]. 2016 [cited 2016 Feb 27]. Available from: http://coda-cerva.be/index.php?option=com_content&view=article&id=136&Itemid=294&lang=en

[30] Studer-Rohr I, Dietrich DR, Schlatter J, Schlatter C. The occurrence of ochratoxin A in coffee. Food Chem Toxicol [Internet]. 1995;33(5):341–55. Available from: www.sciencedirect.com/science/article/B6T6P-3YKM35C-2D/1/bebc42a7963ba82d27f46fe1d7d4d201

[31] Lin L, Chen P, Fu Y, Shih DY. Ochratoxin A Contamination in Coffees , Cereals , Red Wines and Beers in Taiwan. J Food Drug Anal. 2005;13(1):84–92.

[32] Batista LR, Chalfoun SM, Silva CF, Cirillo M, Varga EA, Schwan RF. Ochratoxin A in coffee beans (Coffea arabica L.) processed by dry and wet methods. Food Control [Internet]. Elsevier Ltd; 2009 Sep [cited 2013 Jan 31];20(9):784–90. Available from: http://linkinghub.elsevier.com/retrieve/pii/S0956713508002818

[33] Cox KL. Analysis of Ochratoxin A Using the Hitachi LaChromUltra HPLC System with Fluorescence Detection. Californië: Hitachi; 2010.

[34] Arranz I, Mischke C, Stroka J, Sizoo E, van Egmond H, Neugebauer M. Liquid chromatographic method for the quantification of zearalenone in baby food and animal feed: interlaboratory study. J AOAC Int. 2007;90(6):1598–609.

[35] Diebold GJ, Karny N, Zare RN. Determination of zearalenone in corn by laser fluorimetry. Anal Chem. 1979;51(1):67–9.

[36] Huang ZB, Xu Y, Li LS, Li YP, Zhang H, He QH. Development of an immunochromatographic strip test for the rapid simultaneous detection of deoxynivalenol and zearalenone in wheat and maize. Food Control [Internet]. Elsevier Ltd; 2012;28(1):7–12. Available from: http://dx.doi.org/10.1016/j.foodcont.2012.04.035

[37] Pleadin J, Sokolović M, Perši N, Zadravec M, Jaki V, Vulić A. Contamination of maize with deoxynivalenol and zearalenone in Croatia. Food Control. 2012;28(1):94–8.

Chapter 15

Microprobing Structural Architecture Using Mid-Infrared Vibrational Molecular Spectroscopy

Yuguang Ying and Peiqiang Yu

Additional informaticn is available at the end of the chapter

http://dx.doi.org/10.5772/64927

Abstract

The biofunctions of biopolymers are closely related to their microstructures in the complex plant-based tissue in biological systems. In this chapter, molecular spectroscopy is introduced as an approach to microprobe the structural architecture of plant-based seed tissues. Some recent progresses are made using molecular spectroscopy techniques. The wcrking principles of the techniques, along with the methods of molecular spectral analyses and applications in feed architecture research are described.

Keywords: molecular spectroscopy, chemical functional groups, biopolymers, plant-based feed and food, bio-functions

1. Molecular spectroscopy techniques

According to wavelengths from short to long, the electromagnetic spectrum includes gamma rays, hard X-rays, soft X-rays, ultraviolet, visible light, near infrared, mid-infrared, far infrared, microwaves, and radio waves. Gamma rays have the highest frequency and strongest penetration ability. They are produced by the most energetic objects in the universe, or by nuclear explosions, lightning, and radioactive decay on the earth. X-rays are widely used in medical and science areas, while soft X-rays are also used to analyze the characterization of different layers of plant tissues [1, 2]. Near infrared (NIR), as well as mid-infrared, are effective tools in feed analysis and quality assessment. The mid-IR spectral region (ca. 4000–400 cm^{-1}) is a domain of interest to many scientific areas because many molecules have strong characteristic vibrational transitions, especially in the wave number range of ca. 1800–800 cm^{-1}, which is also called the "fingerprint region" [3–5].

As we know, plants are made up of molecules and the internal molecular energy consists of the electronic, translational, rotational, and vibrational energies. Under normal conditions, the functional groups in organic molecules vibrate independently and only interact weakly with each other. However, the interference from outside, such as electromagnetic radiation, could trigger the nonequilibrium phase and cause energy transitions between the rotational and vibrational energies, which can induce the net change in the electric dipole moment and the absorption of the IR. As the ratio of absorption and transmission IR differs between molecules, nearly every molecular species gives a unique IR absorption spectrum. Hence, IR spectrometry could be used to identify molecular functional groups [5].

2. ATR-FTIR molecular spectroscopy techniques

2.1. Working principles

Attenuated total reflectance Fourier transform infrared (ATR-FTIR) is a globar-sourced FTIR spectroscopy, which could be used to identify the molecular constituents in a wide range of samples in areas such as physics, chemistry, and biology. ATR-FTIR is mainly the combination of the globar light source and a microscope and based on the attenuation effect of light [5, 6].

The core part of a FTIR spectrometer is the Michelson interferometer (**Figure 1**, adapted from [9]).

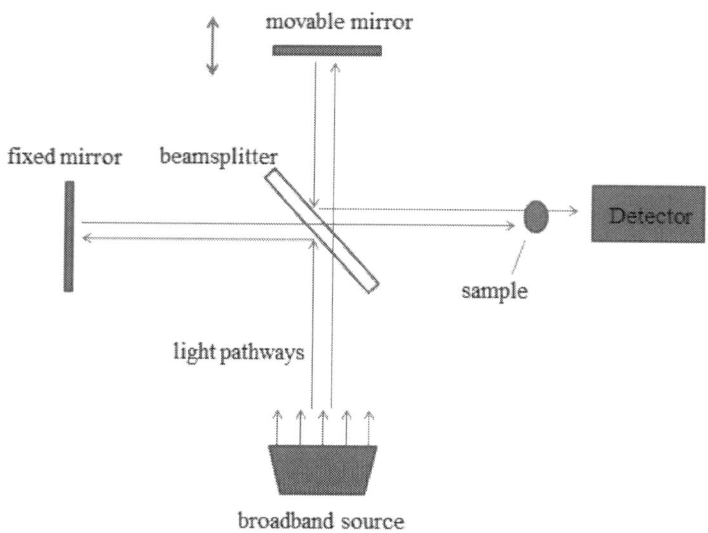

Figure 1. Schematic diagram of a FTIR spectrometer (adapted with permission from McCluskey [9]).

The collimated light from the broadband source travels through the beamsplitter and is split into two beams. One beam travels through the splitter and is reflected by a movable mirror, and the other beam travels to a fixed mirror and is reflected back. A portion of the light finally reaches the sample that may be placed in a liquid-helium cryostat with IR-transparent windows made of ZnSe, KBr, or polypropylene. In an ATR-FTIR instrument, a crystal made of material such as zinc selenide (ZnSe), germanium (Ge), and thallium-iodide is placed under the samples and the incident beam entering at an angle larger than the critical angle, the total reflection could be achieved and only the part of energy that is absorbed by the sample is lost during the process [7, 8]. The part of light that passes through the sample is sensed by the detector, which could be a photoconducting detector such as Ge:Cu placed right behind the sample or a mercury-cadmium-telluride (MCT) mounted in the outside [9]. In this way, all spectral elements are measured simultaneously on the detector and the time consumption (Fellgett advantage or multiplex advantage) depends primarily on the movement of the movable mirror, which could be very short [8, 10].

This technology has a high spectral resolution, a broad measure range and short measure time [9]. At the same time, with no slits to attenuate the infrared light, FTIR has a higher throughput of radiation compared to conventional IR methods (Jacquinot advantage) [10]. Another advantage of ATR-FTIR is it only requires simple sample preparation by finely grinding them and depositing a thin layer on the infrared transparent windows [6].

Nevertheless, the technique also has its shortcomings. Due to the limited brightness, when ATR-FTIR is used to analyze a small region of interest, the decreased aperture would result in the diffraction effects and reduce the signal-to-noise ratio [5]. It is reported that as the plant cell size is normally between 5 and 30 μm, the globar sourced FTIR is not able to obtain a good signal-to-noise ratio within this dimension [5].

2.2. Application of ATR-FTIR techniques in feed research

The ATR-FTIR technique has been proven to be effective in mineral samples in oil shale [11] and biological samples such as cytology and tissue sections, live cells or biofluids [12], and plant samples including transgenic alfalfa [13], hulless barley [14, 15], corn forage [16], DDGS [17, 18], coproducts from bio-oil processing [19–22], heat processing impact [23, 24].

The lipid region, protein region, and carbohydrate region are the three main regions in a spectrum of feed material being analyzed (**Figure 2**). Taking cereal grains as an example, the assessed items included infrared intensity of protein amide I (ca. 1725–1578 cm^{-1}), amide II (ca. 1578–1482 cm^{-1}), amide I peak height (ca. ~1647 cm^{-1}), amide II peak height (ca. ~1537 cm^{-1}), α-helix height (ca. ~1653 cm^{-1}), β-sheet (ca. ~1632 cm^{-1}), lipid (ca. 1798–1709 cm^{-1}) and its peak height (ca. ~1744 cm^{-1}), cellulosic compounds (ca. 1291–1184 cm^{-1}) and its peak height (ca. ~1238 cm^{-1}), total carbohydrates (CHO; ca. 1191–944 cm^{-1}), three major CHO peaks: first peak (ca. 1191–1132 cm^{-1}) and its peak height (ca. ~1150 cm^{-1}), second peak (ca. 1132–1066 cm^{-1}) and its peak height (ca. 67–1078 cm^{-1}), third peak (ca. 1066–944 cm^{-1}) and its peak height (ca. ~1012 cm^{-1}) [25].

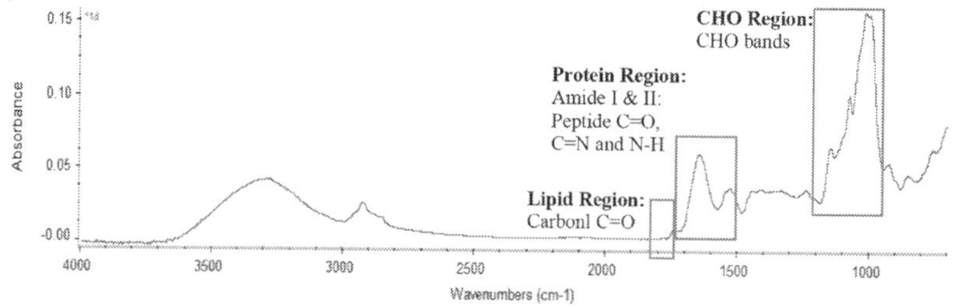

Figure 2. Spectra and fingerprint region and chemical functional groups in plant-based feeds and food.

3. Synchrotron-based molecular spectroscopy techniques (SR-IMS)

3.1. Working principles and advantages

The biggest advantage of SR-IMS is it could preserve the information about the spatial distribution of the objects when detecting the inner structures. This is achieved by the use of synchrotron infrared light source. The nondivergent, intense and extremely fine beamline is created by a giant particle accelerator that turns electrons into light, which is 100–1000 times brighter than the globar source [5]. Therefore, high spatial resolution and signal to noise spectra can be collected at a faster speed [5].

3.2. Novel applications of SR-IMS techniques in feed research

The SR-IMS technology was first applied to animal feed research in 1999 [26]. Since then it has been utilized on several feeds, including transgenic alfalfa [27, 28], hulless barley [5, 15, 29], canola seeds [5, 30], corn seed [28], flaxseeds [28, 30], sorghum seeds [31], wheat [30, 32], wheat DDGS [28], and corn DDGS [28]. In spite of all these applications, this research is still in its infancy.

4. Spectra analysis

Functional groups such as amide I and amide II bonds have certain percentages of C=O, C—N, and N—H stretching vibrations, the wave numbers (cm^{-1}) at which they are absorbed and generally fixed, but they also slightly shift depending on the samples [5]. Some typical IR absorption bands include: amide I (centered at about 1650 cm^{-1}, includes about 80% C=O stretching, 10% C—N stretching and 10% N—H bending), amide II (centered at about 1550 cm^{-1}, includes about 60% N—H bending and 40% C—N stretching), lipid carbonyl C=O (peaks at about 1738 cm^{-1}), and cellulose (at about 1100 cm^{-1}) [33]. Among them, amides I and II are the most dominant vibrational bands of the protein backbone and amide I, due to its high

http://dx.doi.org/10.5772/64927

C=O stretching composition, is the most sensitive and highly related to secondary structural elements of proteins [34].

4.1. Univariate analysis

Using univariate analysis, it is possible to discover quantitative differences in the spectra information, such as the component areas, peak heights, and ratios among different components. Univariate analysis gives very straightforward results in terms of what changes occurred on the mathematical parameters characterizing the spectrum, such as the band intensities, integrated intensities, band frequencies and the band intensity ratios. In addition, this method makes it possible to connect the spectra information to the biological meaning on a mathematical basis [35].

4.2. Multivariate analysis

Multivariate analysis is capable of analyzing multiple variables at same time. Principal component analysis (PCA) and hierarchical cluster analysis (CLA or HCHA) are two of the commonly used methods.

The PCA transforms the original set of variables based on the correlations among them, into a set of independent linear combinations called principal components (PCs), which contain most of the information in the original variables and empirically summarizes their correlations [32, 36]. The first few PCs usually account for more than 95% of the total variation among the variables [27].

The CLA is another data reduction method that calculates a distance matrix, searches for the two most similar objects, and displays the results as dendrograms [32, 37]. In the hierarchical approach, the object or objects are gathered as a group step by step, being nested to the previous groups. Thus, the number of clusters reduces sequentially as the clusters' sizes grow and end up with only one [37].

5. Application

5.1. Application 1: structural responses of functional groups in cereal grains to heat processing methods

The research [25] in processing-induced molecular structure study showed that the sensitivity and responses of functional groups can be detected by both ATR-FTIR and SR-IMS techniques, and different functional groups in cereal grain tissues respond differently to the heating methods, although not all heat-induced structural changes detected by the two mid-IR techniques are highly related to the nutrient availability of cereal grains in ruminants.

Due to the difference in sample-preparation and sampling areas, the results found by the two mid-IR techniques were also different. Similar to the conventional studies, the grains were ground and well-mixed before using the ATR-FTIR technique. The results found in the conventional studies indicated that moist heating had greater impact on nutrient availabilities compared with dry heating. In accordance with such results, ATR-FTIR method also detected stronger influence on spectral peak areas, heights, and ratios (**Figure 3**). These alterations, especially the changes on protein secondary structure, were highly related with the nutrient availability in cereal grains.

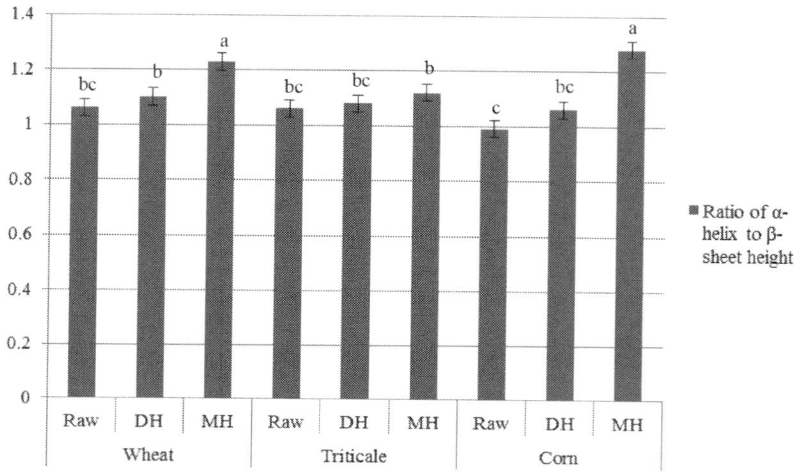

Figure 3. Ratios of modeled α-helix to modeled β-sheet height affected by processing methods detected by ATR-FTIR technique.

The grain seeds were cross-sectioned into thin (6 μm) sections and spectra were collected from the endosperm area. The results discovered by the SR-IMS technique indicated that dry heating also played a big role in changing the secondary structures and functional groups of the grains. As the peak areas and peak heights represent combined information of nutrient amount and molecular structure, with the nutrient contents affected by moist heating, it is very likely that the molecular structures in the endosperm were also changed. Unfortunately, such change was not specified in this part of study. In comparison with results found by the ATR-FTIR technique, less and weaker correlation was discovered between the heat-induced structural changes and the nutrient availability in the endosperm area of cereal grains in ruminants by the SR-IMS technique.

5.2. Application 2: microwave irradiation-induced changes in protein molecular structures of barley grains: relationship with changes in protein chemical profile, protein subfractions, and digestion in dairy cows

These studies [38, 39] aimed to evaluate microwave irradiation (MIR)-induced changes in crude protein (CP) subfraction profiles; ruminal CP degradation characteristics and intestinal digestibility of rumen undegraded protein (RUP); and protein molecular structures in barley

(*Hordeum vulgare*) grains. Samples from hulled and hulless cultivars of barley, harvested in two consecutive years from four replicate plots, were evaluated. The samples were either kept as raw or irradiated in a microwave for 3 min (MIR3) or 5 min (MIR5). Compared with raw grains, MIR5 decreased the contents rapidly degradable CP subfraction (45.2–6.4% CP) and the ruminal degradation rate (8.16–3.53%/h) of potentially degradable subfraction. As a consequence the effective ruminal degradability of CP decreased (55.7–34.1% CP), and RUP supply (43.3–65.9% CP) to the postruminal tract increased. The MIR decreased the spectral intensities of amide 1, amide II, α-helix and β-sheet, and increased their ratios. The changes in protein spectral intensities were strongly correlated with the changes in CP subfractions and digestive kinetics. These results show that MIR for a short period (5 min) with a lower energy input can improve the nutritive value and utilization of CP in barely grains.

6. Summary, implications, and future research areas

6.1. Summary

As different ratios of IR could be absorbed in different molecules when mid-IR is applied, the functional groups have their unique spectra, especially in the "fingerprint region." There have been many applications of ATR-FTIR and SR-IMS on animal feed researches in recent years. The results show that these two advanced mid-IR approaches can effectively detect the microstructural changes in some plant tissues. With the help of the statistical analysis, quantitative differences lie between different spectra could be discovered.

6.2. Implications

Functional groups in different type of plants could have different sensibility and react differently to external changes. Feed processing methods could change the inner structure of the plant tissues, such change can probably be detected by mid-IR techniques such as SR-IMS and ATR-FTIR. Combined with conventional animal nutrition studies, the link between structural changes in spectral areas such as amide, CHO, and cellulosic compounds and nutrient availability of the plant could be found.

6.3. Further research areas

Our further research plans include using ATR-FTIR technique to detect the sensitivity and responses of various chemical functional groups in different types of feed materials to different types of feed processing methods, and building up models using the spectral parameters to estimate the nutrient utilization and availability in ruminants. We would also expand the sampling areas when using the SR-IMS technique and combine methods such as Mid-IR microspectroscopic mapping to better understand the inner structural changes in the plant tissues.

Acknowledgements

The Ministry of Agriculture Strategic Research Chair (PY) research programs have been supported by grants from Natural Sciences and Engineering Research Council of Canada (NSERC—Individual Discovery Grants and CRD grants), Saskatchewan Agricultural Development Fund (ADF), Ministry of Agriculture Strategic Feed Research Chair Program, Western Grain Research Foundation (WGRF), Saskatchewan Forage Network, SaskPulse, SaskCanola, SaskMilk, etc. Parts of this chapter are reproduced with permission from [25].

Author details

Yuguang Ying and Peiqiang Yu*

*Address all correspondence to: peiqiang.yu@usask.ca

Department of Animal and Poultry Science, College of Agriculture and Bioresources, University of Saskatchewan, Saskatoon, Canada

References

[1] Karunakaran, C.; Gaillard, C.; Bouchet, B.; Gnaegi, H.; Buleon, A.; Wang, J. and Hitchcock, A. P. 2009. Characterization of wheat grain tissues by soft X-ray spectromicroscopy. Canadian Light Source. 2010: 50–51.

[2] Yu, P. 2012. Synchrotron soft X-ray and infrared microspectroscopy contributions to advances in feed chemistry and feed science technology. In: Méndez-Vilas, A., ed. Current microscopy contributions to advances in science and technology. Formatex Research Center, Barcelona, Spain, pp. 1504–1510.

[3] Liu, N. 2009. Ruminal nutrient availability and inherent structural features of six barley varieties using in situ technique and mid-IR spectroscopy. MSc Thesis, University of Saskatchewan, Saskatoon, SK, 116 pp.

[4] Schliesser, A.; Pecque, N. and Hänsch. 2012. Mid-infrared frequency combs. Nat Photonics. 6: 440–449.

[5] Yu, P. 2004. Application of advanced synchrotron-based Fourier transform infrared microspectroscopy (SR-FTIR) to animal nutrition and feed science: a novel approach. Br J Nutr. 92: 869–885.

[6] Kazarian, S. G. and Chan, K. L. A. 2013. ATR-FTIR spectroscopic imaging: recent advances and applications to biological systems. Analyst. 138: 1940–1951.

[7] Ochiai, S. 2015. Attenuated total reflection measurements. In: Tasumi, M., ed. Introduction to experimental infrared spectroscopy: fundamentals and practical methods. John Wiley & Sons, Ltd., West Sussex, UK, pp. 179–198.

[8] Stuart, B. 2004. Infrared spectroscopy: fundamentals and applications. England, John Wiley & Sons, Ltd., West Sussex, UK, 242 pp.

[9] McCluskey, M. D. 2000. Local vibrational modes of impurities in semiconductors. J Appl Phys. 87: 3593–617.

[10] Herres, W. and Gronholz, J. 1984. Understanding FT-IR data processing. Part 1: data acquisition and Fourier transformation. [Online] Available: http://mmrc.caltech.edu/FTIR/Understanding%20FTIR.pdf [2015 April 12]

[11] Palayangoda, S. S. and Nguyen, Q. P. 2012. An ATR-FTIR procedure for quantitative analysis of mineral constituents and kerogen in oil shale. Oil Shale. 29: 344–356.

[12] Baker, M. J.; Trevisan, J.; Bassan, P.; Bhargava, R.; Butler, H. J.; Dorling, K. M.; Fielden, P. R.; Fogarty, S. W.; Fullwood, N. J.; Heys, K. A.; Hughes, C.; Lasch, P.; Martin-Hirsch, P. L.; Obinaju, B.; Sockalingum, G. D.; Sulé-Suso, J.; Strong, R. J.; Walsh, M. J.; Wood, B. R.; Gardner, P. and Martin, F. L. 2014. Using Fourier transform IR spectroscopy to analyze biological materials. Nat Protoc. 9: 1771–1791.

[13] Jonker, A. 2011. Characterization of anthocyanidin-accumulating Lc-alfalfa for ruminants: nutritional profiles, digestibility, availability and molecular structure, and bloat characteristics. Ph.D. Thesis, University of Saskatchewan, Saskatoon, SK, 162 pp.

[14] Yang, L.; McKinnon, J. J.; Christensen, D. A.; Beattie, A. D.; Xin, H. and Yu, P. 2013. Investigating the molecular structural features of hulless barley (*Hordeum vulgare* L.) in relation to metabolic characteristics using synchrotron-based Fourier transform infrared microspectroscopy. J Agric Food Chem. 61: 11250–11260.

[15] Yang, L.; McKinnon, J. J.; Christensen, D. A.; Beattie A. D. and Yu, P. 2014. Characterizing the molecular structure features of newly developed hulless barley cultivars with altered carbohydrate traits (*Hordeum vulgare* L.) by globar-sourced infrared spectroscopy in relation to nutrient utilization and availability. J Cereal Sci. 60: 48–59.

[16] Abeysekara, S.; Christensen, D. A.; Niu, Z.; Theodoridou, K. and Yu, P. 2013. Molecular structure, chemical and nutrient profiles and metabolic characteristics of the proteins and energy in new cool-season corn cultivars harvested as fresh forage for dairy cattle. J Dairy Sci. 96: 6631–6643.

[17] Gamage, I. H.; Jonker, A.; Christensen, D. A. and Yu, P. 2012. Metabolic characteristic of proteins and biomolecular spectroscopic profiles in different batches of feedstock (wheat) and their co-products (wheat dried distillers grain with solubles) from the same bioethanol processing plant. J Dairy Sci. 95: 6695–6715.

[18] Gamage, I.H. and Yu, P. 2013. Short communication: comparison of newly developed DVE/OEB2010 system and NRC2001 model in modeling metabolic characteristics of proteins. J Dairy Sci. 96: 5908–5913.

[19] Yu. P. 2012. Short communication: relationship of carbohydrate molecular spectroscopic features to carbohydrate nutrient profiles in co-products from nioethanol production. J Dairy Sci. 95: 2091–2096.

[20] Zhang, X. and Yu, P. 2012. Differentiation of mixtures of co-product blend with barley grain based on Fourier transform infrared attenuated total reflection molecular spectroscopy: carbohydrate molecular structure spectral profiles and nutritive characteristics in dairy cattle. J Dairy Sci. 95: 6624–6634.

[21] Huang, X.; Khan, N. A.; Zhang, X. and Yu, P. 2015. Effects of canola meal pellet conditioning temperature and time on ruminal and intestinal digestion, hourly effective degradation ratio, and potential N to energy synchronization in dairy cows. J Dairy Sci. In Press (DOI: 10.3168/jds.2014-9295).

[22] Huang, X.; Christensen, C. and Yu, P. 2015. Effects of conditioning temperature and time during the pelleting process on feed molecular structure, pellet durability index, metabolic features of co-products from bio-oil processing in dairy cows. J Dairy Sci. 98: 4869–4881.

[23] Samadi and Yu, P. 2011. Dry and moist heating-induced changes in protein molecular structure, protein subfraction, and nutrient profiles in soybeans. J Dairy Sci. 94: 6092–6102.

[24] Peng, Q.; Khan, N. A.; Wang, Z. and Yu, P. 2014. Moist and dry heating-induced changes in protein molecular structure, protein subfractions, and nutrient profiles in camelina seeds. J Dairy Sci. 97: 446–457.

[25] Ying, Y. 2015. Nutritional and microstructural response in cereal grains to heat-related processing methods. MSc Thesis, University of Saskatchewan, Saskatoon, SK, 126 pp.

[26] Kondo, T.; Ohshita, T.; Kyuma, T.; Touno, E. and Murai, M. 1999. Characterization of soluble lignin released from alfalfa by sheep digestion. Anim Feed Sci Technol. 80: 321–328.

[27] Yu, P.; Jonker, A. and Gruber, M. 2009. Molecular basis of protein structure in proanthocyanidin and anthosynin-enhanced Lc-trasgenic alfalfa in relation to nutritive value using synchrotron-radiation FTIR microspectroscopy: a novel approach. Spectrochim Acta Part A. 73: 846–853.

[28] Yu, P. 2010. Plant-based food and feed protein structure changes induced by gene-transformation, heating and bio-ethanol processing: a synchrotron –based molecular structure and nutrition research program. Mol Nutr Food Res. 54: 1535–1545.

[29] Yang, L. 2013. Effect of carbohydrate traits on nutritional characteristics and spectral features of molecular structure in hulless barley. MSc Thesis, University of Saskatchewan, Saskatoon, SK, 151 pp.

[30] Yu, P. 2005. Protein secondary structures (α-helix and β-sheet) at a cellular level and protein fractions in relation to rumen degradation behaviours of protein: a new approach. Br J Nutr. 94: 655–665.

[31] Yu, P. 2011. Microprobing the molecular spatial distribution and structural architecture of feed-type sorghum seed tissue (*Sorghum Bicolor* L.) using the synchrotron radiation infrared microspectroscopy technique. J Synchrotron Radiat. 18: 790–801.

[32] Yu, P.; Block, H.; Niu, Z. and Doiron, K. 2007. Rapid characterization of molecular chemistry, nutrient make-up and microlocation of internal seed tissue. J Synchrotron Radiat. 14: 382–390.

[33] Jackson, M. and Mantsch, H. H. 1996. Biomedical infrared spectroscopy. In: Mantsch, H. H. and Chapman, D., eds. Infrared spectroscopy of biomolecules. Wiley-Liss, New York, NY, pp. 311–340.

[34] Kong, J. and Yu, S. 2007. Fourier transform infrared spectroscopic analysis of protein secondary structures. Acta Biochim. Biophys Sin. (Shanghai). 39: 549–559.

[35] Liu, B. 2011. In-depth study of the relationship between protein molecular structure and the digestive characteristics of the proteins in dried distillers grains with solubles. MSc Thesis, University of Saskatchewan, Saskatoon, SK, 114 pp.

[36] Tabachnick, B. G. and Fidell, L. S. 2007. Using multivariate statistics. Pearson Education, Needham Heights, MA, 980 pp.

[37] Jobson, J. D. 1992. Applied multivariate data analysis: Volume II: categorical and multivariate methods. Springer-Verlag, New York, NY.

[38] Yan, X. and Yu, P. 2016. Reveal interrelationship between processing-induced molecular structure features and metabolic and digestive characteristics in hulled and hulless barley (*Hordeum vulgare*) grains with altered carbohydrate traits using advanced VMS molecular spectroscopy. J Sci Food Agric. Accepted (JSFA-16-1055).

[39] Yan, X.; Khan, N. A.; Zhang, F.; Yang, L. and Yu, P. 2014. Microwave irradiation induced changes in protein molecular structures of barley grains: relationship to changes in protein chemical profile, protein subfractions, and digestion in dairy cows. J Agric Food Chem. 62: 6546–6555.

Chapter 16

Fluorescence Spectroscopy for the Analysis of Spirit Drinks

Jana Sádecká, Veronika Uríčková and
Michaela Jakubíková

Additional information is available at the end of the chapter

http://dx.doi.org/10.5772/64002

Abstract

There are many prescribed methods for the analysis of important components and parameters of spirit drinks. Nevertheless, there is a continuous search for new rapid and simple alternative methods that can be used together with recommended methods. The aim of the chapter is to make a review about themes such as quantification of individual components in the spirit drinks, classification of spirit drinks, and determination of adulterants. The chapter shows that fluorescence spectroscopy has a significant potential for being used in spirit drink research because many alcoholic beverage products contain intrinsic fluorophores. Fluorescence spectroscopy allows the determination of some compounds at concentration as low as 0.1–1 μg/L often without sample preparation, there is no use of chemicals and the time of analysis can be very short. The combination of fluorescence data with chemometric tools is a promising approach for the classification of spirit drinks and for the detection of spirit drink adulteration.

Keywords: fluorescence spectroscopy, chemometrics, beverage, spirit drink, classification

1. Introduction

Fluorescence, like the other molecular spectroscopies, represents an attractive option for food and beverage analysis because it is rapid, sensitive and non-destructive. The reviews on this matter have been reported [1–3]. According to Regulation (EC) No 110/2008 [4], 'spirit drink' means an alcoholic beverage possessing particular organoleptic qualities, having a minimum alcoholic strength of 15% vol., having been produced: (i) either directly (by the distillation, with or without added flavorings, and/or by the maceration of plant materials in ethyl alcohol of agricultural origin, and/or by the addition of flavorings to ethyl alcohol of agricultural origin), (ii) or by the mixture of a spirit drink with one or more other spirit drinks. The Regulation (EC) No 110/2008 defines 46 different categories of spirit drinks. For the purposes of this review, spirit drinks are divided into two general classes: (1) unaged (vodka, gin, juniper-flavoured spirit drink and fruit spirit) and (2) aged in wooden casks (brandy, whisky, mezcal, tequila, cachaça and calvados). The term age refers to the actual duration of storage, while maturity expresses the degree to which chemical changes occur during storage. Most governments specify storage time for various products.

The major constituents of each spirit drink consist of ethanol and water. The minor or trace constituents are higher alcohols, carbonyl compounds, esters, aldehydes, lactones, organic acids, etc. [5]. However, there are almost the same fluorophores in the different spirits, among others, volatile phenols and anisols in unaged spirits, and phenolic compounds and coumarins in spirits aged in wooden casks.

Fluorescence spectra of distilled spirits are typically composed of broad overlapping fluorescence bands containing chemical, physical and structural information of all sample components. Therefore, conventional fluorescence technique based on recording of single emission or excitation spectra is often insufficient for analysing spirit drinks. In some cases, total luminescence or synchronous scanning fluorescence techniques may improve the analytic potential of fluorescence measurements. The analytical information should be extracted from fluorescence spectra using multivariate and multiway methods, which allow to group samples with similar characteristics, to establish classification methods for unknown samples (qualitative analysis) or to perform methods determining some property of unknown samples (quantitative analysis) [6].

There are many prescribed methods for the analysis of important components and parameters of spirit drinks. The most widely used methods are sensory evaluation, gas chromatography, liquid chromatography, mass spectrometry, ultraviolet–visible (UV/VIS) spectrophotometry and infrared spectrometry [5]. Nevertheless, there is a continuous search for new alternative methods that can be used together with recommended methods.

The aim of the chapter is to make a review about themes such as quantification of individual components in the spirit drinks, classification of spirit drinks and adulteration detection in order to highlight the potential of fluorescence spectroscopy in the beverage analysis.

2. Fluorescence spectra of spirit drinks

Conventional fluorescence spectroscopy uses either a fixed excitation wavelength (λex) to record an emission spectrum or a fixed emission wavelength (λem) to record an excitation spectrum. The broad shape of both the excitation and emission fluorescence bands limits the possibility of finding a unique λex and λem for each potential analyte [7]. Selectivity is often improved through fluorimetric strategies such as total luminescence, synchronous scanning fluorescence or total synchronous scanning fluorescence.

Total luminescence spectrum (TLS) presents simultaneously all the excitation and emission spectra over the range of wavelengths scanned [8] and can be shown as a contour map with λem and λex as *x*- and *y*-axes, respectively, and contours linking points of equal fluorescence intensity (**Figure 1a**). In TLS, two types of scattering peaks can be found: Rayleigh scattering at the λex = λem, and Raman scattering at a distance from the Rayleigh peak that is characteristic for the solvent. Because TLS spectra represent the total fluorescence profiles of the samples, they are particularly useful in pattern recognition of samples characterised by small differences in their composition [9].

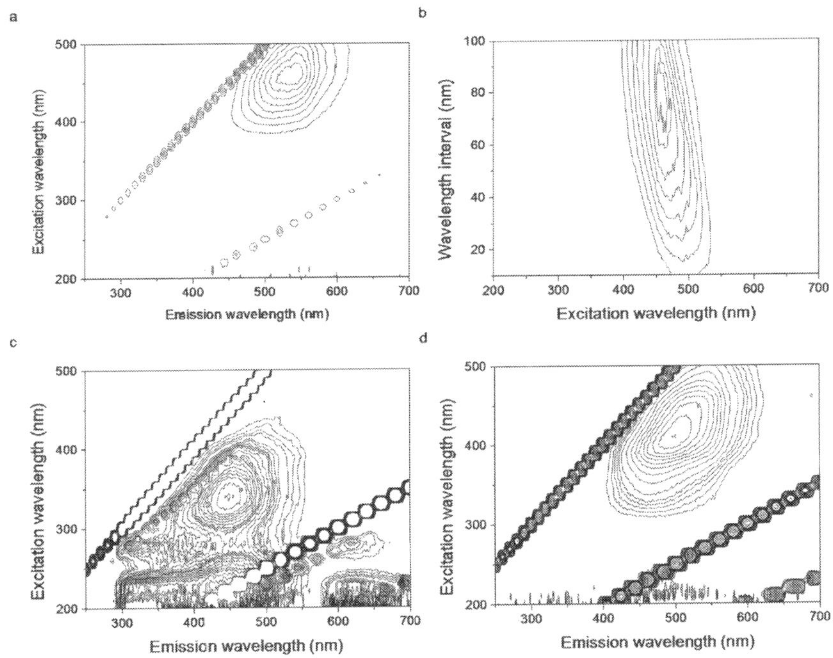

Figure 1. Total luminescence spectra (a,c,d) and TSFS (b) of undiluted (a,b,d) and diluted (c) brandy obtained using right-angled (a,b,c) and front-face geometry (d). **TLS and TSFS were recorded using the Perkin–Elmer LS 50 Luminescence Spectrometer equipped with the Xenon lamp. Samples were placed in 10 × 10 × 45 mm quartz cell. Excitation and emission slits were both set at 5.0 nm. Scan speed was 200 nm.min^{-1}.

In synchronous fluorescence spectroscopy [10], the λex and λem are scanned simultaneously in such a way that a constant wavelength interval Δλ = λem – λex is kept between them. When a value of Δλ is chosen properly, the resulting synchronous fluorescence spectrum (SFS) shows one or a few features that are much more resolvable than those in the conventional fluorescence spectrum because synchronous fluorescence reduces spectral overlaps by narrowing spectral bands and simplifies spectra by amplifying strong fluorescence bands. A choice of $\Delta\lambda$ could be either the difference between the wavelength of emission maximum (λem, max) and the corresponding wavelength of excitation maximum (λex, max) to provide the highest sensitivity, or the particular difference to give a compromise between sensitivity and selectivity [11, 12].

Total synchronous fluorescence spectrum (TSFS) is obtained by plotting fluorescence intensity as a function of the wavelength and $\Delta\lambda$ value (**Figure 1b**) and combine the advantages of TLS and SFS. Because λem is always higher than λex, Rayleigh scattering is not found in TSFS.

Independent of the type of spectrum, the apparent fluorescence intensity and spectral distribution is affected by both the optical density of the sample (**Figure 1a** and **c**) and the geometry of sample illumination (**Figure 1a** and **d**). The most common geometry is right-angle observation of the center of a centrally illuminated cuvette. It is typically used to analyse dilute solutions and other transparent samples (absorbance < 0.1). At high optical densities, signal reaching detector will be significantly disturbed due to the inner filtering effects. In the front-face geometry, the excitation light is focused to the front surface of the samples and then fluorescence emission is collected from the same region at an angle that minimizes reflected and scattered light. Front-face illumination is generally used to decrease the inner filtering effects [7].

2.1. Spirit drinks unaged in wooden casks

The major constituents of each spirit drink, ethanol and water molecules do not exhibit fluorescence. However, when ethanol mixes with water, ethanol and water molecules form molecular clusters by hydrogen bonding and emit different fluorescence photons [13]. When excited by λex = 236 nm, there were eight kinds of luminescence structures in the ethanol–water mixtures, giving the emission bands at 292, 304, 314, 330, 345, 355, 365 and 377 nm, respectively. The fluorescence bands at 355 and 377 nm have maximum intensity when the percent of ethanol is 20%. The other six kinds have maximum intensity for 60% ethanol content [14].

Different flavour exhibits different effect on the fluorescence of 60% ethanol–water mixture characterised by the main band centred at λex/λem = 225/335 nm. The simultaneous addition of eight major flavours (acetaldehyde, ethyl acetate, methanol, propyl alcohol, isobutyl alcohol, isoamyl alcohol, ethyl lactate and acetic acid) make the band at 225/335 nm in excitation/emission disappear and cause the appearance of bands at λex/λem of 285/325 nm as well as at 375/425 nm. The 225/335 nm fluorescence band initially increases and then decreases with increased ethyl acetate or acetate concentration in the 60% ethanol–water mixture. For the Fenjiu samples aged in ceramic containers, the effect of total ester concentration is consistent with the result of ethyl acetate in the 60% ethanol–water mixture, however, the effect of acetic acid differs [15].

Vodka is the simplest distilled spirit, the character of which comes from the ethanol, normally distilled from grain fermentation. Vodka Finlandia (40%) is amongst the purest in the world, its typical TLS and TSFS are shown in **Figure 2**. The short-wavelength band in TLS, which has maximum at λex/λem = 230/335 nm, corresponds to the band at 220–230 nm (Δλ = 90 – 100 nm) in TSFS and can be assigned to luminescence structures in the ethanol–water mixture. It should be noticed that there is no available information or data on the origin of fluorescence of vodka. However, some of the volatile compounds (1,3,5-trimethylbenzene and p-cymene) identified by GC-MS in vodka [16, 17] are known fluorophores. The micro array based on fluorescence dye solutions and their binary mixtures shows vodka pattern with a certain similarity but slightly different from the aqueous ethanol pattern [18].

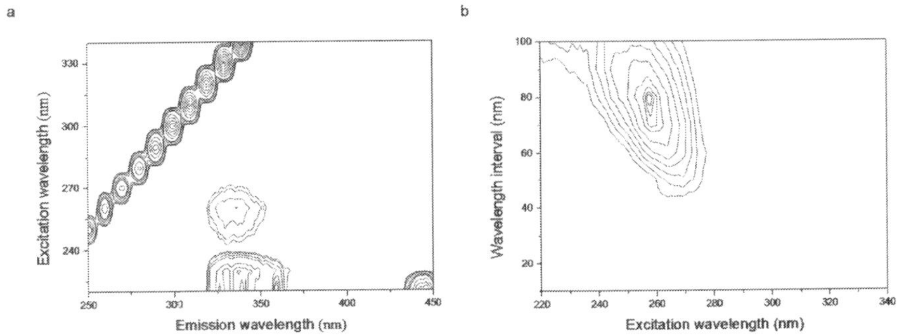

Figure 2. Total luminescence spectrum (TLS) (a) and total synchronous fluorescence spectrum (TSFS) (b) of vodka Finlandia **TLS and TSFS were recorded using the Perkin–Elmer LS 50 Luminescence Spectrometer equipped with the Xenon lamp. Samples were placed in 10 × 10 × 45 mm quartz cell. Excitation and emission slits were both set at 5.0 nm. Scan speed was 200 nm.min^{-1}.

Juniper-flavoured spirit drinks (JFSDs) are produced by flavoring ethyl alcohol of agricultural origin and/or grain spirit with juniper (*Juniperus communis L.* and/or *Juniperus oxicedrus L.*) berries. Eugenol, totarol, *o*-cymene, *p*-cymene, *p*-cymene-8-ol, calamenene, calacorene, phenolic acids, flavonoids, biflavonoids, coumarins, tyrosol and chlorophyll are the best known fluorescent molecules in juniper berries (see references in [19]). Volatile compounds that survived distillation (*o*-cymene, *p*-cymene, calamenene, calacorene [20, 21] and totarol [22], were found in gin, and more than 30 possible fluorophores were detected in JFSDs (mainly substituted benzenes, phenols and anisols) [19, 20]. Many substituted phenols or anisols and diterpenoids show similar fluorescence properties, e.g., λex/λem = 288/315 nm for eugenol [23], λex/λem = 275/315 nm for totarol [24].

The most popular JFSD is gin, which TLS is characterised by the main fluorophores centred at λex = 220 and 304 nm and λem = 337 nm. The first pair of wavelengths (λex/λem = 220/337 nm) is similar to that observed for vodka, the second one (λex/λem = 304/337 nm) is characteristic for London gin. Other JFSDs show band with excitation at about 250–290 nm and emission at about 330–340 nm. Moreover, Belgian and Czech JFSDs show additional band at longer wavelength (**Table 1**). Modelling of TLS allowed relating the fluorescence bands of drinks to 2-phenylethanol, eugenol, carvacrol, 4-allylanisole, *p*-cymene and coumarin derivatives [19]. JFSDs are sometimes marketed in the glass bottles containing dried berries or twig inside. Such

JFSDs had abnormally high fluorescence intensity at about 260/335 nm in excitation/emission, which could be attributed to compounds extracted from berries or twig [25].

Spirit drink	$\lambda_{ex,max}$ (nm)	$\lambda_{em,max}$ (nm)	Reference
Unaged in wooden casks			
Vodka Finlandia	230, 260	335	This work
London Gin	220, 302–306	335–340	[19]
Juniper-flavoured (Slovak)	277–290	330–343	[19]
Juniper-flavoured (Belgian)	280–290	330–340	[19]
	317–350	426–447	
Juniper-flavoured (German)	250–260	331–333	[19]
Juniper-flavoured (Czech)	260–265	335–337	[19]
	300–304	400–402	
Apple	250	327	[26]
	300	419	
Apricot	290	317	[26]
Pear	235	349	[26]
	302	420	
Plum	266	330	[26]
	304	417	
Apricot spirit with fruit	270	360	[26]
	350	443	
Pear spirit with fruit	260	366	[26]
	330	423	
Mixed wine	390–400	480–500	[27]
Aged in wooden casks			
Brandy	450–460	520–540	[27]
Mezcal	514	580	[35]
Tequila	337	430	[38]
Tequila	255	470	[39]
	330	460	[39]
	365	460	[39]
	405	510	[39]
Cachaça (amendoim)	330	400	[44]
Cachaça (balsam)	340	480	[44]
Cachaça (oak)	280	320	[44]
Cachaça (jequitibá)	260	370	[44]
Cachaça (umburana)	380	450	[44]
Calvados	410	506	[26]

$\lambda_{ex,max}$ wavelength of excitation maximum, $\lambda_{em,max}$ wavelength of emission maximum.

Table 1. Fluorescent properties of bulk spirit drinks obtained using right-angled geometry.

Spirit drink	Dilution	$\lambda_{ex,max}$ (nm)	$\lambda_{em,max}$ (nm)	Reference
Apple	40-fold	-	-	[26]
Apricot	40-fold	278	337	[26]
Apricot with fruit	40-fold	280	334	[26]
Pear	40-fold	278	318	[26]
Pear with fruit	40-fold	280	322	[26]
Plum	40-fold	259	329	[26]
Mixed wine	100-fold	280	350	[28]
		330	430	
Brandy	100-fold	280	360–370, 450–460	[28]
		340	450	
Calvados	40-fold	340	450	[26]

$\lambda_{ex,max}$ wavelength of excitation maximum, $\lambda_{em,max}$ wavelength of emission maximum

Table 2. Fluorescent properties of diluted spirit drinks.

Fruit spirits are made of different varieties of fruits by the alcoholic fermentation and distillation. They are usually aged in glass containers, marketed as 'pure' beverages or in the bottles containing a whole dried fruit. **Table 1** shows the characteristic λex,max and λem,max corresponding to the four types of fruit spirits. Bulk apple, pear and plum spirits exhibit two fluorescent bands, one with fluorescent maximum between 250 and 290 nm in the λex and between 330 and 350 nm in the λem range, whose exact position depended on the fruit type, and the second with excitation maximum at about 300 nm and emission at about 420 nm. In contrast, bulk apricot spirit exhibits only the short-wavelength band. Bulk spirits containing fruit show two fluorescent maxima at longer wavelengths (**Table 1**). The UV absorption of bulk fruit spirits is from 2 up to 4 absorbance units when scanning from 225 to 300 nm, and therefore the inner filter phenomena affect the right-angle spectra considerably. One way to reduce the inner filter effects is to dilute the sample with an appropriate solvent. On the other hand, dilution can reduce concentration of some components bellow limit of detection. As an example, **Table 2** shows the λex,max and λem,max of fruit sprits upon dilution. Apple spirits exhibited no reasonable fluorescence upon 40-fold dilution. Both diluted apricot and pear spirits exhibit a band with a maximum fluorescence at λex = 280 nm. The different position of emission band for apricot and pear spirits enables us to distinguish between them. In addition, λex,max and λem,max of diluted plum spirits are different from the other fruit spirits. The compounds such as 1-phenylethanol, 2-phenylethanol, eugenol, 4-allylanisole, 4-vinylanisole,

4-ethylphenol, 4-ethylguaiacol and p-cymene can be detected using λex /λem of 280/320 nm after separation by HPLC. In the case of spirits containing fruit, there is a wider variety of fluorescent compounds, including not only those found in pure spirits but also benzoic and cinnamic acids and their aldehydes [26].

Mixed wine spirits are wine distillates diluted with ethanol from other sources, frequently blended with sugar, brandy aroma and caramel. Some mixed wine spirits contain honey or colourants. TLS contours of bulk mixed wine spirits are concentrated in the λem region from 460 to 530 nm and the λex between 380 and 420 nm [27]. The spectra recorded in right-angled geometry are distorted due to inner-filter effect. Diluted wine distillates exhibit two fluorescence bands centered at the λex /λem pairs of 280/350 nm and 330/430 nm, respectively (**Table 2**). The short-wavelength band is similar to the one observed in the fluorescence spectra of other distilled spirits and it may partly originate from compounds of the grape distillate. The long-wavelength band originates mainly from caramel [28].

2.2. Spirit drinks aged in wooden casks

Freshly distilled spirits are colourless and possess only the flavour and aroma of the grain and the alcohol. Many producers use "ageing wooden barrels" to mature distilled spirits like brandy, Calvados, whisky, mezcal, cachaça and tequila. Barrels are typically made of French or American oak, but chestnut and redwood are also used. The ageing involves several processes: lignins decompose with formation of phenolic compounds (vanillin, syringaldehyde, coniferaldehyde, sinapaldehyde, cinnamic and benzoic acids), hydrolysable tannins and their products (gallic and ellagic acids) and coumarins (particularly scopoletin) are extracted from wood, and reactions may occur between components of wood and spirit. These processes and their products are very important for the quality of the matured spirits (taste, flavour and colour) [29]. In addition, phenolic compounds and coumarins are well-known fluorophores.

Brandy is a spirit drink produced from wine spirit, whether or not blended with a wine distillate. Types of brandies, originally at least, tended to be location-specific. Brandy has to be aged for a certain period in oak casks. Using right-angled geometry, the TLS contours for bulk brandy are concentrated in the λem region from 510 to 570 nm and λex region from 430 to 480 nm [27]. Using front-face geometry, the total luminescence contours for bulk brandy are concentrated in the λem region from 470 to 520 nm and λex region from 390 to 430 nm [30]. Undiluted brandy exhibits a high UV/VIS absorption, thus the fluorescence recorded on the bulk brandy is severely distorted due to the inner filter effects. The short-wavelength fluorescence, with λex,max = 280 nm and λem,max =370 and 450 nm, is clearly observed for diluted brandy samples, along with the longer-wavelength fluorescence, with excitation at 340 nm and emission at 450 nm (**Table 2**). The former band is preliminary attributed to the tryptophol, tyrosol and phenolic acids, the latter band to cinnamic acids, coumarins, tannins and other unknown fluorescent compounds [28].

Whisky (whiskey) is spirit-based drink made from malted or saccharified grains, which should mature for at least 3 years in wooden barrels. Plain spirited caramel of a specific grade is added simply in order to adjust the consistency of the colour [31]. Regarding bulk whisky, front-face fluorescence spectrum recorded at λex = 404 nm exhibit a wide emission band in the 450–700 nm range with maximum at 520 nm. The fluorescent band arises from the caramel, coumarins, tannins and other fluorescent compounds originating from wooden casks [32]. Tequila and mezcal are two traditional Mexican distilled beverages with similar production phases. Tequila must be made exclusively from Agave tequilana Weber blue variety, whereas mezcal is made from different agave species, among them A. salmiana, A. angustifolia and A. potatorum [33]. Maturation of mezcal and tequila is optional, contributing flavour in a similar way to all the other wood-matured spirits. Using liquid chromatography with ion trap mass spectrometry detection, ten phenolic acids were quantified in tequilas [34]. Fluorescence spectra of bulk mezcal obtained using right-angled geometry have emission maximum at about 580 nm (λex = 517 nm). White/young mezcal exhibit spectra similar to ethanol. On the other hand, aged mezcal, and the other types of mezcal differ in the intensity of the emission spectra due to the higher concentration of organic molecules extracted from the wood cask [35, 36]. Using the fluorescent background of Raman spectra, it has been possible to distinguish tequila blanco (unaged) from aged tequila [37]. Later fluorescence between 370 and 510 nm of bulk tequila excited at 337 nm has been observed [38]. Recently reference [39] reported the right-angled fluorescence spectra recorded at four λex (255, 330, 365 and 405 nm) by original tequilas and counterfeit tequilas.

Cachaça, the most popular distilled alcoholic beverage in Brazil, is a distilled spirit made from sugarcane juice. It can be aged in barrels of amendoim (*Pterogyne nitens*), balsam (*Myroxylon peruiferum*), jequitibá (*Cariniana estrellensis*), umburana (*Amburana cearensis*) and oak (*Quercus sp.*). Using HPLC-ESI-MS(n), 14 phenolic compounds and two coumarins were detected in sugarcane spirit extracts of six different Brazilian woods and oak, commonly used by cooperage industries for ageing cachaça [40]. TLSs of bulk cachaça exhibit excitation and emission maxima in the range 260–380 nm and 320–480 nm (**Table 1**), respectively, corresponding to phenolic acids (gallic, syringic, vanillic and ellagic), phenolic aldehydes (sinapaldehyde, coniferaldehyde, syringaldehyde and vanillin) and coumarin [41].

Calvados is an apple-brandy of France. Fluorescent compounds such as 4-vinylanisole, 4-methylguaiacol, methyleugenol, 4-ethylguaiacol, eugenol, 4-ethylphenol and 4-vinylguaiacol are found in freshly distilled Calvados [42], while 2-phenylethanol, 4-methylguaiacol, methyleugenol, 4-ethylguaiacol, eugenol and 4-ethylphenol [43] in matured Calvados. Bulk Calvados is easily distinguishable from the other fruit drinks because its λex,max and λem,max are considerably higher. Diluted Calvados revealed the same fluorescence band as that observed for diluted grape brandies—wine spirits aged in oak barrels. The band could be due to the presence of phenolic compounds extracted from wood [26].

The absorption of undiluted aged spirit samples is from 1 up to 5 absorbance units, thus, brandies, cachaças and mezcals have by far the highest absorbances, regardless of wavelength [33, 44, 45]. Therefore, the analysis of spectra recorded using right-angled geometry, which are affected by inner filter effects, may lead to spectral misinterpretation and invalid assignments of origin of some fluorescent bands. So far, fluorescence spectra unaffected by inner filter effects are available only for diluted brandy, mixed wine spirit and Calvados.

3. Applications of fluorescence spectroscopy

3.1. Quantification of individual components

3.1.1. Naturally occurring components

To determine alcohol, several fluorescence biosensors have been produced by integrating alcohol oxidase or alcohol dehydrogenase enzymes with optical fibers. The utility of enzyme biosensors is restricted due to their low stability and short lifetime determined mainly by enzyme kinetics, the necessity to add the coenzyme to the solution and the temperature [46–48].

Chemosensors are another big group of devices for the determination of alcohol. The application of a fluorescent reagent, fluorescein octadecyl ester, in a fiber optic sensor for the determination of aliphatic alcohols in a range of 10–60 v/v % has been reported [49]. Fluorescence intensity was enhanced due to the formation of hydrogen bonds between alcohol and the hydroxyl group of fluorescein octadecyl ester [49]. The fluorescence quenching of the 5,10,15,20-tetraphenyl porphyrin doped on polyvinyl chloride film by ethanol showed a linear response over the ethanol concentration in the range of 1–75 v/v % with a detection limit of 0.05 v/v % [47]. Using admixture of terphenyl-ol and sodium carbonate, which exhibited bright sky-blue fluorescence in the solid state upon addition of small quantities of ethanol, detection limit at about 5 v/v % of ethanol was demonstrated [50]. A simple visual test has been developed to check the ethanol content of drinks and to detect counterfeit beverages containing methanol. When imidazolium-based dication C10(mim)2 and dianionic 2,2′-azino-bis (3-ethylbenzothiazoline-6-sulfonic acid) are mixed together, they self-assemble into a supramolecular ionic material (SIM). The product is capable of encapsulating the fluorescent dye Rhodamine 6G (R6G) to form SIM-R6G. The addition of ethanol destructs the R6G-SIM structure, resulting in the release of R6G. Alcohol content can be determined by measuring the fluorescence line of R6G on a thin-layer chromatography (TLC) plate within a concentration range from 15 to 40%. The addition of a trace amount of methanol leads to a large increase of the length of R6G on TLC plates [51]. Another supramolecular material has been prepared with 1,4-bis(imidazol-1-ylmethyl)benzene (bix) as the ligand, Zn^{2+} as the central metal ion and encapsulated fluorescent dye Rhodamine B (RhB). The formed RhB/Zn(bix) is stable in ethanol, however, the addition of water results in the release of RhB, allowing the determination of alcohol content within a linear range from 20 to 100 v/v % [52].

The appropriateness of both spectrofluorimetry and HPLC to determine the level of individual coumarins (umbelliferone, scopoletin and 4-methylumbelliferone) in commercial white rum samples has been demonstrated [53]. Recently a simple multivariate calibration spectrofluorimetric method has been developed for the simultaneous determination of gallic, vanillic, syringic and ferulic acids and scopoletin in brandy samples, providing comparable results with those obtained by HPLC method [12].

Ellagic acid is the most explored phenolic acid compound, probably due to direct extraction of free ellagic acid and hydrolysis of wood ellagitannins [54]. Two spectrofluorimetric methods have been developed for the rapid determination of ellagic acid in brandy samples. The first method was based on the complex formation between ellagic acid and borax in methanol solution (λex/λem = 383/456 nm). In the second method, the complex was formed between ellagic acid and boric acid in ethanol solution (λex/λem = 387/447 nm). The limit of determination was at about 0.3 μg/L. The results were found to be in good agreement with those obtained by HPLC method [55]. The potential of SFS (Δλ = 40 nm) has been demonstrated to differentiate caramel from oak wood extract. The method was selective for the determination of caramel in the presence of common components of brandies (gallic acid, syringic acid, vanillic acid, caffeic acid, ferulic acid, p-coumaric acid, vanillin, syringaldehyde, coniferaldehyde, sinapaldehyde, furfural, 5-hydroxymethylfurfural and scopoletine). The limit of determination was 5 mg/L for caramel [56].

3.1.2. *Contaminants*

AMPHORA project, which assessed the quality of illegally and informally produced alcohol in the European Region, reports that compared to the health effects of ethanol, the contamination problems may be of minor importance as exposure will only in worst-case scenarios reach tolerable daily intakes of the substances as ethyl carbamate, copper manganese, acetaldehyde, methanol, higher alcohols and phthalates [57]. The incidence of the aldehydes, especially of formaldehyde, in the Asian samples was considerably higher than that found in European alcoholic beverages [58].

Fluorimetry with Hantzsch reaction is commonly used for the determination of formaldehyde. Cyclohexane-1,3-dione (CHD) [59] and 4-amino-3-penten-2-one (Fluoral-P) [60, 61] have been used as Hantzsch reaction reagents. The Fluoral-P method is based on the reaction of 4-amino-3-penten-2-one with formaldehyde, producing 3,5-diacetyl-1,4-dihydrolutidine, which fluoresces at 510 nm when excited at 410 nm. The method is specific for formaldehyde, allowing for the determination of this analyte even in the presence of acetaldehyde concentrations 1000 times higher than formaldehyde [60]. Limit of detection was 3 μg/L for formaldehyde in cachaça, rum and vodka [61]. Aldehydes, such as formaldehyde, acetaldehyde, propionaldehyde and *n*-butyraldehyde, can completely react with CHD to form fluorescence derivatives (9-substituted decahydroacridine-l,8-diones) in the presence of ammonium acetate in 1 h at 60°C. The application of microwave irradiation accelerates considerably the derivatisation reaction of formaldehyde with CHD and allows attaining a limit of detection 0.02 μg/L for formaldehyde in a shorter derivatisation reaction time. Tolerated ratio of acetaldehyde was 100 in the determination of 0.500 mg/L formaldehyde [59].

Based on the fluorescence properties of 2,4-(1H,3H)-quinazolinedione (λex/λem = 310/410 nm), a product of the reaction between cyanate and 2-aminobenzoic acid, a method for the determination of cyanate was developed with a limit of detection 4 μg/L. A correlation between the cyanate and ethyl carbamate concentrations in the sugar cane spirit was observed [62].

Fluorescent molecularly imprinted polymer (fluorescein 5(6)-isothiocyanate-3-aminopropyltriethoxysilane /SiO_2 particles) has been used for the selective recognition and the determination of λ-cyhalothrin (pesticide) in Chinese spirits. Based on fluorescence quenching, the limit of detection 4 μg/L was obtained [63].

Recently, a rapid methodology has been proposed for simultaneous quantification of five PAHs (acenaphten, anthracene, benzo[*a*]pyrene, fluoranthene and pyrene) in three types of spirits (rum, cachaça and vodka) [64].

3.1.3. Drugs

Three most commonly used drugs in drink spiking are ketamine, benzodiazepines, including diazepam and flunitrazepam, and gamma-hydroxybutyric acid (GHB).

The determination of diazepam in commercial beverages, previously spiked with drug, has been implemented through photo degradation of diazepam and detection of degradation products at λem= 463 nm (λem = 262 nm). The limit of detection was 2 mg/L [65]. A screening method for flunitrazepam in colourless alcoholic beverages is based on emission at 472 nm of protonated drug given the limit of detection 1 mg/L [66].

Zhai group has recently reported the first fluorescent sensor for gamma-butyrolactone (GBL), the pro-drug of GHB. GBL sensor was named Green Date and required an extraction to eliminate alcohol effects for GBL detection in real drinks [67]. The team also found that an orange fluorescent compound named GHB Orange is capable of detecting GHB in different beverages with explicit intensity change under the irradiation of a hand-held 365 nm lamp [68].

3.2. Classification of spirit drinks

Visual inspection of fluorescence spectra seldom shows that they fall naturally into a number of groups [25, 39]. Thus, pattern recognition methods are usually required to gain significant meaningful information from the spectrometric data (**Table 3**). Non-supervised pattern recognition methods as hierarchical cluster analysis (HCA) or principal component analysis (PCA) discover, previously unknown, the group structure in the data. With supervised pattern recognition methods, the number of groups is known in advance and representative samples of each group are available. This information is used to develop a suitable discriminating rule or discriminate function with which new, unknown samples can be assigned to one of the groups. Supervised pattern recognition methods as linear discriminant analysis (LDA), general discriminant analysis (GDA), k-nearest neighbour (kNN), support vector machine (SVM) and partial least squares discriminant analysis (PLS-DA) can be used. The choice of the chemometric method often depends on preference of the analyst and the complexity of the data. LDA requires the number of variables (wavelengths) smaller than the number of samples in each group. Consequently, large spectral datasets with few samples cannot be analysed using LDA. As PCA is a dimensionality reduction method, combining LDA with a PCA overcomes this problem. On the other hand, PLS-DA is well suited to deal with a much larger number of variables than samples [6]. Parallel factor analysis (PARAFAC) is commonly used for modeling fluorescence excitation-emission data. PARAFAC decomposition gives the loading and the score profiles of the components. The comparison of loading profiles of component with the fluorescence spectra for a standard of the analyte often leads to the identification of the fluorophore. Calibration model can be obtained by PLS regression between the scores related

http://dx.doi.org/10.5772/64002

to the fluorophore and the reference concentrations of the fluorophore in the calibration samples [69].

Sample	Spectral region[a]	Multivariate analysis[b]	Purpose of analysis and quality of method (the percentage of correct classification in the prediction step)[a, b, c]	Reference
Quality				
Brandy	EX (225–425 nm), λem = 440 nm; EM (360–650 nm), λex = 350 nm; SFS (200–700 nm), Δλ=90 nm	PCA, HCA	Classification of bulk brandies and mixed wine spirits using front-face geometry	[30]
	EX (225–460 nm), λem=470 nm; EM (400–650 nm), λex=390 nm; SFS (200–700 nm), Δλ=80 nm	PCA-LDA, HCA	Classification of bulk brandies and mixed wine spirits using right-angled geometry; SFS (PCA-LDA): 99.6% classification	[27]
	EX (240–380 nm), λem= 450 nm; EM (400–470 nm), λex=340 nm	PCA, HCA	Classification of diluted brandies and mixed wine spirits	[28]
	SFS 220-700 nm, Δλ = 40 nm	PCA-LDA, HCA	Classification of diluted brandies and mixed wine spirits; SFS (PCA-LDA): 99.2% classification	[70]
Mezcas	EM (540-800 nm), λex=514 nm	PCA	Classification of the group including white mezcals (non-maturated) and ethanol from the group including rested (matured 2 months in wood casks), abocado (white or young mezcal artificially coloured and flavoured) and distilled mezcals (coloured white mezcal)	[35]
Fruit	SFS (200-500 nm). Δλ=10, 90 and 100 nm	PCA-LDA, GDA	Classification of apple, apricot, pear, and plum spirits ; PCA-LDA: 100, 90 and 90% classification for Δλ = 10, 90 and 100 nm, resp.; GDA: 100% classification regardless Δλ used	[26]
Tequila	EM (250-800 nm). λex = 255, 330, 365 and 405 nm		Discrimination adulterated and counterfeit tequilas from the genuine ones (λex = 255 nm), and aged, rested, and mixed tequilas from fake ones (λex = 330, 365, and 405 nm).	[39]
Cachaça	UV/VIS (190-500 nm); EM (260-600 nm). λex=250, 280, 330, 360, and 450 nm; fusion of the UV/VIS and EM	PLS-DA, NPLS-DA	Prediction of the wood used in the ageing of commercial cachaças; UV/VIS (PLS-DA): 56–89% classification; EM (NPLS-DA): 37–91% classification; Low-level fused UV/VIS and EM data (PLS-DA): 60–94% classification	[44]
Whisky	UV-Vis (290-600 nm); NIR (1200-1880 nm); EM (450-700 nm), λex= 404 nm	PCA-LDA	Distinguishing between the single-malt whiskies and the commercial-grade blended whiskies; 100% classification	[32]

Sample	Spectral region[a]	Multivariate analysis[b]	Purpose of analysis and quality of method (the percentage of correct classification in the prediction step)[a, b, c]	Reference
Region				
Whisky	UV-Vis (290-600 nm); NIR (1200-1880 nm); EM (450-700 nm), λex= 404 nm	PCA-LDA	Classification of single-malt whiskies come from two main production areas, the islands and the highlands, respectively: 89% classification	[32]
JFSD	UV (250–325 nm); SFS (250–450 nm), Δλ = 10 nm; TLS (λem = cx275–490 nm, λex = 250–400 nm)	PCA-LDA, PARAFAC-LDA	Distinguishing between Slovak, Belgian, German, Czech and British JFSDs; UV (PCA-LDA) 88 %, SFS (PCA-LDA) 97 %, TLS (PARAFAC-LDA) 88 %	[19]
Plum	SFS (230–550 nm), Δλ = 60 nm	PCA-LDA	Differentiation of Czech, Hungarian and Slovak plum spirits; 100 % classification	[72]
Producer				
JFSD	SFS (250–350 nm), Δλ = 10 nm	PCA-LDA, GDA, *k*NN, SVM	Distinguishing between (1) drinks from different producers and (2) distillates of different geographical indications and others; GDA: 100 % classification	[25]
Adulteration				
Brandy	TLS (λem = 485–580 nm, λex = 363–475 nm)	PARAFAC-PLS	Determination of the mixed wine spirit in adulterated brandy; RMSEP: 1.9%, R^2Pred: 0.995.	[73]
Brandy	TLS (λem = 510–600 nm, λex = 393–497 nm)	PARAFAC-PLS	Determination of the adulterants (water, ethanol, methanol) in adulterant-brandy blends; RMSEP: 0.24%, 0.20% and 0.22%, R^2 Pred 0.993, 0.997 and 0.995 for water, ethanol and methanol, respectively.	[74]
Fruit	TLS (λem = 315–450 nm, λex = 240–305 nm)	PARAFAC-PLS	Determination of water or ethanol in adulterant-fruit spirit blends; apple spirit: RMSEP: 1,8% and 1.9%, R^2 Pred 0.92 and 0.90, for ethanol and water, respectively; plum spirit: RMSEP: 3.5% and 0.7%, R^2 Pred 0.66 and 0.99, for ethanol and water, respectively.	[75]

[a] EX excitation wavelength, EM emission wavelength, SFS synchronous fluorescence spectrum, Δλ wavelength interval, UV/VIS ultraviolet/visible, NIR near infrared, TLS total luminescence spectrum.
[b] PCA principal component analysis, HCA hierarchical cluster analysis, LDA linear discriminant analysis, GDA general discriminant analysis, PLS-DA partial least squares regression discriminant analysis, NPLS-DA multi-way partial least squares discriminant analysis, PARAFAC-LDA parallel factor analysis-linear discriminant analysis, *k*NN k-nearest neighbour, SVM support vector machine.
[c] RMSEP root mean square error of prediction, R^2Pred coefficient of determination of prediction.

Table 3. Application of fluorescence spectroscopy and pattern recognition methods.

3.2.1. *Classification of spirit drinks according to the quality*

Spirit drinks can be sometimes adulterated in the flavour and colour to imitate the sensorial and visual characteristics of the authentic matured beverages. Thus, one way of classifying spirit drinks is as aged or unaged—for example, brandy or less expensive mixed wine spirit. The λex/λem values of the major peaks of the bulk brandies are generally longer than those recorded for bulk mixed wine spirits. Thus, both PCA and HCA carried out on the front-face emission spectra recorded at λex = 350 nm and SFSs collected at $\Delta\lambda$ = 90 nm provided very good differentiation between brandies and mixed wine spirits. Less good classification was obtained using excitation spectra recorded at λem = 440 nm [30]. Right-angle fluorescence spectroscopy can be used as an alternative method to front-face fluorescence technique, exigent of special front surface accessory, as both the techniques provide similar classification [27]. Regardless of fluorescence technique used, scattering is much more intense and/or heterogeneous for mixed wine spirits than for brandies and can result from the presence of the colloids in mixed wine spirits. Although the phenomenon was not studied in detail, the differences between brandy and mixed wine spirit are also due to scatter bands [27, 30]. Regarding classification of diluted samples, again better results were obtained from excitation and synchronous fluorescence spectra [28, 70].

UV-absorption and fluorescence spectroscopy have been compared for the evaluation of the authenticity of matured mezcal. The results showed that PCA conducted over a set of UV absorption spectra allows a reliable discrimination between artificially and naturally maturated mezcals. On the other hand, PCA conducted over fluorescence spectra allowed the identification of two main groups, not necessarily correlated with maturation in the wood casks (**Table 3**) [35].

Raman spectroscopy has been able to distinguish unaged (silver) tequila from aged tequilas by the application of a PCA to the fluorescence background of the Raman spectra [37]. The same authors observed that the lower and highest fluorescence background of the Raman spectra corresponds to the Herradura tequila and Rancho Escondido distillated of the given samples, respectively. It is supposed that this fluorescence background behaviour is related with the production processes of the samples [37]. PCA performed on the combination of Raman spectra and the fluorescent background information has been used to classify various brands of whiskies based on flavour, age and type of cask. The fluorescence decay constant can be also used as another parameter to distinguish whisky types which are otherwise non-distinguishable [71].

The character and potential nutritional value of spirits is reliant, among others, on the type of wood used for the barrel in which spirits may be aged. UV-Vis spectrophotometry and fluorescence spectrometry have been compared for the discrimination of the cachaças according to the wood used in their ageing. It was observed that the PLS-DA based on UV-Vis spectrophotometry provided better results for two classes of aged cachaça, amendoim and jequitibá, whereas NPLS-DA of emission spectra recorded at λex = 250, 280, 330, 360, and 450 nm provided better results for the other two classes of aged cachaças, balsam and oak. For the class of cachaça aged in umburana, both models provided similar and good results. Consequently, a fused PLS-DA model based on the UV-Vis and emission spectra was developed, providing highest classification for four out of the five analysed classes. The only exception was the class of samples aged in oak, better classified using emission spectra and NPLS-DA [44].

Using the combination of absorption (UV/VIS, NIR) and fluorescence spectroscopic data, it has been possible to distinguish the single-malt whiskies from the commercial-grade blended whiskies. First, PCA was applied to each data-block. Next a joint-data matrix containing PC1 and PC2 scores from UV/VIS data, PC1, PC2 and PC3 scores from NIR data and PC1 scores from fluorescence data was created. Then, LDA was applied to this matrix, and 100% classification was obtained [32].

3.2.2. Classification of spirit drinks according to the region of production

A few papers have been published on the use of fluorescence to classify spirit drinks according to the region of production. UV absorption spectra, TLS and SFS combined with PCA, PARAFAC and LDA were applied to distinguish between Slovak, Belgian, German, Czech and British JFSDs. PCA-LDA performed on the UV spectral data showed a good discrimination of Slovak, British, German and Czech drinks; however, the UV spectra failed to discriminate Belgian samples. LDA applied to the PARAFAC components calculated on TLS showed correct classification for German, Czech and Belgian drinks, whereas British samples were classified as belonging to Slovak group. PCA-LDA performed on the SFS data lead to the best discrimination as only one Slovak sample was classified as Belgian in the prediction step [19].

SFS combined with PCA-LDA have been used for the differentiation of plum spirits according to their geographical origin. The samples were divided in two categories: colourless and coloured. All colourless and Czech and Hungarian coloured samples were properly classified in both calibration and prediction sets. A group of Slovak coloured was classified as belonging to the Hungarian group in the calibration set; however, it was correctly classified in the prediction step [72].

SFS and pattern-recognition methods have been used for searching the natural grouping among Slovak JFSDs. LDA was applied to the first PCs; however, GDA, *k*NN and SVM were performed on the whole SFS. Regarding different producers, both GDA and SVM resulted in 100% correct classification. Regarding geographical indication, 100% correct classification was obtained using GDA [25].

3.3. Determination of adulterants

TLS and PARAFAC-PLS have been used for the determination of the adulterants (mixed wine spirits, water, ethanol and methanol) in adulterant-brandy blends [73, 74]; the best results were obtained for ethanol (RMSEP = 0.20% and R^2 Pred = 0.997). A comparison with UV/VIS absorption and NIR spectroscopy showed that the fluorescence method is slightly less sensitive than UV/VIS absorption, but more sensitive than the NIR technique in the process of determining the percentage of adulterant (water, ethanol and methanol) in the adulterant–brandy blend. NIR technique showed the best discrimination of the adulterant type [74]. Regarding determination of water or ethanol in adulterant-fruit spirit blends, PARAFAC-PLS provided a model with very limited predictive ability for ethanol-plum spirit blends (RMSEP = 3.5% and R^2 Pred = 0.66) [75].

http://dx.doi.org/10.5772/64002

4. Conclusions

Our literature survey revealed that the intrinsic fluorescence from spirit drinks contains valuable information on the quality and origin of such products. Many of the reported studies examining the potential of fluorescence spectroscopy to classify spirit drinks and/or quantify adulterants in spirit drinks until now have been preliminary or feasibility studies, performed on a limited number of samples. This was mainly due to the price and complexity of collecting an adequate number of samples with sufficient variation within the sample set. Therefore, appropriate verification should always be performed before implementation of any such method. The results presented were usually achieved using a conventional spectrophotometer, which can be replaced by diode lasers or bright light-emitting diodes as good alternative light sources. This reduces hardware complexity and can lead to a compact portable device to be used for authentication or fraud detection. The increasing research work is needed to better explore the connection between chemical composition and fluorescence spectra, which in most cases is not fully described. Instead, the tentative assignments of fluorophores are suggested in the application studies. Thus, fluorescence spectroscopy presents several opportunities for future research with potential application in spirit drink analysis.

Acknowledgements

This research was supported by the Scientific Grant Agency of the Ministry of Education of Slovak Republic and the Slovak Academy of Sciences VEGA No 1/0499/14.

Author details

Jana Sádecká*, Veronika Uríčková and Michaela Jakubíková

*Address all correspondence to: jana.sadecka@stuba.sk

Institute of Analytical Chemistry, Faculty of Chemical and Food Technology, Slovak University of Technology, Bratislava, Slovak Republic

References

[1] Sádecká J., Tóthová J. Fluorescence spectroscopy and chemometrics in the food classification. Czech Journal of Food Sciences. 2007;25:159–173.

[2] Karoui R., Blecker Ch. Fluorescence spectroscopy measurement for quality assessment of food systems—a review. Food Bioprocess Technol. 2011;4:364–386.

[3] Al-Rawashdeh N.A.F. Current achievement and future potential of fluorescence spectroscopy. In: Uddin J., editor. Macro To Nano Spectroscopy. InTech; 2012. pp. 209–250. DOI: 10.5772/48034.

[4] Regulation (EC) No. 110/2008 of the European Parliament and of the Council of 15 January 2008 on the definition, description, presentation, labelling and the protection of geographical indications of spirit drinks and repealing Council Regulation (EEC) No. 1576/89. Official Journal of the European Union. 2008;L39:16–54.

[5] Buglass A.J., editor. Handbook of Alcoholic Beverages: Technical, Analytical and Nutritional Aspects, 2 Volume Set. Chichester: Wiley; 2011. 1204 p. ISBN: 978-0-470-51202-9.

[6] Otto M. Chemometrics. 2nd ed. Weinheim: Wiley-Vch; 2007. 343 p. ISBN: 978-3-527-31418-8.

[7] Lakowicz J.R., editor. Principles of Fluorescence Spectroscopy. 3rd ed. Springer US; 2006. 954 p. DOI: 10.1007/978-0-387-46312-4.

[8] Guilbault G.G., editor. Practical Fluorescence. 2nd ed. Marcel Dekker; 1990. 826 p. ISBN: 978-0824783501.

[9] Airado-Rodríguez D., Durán-Merás I., Galeano-Díaz T., Wold J.P. Front-face fluorescence spectroscopy: A new tool for control in the wine industry. Journal of Food Composition and Analysis. 2011;24:257–264.

[10] Lloyd J. B. F. Synchronized excitation of fluorescence emission spectra. Nature. 1971;231:64-65. DOI: 10.1038/10.1038/physci231064a0.

[11] Divya O., Mishra A.K. Understanding the concept of concentration-dependent red-shift in synchronous fluorescence spectra: Prediction of λmax, SFS and optimization of $\Delta\lambda$ for synchronous fluorescence scan. Analytica Chimica Acta. 2008;630:47–56. DOI: 10.1016/j.aca.2008.09.056.

[12] Žiak Ľ., Sádecká J., Májek P., Hroboňová K. Simultaneous determination of phenolic acids and scopoletin in brandies using synchronous fluorescence spectrometry coupled with partial least squares. Food Analytical Methods. 2014;7:563-570. DOI: 10.1007/s12161-013-9656-y.

[13] Liu Y., Luo X.S., Shen Z.H., Lu J., Ni X.W. Studies on molecular structure of ethanol-water clusters by fluorescence spectroscopy. Optical Review. 2006;13:303-307.

[14] Bin W., Ying L., Caiqin H., Xiaosen L., Jian L., Xiaowu N. Derivative fluorimetry analysis of new cluster structures formed by ethanol and water molecules. Chinese Optics Letters. 2009;7:159-161. DOI: 10.3788/COL20090702.0159.

[15] Hua Q., Shengwan Z., Wei W. Fluorescence spectroscopic and viscosity studies of hydrogen bonding in Chinese Fenjiu. Journal of Bioscience and Bioengineering. 2013;115:405-411. DOI: 10.1016/j.jbiosc.2012.10.023.

[16] Lachenmeier D.W., Anh P.T.H., Popova S., Rehm J. The quality of alcohol products in Vietnam and its implications for public health. International Journal of Environmental Research and Public Health. 2009;6:2090–2101. DOI: 10.3390/ijerph6082090.

[17] Siříšťová L., Přinosilová Š., Riddellová K., Hajšlová J., Melzoch M. Changes in quality parameters of vodka filtered through activated charcoal. Czech Journal of Food Science. 2012;30:474–482.

[18] Thete A.R., Henkel T., Göckeritz R., Endlich M., Köhler J.M., Gross G.A. A hydrogel based fluorescent micro array used for the characterization of liquid analytes. Analytica Chimica Acta. 2009;633:81–89. DOI: 10.1016/j.aca.2008.11.030.

[19] Sádecká J., Uríčková V., Hroboňová K., Májek P. Classification of juniper-flavoured spirit drinks by multivariate analysis of spectroscopic and chromatographic data. Food Analytical Methods. 2015;8:58–69. DOI: 10.1007/s12161-014-9869-8.

[20] Janáčová A., Jakubík T., Pažitná A., Špánik I. The comparison of VOC composition of juniper favored spirit drinks from various EU countries using GC-TOFMS. In: Joint Congress 2011 including 35th ISCC, 26th MSB and 8th GCxGC Symposium; 1–5 May 2011; San Diego.

[21] Robbat Jr. A., Kowalsick A., Howell J. Tracking juniper berry content in oils and distillates by spectral deconvolution of gas chromatography/mass spectrometry data. Journal of Chromatography A. 2011;1218:5531–5541. DOI: 10.1016/j.chroma.2011.06.053.

[22] Vichi S., Riu-Aumatell M., Buxaderas S., López-Tamames E. Assessment of some diterpenoids in commercial distilled gin. Analytica Chimica Acta. 2008;628:222–229. DOI: 10.1016/j.aca.2008.09.005.

[23] Zhan H., Jiang Z.T., Wang Y., Li R., Dong T.S. Molecular microcapsules and inclusion interactions of eugenol with β-cyclodextrin and its derivatives. European Food Research and Technology. 2008;227:1507–1513. DOI 10.1007/s00217-008-0873-3.

[24] Mateo C.R., Prieto M., Micol V., Shapiro S., Villalain J. A fluorescence study of the interaction and location of (+)-totarol, a diterpenoid bioactive molecule, in model membranes. Biochimica et Biophysica Acta. 2000;1509:167–175. DOI: 10.1016/S0005-2736(00)00291-1.

[25] Uríčková V., Sádecká J., Májek P. Classification of Slovak juniper-flavoured spirit drinks. Journal of Food and Nutrition Research. 2015;54:298–307.

[26] Tomková M., Sádecká J., Hroboňová K. Synchronous fluorescence spectroscopy for rapid classification of fruit spirits. Food Analytical Methods. 2015;8:1258–1267. DOI: 10.1007/s12161-014-0010-9.

[27] Uríčková V., Sádecká J., Májek P. Right-angle fluorescence spectroscopy for differentiation of distilled alcoholic beverages. Nova Biotechnologica et Chimica. 2013;12:83–92. DOI: 10.2478/nbec-2013-0010.

[28] Tóthová J., Sádecká J., Májek P. Total luminescence spectroscopy for differentiating between brandies and wine distillates. Czech Journal of Food Sciences. 2009;27:425–432.

[29] Mosedale J.R., Puech J.L. Wood maturation of distilled beverages. Trends in Food Science & Technology. 1998;9:95–101. DOI: 10.1016/S0924-2244(98)00024-7.

[30] Sádecká J., Tóthová J., Májek P. Classification of brandies and wine distillates using front face fluorescence spectroscopy. Food Chemistry. 2009;117:491–498. DOI: 10.1016/j.foodchem.2009.04.053.

[31] The Scotch Whisky Regulations 2009, UK Government [Internet]. 2009. Available from: http://www.legislation.gov.uk/uksi/2009/2890/pdfs/uksi_20092890_en.pdf [Accessed: 2016-02-01]

[32] Mignani A.G., Ciaccheri L., Gordillo B., Mencaglia A.A., González-Miret M.L., Heredia F.J., et al. Identifying the production region of single-malt Scotch whiskies using optical spectroscopy and pattern recognition techniques. Sensors and Actuators B. 2012;171–172:458–462. DOI: 10.1016/j.snb.2012.05.01.

[33] Muñoz-Muñoz A.C., Pichardo-Molina J.L., Ramos-Ortíz G., Barbosa-García O., Maldonado J.L., Meneses-Nava M.A., et al. Identification and quantification of furanic compounds in tequila and mezcal using spectroscopy and chemometric methods. Journal of the Brazilian Chemical Society. 2010;21:1077–1087. DOI: 10.1590/S0103-50532010000600018.

[34] Magana A.A., Wrobel K., Elguera J.C.T., Escobosa A.R.C., Wrobel K. Determination of small phenolic compounds in tequila by liquid chromatography with ion trap mass spectrometry detection. Food Analytical Methods. 2015;8:864–872. DOI: 10.1007/s12161-014-9967-7.

[35] Araujo-Andrade C., Frausto-Reyes C., Gerbino E., Mobili P., Tymczyszyn E., Esparza-Ibarra E.L., Ivanov-Tsonchev R., Gómez-Zavaglia A. Application of principal component analysis to elucidate experimental and theoretical information. In: Sanguansat P, editor. Principal Component Analysis. InTech; 2012. pp. 23–48. DOI: 10.5772/36970.

[36] Ávila-Reyes J.A., Almaraz-Abarca N., Delgado-Alvarado E.A., González-Valdez L.S., del Toro G.V., Páramo E.D. Phenol profile and antioxidant capacity of mescal aged in oak wood barrels. Food Research International. 2010;43:296–300. DOI: 10.1016/j.foodres.2009.10.002.

[37] Frausto-Reyesa C., Medina-Gutiérrez C., Sato-Berrú R., Sahagún L.R. Qualitative study of ethanol content in tequilas by Raman spectroscopy and principal component analysis. Spectrochimica Acta Part A. 2005;61:2657–2662. DOI: 10.1016/j.saa.2004.10.008.

[38] Arvizu A.R. Detection of compounds by laser-induced fluorescence. Unidad Profesional Adolfo López Mateos Edificio N°7: Instituto Politécnico Nacional; 2007.

[39] Vázquez J.M.R., Fabila-Bustos D.A., Quintanar-Hernández L.F.J., Valor A., Stolik S. Detection of counterfeit tequila by fluorescence spectroscopy. Journal of Spectroscopy. 2015;2015:1–7. DOI: 10.1155/2015/403160.

[40] da Silva A.A., do Nascimento E.S., Cardoso D.R., Franco D.W. Coumarins and phenolic fingerprints of oak and Brazilian woods extracted by sugarcane spirit. Journal of Separation Science. 2009;32:3681–3691. DOI: 10.1002/jssc.200900306.

[41] Aquino F.W.B , Rodrigues S., Nascimento R.F., Casimiro A.R.S. Simultaneous determination of ageing markers in sugar cane spirits. Food Chemistry. 2006;98:569–574. DOI: 10.1016/j.foodchem.2005.07.034 .

[42] Ledauphin J., Saint-Clair J.F., Lablanquie O., Guichard H., Founier N., Guichard E., Barillier D. Identification of trace volatile compounds in freshly distilled Calvados and Cognac using preparative separations coupled with gas chromatography-mass spectrometry. Journal of Agricultural and Food Chemistry. 2004;52:5124–5134. DOI: 10.1021/jf040052y.

[43] Ledauphin J., Le Milbeau C., Barillier D., Hennequin D. Differences in the volatile compositions of French labeled brandies (Armagnac, Calvados, Cognac, and Mirabelle) using GC-MS and PLS-DA. Journal of Agricultural and Food Chemistry. 2010;58:7782–7793. DOI: 10.1021/jf9045667.

[44] Bernardes C.D., Barbeira P.J.S. Different chemometric methods for the discrimination of commercial aged Cachaças. Food Analytical Methods. 2016;9:1053–1059. DOI: 10.1007/s12161-015-0284-6.

[45] Savchuk S.A., Vlasov V.N., Appolonova S.A., Arbuzov V.N., Vedenin A.N., Mezinov A.B., et al. Application of chromatography and spectrometry to the authentication of alcoholic beverages. Journal of Analytical Chemistry. 2001;56:214–231. DOI: 10.1023/A:1009446221123.

[46] Kuswandi B., Ahmad M. Recent progress in alcohol biosensors. OA Alcohol. 2014;2:1–8.

[47] Bozkurt S.S., Merdivan E., Benibol Y. A fluorescent chemical sensor for ethanol determination in alcoholic beverages. Microchimica Acta. 2010;168:141–145. DOI: 10.1007/s00604-009-0271-y.

[48] Bosch M.E., Sánchez A.J.R., Rojas F.S., Ojeda C.B. Recent development in optical fiber biosensors. Sensors. 2007;7:797–859. DOI: 10.3390/s7060797.

[49] Zeng H.H., Wang K.M., Li D., Yu R.Q. Development of an alcohol optode membrane based on fluorescence enhancement of fluorescein derivatives. Talanta. 1994;41:969–975. DOI: 10.1016/0039-9140(94)E0103-X.

[50] Akamatsu M., Mori T., Okamoto K., Komatsu H., Kumagai K., Shiratori S., et al. Detection of ethanol in alcoholic beverages or vapor phase using fluorescent molecules

embedded in a nanofibrous polymer. ACS Applied Materials & Interfaces. 2015;7:6189–6194. DOI: 10.1021/acsami.5b00289.

[51] Zhang L., Qi H., Wang Y., Yang L., Yu P., Mao L. Effective visualization assay for alcohol content sensing and methanol differentiation with solvent stimuli-responsive supramolecular ionic materials. Analytical Chemistry. 2014;86:7280–7285. DOI: 10.1021/ac5014546.

[52] Deng J., Ma W., Yu P., Mao L. Colorimetric and fluorescent dual mode sensing of alcoholic strength in spirit samples with stimuli-responsive infinite coordination polymers. Analytical Chemistry. 2015;87:6958–6965. DOI: 10.1021/acs.analchem.5b01617.

[53] Fernández Izquierdo M.E., Quesada Granados J., Villalón Mir M., López Martinez M.C. Comparison of methods for determining coumarins in distilled beverages. Food Chemistry. 2000;70:251–258. DOI: 10.1016/S0308-8146(00)00071-6.

[54] Canas S., Belchior A.P., Spranger M.I., Bruno-de-Sousa R. High-performance liquid chromatography method for analysis of phenolic acids, phenolic aldehydes, and furanic derivatives in brandies. Development and validation. Journal of Separation Science. 2003;26:496–502. DOI: 10.1002/jssc.200390066.

[55] Sádecká J., Tóthová J. Spectrofluorimetric determination of ellagic acid in brandy. Food Chemistry. 2012;135:893–897. DOI: 10.1016/j.foodchem.2012.06.019.

[56] Sádecká J., Tóthová J. Determination of caramel in non-aged mixed wine spirits by synchronous fluorescence spectroscopy. European Food Research and Technology. 2010;230:797–802. DOI 10.1007/s00217-010-1221-y.

[57] Lachenmeier D.W., Leitz J., Schoeberl K., Kuballa T., Straub I., Rehm J. Quality of illegally and informally produced alcohol in Europe: results from the AMPHORA project. Adicciones. 2011;23:133–140.

[58] Lachenmeier D.W., Monakhova Y.B., Rehm J., Kuballa T., Straub I. Occurrence of carcinogenic aldehydes in alcoholic beverages from Asia. The International Journal of Alcohol and Drug Research. 2013;2:31–36. DOI: 10.7895/ijadr.v2i2.88.

[59] Zhao X.-Q., Zhang Z.-Q. Microwave-assisted on-line derivatization for sensitive flow injection fluorometric determination of formaldehyde in some foods. Talanta. 2009;80:242–245. DOI: 10.1016/j.talanta.2009.06.066.

[60] de Andrade J.B., Bispo M.S., Rebouças M.V., Carvalho M.L.S.M., Pinheiro H.L.C. Spectrofluorimetric determination of formaldehyde in liquid samples. American Laboratory. 1996;8:56–58.

[61] de Oliveira F.S., Sousa E.T., de Andrade J.B. A sensitive flow analysis system for the fluorimetric determination of low levels of formaldehyde in alcoholic beverages. Talanta.2007;73:561–566. DOI: 10.1016/j.talanta.2007.04.027.

[62] Ohe T.H.K., da Silva A.A., Rocha T.daS., de Godoy F.S., Franco D.W. A fluorescence-based method for cyanate analysis in ethanol/water media: correlation between cyanate presence and ethyl carbamate formation in sugar cane spirit. Journal of Food Science. 2014;79:C1950–C1955. DOI: 10.1111/1750-3841.12587.

[63] Wang J., Gao L., Han D., Pan J., Qiu H., Li H., et al. Optical detection of λ-cyhalothrin by core-shell fluorescent molecularly imprinted polymers in Chinese spirits. Journal of Agricultural and Food Chemistry. 2015;63:2392–2399. DOI: 10.1021/jf5043823.

[64] da Silva A.C. Second-order calibration model for spectrofluorimetric determination of polycyclic aromatic hydrocarbons in distilled drinks. Cidade Universitária: Federal University of Paraíba; 2015

[65] Ribeiro D.S., Prior J.A., Santos J.L., Lima J.L. Automated determination of diazepam in spiked alcoholic beverages associated with drug-facilitated crimes. Analytica Chimica Acta. 2010;668:67–73. DOI: 10.1016/j.aca.2010.01.016.

[66] Leesakul N., Pongampai S., Kanatharana P., Sudkeaw P., Tantirungrotechaid Y., Buranachaic C. A new screening method for flunitrazepam in vodka and tequila by fluorescence spectroscopy. Luminescence. 2013;28:76–83. DOI: 10.1002/bio.2348.

[67] Zhai D., Agrawalla B.K., Eng P.S., Lee S.C., Xu W., Chang Y.T. Development of a fluorescent sensor for an illicit date rape drug–GBL. Chemical Communications. 2013;49:6170–6172. DOI: 10.1039/c3cc43153c.

[68] Zhai D., Tan Y.Q.E., Xua W., Chang Y.-T. Development of a fluorescent sensor for illicit date rape drug GHB. Chemical Communications. 2014;50:2904–2906. DOI: 10.1039/C3CC49603A.

[69] Murphy K.R., Stedmon C.A., Graeber D., Bro R. Fluorescence spectroscopy and multi-way techniques. PARAFAC. Analytical Methods. 2013;5:6557–6566. DOI: 10.1039/C3AY41160E.

[70] Píš Ľ., Májek P., Sádecká J. Synchronous fluorescence spectroscopy for differentiating between brandies and wine distillates. Acta Chimica Slovaca. 2011;4:47–581.

[71] Ashok P.C., Praveen B.B., Dholakia K. Near infrared spectroscopic analysis of single malt Scotch whisky on an optofluidic chip. Optics Express. 2011;19:22982–22992. DOI: 10.1364/OE.19.022982.

[72] Sádecká J., Jakubíková M., Májek P., Kleinová A. Classification of plum spirit drinks by synchronous fluorescence spectroscopy. Food Chemistry. 2016;196:783–790. DOI: 10.1016/j.foodchem.2015.10.001.

[73] Markechová D., Májek P., Sádecká J. Fluorescence spectroscopy and multivariate methods for the determination of brandy adulteration with mixed wine spirit. Food Chemistry. 2014;159:193–199. DOI: 10.1016/j.foodchem.2014.02.085.

[74] Markechová D., Májek P., Kleinová A., Sádecká J. Determination of the adulterants in adulterant-brandy blends using fluorescence spectroscopy and multivariate methods. Analytical Methods. 2014;6:379–386. DOI: 10.1039/C3AY41405A.

[75] Jakubíková M., Sádecká J., Májek P. Determination of adulterants in adulterant-fruit spirit blends using excitation-emission matrix fluorescence spectroscopy. Acta Chimica Slovaca. 2015;8:52–58. DOI: 10.1515/acs-2015-0010.

Section 5

Some Molecular Spectroscopic Methodologies

Chapter 17

Laser Spectroscopy in Hollow-Core Fibers: Principles and Applications

Philip G. Westergaard

Additional information is available at the end of the chapter

http://dx.doi.org/10.5772/63536

Abstract

The development of hollow-core photonic crystal fiber (HC-PCF) technology over the past decade has opened up a vast array of possibilities for new applications. When the hollow core is filled with gas, the HC-PCF is ideal for molecular spectroscopy applications that require long path length interaction. When light is coupled into the HC-PCF, the overlap between light and the molecules inside the hollow core is excellent all along the length of the fiber, which can be hundreds of meters long. Coiling the fiber up provides a compact, low-weight gas cell at the same time featuring a high level of interaction between laser light coupled through the fiber and the molecules inside.

This chapter presents some theoretical background, different applications, and selected results for molecular spectroscopy using gas-filled hollow-core fibers. The applications include frequency stabilization of a laser to a molecular transition and stimulated Raman scattering in a hollow-core fiber.

Keywords: photonic crystal fiber, hollow-core, HC-PCF, CO2, laser stabilization, stimulated Raman scattering

1. Introduction

Spectroscopy of weak molecular lines or weak scattering processes (such Raman scattering) is typically troubled by a small signal strength caused by insufficient interaction length or optical power, depending on the type of spectroscopy. One possible solution is to use multipass cells for an effective interaction length of up to a few hundred meters. Another, more compact solution is to fill gas into the hollow core of specially designed optical fibers —the so-called hollow-core photonic crystal fibers (HC-PCFs). Here, an interaction length of more than 100 m can be coiled up to a diameter of less than 10 cm. The hollow core ranges from around 10 to 100 microns in diameter, thus providing high intensity of the probe light propagating through the fiber and interacting with the molecules confined to the hollow core inside the fiber.

Applications of gas-filled HC-PCFs include high-accuracy laser stabilization, which can be useful in determining, e.g., the CO_2 content in the atmosphere using remote sensing techniques. These sensing techniques require stabilized laser systems for optimum sensitivity, and using a gas-filled HC-PCF for spectroscopy to stabilize the laser frequency is well suited for this purpose.

Hollow-core fibers with their high optical intensity inside the hollow core and long interaction length are also well suited for observing weak processes such as Raman scattering. This can be carried out by either observing the spontaneously generated stokes light at the output of the fiber, or by stimulating the Raman transition coherently using two input laser beams. Stimulating the Raman transition gives orders of magnitude higher signal and a possibility for higher spectral resolution.

2. Hollow-core fibers

2.1. Guiding light through a hollow core

Conventional optical fibers developed in the beginning of the 1970s [1] guide light inside a core with higher index of refraction than the surrounding layers (the cladding). These fibers are known as step index fibers and guide the light via classical refraction mechanisms such as total internal reflection.

In the beginning of the 1990s, a new technique for guiding light was envisioned [2]. Instead of relying on the mechanism that results from Maxwell's equations for the interface between two infinite mediums, this new technique exploits the constraints on the propagation of the electromagnetic field that arises from a periodic lattice structure of varying index of refraction. This lattice—or crystal—structure gives rise to a band gap in the energy spectrum in which light cannot propagate. This effect is also known from solid-state physics where it is observed for electrons in various periodic structures such as metals and semiconductors. By introducing a regular pattern of tiny holes in the silica structure that makes up a regular fiber (see **Figure 1**, left), an effective lattice pattern emerges and the light will be guided inside the lattice structure according to the band gap equations.

Most notably, it became possible to guide light in a central large hole inside the fiber instead of in a high index medium. This type of fiber is called hollow-core photonic crystal fiber (HC-PCF) and a first realization of this type was produced in 1999 [3]. Confining light in free space (as opposed to a solid core) with surrounding silica structure meant that now it was possible to fill molecules into this central empty part of the fiber and have the light interact with these molecules as it traverses the length of the fiber. This has several advantages. Since the central hollow core has a diameter of typically 10–20 microns[1] depending on the type (**Figure 1**), the

http://dx.doi.org/10.5772/63536

mode of the guided light (which has a shape close to Gaussian) will have a waist diameter around the same size as the core diameter. Such small waists give high intensity of the light, and not just in a small focused spot limited to the corresponding Rayleigh length for Gaussian beams ($\pi w_0^2/\lambda$, where w_0 is the waist radius and λ the wavelength of the light), but over the entire length of the fiber. One can imagine that this potentially allows for a very high degree of interaction between the molecules and the light.

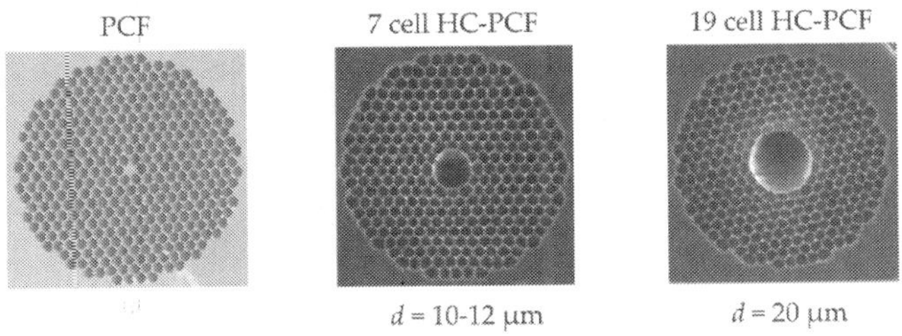

Figure 1. Different examples of photonic crystal fibers. The figure shows microscope images of the fiber tips. Left: A solid-core fiber with a periodic lattice structure of small holes. Middle: The seven-cell HC-PCF has a large central hole where seven rods that comprise the surrounding small holes have been removed. Similarly for the 19 cell HC-PCF (right).

For applications that require compact and lightweight components, such as operation in space, the HC-PCF is an excellent alternative to a bulky glass cell.

2.2. Filling and coupling

When light is transmitted through the HC-PCF, it interacts with the molecules that happen to be situated inside the hollow core. If interaction with molecules other than air is desired, the core of the HC-PCF should first be evacuated and then filled with the molecules of choice. This is done by mounting the fiber ends (or the entire fiber) inside a vacuum system. The vacuum system should also allow optical access for coupling light into the fiber. This is not easily achieved but various techniques exist. If the fiber ends are not enclosed in a vacuum system at all times, the HC-PCF must be sealed in some way to keep the desired molecules inside the fiber.

Basically, two techniques have been used in the literature to seal the gas-filled HC-PCFs. The first technique is based on splicing the ends to standard single-mode fibers. This is relatively easy to do for one end of the HC-PCF before the gas filling, but the single-ended filling time will be rather large for long fibers. Furthermore, Fresnel reflections at the splice interface between the two fibers often introduce interference effects that can affect the transmission properties of the fiber in an unpredictable way [4]. Losses at the splice interface due to mode mismatch can be significant [5], in particular for a (nearly) single-mode HC-PCF. It is also

[1] Other types of hollow-core fibers, such as the Kagomé type, can have several times larger core diameters. These types of fibers will not be discussed here.

possible to splice the HC-PCF to a single-mode fiber after gas filling as demonstrated in [6]. Here, the gas filling is followed by filling with He gas at high pressure (above 1 atm). The high He pressure prevents air from entering the HC-PCF during the splice process. Subsequently, He escapes by diffusion through the HC-PCF silica. With both ends of the HC-PCF spliced to single-mode fiber, strong interference fringes are generally seen due to Fresnel reflections at both splice interfaces [6, 7]. It has been shown that these interferences can be suppressed by angling one splice [7, 8]. However, this technique increases losses, is not yet robust or very reproducible and is susceptible to gas leaks and contamination [7].

An alternative technique uses a compact bulk cell around one or both ends of the HC-PCF [9, 10]. **Figure 2** shows a design recently developed at Danish Fundamental Metrology (DFM) [11] that allows for a compact solution for both coupling of the light and gas filling into the fiber in the same device.

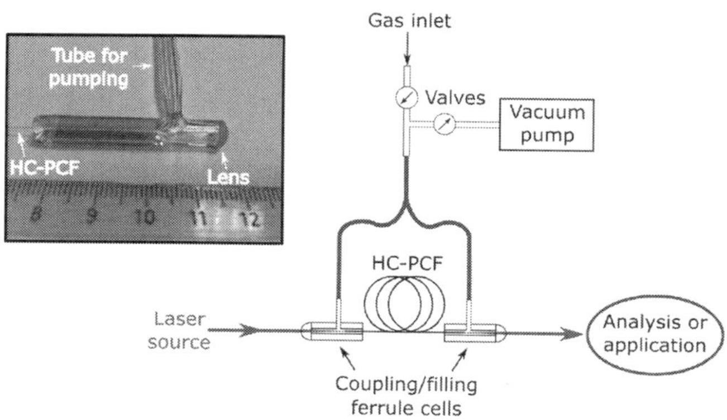

Figure 2. Schematic overview of filling and coupling of the HC-PCF. The hollow core in the fiber must first be evacuated with a vacuum pump, and then it can be filled with a gas. Inset: The ferrule design used for gas filling and coupling light into the HC-PCF. The HC-PCF is inserted into the ferrule from the left and a lens is glued on the right. Evacuating and filling of the hollow core is attained through the tube connected from the top.

In the DFM design, the HC-PCF is inserted into the back of a glass ferrule and is glued in place to avoid leaks and misalignment. The ferrule has a lens glued to the front face for optical coupling, that is, focusing the input light into the core of the HC-PCF. The cell is evacuated and filled with gas via a second smaller tube attached to the side of the outer tube. The cell is sealed off from the pump and filling system after it has been filled with the desired gas. The output of the compact glass cell can be coupled into a single-mode fiber using standard fiber coupling optics.

When the HC-PCF has been set up for gas filling, the system is first evacuated for an extended period, usually several days. After that the system is filled with the relevant gas, which enters the fiber by diffusion. The filling time in the hydrodynamic regime, which is the regime for the relevant gas pressures used here, can be calculated from molecular parameters and fiber geometry as [12]

$$g_V(f, f_0, p, T, m_i) = \frac{\sqrt{\frac{\ln(2)}{\pi}}\mathrm{Re}\left(\exp\left(-z^2\right)\mathrm{erfc}\left(-iz\right)\right)}{\Delta f_D}, \tag{1}$$

Considering filling with CO_2 molecules as an example, typical parameters are fiber lengths of up to L = 100 m, d = 10 μm (HC-PCF core diameter), $p_0 \simeq 100$ hPa (gas pressure), and η = $1.50 \cdot 10^{-5}$ kg/(m·s) (CO_2 gas viscosity). The filling time is then calculated to be 28 days. This assumes filling from both ends of the fiber. If the fiber is only filled from one end, the filling time will be four times longer (112 days).

The actual filling time does have some uncertainty. First, it can be reduced by starting with a somewhat higher pressure and then lowering the pressure in the vacuum system to the final pressure at the end of the filling period. It has previously been shown that in the low-pressure regime, the filling time can be reduced to one-sixth of t_{fill} in Eq. (1) by starting out with a pressure 20 times higher than the final pressure [13]. On the other hand, CO_2 tends to adsorb to glass surfaces, and this may increase the filling time [14].

Since the fiber length enters as L^2 in Eq. (1), the filling time can be reduced significantly by reducing the fiber length. Typically, a trade-off between filling time and signal strength must be considered.

3. Spectroscopic applications for laser stabilization

3.1. Molecular absorption

Spectroscopy necessarily relies on the absorption of light by the molecules that are investigated. The amount of absorption depends on molecular parameters (line strength, pressure and velocity distribution), laser parameters (frequency and in some cases optical power), and the interaction length between light and molecules. Power-dependent absorption involves a nonlinear process, such as Raman excitation, which is discussed in Section 4.

When the molecules are held at a given temperature T, their velocity distribution will be given by the Maxwell-Boltzmann equation. Through the Doppler effect, molecules with different velocities will absorb light at different frequencies. This affects the molecular line profile, that is, the absorption (or, correspondingly, the transmission) as a function of laser frequency. Even at zero velocity, the molecular line has a natural profile—the Lorentz profile. When the molecules are moving due to finite temperature, the Lorentz profile is changed into a so-called Voigt profile which is broader and has less absorption on resonance. The effect is illustrated in **Figure 3** where a CO_2 transition at 2051 nm is plotted for different temperatures. For zero temperature, the line profile is Lorentzian.

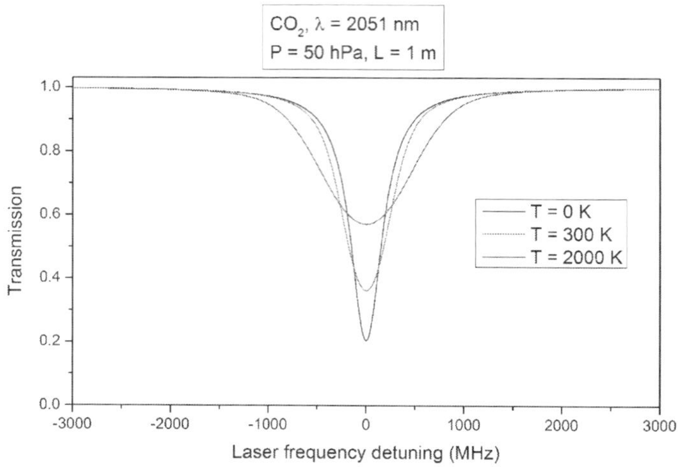

Figure 3. The line profile as a function of laser frequency detuning from molecular resonance for different temperatures of the molecular sample, in this case CO_2 at 50 hPa pressure and 1 m interaction length. The line is broadened by thermal motion at increasing temperatures due to the Doppler effect. The line strength of the transition shown here is 1.504×10^{-22} cm/molecule.

Considering normal excitation of a molecular transition, the amount of absorption can be quantified by the absorption coefficient α_i given by [15]

$$\alpha_i(f, f_0, p, T, m_i) = N(p) \cdot S_i \cdot g_V(f, f_0, p, T, m_i), \tag{2}$$

where i denotes the molecular species, p is the pressure, T is the temperature, f is the laser frequency, f_0 is the center frequency of the molecular line and m_i is the mass of the molecule. $N(p)$ is the number of molecules per volume, given by the ideal gas law $N(p) = \frac{p}{k_B T}$, with k_B being the Boltzmann constant. S_i is the line strength of the transition given in SI units, $[S_i] = \mathrm{m^2Hz}$. In the literature, the line strength is often given in units of [cm/molecule]. The relation between the value $S_{i,c}$ in units of [cm/molecule] and the SI line strength $S_{i,c}$ is given by [15]

$$S_i = cS_{i,c}$$

with c being the speed of light.

The factor $g_V(f, f_0, p, T, m_i)$ in Eq. (2) accounts for the Voigt profile of the transition and is given on resonance by [15]

http://dx.doi.org/10.5772/63536

$$g_V\left(f,f_0,p,T,m_i\right)=\frac{\sqrt{\frac{\ln(2)}{\pi}}\mathrm{Re}\left(\exp\left(-z^2\right)\mathrm{erfc}\left(-iz\right)\right)}{\Delta f_D},\tag{3}$$

where $z = -x + iy$ with $x = \sqrt{\ln(2)}\frac{f - f_0}{\Delta f_D}$ and $y = p\frac{\mathrm{PBC}(p)}{\Delta f_D}\sqrt{\ln(2)}$ with PBC(p) being the pressure broadening coefficient for a pressure p and $\Delta f_D = \frac{f_0}{c}\sqrt{\frac{2\ln(2)k_BT}{m_i}}$ is the Doppler width of the transition. The function erfc(y) is the complex complementary error function.

Equations (2) and (3) can be used to calculate the absorption profile of an arbitrary molecular gas if the line strength, pressure, temperature, and pressure-broadening coefficient are known. In the following section, we will see how this is applied to laser frequency stabilization to a molecular transition.

3.2. Modulation spectroscopy and frequency stabilization

Being an indispensable prerequisite for high-performance remote sensing such as LIDAR (see Section 3.3), frequency stabilization of the laser used for the sensing measurement hinges upon a frequency discriminator that serves as reference for the "correct" value of the laser frequency. Typically, the discriminator is composed of an optical cavity, an atomic line, or—as in the case of molecular sensing—a molecular transition.

Once a suitable frequency discriminator is found, stabilization (otherwise known as locking) of the laser frequency to this reference is accomplished by generation of an electronic signal, known as the error signal, which can be used to steer the laser frequency toward the desired frequency—the laser is then locked. The error signal typically has a large slope around the reference (cavity, molecular, or other) resonance and changes sign when the laser frequency crosses the resonance. This is useful when designing simple electronic circuits (typically PID circuits) that can be used for controlling the laser frequency and locking it to the reference resonance. The standard approach to generate the error signal is by modulation and subsequent demodulation of the laser frequency or phase after the laser light has interacted with the discriminator reference. There are several ways of doing this. In the following section, the frequency modulation spectroscopy (FMS) approach will be discussed for the case of laser frequency stabilization to a molecular transition.

3.2.1. Frequency modulation spectroscopy error signal

For a given set of parameters, the error signal can be simulated from the Voigt profile (Eq. (2)) of the molecular transition. Here, we consider phase modulation of the light and subsequent demodulation to obtain an error signal.

To phase modulate laser light an electro-optic modulator is the most common choice. To see how this affects the light, we consider the electric field component E of the light, which can be written in complex notation as $E = E_0 e^{i\omega t}$ where ω is the (angular) frequency of the light and E_0 is the amplitude. After modulation, the field becomes (to first order in the Bessel expansion)

$$E_{mod} = E_0 e^{i(\omega t - \beta \sin(\Omega t))} \cong E_0 \left(J_0(\beta) e^{i\omega t} + J_1(\beta) e^{i(\omega+\Omega)t} - J_1(\beta) e^{i(\omega-\Omega)t} \right), \tag{4}$$

where Ω is the modulation (angular) frequency, β is the phase modulation index, and $J_n(\beta)$ is the Bessel function of order n. Spectrally, the laser light is now composed of three components; the carrier at frequency ω and two modulation sidebands at $\omega \pm \Omega$. After the light has passed through the molecular gas (which is here acting as the frequency discriminator), it will have experienced absorption and a phase shift, which both depend on the frequency ω of the light. Introducing the amplitude coefficient of absorption $\alpha_E(\omega)$, the field after passing through the molecular gas becomes[2]

$$E_{out} = E_0 \left[J_0(\beta) e^{i\omega t - \alpha_E(\omega) l} + J_1(\beta) e^{i(\omega+\Omega)t - \alpha_E(\omega+\Omega) l} - J_1(\beta) e^{i(\omega-\Omega)t - \alpha_E(\omega-\Omega) l} \right], \tag{5}$$

where l is the length of the interaction.

When detecting the light, the quantity measured is the intensity $I \propto \left| E_{out} \right|^2$. After detection, the signal is demodulated by mixing (multiplying) the measured signal with the modulation signal and low-pass filtering the mixed output to get a DC error signal. Thus, with the field in Eq. (5), the DC error signal becomes after some algebra

$$\mathrm{Err}(\omega, \Omega, \beta) = \frac{\Omega}{2\pi} \int_0^{\frac{2\pi}{\Omega}} \kappa \sin(\Omega t) \left| E_{out} \right|^2 \mathrm{d}t = \chi e^{-\alpha_E(\omega) l} \left(e^{-\alpha_E(\omega+\Omega) l} - e^{-\alpha_E(\omega-\Omega) l} \right) \tag{6}$$

where $\chi = \kappa E_0^2 J_0(\beta) J_1(\beta)$ and κ is the quantum efficiency of the detector. We note that $\alpha_E(\omega)$ is the absorption coefficient for the field amplitude and that the transmission $T_r(\omega) = e^{-\alpha_i(\omega) l}$ through the gas is measured for the intensity of the light using α_i from Eq. (2). The two quantities can be related to via the relation

---

[2] Here, we neglect the phase shift imparted on the light by the molecules. For typical parameters, this phase shift will be very small and have a negligible contribution to the error signal. For a thermal ensemble, molecules with oppositely directed velocities will contribute equally to the absorption but their contribution to the phase shift will cancel each other out.

$$e^{-\alpha_E(\omega)l} = \sqrt{T_r(\omega)} = \sqrt{e^{-\alpha_i(\omega)l}}. \tag{7}$$

Now, we can model the error signal[3] by inserting Eqs. (2), (3), and (7) in Eq. (6). The transmission $T_r(\omega)$ and the corresponding error signal is shown in **Figure 4** for a CO_2 transition at 2051 nm with a pressure of 5 hPa and 10 m interaction length.

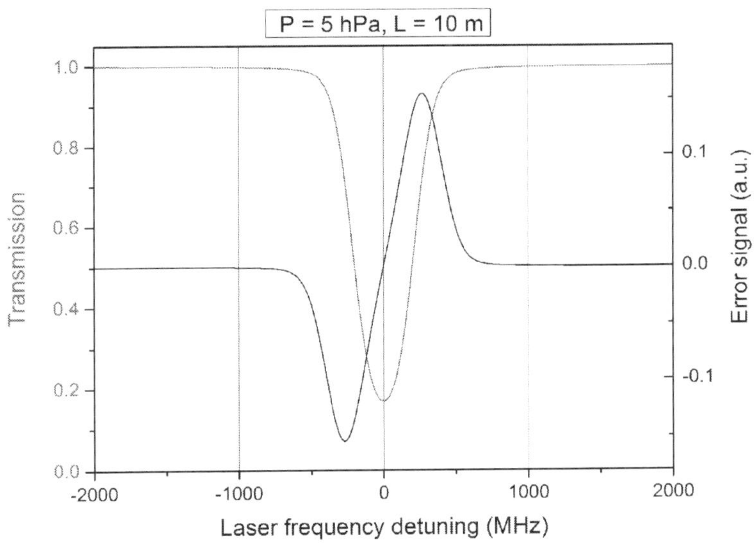

Figure 4. Modeled transmission (red, left scale) and error signal (black, right scale) calculated from Eqs. (2), (3), (6), and (7) for a CO_2 transition at 2051 nm with a pressure of 5 hPa and 10 m interaction length.

The thing to note here is the very linear behavior and zero crossing of the error signal around resonance (i.e., zero detuning). This is ideal as input to an electronic locking mechanism, such as a PID circuit. The long interaction length required to get a large error signal can be difficult to obtain and/or bulky with standard single- or multipass cells. Here, the HC-PCF is ideal as a gas container, since it can have lengths of up to 100 m while being very compact and light weight.

3.3. Applications in remote sensing (LIDAR, DIAL)

The principle of remote sensing (or LIDAR, from LIght Detection And Ranging) using a laser is shown in **Figure 5**. Here, a laser beam is sent through the molecular gas under test and the

[3] In the model presented, here, we do not take into account power broadening because usual power levels are much smaller than the typical saturation power.

absorption from scattered light is detected. For atmospheric sensing the laser source can be satellite based. Important molecules in this respect include the greenhouse gasses CO_2, CH_4, and N_2O.

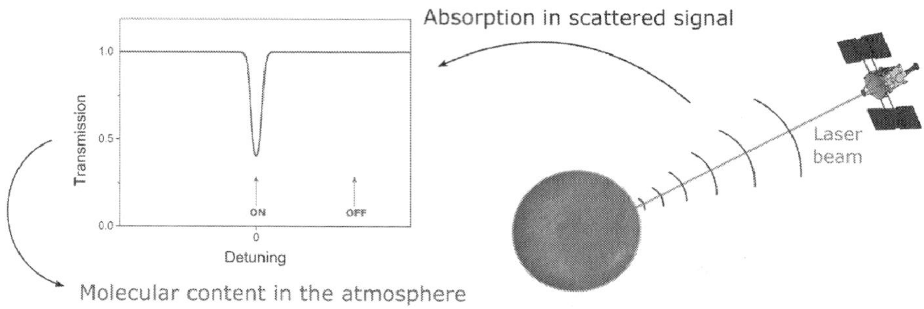

Figure 5. The principle of satellite-based differential atmospheric sensing using LIDAR.

Depending on where the laser frequency is tuned with respect to the molecular transition, the light will experience some absorption by the molecules. If the laser detuning from resonance is well known, this absorption can be directly related to the molecular concentration in the sample gas ($N(p)$ in Eq. (2)). Typically, the laser transmission on resonance (or close to it) is compared to an "off" frequency, where absorption is negligible, which serves as a reference background signal. This differential approach—usually known as DIAL (DIfferential Absorption Lidar)—is one of the most accurate for determining absolute concentrations of molecules in the atmosphere.

One of the main factors determining the accuracy and reliability of the DIAL technique is the frequency stability of the laser used for the absorption measurements. If the laser frequency drifts, the absorption from the molecular transition will change and give an error in the determination of the molecular concentration. Frequency stability down to the level of ±0.3 MHz may be necessary to achieve sufficient accuracy for CO_2 measurements [16]. Most free-running (i.e., not stabilized) laser sources show frequency fluctuations and drift on the order of tens or even hundreds of MHz over the course of minutes to hours. It is thus imperative to lock the laser frequency to a known reference when doing DIAL measurements.

Setting the "on" frequency to the exact center of the molecular resonance does not provide the best possible sensitivity, however. Detecting the transmission at the slope of the transmission curve will give a much higher sensitivity to changes in concentration (see, e.g., **Figure 4**), since pressure-broadening effects render this position highly sensitive to pressure/concentration changes. Therefore, it is desirable to have the possibility to lock the laser "on" frequency at a fixed detuning from the molecular resonance. This can either be performed with additional slave lasers or with just a single laser [17]. For satellite-based operation, it is advantageous to reduce the weight and size of the laser system, and a single laser locked off resonance is preferable if the laser can live up to the stability requirements.

In recent years, efforts have been carried out to obtain sufficiently stable satellite-based laser sources for CO_2 measurements. Among others, NASA has realized stabilization of a fiber-coupled DBF laser to the 1572.3 nm line of CO_2 using a multipass cell [18, 19]. The stability obtained here was at the level of 5.7 kHz up to 1000 s integration time. In the interest of probing atmospheric CO_2 with a wide tuning range about the line center, six DFB slave lasers were offset locked to the master laser at different offsets, and the output light was switched in pulses between the six different slaves. Of course, this type of setup can be bulky and heavy and the large number of components drives up the costs.

To reduce the size and weight, instead of using a multi-pass cell, in 2010 a fiber-coupled Tm:Ho:YLF laser was stabilized to the line center of the CO_2 transition at 2051 nm using a CO_2-filled 10 m long HC-PCF [20]. The resulting standard deviation of the locked frequency was at the level of 2.4 MHz. Also, acetylene and iodine have been used in hollow-core fibers for frequency stabilization, mostly using saturated absorption spectroscopy which can give a more accurate and stable lock, but less opportunity for tuning of the frequency. The most recent experiments report a long-term stability of around 800 Hz [4, 21].

Recently, [17], a compact system was demonstrated using a 10 m HC-PCF for stabilization of a DFB laser to CO_2 at 2051 nm. The setup was designed to be compact and lightweight. This included offset locking away from resonance with just a single laser. **Figure 6** shows experimental data of the transmission and error signal from a 10 m long HC-PCF filled with CO_2 to a pressure of around 20 hPa.

The HC-PCF in [17] was coiled up to a diameter of 8 cm on a copper mount. The control software developed for this system features an automated calibration sequence, which makes it possible to achieve an accurate and stable lock of the laser away from resonance without any additional lasers.

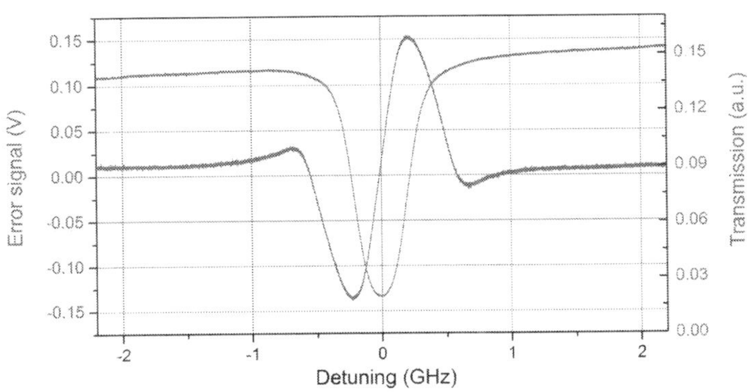

Figure 6. Experimental data from a CO_2-filled hollow-core fiber of length 10 m. The figure was adapted from [17].

To characterize the frequency stability of a system, a quantity called the Allan deviation is typically used. The Allan deviation gives information about how well the system averages out noise over time and the type of noise present at different time scales. For the system in [17], the Allan deviation of the locked laser is shown in **Figure 7**.

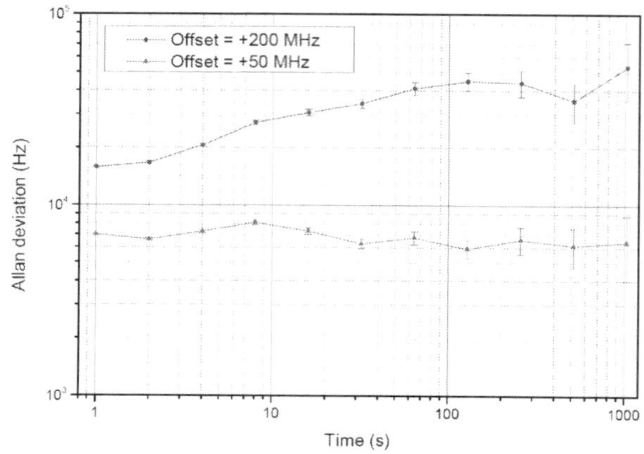

Figure 7. The Allan deviation for the locked laser system at two different offset frequencies in [17].

Here, the laser frequency stability at two different offset frequencies was measured. The figure shows that smaller offset frequency demonstrates better stability (lower Allan deviation) and that the frequency stability is dominated by flicker noise (1/*f*-type noise). This can be seen from the fact that the Allan deviation does not decrease over time. For white noise, the Allan deviation decreases as $t^{-1/2}$, where *t* is the time. Even with the presence of flicker noise, the level of stability here is better than 10 kHz for the low-frequency offset (50 MHz) until at least 1000 s integration time. The typical duration of the LIDAR measurement is only on the order of 10 seconds, so this stability is more than sufficient to meet the requirements.

Thus, in the study of Westergaard et al. [17], a hollow-core fiber has successfully been used for molecular spectroscopy applied to frequency stabilization of a DFB laser with the aim of satellite-based DIAL measurements.

4. Raman spectroscopy

Raman spectroscopy is a versatile technique that finds applications in physics, chemistry, biology, and in a number of different industrial areas. Raman spectroscopy typically explores transitions between motional states in molecules that have a unique energy spectrum for each molecule—a "fingerprint"—which therefore allows identification of that molecule [22].

Raman scattering is a weak process that requires high optical intensity in the excitation beam to take place. This places some limitations on the laser sources and the damage threshold of

http://dx.doi.org/10.5772/63536

the sample under test. Traditionally, the signal from the Raman scattering process has been obtained via a spontaneous process. The principle of the process is illustrated in **Figure 8**.

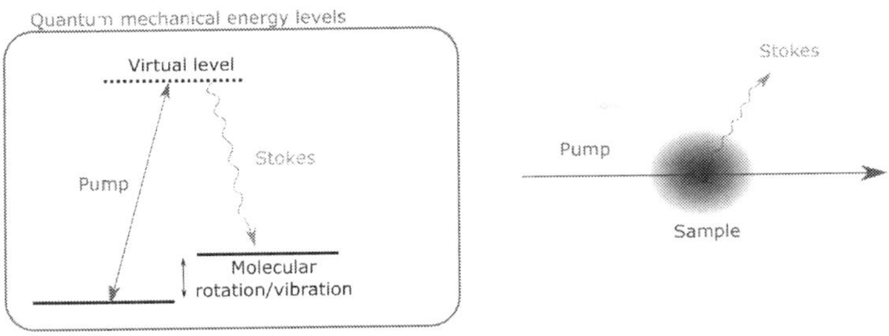

Figure 8. Spontaneous Raman scattering. A sample of molecules is excited by a pump laser, and when the molecules decay, a signal with lower (Stokes) or higher (anti-Stokes) energy is emitted from the sample in an arbitrary direction.

Here, a molecular sample is excited with a powerful pump laser to a virtual level; something which is allowed by quantum mechanics—but not very likely to happen for any given pump photon, which is why a powerful laser with a large number of photons is used. After the excitation, the molecules quickly decay to a real level, which typically is excited motionally (i.e., with more rotational or vibrational energy) with respect to the initial state.

In the decay process, the molecule emits a photon corresponding to the energy difference between the virtual level and the excited motional level. This photon contains information about the motional energy levels of the molecule, thus providing the spectroscopic fingerprint of the technique. When the level to which the molecules decay has a higher energy than the initial level (as illustrated in **Figure 8**), the emitted radiation is known as the Stokes signal. If the initial level has a higher energy, the radiation is called the anti-Stokes signal.

To achieve a larger Raman signal, a different approach can be applied, where the Raman transition is stimulated coherently using two different laser sources at the same time, with the

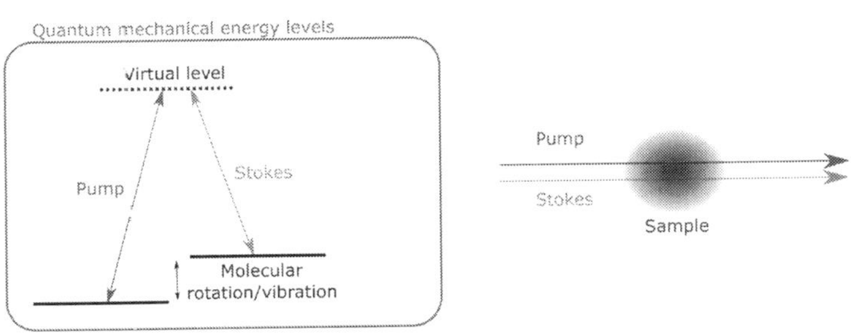

Figure 9. Stimulated Raman scattering. Here, two lasers are used to stimulate the Raman transition. The two laser beams must overlap spatially for the process to take place. Here, they are displaced slightly for clarity.

frequency difference between the two lasers corresponding to the energy difference between the two molecular levels involved in the Raman transition [22]. The situation is illustrated in **Figure 9**.

Stimulating the Raman transition in this way instead of relying on a spontaneous process has several advantages. First of all, it provides orders of magnitude larger signal [23] and makes it possible to observe the Raman signal with continuous-wave (CW) lasers with moderate (sub-Watt) optical power, which otherwise for spontaneous excitation would require Watt levels of CW power in the best case [24, 25]. Secondly, when the Raman process takes place, photons are effectively transferred from one laser beam to the other. Therefore, the signal is contained in a laser beam with a well-defined direction, in contrast to the spontaneous case where the Stokes radiation is emitted in an arbitrary direction, and the collection of the stimulated signal can be much more effective.

When the frequency difference between the two lasers matches that of an allowed Raman transition, the pump laser (typically the high-frequency laser) will be depleted, while the probe laser (typically the low-frequency laser) will experience a gain in intensity. As opposed to spontaneous scattering, which is proportional to only the pump intensity, the stimulated Raman signal is proportional to the product of the pump and probe intensity,

$$\delta I\left(\nu_{probe}\right) \propto I\left(\nu_{pump}\right) I\left(\nu_{probe}\right).$$

This proportionality gives the possibility of boosting the signal by many orders of magnitude.

By containing molecules inside a hollow-core fiber, the Raman signal is increased even further. This was first achieved around 15 years ago [23, 26]. The intrafiber interaction ensures high intensity between the light and the molecules over a long distance. It has been shown that HC-PCFs are very efficient for enhancing the Raman signals for both spontaneous [24, 27] and stimulated Raman transitions [23, 28, 29].

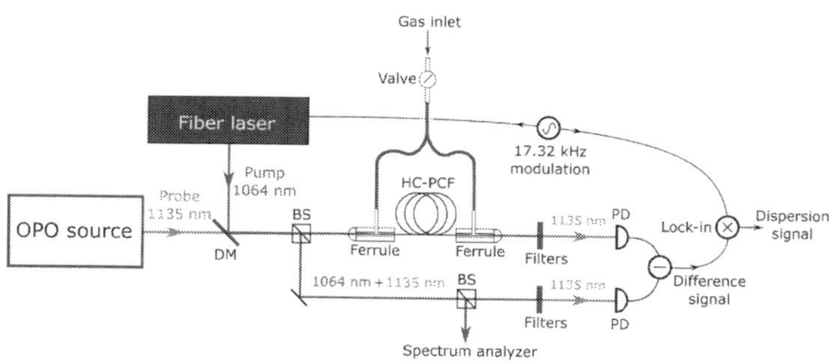

Figure 10. The setup in [29] for differential detection of stimulated Raman scattering. DM: dichroic mirror, PD: photo diode, BS: beam splitter, OPO: optical parametric oscillator.

Most recently, in [29], measurements were performed on the ortho-$S_0(1)$ transition of molecular hydrogen at 587 cm^{-1}, which is the most intense rotational line at room temperature for low wavenumbers. For a pump wavelength of 1064 nm, the Raman transition corresponds to probe (Stokes) radiation at 1135 nm. The pump light at 1064 nm was obtained from a commercial fiber laser and the light at 1135 nm was generated via a nonlinear process in a crystal—so-called optical parametric oscillator (OPO). The signal-to-noise ratio (SNR) can be further enhanced by using modulation detection [28, 29], where the pump laser is frequency modulated and this modulation is transferred to the probe laser via the Raman transition and demodulated with a lock-in amplifier giving the Raman signal. In [29], the setup was further improved by employing differential measurement detection of the probe laser, while suppressing the pump laser (see **Figure 10**). With this technique, any classical noise in the probe beam can be rejected, thus enhancing the SNR of the Raman signals. This resulted in spectroscopic data such as shown in **Figure 11**. Here, the Raman transition is probed with high sensitivity; the achieved SNR was around 1600 using only around 200 mW and 2 mW of optical power in the pump and probe laser, respectively, and a H_2 pressure of 867 hPa in 4.5 m of HC-PCF. Using two spectrally narrow light sources (such as the two lasers used here) additionally allows a for spectral resolution of features of Raman transitions down to around 5 MHz or 1.6×10^{-4} cm^{-1}, provided there are no other broadening mechanisms (which there typically are at room temperature, however, as described in Section 3.1).

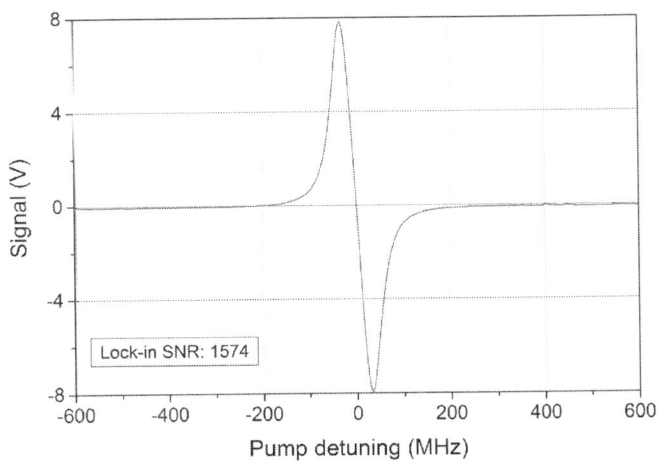

Figure 11. The signal from H_2 with around 2 mW of power in the 1135 nm (probe) laser and 200 mW in the 1064 nm (pump) laser. The Raman transition is detected using a lock-in technique, where the phase quadrature is recorded giving a dispersion-shaped curve instead of the usual absorption curve when the pump laser frequency is scanned across the Raman transition.

The measurements presented in [29] demonstrate that the use of a HC-PCF for enhancing the Raman signal makes it possible to achieve much higher SNR than without the hollow-core fiber and in principle the system is able to measure Raman transitions with only a few mW of power in the pump and probe beams at ambient pressure with a high spectral resolution.

Author details

Philip G. Westergaard

Address all correspondence to: pgw@dfm.dk

Danish Fundamental Metrology (DFM), Lyngby, Denmark

References

[1] F. P. Kapron, D. B. Keck and R. D. Maurer, "Radiation losses in glass optical waveguides," *Applied Physics Letters,* vol. 17, p. 423, 1970.

[2] P. Russell, "Photonic crystal fibers," *Science,* vol. 299, pp. 358–362, 2003.

[3] R. F. Cregan, B. J. Mangan, J. C. Knight, T. Birks, P. Russell, P. J. Roberts and D. C. Allan, "Single-mode photonic band gap guidance of light in air," *Science,* vol. 285, pp. 1537–1539, 1999.

[4] M. Triches, M. Michieletto, J. Hald, J. Lyngsø, J. Lægsgaard and O. Bang, "Optical frequency standard using acetylene-filled hollow-core photonic crystal fibers," *Optics Express,* vol. 23, pp. 11227–11241, 2015.

[5] R. Thapa, K. Knabe, K. L. Corwin and B. R. Washburn, "Arc fusion splicing of hollow-core photonic bandgap fibers for gas-filled fiber cells.," *Optics Express,* vol. 14, no. 21, pp. 9576–9583, 2006.

[6] P. S. Light, F. Couny and F. Benabid, "Low optical insertion-loss and vacuum-pressure all-fiber acetylene cell based on hollow-core photonic crystal fiber," *Optics Letters,* vol. 31, no. 17, pp. 2538–2540, 2006.

[7] K. Knabe, "Using saturated absorption spectroscopy on acetylene-filled hollow-core fibers for absolute frequency measurements," Ph.D. thesis, Kansas State University, 2010.

[8] F. Couny, F. Benabid and P. Light, "Reduction of Fresnel back-reflection at splice interface between hollow core PCF and single-mode fiber," *IEEE Photonics Technology Letters,* vol. 19, no. 13, pp. 1020–1022, 2007.

[9] P. Marty, J. Morel and T. Feurer, "All-fiber multi-purpose gas cells and their applications in spectroscopy," *Journal of Lightwave Technology,* vol. 28, no. 8, pp. 1236–1240, 2010.

[10] P. J. Meras, I. Y. Poberezhskiy, D. H. Chang, J. Levin and G. D. Spiers, "Laser frequency stabilization for coherent lidar applications using novel all-fiber gas reference cell fabrication technique," in *24th International Laser Radar Conference (ILRC)*, Boulder, Colorado, 2008.

[11] M. Triches, A. Brusch and J. Hald, "Portable optical frequency standard based on sealed gas-filled hollow-core fiber using a novel encapsulation technique," *Applied Physics B*, vol. 121, pp. 251–258, 2015.

[12] J. Henningsen and J. Hald, "Dynamics of gas flow in hollow core photonic bandgap fibers.," *Applied Optics*, vol. 47, no. 15, pp. 2790–2797, 2008.

[13] J. Hald, J. Henningsen and J. C. Petersen, "Spectroscopy on Slow Molecules in Hollow-Core Photonic Bandgap Fibers," Frontiers in Optics 2007 (OSA's 91st Annual Meeting / Laser Science XXIII), San Jose, USA, 2007.

[14] R. Baptist and F. Levy, "Carbon dioxide adsorption on glass," *Vacuum*, vol. 43, no. 3, pp. 213–214, 1992.

[15] L. Nielsen, "Optical line profiles for use in gas monitoring," *EMRP Eumetrispec Report D1-03-01*, 2013.

[16] G. Ehret, C. Kiemle, M. Wirth, A. Amediek, A. Fix and S. Houweling, "Space-borne remote sensing of CO2, CH4, and N2O by integrated path differential absorption lidar: a sensitivity analysis," *Applied Physics B*, vol. 90, pp. 593–608, 2008.

[17] P. G. Westergaard, J. W. Thomsen, M. Henriksen, M. Michieletto, M. Triches, J. K. Lyngsø and J. Hald, "Compact, CO2-stabilized tuneable laser at 2.05 microns," *Optics Express*, vol. 24, no. 5, pp. 4872–4880, 2016.

[18] K. Numata, J. R. Chen, S. T. Wu, J. B. Abshire and M. a. Krainak, "Frequency stabilization of distributed-feedback laser diodes at 1572 nm for lidar measurements of atmospheric carbon dioxide.," *Applied Optics*, vol. 50, no. 7, pp. 1047–1056, 2011.

[19] K. Numata, J. R. Chen and S. T. Wu, "Precision and fast wavelength tuning of a dynamically phase-locked widely-tunable laser.," *Optics Express*, vol. 20, no. 13, pp. 14234–14243, 2012.

[20] P. Meras, I. Y. Poberezhskiy, D. H. Chang and G. D. Spiers, "Frequency stabilization of a 2.05 μm laser using hollow-core fiber CO2 frequency reference cell," Proc. of SPIE vol. 7677, p. 767713, 2010.

[21] A. Lurie, F. N. Baynes, J. D. Anstie, P. S. Light, F. Benabid, T. M. Stace and A. N. Luiten, "High-performance iodine fiber frequency standard," *Optics Letters*, vol. 36, pp. 4776–4778, 2011.

[22] N. B. Colthup, L. H. Daly and S. E. Wiberley, "Introduction to infrared and Raman spectroscopy," 2nd ed., Academic Press New York, 1975.

[23] F. Benabid, J. C. Knight, G. Antonopoulos and P. S. J. Russel, "Stimulated Raman scattering in hydrogen-filled," *Science,* vol. 298, pp. 399–402, 2002.

[24] F. Couny, F. Benabid and P. S. Light, "Subwatt threshold cw Raman fiber-gas laser based on H2-filled hollowcore," *Physical Review Letters,* vol. 99, p. 143903, 2007.

[25] F. Couny, O. Carraz and F. Benabid, "Control of transient regime of stimulated Raman scattering using hollow-core PCF," *Journal of the Optical Society of America B,* vol. 26, pp. 1209–1215, 2009.

[26] F. Benabid, F. Couny, J. C. Knight, T. A. Birks and P. S. J. Russell, "Compact, stable and efficient all-fibre gas cells using hollow-core photonic crystal fibres," *Nature,* vol. 434, pp. 488–491, 2005.

[27] M. P. Buric, K. P. Chen, J. Falk and S. D. Woodruff, "Enhanced spontaneous Raman scattering and gas composition analysis using a photonic crystal fiber," *Applid Optics,* vol. 47, pp. 4255–4261, 2008.

[28] J. L. Doménech and M. Cueto, "Sensitivity enhancement in high resolution stimulated Raman spectroscopy of gases with hollow-core photonic crystal fibers," *Optics Letters,* vol. 38, p. 4074–4077, 2013.

[29] P. G. Westergaard, M. Lassen and J. C. Petersen, "Differential high-resolution stimulated CW Raman spectroscopy of hydrogen in a hollow-core fiber," *Optics Express,* vol. 23, pp. 16320–16328, 2015.

Chapter 18

Enhanced Molecular Spectroscopy via Localized Surface Plasmon Resonance

Lu Sun, Ping Chen and Lie Lin

Additional information is available at the end of the chapter

http://dx.doi.org/10.5772/64380

Abstract

Numerous novel spectroscopy techniques have been developed to perform detection and characterization at molecular level. Nevertheless, the resolution of spectroscopy remains to be the bottleneck, and local electric field is involved to solve this issue. Localized surface plasmon resonance (LSPR) occurred at the surface of noble metal nanoparticles is a major source of enhanced local electric field which provide notable enhancement factor of spectroscopy applying fluorescence and the Raman scattering. In this chapter, we will firstly present the physics of localized surface plasmon resonance to gain a basic understanding. Several current techniques to prepare a wide variety of nanoparticles and localized surface plasmon resonance detector are subsequently introduced. We further illustrate two examples taking advantage of experiments and modeling to elaborate the effect of localized surface plasmon resonance on spectroscopy under different circumstances. The combination of experimental and theoretical approaches elucidates the influence of each factor and promotes the design of localized surface plasmon resonance detector used in spectroscopy.

Keywords: spectroscopy, localized surface plasmon resonance, nanoparticle, detection, enhancement factor, finite-difference time-domain

1. Introduction

The advance of science and technology has drawn people's attention to the molecular level, and characterization of molecular configuration is among the most significant challenges. Spectroscopy takes advantage of the interaction between electromagnetic radiation and matter and records the response of interest. The resulting spectrum containing the fingerprint of the analyte sheds light on specific structural details of a single molecule.

State-of-the-art spectroscopy techniques employing fluorescence [1], the Raman scattering [2], X-ray [3], NMR [4], etc. have been successfully utilized to illustrate the conformations of biomolecules such as protein, DNA, and RNA. Moreover, utilizing lasers with impulse interval at femtosecond as excitation power has accomplished ultrafast detections. For example, the instantaneous structures of mRNA-tRNA translocation intermediates have been characterized through single-molecule fluorescence resonance energy transfer method, the achievement of which is a huge step toward the comprehensive mechanism of in vivo protein synthesis [5].

Despite the pronounced temporal resolution achieved during the past two decades, the spatial resolution is another key issue to fulfill detection and characterization at the single-molecule level. For instance, the cross section of non-resonant Raman scattering is typically ranging from 10^{-30} to 10^{-25} cm^2 per molecule, a value so weak that a notable amount of analyte molecules is demanded to convert the incident photon to the Raman photon [6]. Although the laser power still has the potent to be augmented, the loss during transmission is too dramatic to exert a distinct influence by simply replacing an intensified laser. Local electromagnetic field is therefore more applicable in enhancing the resolution of spectroscopy.

The progress of nanoparticles' localized surface plasmon resonance (LSPR) is becoming a major solution to enhance the intensity of engendered signals through the highly localized electromagnetic field. It has been discovered that the electrons within the conduction band can be excited collectively at noble metal surface and the consequential oscillation of the excited electrons would be localized instead of propagating on a rough surface [7].

Extensive studies have been conducted to manipulate LSPR at the surface of different kinds of nanoparticles, and LSPR has displayed distinct properties by regulated size, shape, structure, material, and other factors [8]. For example, it has been shown that the wavelength of plasmon varies with the particle radius [9], two distinguished plasmonic radiations have been found in nano-rod [10], and aggregates of nanoparticles show more localized optical field with hot spots and cold zones compared with isolated nanoparticles [11].

Together with the significant improvement in the fabrication of a variety of nanostructures during last decade, gold- and silver-nanostructures generating LSPR are nowadays applicable and have been integrated into sensors and detectors [12, 13] Surface-enhanced Raman spectroscopy has recently made detection of biological and chemical analytes with concentration as low as nanogram and femtogram feasible [14]. For different nanostructures, enhancement effect of LSPR cannot be predicted instinctively but through theoretical methods such as finite-difference time-domain (FDTD), discrete dipole approximation (DDA), and finite element method (FEM) [15].

In this chapter, we elaborate the efficacy of researches applying experimental techniques and computation modeling to enhance spectroscopy through LSPR. We first interpret the physics that originated LSPR. Since LSPR occurs at the surface of nanoparticles, different ways to fabricate nanostructure in order to generate LSPR and how nanoparticles are used as detector are thereafter introduced. Finally, examples of studies applying LSPR to enhance molecular spectroscopy and interpretations through finite-difference time-domain simulations are illustrated.

http://dx.doi.org/10.5772/64380

2. Physics of localized surface plasmon resonance

The observation of surface plasmon could be dated back to the beginning of the last century when Wood observed the anomalous light diffraction on a metallic diffraction grating, a phenomenon later proved to be correlated with the excitation of electromagnetic waves on the surface of the diffraction grating [16]. The plasmon generated from the collective oscillation of the free electrons can be described by the classical Maxwell's equation. We can treat the plasmon as the mechanical oscillations of the electron gas of a metal resulted from an external electric field. For the bulk system with size larger than the wavelength of the incident light, the oscillations occur at the plasma frequency with the energy:

$$E_{\mathrm{p}} = \hbar\sqrt{\frac{ne^2}{m\varepsilon_0}} \tag{1}$$

where n denotes the electron density, e is the electron charge, m is the electron mass, and ε_0 represents the permittivity of free space.

Under this circumstance, the oscillations of electrons are simply called surface plasmons. Surface plasmons can be excited through incident light. A light can couple with a surface plasmon at a metal-dielectric interface only if the incidence angle meets the criteria, because the wavevector of the incident light should accord with the propagation constant of the plasmon so that the oscillating electric field of the incident light is capable of exciting surface plasmons. The application of surface plasmon is therefore limited.

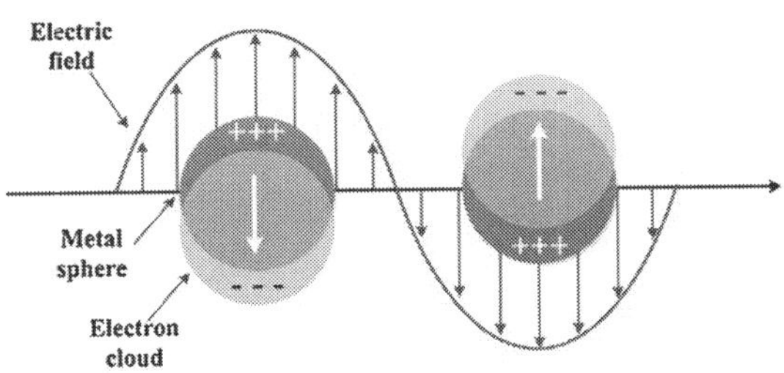

Figure 1. Illustration of the excitation of localized surface plasmon resonance [17].

However, when a surface plasmon is excited at the surface of a metallic nanoparticle with the size comparable to the wavelength of light, the free electrons are confined and take parts in the collective oscillations. This kind of oscillation is thus termed as localized surface plasmon (LSP), with the oscillation shown in **Figure 1**. Since the oscillation is collective, the LSP has taken advantage of significant enhancement at the surface. It is worth noticing that the magnitude of the field attenuates drastically with the distance to the surface of the nanoparticle.

Moreover, the size of nanoparticle makes the frequency of LSP, which also depends on the refractive index of the medium, at visible wavelengths for noble metal nanoparticles.

Since the size of nanoparticle is comparable to the wavelength of light, the Mie theory for light scattering would be considered. Through the analytical solution to Maxwell's equation in the Mie theory, the scattering, extinction, and absorption cross sections are solved as

$$C_{sca} = \frac{2\pi}{k^2}\sum_{N=1}^{\infty}(2N+1)\left(|a_n|^2 + |b_n|^2\right) \tag{2}$$

$$C_{ext} = \frac{2\pi}{k^2}\sum_{N=1}^{\infty}(2N+1)\mathrm{Re}\{a_n + b_n\} \tag{3}$$

$$C_{\mathrm{abs}} = C_{ext} - C_{sca} \tag{4}$$

where k is the incident wavevector, N is an integer representing the dipole, quadrupole, and higher multipoles of the scattering, a_L and b_L are the parameters represented by the Riccati-Bessel functions ψ_L and ξ_L expressed below:

$$a_n = \frac{m\psi_n(mx)\psi_n'(x) - \psi_n(x)\psi_n'(mx)}{m\psi_n(mx)\xi_n'(x) - \xi_n(x)\psi_n'(mx)} \tag{5}$$

$$b_n = \frac{\psi_n(mx)\psi_n'(x) - m\psi_n(x)\psi_n'(mx)}{\psi_n(mx)\xi_n'(x) - m\xi_n(x)\psi_n'(mx)} \tag{6}$$

$$\mathrm{m} = \frac{\mathrm{n_p}}{\mathrm{n_m}} \tag{7}$$

where $\mathrm{n_p}$ is the complex refractive index of the metal utilized and is equivalent to n_r+in_i, $\mathrm{n_m}$ is the real refractive index of the medium, and x equals to $k_m r$ (r is the radius of the particle, k_m=2π/ λ_{m} indicates the wave number in the medium).

To simplify the equation, Riccati-Bessel functions can be approximated by power series if we assume the nanoparticle is much smaller compared to the wavelength (i.e., $x \ll 1$). By truncating terms after the order of x^3, we have

$$\mathrm{a}_1 \approx -\frac{i2x^3}{3}\frac{m^2-1}{m^2+2} \tag{8}$$

$$b_1 \approx 0 \tag{9}$$

The real part of a_1 required to calculate the cross section of extinction can be found by replacing $m = \frac{n_r + in_i}{n_m}$ into a_1 as

$$a_1 \approx -\frac{i2x^3}{3}\frac{m^2-1}{m^2+2} = -\frac{i2x^3}{3}\frac{n_r^{\ 2} - n_i^{\ 2} + i2n_r n_i - n_m^{\ 2}}{n_r^{\ 2} - n_i^{\ 2} + i2n_r n_i + 2n_m^{\ 2}} \tag{10}$$

Further substituting the dielectric function of metal with the complex form

$$\varepsilon_p = \varepsilon_1 + i\varepsilon_2 \tag{11}$$

$$\varepsilon_1 = n_r^{\ 2} - n_i^{\ 2} \tag{12}$$

$$\varepsilon_2 = 2n_r n_i \tag{13}$$

and replacing the dielectric function of medium, $\varepsilon_m = n_m^2$, will result in the following relation as

$$a_1 = -\frac{i2x^3}{3}\frac{\varepsilon_1^{\ 2} + \varepsilon_1\varepsilon_m - 3\varepsilon_2\varepsilon_m + \varepsilon_2^{\ 2} - 2\varepsilon_m^{\ 2}}{\left(\varepsilon_1 + \varepsilon_m\right)^2 + \varepsilon_2^{\ 2}} \tag{14}$$

Substituting the above equation into the extinction cross section and only taking the dipole term, we can get the most quoted expression for LSPR as

$$C_{ext} = \frac{18\pi\varepsilon_m^{\frac{3}{2}}V}{\lambda}\frac{\varepsilon_2(\lambda)}{\left[\varepsilon_1(\lambda) + 2\varepsilon_m\right]^2 + \varepsilon_2(\lambda)^2} \tag{15}$$

In here, V represents the volume of the particle. Similarly, the scattering cross section can be expressed as

$$C_{sca} = \frac{34\pi^4\varepsilon_m^{\ 2}V^2}{\lambda^4}\frac{\left(\varepsilon_1 - \varepsilon_m\right)^2 + \varepsilon_2^{\ 2}}{\left(\varepsilon_1 + 2\varepsilon_m\right)^2 + \varepsilon_2^{\ 2}} \tag{16}$$

Because we supposed that the nanoparticles are small enough to use the approximation, the equation above would be strictly applied to particles with diameter smaller than 10 nm. Nevertheless, it is noteworthy that the expression will give certain accuracy for larger particles as well [18].

The functional form of the LSPR peak wavelength is dependent on the dielectric function of the medium [19], and the dependence can be derived by the following access.

The frequency-dependent dielectric constant for ε_1 according to the Drude model of the electronic structure of metal would be

$$\varepsilon_1 = 1 - \frac{\omega_p^{\ 2}}{\omega^2 + \gamma^2} \tag{17}$$

in which ω_p denotes the plasma frequency and γ represents the damping parameter of the bulk metal. It is notable the Drude model is a classical model of electronic transport in conductors and describes the collisions between freely moving electrons and the lattice of heavy, stationary ionic cores. The model is a very good approximation for the conductivity of noble metals. For visible and near-infrared frequencies, where $\gamma \ll \omega_p$, the above relation would be reduced to the following form as

$$\varepsilon_1 = 1 - \frac{\omega_p^{\ 2}}{\omega^2} \tag{18}$$

Substituting the above expression for ε_1 and setting $\varepsilon_1 = -2\varepsilon_m$ as the resonance condition, we can obtain the maximum peak of the LSPR frequency as

$$\omega_{max} = \frac{\omega_p}{\sqrt{2\varepsilon_m + 1}} \tag{19}$$

Because the relation between frequency and wavelength is denoted as $\lambda = 2\pi c/\omega$, the wavelength of LSPR can be expressed after replacing the dielectric constant with the refraction $\varepsilon_m = n^2$:

$$\lambda_{max} = \lambda_p \sqrt{2n_m^{\ 2} + 1} \tag{20}$$

in which λ_{max} is the peak wavelength of LSPR while λ_p represents the corresponding wavelength to the plasma frequency of the bulk metal.

Because the nanoparticles generating LSPR are generally not strictly spherical, Richard Gans further complemented the Mie theory to spheroidal particles of any aspect ratio in the small particle approximation. The absorption cross section for a prolate spheroid (nanorod structure) is found analogous to that of the spherical nanoparticles, as

$$C_{\text{abs}} = \frac{\omega}{3c}\varepsilon_{\text{m}}^{\frac{3}{2}}V\sum_{j}\frac{\left(1/P_j^2\right)\varepsilon_2}{\left[\varepsilon_1+\frac{\left(1-P_j\right)\varepsilon_{\text{m}}}{P_j}\right]^2+\varepsilon_2^{\ 2}} \tag{21}$$

The sum over j infers to the three dimensions of the nanoparticle. P_j denoting the depolarization factors has three components, P_A, P_B, and P_C, along each axis. For a prolate spheroid with aspect ratio A > B = C, the depolarization factors alter the dielectric constant ε_1 and ε_2 anisotropically. Therefore, the corresponding LSPR peak frequencies are different at different directions. The depolarization factors are expressed as

$$P_A = \frac{1-e^2}{e^2}\left[\frac{1}{2e}\ln\left(\frac{1+e}{1-e}\right)-1\right] \tag{22}$$

$$P_B = P_B = \frac{1-P_A}{2} \tag{23}$$

where e is the ellipticity factor that includes the particle aspect ratio R:

$$\text{e} = \sqrt{1-\left(\frac{B}{A}\right)^2} = \sqrt{1-\left(\frac{1}{R}\right)^2} \tag{24}$$

The extinction spectrum resulting from nanorod has two peaks, one corresponding to the transverse plasmon mode and the other corresponding to the longitudinal plasmon mode (**Figure 2**).

For example, the absorption spectra of nanoparticle with different aspect ratios have been simulated, and it is shown that the increase of aspect ratio would dramatically increase the wavelength. From the result by EL-Sayed et al., the maximum peak of longitude plasmon band displayed red shift from 650 to 800 nm after altering the ratio aspect from 2.6 to 3.6 [21]. For nanoparticles beyond these spheres and spheroids, particle shape plays a significant role in determining the LSPR spectrum.

The plasmonic spectrum would also rely on many other factors, such as the local medium surrounding. The LSPR wavelength shift is in accordance with the refractive index change and follows the relation as

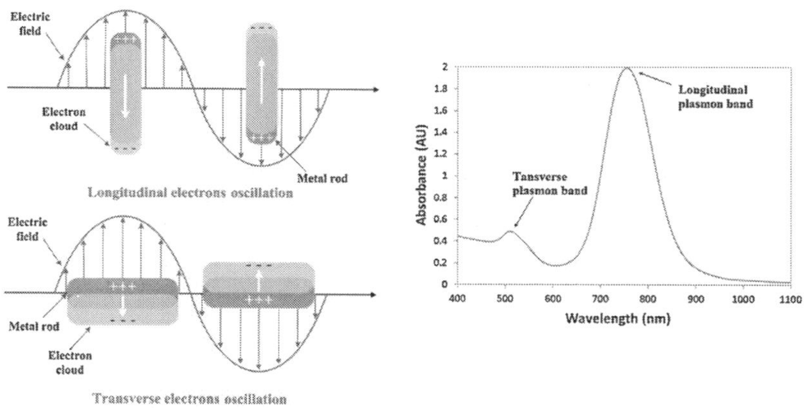

(a) (b)

Figure 2. An illustration of LSPR excitation for prolate spheroid. The discrepant oscillation of electrons at longitudinal and transverse plasmon bands results in different plasmonic spectra [20].

$$\Delta\lambda = \mathrm{m}\Delta n\left[1-\exp\left(\frac{-2d}{l_d}\right)\right] \tag{25}$$

where m represents the sensitivity factor (measured in nm per refractive index unit), Δn is the change of the refractive index, d indicates the effective thickness of the absorbed layer, and l_d denotes the characteristic electromagnetic field decay.

The complication of LSPR contributes to its potential applications only if we have gained a thorough understanding. Currently, there are several numerical methods for simulation of LSPR occurring at the surface of nanoparticles, including finite-difference time-domain (FDTD), discrete dipole approximation (DDA), and finite element method (FEM) [15]. An example of FDTD simulation will be illustrated in Section 5.

3. Fabrication of nanoparticles

The extensive studies on the fabrication of nanoparticles have promoted the potential application of LSPR. A reliable and reproducible synthesis method of nanostructures is the basis for LSPR detector. While the spherical nanoparticle of a noble metal is most readily prepared, it takes efforts to fabricate other desired structures. Several ways to fabricate desired nanostructure are illustrated in this section [22].

3.1. Citrate reduction

Citrate reduction is the most widely applied method for producing nanoparticles. The delicate addition of a calculated amount of citrate solution into the boiling metallic salt (such as $HAuCl_4$) solution generates solutions containing nanoparticles. The size of the nanoparticle would be controlled through the ratio between the citrate and the gold salt, the reaction temperature, and the reaction time [23]. The simplicity of these reductions makes it the most popular method to form nanoclusters.

3.2. Electrochemical method

The preparation of high yield of nanoparticles was initially proposed through applying an electrochemical method which can produce both nanocluster and nanorod structures [24, 25]. While preparing the cluster structure is easier, the nanorod structure was first synthesized through this method by Wang et al. in the late 1990s [26]. The synthesis of nanorod was carried out using a two-electrode-type electrochemical cell containing a gold metal plate as the sacrificial anode and a platinum plate as cathode and the electrolytic solution containing a rod-inducting cationic surfactant cetyltrimethylammonium bromide (CTAB) and a cationic co-surfactant tetradodecylammonium bromide (TCAB). During the electrolysis, the gold metal anode was oxidized into $AuBr_4$ and subsequently formed complexes with the cationic surfactants. The gold nanoparticle was generated through the reduction process after the complexes migrated to the cathode. In order to control the aspect ratio, a silver plate was placed in a position behind the platinum cathode. The aspect ratio was found to be dependent on the concentration and the release rate of the silver ions.

3.3. Electron beam lithography method

Electron beam lithography method is another common way used to generate metallic nanostructures. This method takes advantage of the precise control of the size, shape, and spatial distribution of the nanoparticles synthesized [27, 28]. Nevertheless, the lithography applied in this method makes it highly time-consuming owing to the small region processed.

3.4. Seed-mediated growth method

For the synthesis of nanostructure with specific aspect ratio, the seed-mediated growth method is the most used and has been extensively utilized [29, 30]. The method possesses merits such as the simplicity of the experiment, the high yield and high quality of produced nanorods, the convenience of size control, and the flexibility in structural modifications [31]. Lately, nanostructures such as 2D gold nanorings [32], composite core-shell nanorod [33], and branched gold nanodendrites [34] have been reported applying this experimental approach.

The seed-mediated growth method for nanorod structure was initially demonstrated by Jana et al. [35]. In their early experiment, the seed solution was prepared by the reduction of gold salt ($HAuCl_4$) with $NaBH_4$ in the presence of sodium citrate. The produced nanoparticles usually have a diameter of 3–4 nm and were used as the seeds by being added to the so-called growth solution, which was composed of $HAuCl_4$, cetyltrimethylammonium bromide (CTAB), ascorbic acid, and $AgNO_3$. The latter three compositions acted as the template, the reducing agent, and the shape induction agent, respectively. The nanoparticle can subsequently grow into various aspect ratios by controlling the ratio of seed solution to the growth solutions.

The method was further improved later by the same group to obtain larger aspect ratio [36]. The seed and the growth solutions were prepared similarly except for adding $AgNO_3$. After adding the seed solution to the growth solution, the generated nanostructure was again used as seed with a repeated step. Despite nanostructures with larger aspect ratio acquired, by-products resulted from the reaction become significant, and more difficult purification processes are required in this process [37].

Based on the prototype, a wealth of significant improvements has been thereafter accomplished. For example, Nikoobakht et al. synthesized high-quality and high-yield nanorods by replacing the sodium citrate and adjusting the concentration of silver ions [38]. Ye et al. employed aromatic additives and a low CTAB concentration to achieve a broadly tunable localized plasmon band with higher purity [39].

Above all, mastering nanostructure is becoming more and more feasible, which has significantly advanced the applications of LSPR in molecular spectroscopy.

4. Nanoparticle LSPR sensor and detector

There are a wide variety of designs of LSPR detectors because the LSPR is more readily to be excited compared to the surface plasmon on a planar surface. The LSPR detector can be typically designed as either substrate-based or solution-based. We will hereby introduce three most widely studied structures: chip-based, optical-fiber-based, and solution-phase-based.

4.1. Chip-based LSPR sensor and detector

The chip-based LSPR detector can be fabricated by immobilizing nanostructures on the surface of a substrate, such as a glass slide and cover slip. The detector chip is easily achieved when nanostructures are produced through electron beam lithography technique or are grown on the substrate.

If the nanoparticles are produced in solution, such as citrate reduction, they can be immobilized on the surface through electrostatic force [40]. For instance, a clean glass substrate can be coated with polyelectrolyte that shows opposite charge to the surface charge of the generated nanoparticles. The charged substrate is subsequently immersed into the nanoparticle solution to attract nanoparticles by electrostatic force. However, the LSPR detector prepared in this way suffers from poor stability and poor uniformity.

Another method is based on the SAM technique [41]. A clean substrate is dipped into an alkylsilane solution, such as MPTMS, to form a thiol-terminated silane membrane on the surface. The silanized substrate surface would subsequently form covalent bonds with single layer of nanoparticles.

4.2. Optical fiber-based LSPR sensor and detector

Optical fiber-based LSPR detector is typically fabricated by immobilizing nanoparticles on the decladed fiber core of a multimode or single-mode optical fiber [42, 43]. Optical fiber-based LSPR detector shows advantages such as small sample volume, simple optical design, and minor electromagnetic interference [44].

4.3. Solution-phase-based LSPR sensor and detector

Solution-phase-based LSPR detectors are the nanoparticles suspended in solution rather than immobilized on a substrate, which makes the detection process of analytes inside the solution [45]. During the detection, the functional molecules should mix evenly with the nanoparticle solution and be closed enough to the surface of nanoparticles because of the decaying electric field.

5. LSPR-enhanced molecular spectroscopy

The utilization of localized surface plasmon resonance to enhance molecular spectroscopy has achieved prodigious enhancement factors. Applications of surface plasmon polariton include enhanced Raman spectrum [46], enhanced fluorescence [47], enhanced optical nano-devices [48], etc. Examples of the related experimental achievements are introduced and further elucidated through theoretical modeling applying Maxwell's equation.

5.1. Metal-enhanced fluorescence spectroscopy detecting polycyclic aromatic hydrocarbons

Oil spills are the major sea pollutants originating from tankers, offshore drilling rigs, etc. [49]. Spilled oil is toxic to living organisms, and even one spill oil would cause large mortality in the marine ecosystem [50]. The monitor of the water quality is crucial and demands trace amount detection technique. Polycyclic aromatic hydrocarbons are a major component of crude oil and would be selected as the surveillance target.

The easy-to-handle fluorescence spectroscopy is enhanced by LSPR during the detections of crude oil [51]. The silver nanoparticle solution was prepared with the citrate reduction method shown previously. The glass-based detector was prepared by (3-aminopropyl)trimethoxysilane-coated quartz substrates. Through the SEM image, an average grain size of 80 nm was analyzed over 200 particles, as shown in **Figure 3(a)**. The characteristic absorption spectrum of the silver nanoparticles has its peak at 405 nm as displayed in **Figure 3(b)**, resulting in LSPR at the surface of the silver nanoparticles.

The fluorescence spectrum of artificial diesel oil polluted seawater emulsions was measured in the presence and in the absence of the glass-based nanoparticles, as shown in **Figure 3(c)**. It is distinct that the maximum peak is enhanced with the intensity rising from 6×10^4 to more than 30×10^4, indicating an enhancement factor of 5.44.

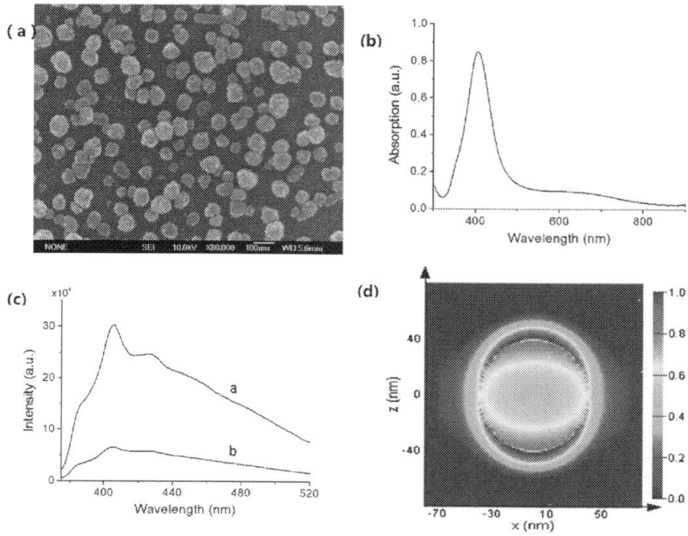

Figure 3. (a) SEM image shows the structure of the applied silver nanoparticles, (b) the absorption spectrum of silver nanoparticles at room temperature, (c) fluorescence spectra of diesel oil emulsions in artificial seawater before and after enhancement (emitted at 355 nm), and (d) the electric field enhancements of silver nanoparticles at 355 nm.

To understand the enhancement, the electric field effect could be modeled through FDTD method, as shown in **Figure 3(d)** with the near-field plotted. During the simulation, the excitation field is incident in the positive x-direction and polarized along the z-axis. The dipole resonance mode is the key to the enhancement in this case, which explains the enhanced fluorescence is resulted from the increased electric fields.

5.2. Enhanced Raman scattering

The enhancement effects of LSPR on Raman spectroscopy are generally attributed to the presence of hot spots on the rough particle surface, but there are more factors involved during this process, such as the complicated intraparticle coupling [52, 53]. Combined experiments and simulations are performed to interpret this issue.

The silver nanoparticles on quartz substrates were first synthesized through citrate reduction and applied to enhance the Raman spectrum of methylene blue. From the obtained spectra shown in **Figure 4**, it is obvious that the involvement of LSPR improves the resolution, where characteristic vibrational peaks are distinctly displayed. The enhancement factors are 3.2×10^5 and 1.3×10^7 for SERS excited by a 514.5 nm Ar-ion laser and a 785 nm diode laser, respectively. Since the nanoparticles were attracted by (3-aminopropyl)trimethoxysilane with SAM method, it is estimated that they lay as one layer instead of taking the form of aggregation. However, it is notable that the enhancement factor was higher than theoretically estimated for spherical particles [54]. More sophisticated explanation should be accounted.

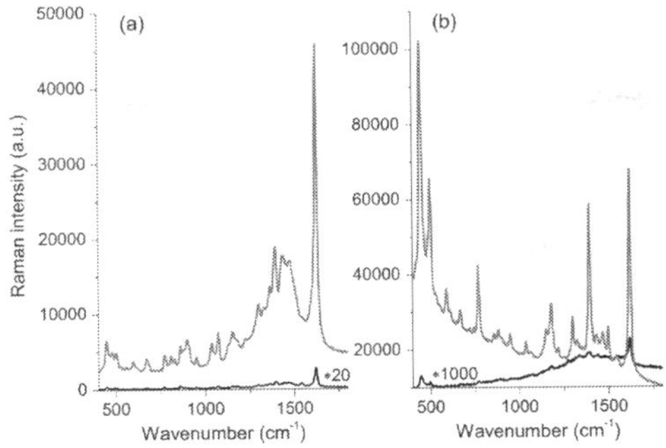

Figure 4. (a) The Raman spectra of methylene blue excited by 514.5 nm in the absence (black) and the presence (red) of SERS substrates and (b) the Raman spectra of methylene blue excited by 785 nm in the absence (black) and the presence (red) of SERS substrates.

SEM image shown in **Figure 5(a)** indicates that the submicrometer silver particles are flower-like with distinct surface protrusions. The average diameter of the silver particles is analyzed to be about 500 nm. Three-dimensional finite-difference time-domain (FDTD) method was employed to calculate both far- and near-field optical properties of the submicrometer silver particles. The contrcl structure, i.e., the smooth spherical structure, and the mimicking structure, i.e., a large particle (D = 400 nm) with 26 small spherical particles (D = 100 nm) evenly distributed on the exterior surface, were modeled.

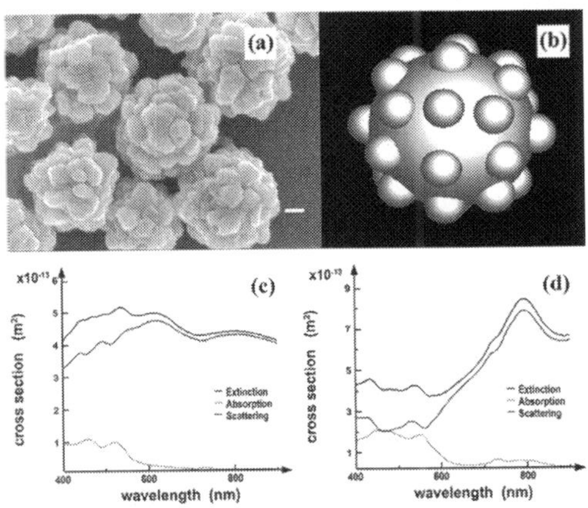

Figure 5. (a) The SEM image of rough submicrometer silver particles. Scale bar: 100 nm. (b) Schematic diagram of the rough submicrometer silver particle model. The 26 small peripheral particles were submerged into the large core particle, effectively generating many hemispheres on the surface. (c) Calculated extinction, absorption, and scattering spectra of the smooth silver particles model. (d) Calculated extinction, absorption, and scattering spectra of the rough silver particle model.

The characteristic spectra of both models are illustrated in **Figure 5(c)** and **(d)**. It is noteworthy that the extinction spectrum of the smooth particle shows featured bands originated from dipole resonance (ca. 800 nm) and higher-order multipole resonances, such as quadrupole resonance (ca. 620 nm), while the featured band of rough particles located at ca. 800 nm. The dipole field at the surface of rough particles is intensified through scattering. The augmented enhancement factor at 785 nm in the SERS spectrum is thus interpreted.

Understanding the wavelength dependence of SERS requires the distribution of electric fields of metal particles, as shown in **Figure 6**. The different electric distributions at different wavelengths for the smooth surface point out that the enhancement effect under the shorter wavelength (514.5 nm) originates from the multipole effect, while the longer wavelength (785 nm) is resulted from the dipole effect. The electric distribution for the rough surface denotes the prominent near-field enhancement at the rough surface because of a more localized distribution of the particle's conduction electrons.

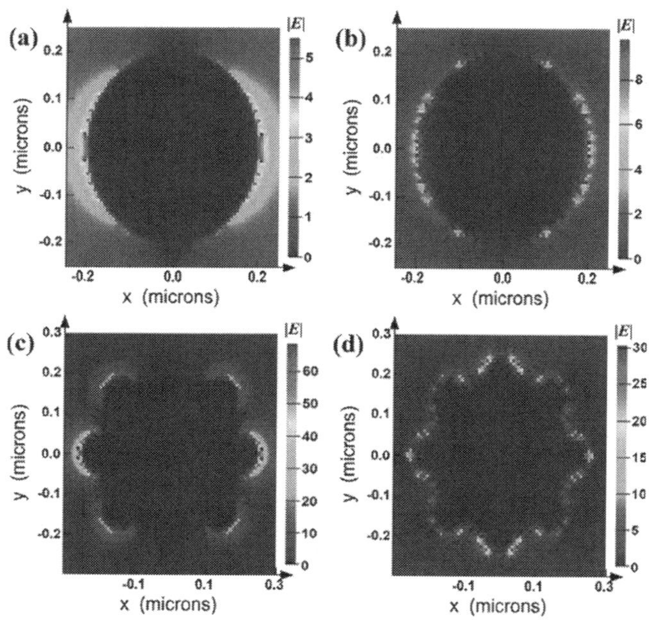

Figure 6. The distribution of electric fields around the silver particle model through FDTD calculation. The excitation field is incident in the positive z-direction. (a) and (b) The smooth silver particle model excited by 785 and 514.5 nm, respectively. (c) and (d) The rough silver particle model excited by 785 and 514.5 nm, respectively.

Further verification of the theoretical computed electric field is accomplished through near-field scanning optical microscopy (NSOM), which can measure the intensity distribution of optical fields of the rough submicrometer silver particles on a quartz slide, as shown in

Figure 7. The rough submicrometer silver particle shows a strong optical field distribution, shedding light on the stronger enhancement factor.

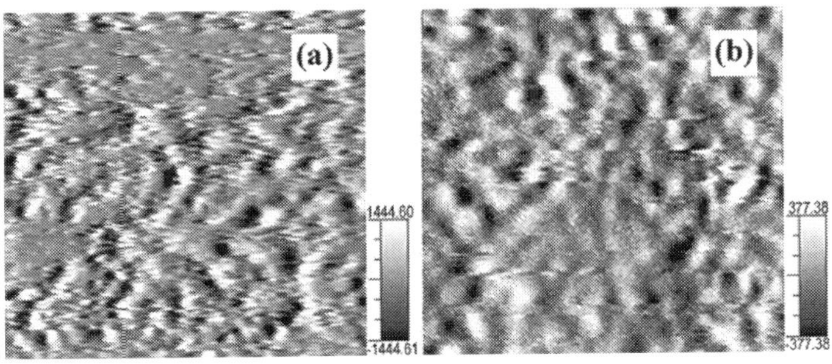

Figure 7. The measured intensity distribution of optical fields on the surface of nanoparticles by NSOM under 785 nm excitation. (a) Rough submicrometer silver particles and (b) spherical silver nanoparticles.

In order to illustrate the enhancement of electric field with a formula, the total electric field ($\vec{E}_{Total}$) of LSPR is accounted as the sum of radiation field produced by LSPR of the core particle ($\vec{E}_{o}$) and the peripheral particles ($\vec{E}_{i}$), as

$$\vec{E} = \vec{E}_o + \sum_{i=1}^{n} \chi_i(\mathrm{r}_i)\vec{E}_i \tag{26}$$

where $\chi_i(\mathrm{r}_i)$ is the weighing factor for the *ith* peripheral particles at r_i.

Based on the equation of the radiation field produced by the dipole resonance, the constant phase difference could be found for the radiation waves of large core particle and small peripheral for a direction r ($R\cos\theta_i$). Electric field from the core particle ($\vec{E}_{o}$) and the peripheral particles ($\vec{E}_{i}$) can be demonstrated and $\vec{E}_{Total}$ can be rewritten as

$$\vec{E}_o = \frac{\mathrm{k}^2 P_o \sin(\alpha)}{4\pi\varepsilon_0 \mathrm{r}} e^{ikr} \tag{27}$$

$$\vec{E}_i = \frac{\mathrm{k}^2 P \sin(\alpha)}{4\pi\varepsilon_0 \mathrm{r}} e^{ik(r + R\cos\theta_i)} \tag{28}$$

$$\vec{E}_{Total}^2 = E_o{}^2 + \chi_i{}^2 E_i{}^2 + 2\chi_i E_o E_i \cos(kR\cos\theta_i) \tag{29}$$

where ε_0 represents the permittivity of vacuum, P_o denotes the dipole moment of the large core particle, k expresses the wave number of the radiation field, and α is the angle between the incident and radiation fields. The effect of the roughness of the surface would be described by different induced dipole moments, as

$$P = E_{ex} a^3 \frac{\varepsilon_{np} - \varepsilon_0}{\varepsilon_{np} + 2\varepsilon_0} \tag{30}$$

In here, a is the radius of the particle, E_{ex} represents the incident electric field, and ε_{np} and ε_o denote the relative permittivity of the particle and the surrounding medium, respectively. The dependence of dipole enhancement with the grain is thus illustrated. A more distinct demonstration can be achieved by simulation through controlling the different sizes of peripheral particles, as shown in **Figure 8**, which indicates the enhanced induced dipole moment with larger peripheral particles.

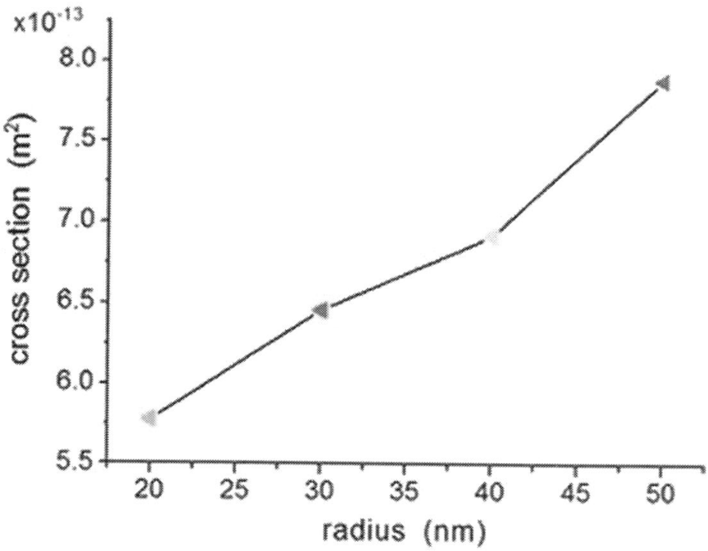

Figure 8. Enhanced dipole mode of LSPR with the increasing radius of the peripheral particles.

The effects of a rough surface are thus thoroughly studied combining experiment and theoretical calculation. The enlightening result indicating how small particles affect the enhancement factor helps further design more advanced nanoparticle detectors.

6. Conclusion

The localized surface plasmon resonance taking advantage of easy excitation is becoming increasing popular in the application of molecular spectroscopy, which has improved the resolution of spectroscopy and makes detection limit as low as femtogram. As we show in this chapter, there are numerous techniques to synthesize the desired nanostructures nowadays, and LSPR derived from those nanoparticles has demonstrated considerable enhancement factor to improve the applicability of molecular spectroscopy involving fluorescence, the Raman scattering, etc. On the one hand, it is to design better nanoparticles that arise localized surface plasmon; on the other hand, the mechanism of resulting electric field on the surface of nanoparticle needs to be accounted for different nanostructures. The multifactor-determined LSPR can now be currently elaborated through FDTD method. The joint experimental data with theoretical perspective are beneficial for a better understanding of the characteristic of LSPR and the resulting enhancement factor. Further intensive studies on LSPR combining experiments and modeling will broaden the application of LSPR and favor spectroscopies at molecular precision.

Acknowledgements

This work was supported by the Tianjin Municipal Science and Technology Commission (No. 15ZCZDGX00250 and No. 08ZCDFGX09400), the Doctoral Fund of Ministry of Education of China (No. 20110031110035), the National Natural Science Foundation of China (No. 60508004 and No. 60778043), and the National High Technology Research and Development Program of China ("863" Program, No. 2011AA030205).

Author details

Lu Sun, Ping Chen* and Lie Lin

*Address all correspondence to: chping@nankai.edu.cn

Key Laboratory of Optoelectronic Information Science and Technology, Ministry of Education of China, Institute of Modern Optics, Nankai University, Tianjin, China

References

[1] Wozniak, A.K., Schroder, G.F., Grubmuller, H., Seidel, C.A., Oesterhelt, F. Single-molecule FRET measures bends and kinks in DNA. Proc Natl Acad Sci U S A. 2008;105(47):18337-18342. DOI: 10.1073/pnas.0800977105

[2] Fujisawa, T., Kuramochi, H., Hosoi, H., Takeuchi, S., Tahara, T. Role of coherent low frequency motion in excited-state proton transfer of green fluorescent protein studied by timeresolved impulsive stimulated raman spectroscopy. J Am Chem Soc. 2016;138(12):3942–3945. DOI: 10.1021/jacs.5b11038 19

[3] Shi, X., Beauchamp, K.A., Harbury, P.B., Herschlag, D From a structural average to the conformational ensemble of a DNA bulge. Proc Natl Acad Sci U S A. 2014;111(15): 1473-1480. DOI: 10.1073/pnas.1317032111

[4] Salmon, L., Nodet, G., Ozenne, V., Yin, G., Jensen, M.R., Zweckstetter, M., Blackledge, M. NMR characterization of long-range order in intrinsically disordered proteins. J Am Chem Soc. 2010;132(24):8407-8408. DOI: 10.1021/ja101645g

[5] Agirrezabala, X., Liao, H.Y., Schreiner, E., Fu, J., Ortiz-Meoz, R.F., Schulten, K., Green, R., Frank, J. Structural characterization of mRNA-tRNA translocation intermediates. Proc Natl Acad Sci U S A. 2012;109(16):6094-6099. DOI: 10.1073/pnas.1201288109

[6] Kneipp, K., Kneipp, H., Kneipp, J. Surface-enhanced Raman scattering in local optical fields of silver and gold nanoaggregates—from single-molecule Raman spectroscopy to ultrasensitiveprobing in live cells. Acc Chem Res. 2006;39(7):443-450. DOI: 10.1021/ar050107x

[7] El-Sayed, M.A. Some interesting properties of metals confined in time and nanometer space of different shapes. Acc Chem Res. 2001;34(4):257-264. DOI: 10.1021/ar960016n

[8] Rosi, N.L., Mirkin, C.A. Nanostructures in biodiagnostics. Chem Rev. 2005;105(4): 1547-1562. DOI: 10.1021/cr030067f

[9] Chen, H.J., Kou, X.S., Yang, Z., Ni, W.H., Wang, J.F. Shape- and size-dependent refractive index sensitivity of gold nanoparticles. Langmuir. 2008;24(10):5233-5237. DOI: 10.1021/la800305j

[10] Hu, M., Chen, J.Y., Li, Z.Y., Au, L., Hartland, G.V., Li, X.D., Marquez, M., Xia, Y.N. Gold nanostructures: engineering their plasmonic properties for biomedical applications. Chem Soc Rev. 2006;35(11):1084-1094. DOI: 10.1039/B517615h

[11] Zhang, P., Haslett, T.L., Douketis, C., Moskovits, M. Mode localization in self-affine fractal interfaces observed by near-field microscopy. Phys Rev B. 1998;57(24): 15513-15518. DOI: 10.1103/PhysRevB.57.15513

[12] Reinhard, B.M., Sheikholeslami, S., Mastroianni, A., Alivisatos, A.P., Liphardt, J. Use of plasmon coupling to reveal the dynamics of DNA bending and cleavage by single EcoRV restriction enzymes. Proc Natl Acad Sci U S A. 2007;104(8):2667-2672. DOI: 10.1073/pnas.0607826104

[13] Sonnichsen, C., Reinhard, B.M., Liphardt, J., Alivisatos, A.P. A molecular ruler based on plasmon coupling of single gold and silver nanoparticles. Nat Biotechnol. 2005;23(6): 741-745. DOI: 10.1038/nbt1100

[14] Yu, W.W. and White, I.M. Inkjet-printed paper-based SERS dipsticks and swabs for trace chemical detection. Analyst. 2013;138(4):1020-1025. DOI: 10.1039/c2an36116g

[15] Zhao, J., Pinchuk, A.O., Mcmahon, J.M., Li, S.Z., Ausman, L.K., Atkinson, A.L., Schatz, G.C. Methods for describing the electromagnetic properties of silver and gold nanoparticles. Acc Chem Res. 2008;41(12):1710-1720. DOI: 10.1021/ar800028j

[16] Wood, R.W. On a remarkable case of uneven distribution of light in a diffraction grating spectrum. Proc Phys Soc Lond 1902;18:269. DOI: 10.1088/1478-7814/18/1/325

[17] Willets, K.A. and Van Duyne, R.P.. Localized surface plasmon resonance spectroscopy and sensing. Annu Rev Phys Chem 2007;58:267-297. DOI: 10.1146/annurev.physchem.58.032806.104607

[18] Anderson, L.J.E., Mayer, K.M., Fraleigh, R.D., Yang, Y., Lee, S., Hafner, J.H. Quantitative measurements of individual gold nanoparticle scattering cross sections. J Phys Chem C. 2010;114(25):11127-11132. DOI: 10.1021/jp1040663

[19] Jensen, T.R., Duval, M.L., Kelly, K.L., Lazarides, A.A., Schatz, G.C., Van Duyne, R.P. Nanosphere lithography: effect of the external dielectric medium on the surface plasmon resonance spectrum of a periodic array of sliver nanoparticles. J Phys Chem B. 1999;103(45):9846-9853. DOI: 10.1021/Jp9926802

[20] Cao, J., Sun, T., Grattan, K.T.V. Gold nanorod-based localized surface plasmon resonance biosensors: a review. Sens Actuat B Chem. 2014;195:332-351. DOI: 10.1016/j.snb.2014.01.056

[21] Link, S., Mohamed, M.B., El-Sayed, M.A.. Simulation of the optical absorption spectra of gold nanorods as a function of their aspect ratio and the effect of the medium dielectric constant. J Phys Chem B. 1999;103(16):3073-3077. DOI: 10.1021/Jp990183f

[22] Petryayeva, E., Krull, U.J. Localized surface plasmon resonance: nanostructures, bioassays and biosensing—a review. Anal Chim Acta. 2011;706(1):8-24. DOI: 10.1016/j.aca.2011.08.020

[23] Daniel, M.C., Astruc, D. Gold nanoparticles: assembly, supramolecular chemistry, quantum size-related properties, and applications toward biology, catalysis, and nanotechnology. Chem Rev. 2004;104(1):293-346. DOI: 10.1021/cr030698+

[24] Hammons, J.A., Ustarroz, J., Muselle, T., Torriero, A.A.J., Terryn, H., Suthar, K., Ilavsky, J. Supported silver nanoparticle and near-interface solution dynamics in a deep eutectic solvent. J Phys Chem C. 2016;120(3):1534-1545. DOI: 10.1021/acs.jpcc.5b09836

[25] Lu, W.S., Wang, H.F., Zhang, J.P., Jiang, L. Gold nanorods: synthesis, growth mechanism and purification. Prog Chem. 2015;27(7):785-793. DOI: 10.7536/PC150111

[26] Chang, S.S., Shih, C.W., Chen, C.D., Lai, W.C., Wang, C.R.C. The shape transition of gold nanorods. Langmuir. 1999;15(3):701-709. DOI: 10.1021/La980929l

[27] Boltasseva, A. Plasmonic components fabrication via nanoimprint. J Opt A Pure Appl Opt. 2009;11(11):114001. DOI: 10.1088/1464-4258/11/11/114001

[28] Zhang, X.Y., Whitney, A.V., Zhao, J., Hicks, E.M., Van Duyne, R.P. Advances in contemporary nanosphere lithographic techniques. J Nanosci Nanotechnol. 2006;6(7): 1920-1934. DOI: 10.1166/jnn.2006.322

[29] Feng, L.L., Xuan, Z.W., Ma, J.B., Chen, J., Cui, D.X., Su, C.W., Guo, J.M., Zhang, Y.J. Preparation of gold nanorods with different aspect ratio and the optical response to solution refractive index. J Exp Nanosci. 2015;10(4):258-267. DOI: 10.1080/17458080.2013.824619

[30] Satija, J., Tharion, J., Mukherji, S. Facile synthesis of size and wavelength tunable hollow gold nanostructures for the development of a LSPR based label-free fiber-optic biosensor. RSC Adv. 2015;5(86):69970-69979. DOI: 10.1039/c5ra13941d

[31] Zhang, D.J., Zhang, C.Y., Lu, Y., Hao, Y.W., Liu, Y.Q. Preparation of Au@ Ag core-shell nanoparticles through seed-mediated growth method. Prog Chem. 2015;27(8): 1057-1064. DOI: 10.7536/PC150140

[32] Jang, H.J., Ham, S., Acapulco, J.A.I., Song, Y., Hong, S., Shuford, K.L., Park, S. Fabrication of 2D Au nanorings with Pt framework. J Am Chem Soc. 2014;136(50):17674-17680. DOI: 10.1021/ja510916y

[33] Ma, Y.N., Zhou, J., Shu, L., Li, T.H., Petti, L., Mormile, P. Optimizing Au/Ag core-shell nanorods:purification, stability, and surface modification. J Nanopart Res. 2014;16(6): 2439. DOI: 10.1007/S11051-014-2439-6

[34] Jia, W.F., Li, J.R., Jiang, L. Synthesis of highly branched gold nanodendrites with a narrow size distribution and tunable NIR and SERS using a multiamine surfactant. ACS Appl Mater Inter. 2013;5(15):6886-6892. DOI: 10.1021/am401006b

[35] Jana, N.R., Gearheart, L., Murphy, C.J. Seed-mediated growth approach for shape-controlled synthesis of spheroidal and rod-like gold nanoparticles using a surfactant template. Adv Mater. 2001;13(18):1389-1393. DOI: 10.1002/1521-4095

[36] Jana, N.R., Gearheart, L., Murphy, C.J. Wet chemical synthesis of high aspect ratio cylindrical gold nanorods. J Phys Chem B. 2001;105(19):4065-4067. DOI: 10.1021/Jp0107964

[37] Huang, X.H., Neretina, S., El-Sayed, M.A. Gold nanorods: from synthesis and properties to biological and biomedical applications. Adv Mater. 2009;21(48):4880-4910. DOI: 10.1002/adma.200802789

[38] Nikoobakht, B., El-Sayed, M.A. Preparation and growth mechanism of gold nanorods (NRs) using seed-mediated growth method. Chem Mater. 2003;15(10):1957-1962. DOI: 10.1021/cm0207321

[39] Ye, X.C., Jin, L.H., Caglayan, H., Chen, J., Xing, G.Z., Zheng, C., Doan-Nguyen, V., Kang, Y.J., Engheta, N., Kagan, C.R., Murray, C.B. Improved size-tunable synthesis of

monodisperse gold nanorods through the use of aromatic additives. ACS Nano. 2012;6(3):2804-2817. DOI: 10.1021/nn300315j

[40] Gole, A., Murphy, C.J. Polyelectrolyte-coated gold nanorods: synthesis, characterization and immobilization. Chem Mater. 2005;17(6):1325-1330. DOI: 10.1021/cm048297d

[41] Marinakos, S.M., Chen, S.H., Chilkoti, A. Plasmonic detection of a model analyte in serum by a gold nanorod sensor. Anal Chem. 2007;79(14):5278-5283. DOI: 10.1021/ac0706527

[42] Cheng, S.F., Chau, L.K. Colloidal gold-modified optical fiber for chemical and biochemical sensing. Anal Chem. 2003;75(1):16-21. DOI: 10.1021/ac020310v

[43] Tang, J.L., Cheng, S.F., Hsu, W.T., Chiang, T.Y., Chau, L.K. Fiber-optic biochemical sensing with a colloidal gold-modified long period fiber grating. Sens Actuat B Chem. 2006;119(1):105-109. DOI: 10.1016/j.snb.2005.12.003

[44] Cao, J., Galbraith, E.K., Sun, T., Grattan, K.T.V. Cross-comparison of surface plasmon resonance-based optical fiber sensors with different coating structures. IEEE Sens J. 2012;12(7):2355-2361. DOI: 10.1109/Jsen.2012.2188713

[45] Yu, C.X., Irudayaraj, J. Multiplex biosensor using gold nanorods. Anal Chem. 2007;79(2):572-579. DOI: 10.1021/ac061730d

[46] Gawinkowski, S., Pszona, M., Gorski, A., Niedziolka-Jonsson, J., Kaminska, I., Nogala, W., Waluk, J. Single molecule Raman spectra of porphycene isotopologues. Nanoscale. 2016;8(6):3337-3349. DOI: 10.1039/c5nr08627b

[47] Shen, Y.B., He, T., Wang, W.H., Zhan, Y.L., Hu, X., Yuan, B.F., Zhou, X.C. Fluorescence enhancement on silver nanoplates at the single- and sub-nanoparticle level. Nanoscale. 2015;7(47):20132-20141. DOI: 10.1039/c5nr06146f

[48] Shen, W.F., Tang, J.G., Wang, D., Yang, R.Q., Chen, W.C., Bao, X.C., Wang, Y., Jiao, J.Q., Wang, Y.X., Huang, Z., Huang, L.J., Liu, J.X., Wang, W., Wu, P.H., Belfiore, L.A. Enhanced efficiency of polymer solar cells by structure-differentiated silver nanodopants in solution-processed tungsten oxide layer. Mater Sci Eng B Adv. 2016;206:61-68. DOI: 10.1016/j.mseb.2015.11.004

[49] Rogowska, J., Namiesnik, J. Environmental implications of oil spills from shipping accidents. Rev Environ Contam Toxicol. 2010;206:95-114. DOI: 10.1007/978-1-4419-6260-7_5

[50] Mearns, A.J., Reish, D.J., Oshida, P.S., Ginn, T., Rempel-Hester, M.A., Arthur, C., Rutherford, N., Pryor, R. Effects of pollution on marine organisms. Water Environ Res. 2015;87(10):1718-1816. DOI: 10.2175/106143015X14338845156380

[51] Zhang, F., Zhang, L., Mao, S.C., Chen, P., Cui, J.C., Tang, Y.G., Wang, K., Lin, L., Qi, X.D. Use of metal-enhanced fluorescence spectroscopy for detection of polycyclic aromatic

hydrocarbons in diesel oil emulsions in artificial seawater. Environ Technol. 2012;33(18):2071-2075. DOI: 10.1080/09593330.2012.660643

[52] Zhang, F., Chen, P., Zhang, L., Mao, S.C., Lin, L., Tang, Y.G., Cui, J.C., Qi, X.D., Yang, J.H., Ma, Y.F. Enhancement of Raman scattering by field superposition of rough submicrometer silver particles. Appl Phys Lett. 2012;100(17):4705083. DOI: 10.1063/1.4705083

[53] Bok, H.M., Shuford, K.L., Jeong, E., Park, S. Intra-surface plasmon coupling between smooth and nanoporous blocks in a gold nanorod. Chem Commun. 2010;46(6):982-984. DOI: 10.1039/b918510k

[54] Kerker, M. Estimation of surface-enhanced Raman scattering from surface-averaged electromagnetic intensities. J. Colloid Interface Sci.. 1987;18:417-421. DOI: 10.1016/0021-9797(87)90477-2

Chapter 19

Using Raman Spectroscopy for Characterization of Aqueous Media and Quantification of Species in Aqueous Solution

Ivana Durickovic

Additional informat on is available at the end of the chapter

http://dx.doi.org/10.5772/64550

Abstract

In this chapter, the use of Raman spectroscopy (RS) for studies of aqueous solutions is shown. This technique is mainly used for the characterization of solid samples, but presents numerous features permitting its use for the analysis of aqueous media. Indeed, it possesses all the advantages of optical methods (versatility, rapidity, contact-less non-destructive measurement, etc.), but also offers possibilities for *in situ* measurements. The Raman spectrum will be influenced by several parameters such as the solution concentration or its temperature-phase. Thus, the analysis of a set of aqueous solutions of different concentrations in a certain temperature range can permit the identification of the specific effect of salt and temperature. A proper analysis based on the follow-up of the specific peak areas or intensities can permit the determination of the salt concentration or the phase transition of the studied solution. The analysis can be focused on the salt direct effect on the spectrum, analysis of the salt signature itself, or on its indirect effect on the water signature. The method for the characterization of aqueous solutions of some salts is presented: elaboration of calibration curves and concentration determination. As an application example, a special attention is devoted to aqueous solutions that are used in the winter maintenance domain (solution of acetates, formates, or chlorides), which are very relevant examples of aqueous solution behavior. A specific analysis set to determine the solution solid-liquid phase transitions is presented as well as the thus-constructed phase diagram.

Keywords: aqueous solutions, Raman spectroscopy, characterization, quantification, phase transition

INTECH
open science | open minds

1. Introduction

In this chapter, the possibility of using Raman spectroscopy (RS) for studies of aqueous solutions is shown. Section 2 is devoted to the description of the general principle and characteristics of Raman spectroscopy (RS) as a technique for the investigation of molecular structure. Since Raman spectra contain information not only on intra-molecular vibrations but also on vibrations of the crystal lattice and other solid movements, RS is a well-established technique for the studies on solid materials [1].

The aim of this chapter is to show the possibility of its use to study aqueous solutions and chemical species dissolved in water. The application of RS on the study of aqueous media is more complicated than in solids, but is still very efficient if a proper signal treatment is applied. In order to show the different possibilities of RS for the characterization of water media, a special attention is devoted to one application example: aqueous solutions that are used in the winter maintenance domain (solution of acetates, formates, or chlorides). The possibility of its use for the detection of some water pollutants is also discussed.

In the case of water quality, there is a need for a device presenting a high polyvalence degree to detect and to quantify several chemical species in aqueous solutions but also to discriminate them from each other. There is an increasing need for a technique that could permit *in situ* measurements and continuous monitoring of water.

However, the techniques actually used for the detection and quantification of chemicals in water often do not respond to these criteria [2]. Some techniques, such as the widely used solution conductometry, are fully dedicated to one chemical species or are not able to make an accurate discrimination between different chemical species. In other cases, the quantification of chemicals in water media usually needs the combination of several techniques such as ionic chromatograph and electrical conductivity. This approach is then not appropriate for *in situ* measurements and are time consuming, inducing a time delay between the sample collection and analysis. The same remark concerns the inductively coupled plasma atomic emission spectroscopy, which permits to detect chemical atoms.

On the other hand, previous studies indicated the great potential of the optical and spectroscopic tools in the detection of salts in solutions and, more specifically, RS, which could then permit to avoid the inconveniences of the techniques mentioned above [3]. Indeed, as it is shown, RS can permit us to determine the nature and the quantity of chemical species present in water media, offering the possibility to discriminate the different species present. In addition, the RS is an all-optical method, so it possesses all the advantages of optical methods (versatility, rapidity, contactless non-destructive measurement, etc.), as it will be detailed further on. Moreover, RS is generally more versatile and easier to set up *in situ*.

2. General principle of Raman spectroscopy

RS provides numerous information about the structure and chemical composition of the sample, information obtained by an illumination of the sample with a monochromatic radiation (laser beam) that excites the vibrational structure of molecules; RS is thus called a vibrational spectroscopy. As illustrated in **Figure 1**, the radiation coming to the sample undergoes two types of scattering: Rayleigh scattering, where the radiation is at the exact same frequency as the excitation laser line (υ), and Raman scattering where the scattered radiation has a different frequency (υ'). The difference between the two frequencies, called "Raman shift" ($\Delta\upsilon$), is the consequence of the interaction of the radiation with the sample molecules by the means of the vibrations of the chemical bonds present in the sample [4–6]. Note that, because it is a difference value, the Raman shift is consequently totally independent of the frequency of the incident radiation.

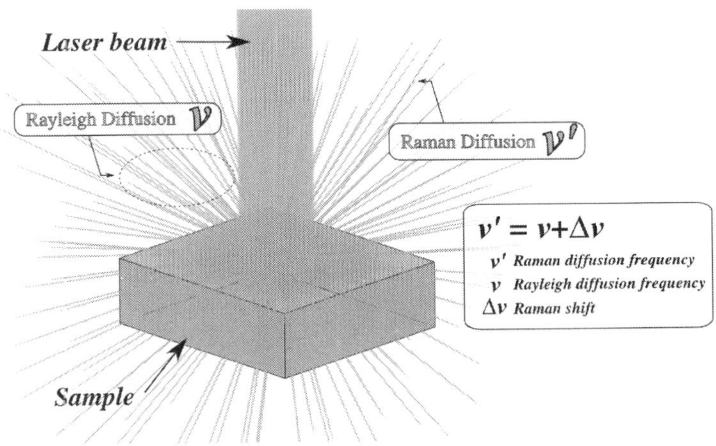

Figure 1. Schematic representation of the Rayleigh and the Raman diffusions.

In order to properly extract the Raman shift, the radiation coming from the Rayleigh diffraction has to be filtered out with a notch or a band pass filter from the total radiation before being collected on the spectrometer sensor (generally a CCD). Thus, only the radiation with a different wavelength than the incident laser beam is analyzed.

2.1. Raman spectrum

The results of the measurements are depicted graphically as Raman spectra. The intensity of the scattered light is plotted for each energy (frequency) of light. The frequency axis represents the Raman shift $\Delta\upsilon$, as it is the shift in energy/frequency of the light that is of particular interest. In vibrational spectroscopy, the frequency is traditionally measured in a unit called the "wavenumber" (number of waves per cm, cm^{-1}), which is directly proportional to energy. Wavenumbers are easily converted into the more familiar wavelength scale by calculating the reciprocal.

A thus-plotted Raman spectrum is composed of peaks, where each peak corresponds to a specific vibration of a chemical bond present in the sample. The more complex the chemical composition, the richer is its Raman spectrum.

When the specific vibration is represented by a wider peak, we are generally talking about a "band" that can be composed of several peaks. In this case, each peak represents the vibration of the same chemical bond, but its surrounding environment is slightly different, provoking a frequency shift. Thus, the partly overlapping peaks form a resulting band that is covering a more or less large range of frequencies.

2.2. Information accessible

By the analysis of the different Raman peaks present, a Raman spectrum can give us numerous qualitative and quantitative information on the sample, as presented in **Figure 2**.

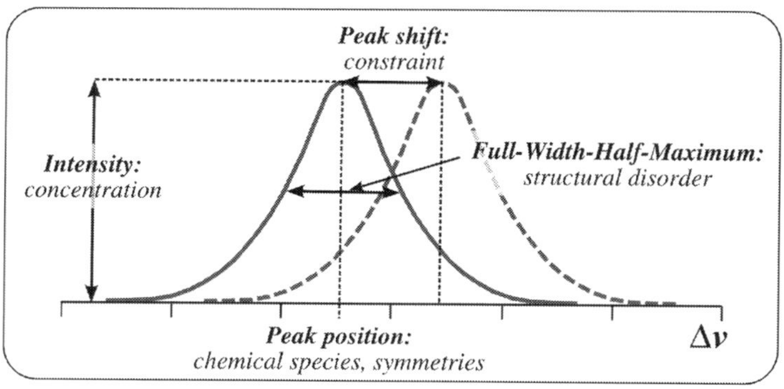

Figure 2. Information accessible from a Raman spectrum [6].

Three spectral parameters can be derived from the analysis of a Raman line.

The **position of the peak** defined by its maximum corresponds to the vibration frequency of the chemical species. Since each chemical bond has its own characteristic vibrations, the position of the peaks lead to the identification of the chemical species. The determination of the peak locations in a Raman spectrum is sensitive enough for the recognition and the specification of chemical species in heterogeneous samples [7].

The **peak intensity** is related to the corresponding chemical species concentration. In order to determine this parameter, it is necessary to use normalization of the integrated intensity of the Raman line as the peak intensity is also sensitive to the laser power. The follow-up of the relative changes in the integrated intensities of peaks is then necessary for excluding the laser influence and to determine correctly the species concentration.

Furthermore, the full width half maximum (FWHM) reflects the order character of the sample: structure with the lower the FWHM, the higher the local order.

Thereby, the peak position, the linewidth, and intensity extracted from a Raman line can be used for the determination of some physico-chemical parameters such as the sample phase or constraint [8]. The analysis of all the peaks present in the spectrum gives us an insight on the sample composition and structure. RS can thus be used to determine various quantities that can affect the vibrational state (modes/peaks) specific to a substance.

In addition, specific Raman peaks are sensitive not only to the above-mentioned composition or concentration of the substance under study but also to external parameters such as the temperature or the pressure that affect the structure of the sample. Thus, the spectroscopic follow-up of a sample can permit to identify which chemical bond is affected by a change of an external parameter and to identify the constraint through the resulting peak shift. This aspect is explained further in this chapter in Section 3.2.

Since a Raman spectrum gives details on the chemical composition, molecular structure, and molecular interactions, it is hence considered as a chemical compound fingerprint at a molecular or crystalline level.

2.3. Advantages

RS is useful for chemical analysis for several reasons [9, 10]. As many optical techniques, it is non-destructive and non-intrusive, permitting to perform contactless measurements. Furthermore, in contrast with most other chemical analysis techniques, it does not require any specific sample preparation. Moreover, this technique requires a small volume of the substance, in the order of 1 μm^3, for the analysis and it is possible to use optical fibers for deported measurement. This kind of set up is particularly well adapted for *on-site* measurements (on the field or in an industrial context). Furthermore, as a Raman spectrum can be acquired in the range of time of seconds, RS permits an almost "real-time" monitoring of chemical reactions. Depending on the experimental set up and on the application aimed, it can offer different spatial resolutions (from 1 μm to 1 cm). An additional advantage of RS is the possibility to analyze samples in solid, liquid, or gaseous state.

It is to note that RS offers the possibility to use lasers in a large spectral domain, going from the ultraviolet to the near infrared (IR). The choice of the most appropriate laser to use will depend on the nature of the sample under study. Indeed, the absorption, the fluorescence, the solidity toward light exposure, and more generally the interaction between light and the sample have to be taken into account. For the specific case of aqueous solutions, the most appropriate laser is in the visible green light (514 or 532 nm), which permits to diminish the fluorescence.[1] Furthermore, the excitation wavelength will have an important impact on experimental capacities; the laser wavelength λ will influence Raman intensity, proportional to λ^{-4}, and spatial resolution, defined by a diameter equal to $1.22\,\lambda / NA$, the numerical aperture of the objective used.

1 In fluorescence, the incident excitation light is completely absorbed, transferring the system to an excited state and producing a photon with a different frequency. The fluorescence photons, which are of lower energy, are emitted after a certain resonance lifetime.

RS presents also some inconveniences and limits, but the technology development offers possibilities to overcome most of them. For instance, Raman signal of some compounds can be weak comparing to the total signal, making it difficult to determine small concentrations. A stronger signal can obviously be obtained by the increase of the laser power, but the influence of the laser on the sample state could be strong enough to be problematic for the analysis or even induce a sample damage. To compensate this detrimental effect, the recent technological development improved significantly the performance of multichannel detectors, leading to a considerable increase of the sensitivity of Raman spectrometers.

Another limiting factor is the fluorescence, which can be much higher than the Raman signal, dominating in intensity the Raman effect, at times diluting it completely in the signal noise. However, as the Raman effect is independent on the excitation frequency, it is often possible to overcome this difficulty by choosing an appropriate laser.

Despite these few restrictions, RS appears well adapted for the analysis of aqueous solutions. Compared to infrared spectroscopy where the spectrum of water is so strong and complex that it interferes with the signatures of chemical species, the Raman spectrum is not sensitive to aqueous absorption bands that are weak and unobtrusive. The analysis of liquids is easier, thanks to the transparency of glass containers in the spectral domains concerned.

3. Spectroscopic analysis of water

3.1. Specific water signature

The structure of water has been studied for decades by both infrared (IR) and RS [11–15]. The normal modes of water are nowadays well known and detailed on **Figure 3**; The spectral band corresponding to the O—H bending (noted v_2) is located at about 1600 cm^{-1}, and to the O—H stretching band around 2900–3700 cm^{-1}. This broad range of the Raman spectrum of liquid water is composed of symmetric (noted v_1) and asymmetric (noted v_3) O—H stretching vibrations [14, 16, 17].

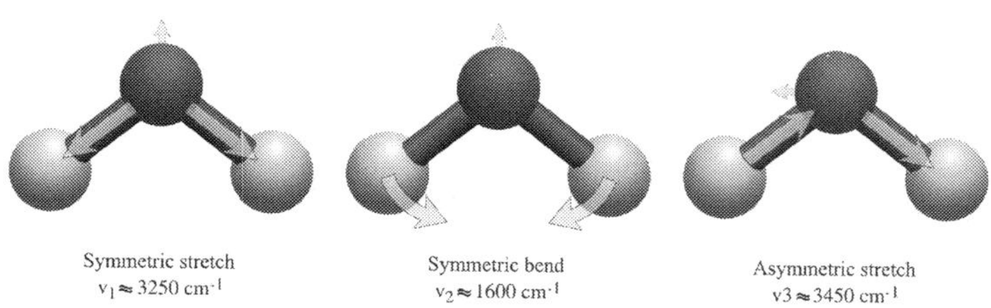

Figure 3. Normal modes of water.

In the literature, a special attention is devoted to the study of the O—H stretching region [18, 19]. The particular interest for this region originates from the fact that it is generally considered to be closely related to the structure of water [20, 21] because it is an indicator of the hydrogen-bonding network [22, 23]. Moreover, even though it is well known that this band constitutes of the symmetric and asymmetric O─H stretching, further designations of these normal modes are still not well elucidated. It is generally admitted that all the bands of water are made up from contributions from different components from water molecules in different hydrogen bonded environments. In order to analyze this region in a more detailed way, it is necessary to proceed to a deconvolution of the band corresponding to this complex contribution of the water spectrum into various peak components, to assign each vibrational mode and to get back to their molecular origins. However, when it comes to bands that are as large as the O─H stretching band, the decomposition is rather complicated and numerous deconvolution models can be found in the literature. Depending on the authors, different number of components as well as different mechanisms involved are considered. The use of two to five components for deconvolution is suggested to correctly describe the O─H stretching band [18, 24–27]. The most commonly proposed deconvolution contains five components [28–30]. An example of such a deconvolution of the O─H stretching band is presented in **Figure 4**.

It is to note that even when using the same number of components for the deconvolution, different results can be found and different attributions proposed. For each deconvolution, different hydrogen bond properties, such as their number, angle or length, are used to define the attributions [18, 27, 30–33]. Generally speaking, lower frequency components are attributed to water molecules with stronger hydrogen bonds and higher frequency components have weaker hydrogen bonds [34].

All these studies show that, besides the intra-molecular O─H pairs, intermolecular O─H linked by hydrogen bonds contribute to the O─H stretching. The role of the hydrogen bonds is therefore of great importance for the understanding of this spectral range. This bond being

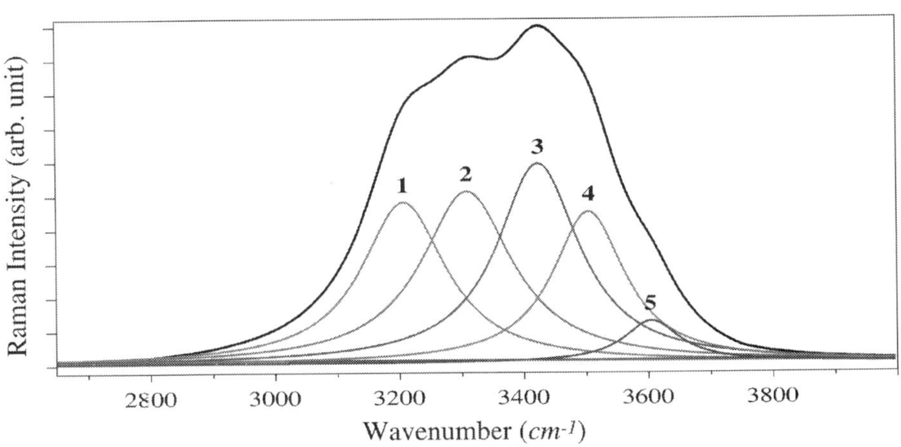

Figure 4. Example of deconvolution of the O─H stretching band into five sub-bands [28].

flexible is sensitive to temperature [35] and the presence of ions. The study of this spectral region can therefore be used for the determination of the water phase or for the detection of ions dissolved in water, as it will be presented thereafter.

3.2. Phase effect

As a molecule, water can be present at three different phases—solid, liquid, and gaseous—the state depending on the environmental conditions, namely the pressure and the temperature. This chapter concentrates on the solid-liquid phase transition, with a special focus on measurements performed on atmospheric pressure, the application example being the winter maintenance domain.

As mentioned earlier, the O—H stretching band is considered to be closely related to the structure of water. It is expected that the morphology of this band will be modified by a change in the temperature or in its chemical composition, as it can be seen on **Figure 5** where normalized spectra of liquid water and ice are compared.

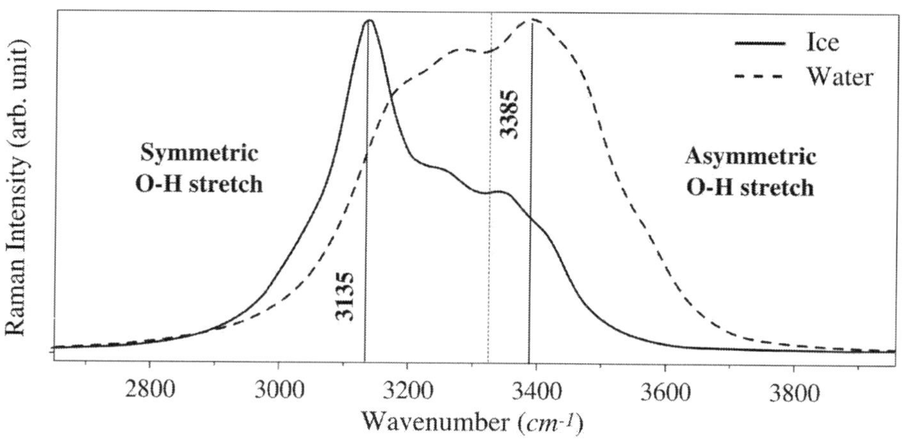

Figure 5. Raman signature of the O—H stretching region for water in its liquid (at 20°C) and solid form (at −20°C) obtained with a 532-nm laser beam at 100 mW.

In the case of liquid water, molecules are in constant movement, provoking continuous creation and break of hydrogen bonds that are rather weak. Water can be considered as a mixture of isolated water molecules and water molecules forming clusters by the hydrogen bonds [14]. Hence, all contributions, in terms of number of hydrogen bonds present, contribute to the O—H stretching band. This leads to a broad spectrum with the symmetric and asymmetric regions of the band that are almost equally represented. An average number of hydrogen bonding for each molecule is found to be 2.75 [29], which is manifested by a slightly greater intensity at 3385 cm^{-1} corresponding to the contribution possessing three H-bonds.

In the case of ice, on the other hand, the decrease of temperature will strengthen the hydrogen bonds, leading to weaker O—H bonds, which will thus vibrate at lower frequencies. Furthermore, besides the shift toward the lower frequencies, due to the increased importance of intermolecular hydrogen bonding at lower temperature, the lower part of the spectrum, related to fully H-bonded atoms, enhances and becomes narrower [36].

3.3. Phase transition determination

Since the spectra of liquid water and ice are significantly different, the combination of Raman spectrometry and micro-thermometry can lead to precise detection of the water phase transition. Spectra obtained in the −3 to 3°C temperature region were collected and their O—H stretching band presented in **Figure 6**. It is reminded that the study is focused on the O—H stretching band, as it is the spectral region considered to be closely related to the structure of water and which is thus expected to be the most influenced by a change of temperature.

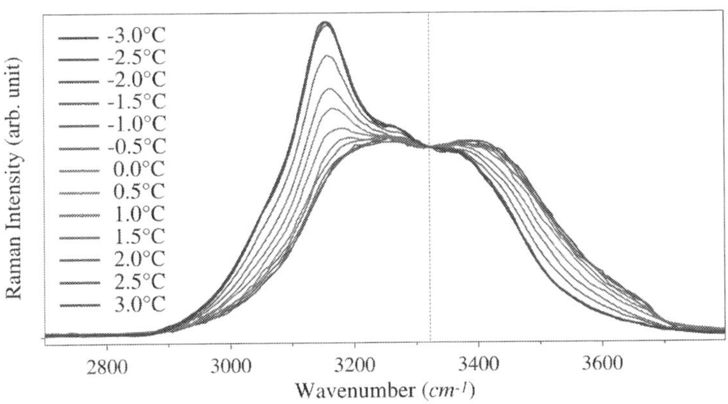

Figure 6. Raman spectra of water between −3 and 3°C, obtained with a 532-nm laser beam at 100 mW.

The center of the O—H stretching bond is located around 3325 cm^{-1}. The temperature modification induces large changes in the shape and the intensity of the O—H stretching region in both the part corresponding to symmetric and asymmetric stretching vibrations. It can be considered that the lower frequency part of the spectrum (between 2900 and 3325 cm^{-1}) roughly corresponds to the ordered solid phase as this contribution is related to fully H-bonded atoms, whereas partly H-bonded and free O—H are expressed in the upper frequency part, characteristic of the liquid phase [30]. It is then possible to follow the phase transition by the relative evolution of these two parts of the O—H stretching region as the evolution of the order/disorder of the water structure can be reflected from the values of their intensities. Thus, a spectral marker S_D, can be defined as the ratio of the asymmetric and the symmetric part of the O–H band, parts centered on 3385 and 3135 cm^{-1}, respectively, in order to detect the phase transition.

A method for the determination of the phase transition based on that principled was developed and tested [37]. That study showed that, for the calculation of the spectral marker, it is possible to use intensities at the peak maximum or integrated intensities of raw spectra.[2] The comparison of curves obtained by the calculation of ratios of simple intensities of the two most intense parts (c) and of integrated intensities of more (a) or less broad (b) spectral areas around the peak maximum is shown in **Figure 7**.

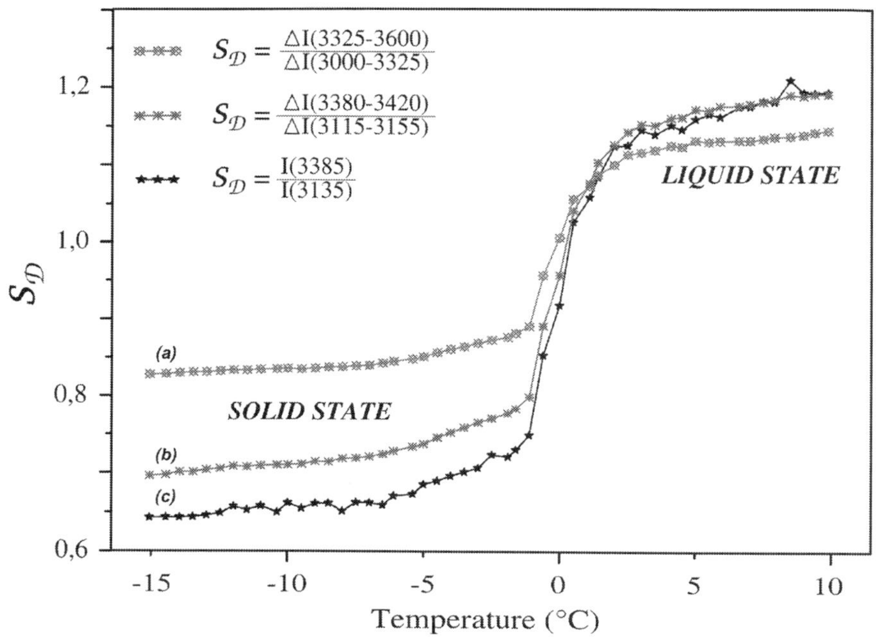

Figure 7. Plots of the spectral marker S_D calculated with intensities (c) or integrated intensities (a, b) as a function of temperature [37].

The temperature of the phase transition is then determined by a simple calculation of the curve inflection point. The uncertainty of the phase temperature thus obtained is dependent on the speed of the temperature change during the measurements. In the case presented here, the speed was set to 0.5°C/minute, and the uncertainty obtained is about 0.5°C for each ratio tested (ratio of simple or integrated intensities).

2 The use of raw spectra permits to overcome the possible controversies about the deconvolution of the O–H stretching region. In this method, only two parts of the spectra that are affected by the temperature and phase change are then analyzed.

4. Spectroscopic analysis of aqueous media

A Raman spectrum of an aqueous solution will present specific the different O–H bond vibrations of water, as well as specific peaks of each chemical/salt diluted. Indeed, each type of salt (acetate, formate, nitrate, etc.) diluted in an aqueous solution presents specific signatures/peaks corresponding to the vibrations of the different chemical bonds constituting it. By the analysis of each peak, it is possible to identify the nature of the chemical present, as well as its concentration. Indeed, the systematic analysis of a set of aqueous solutions of different concentrations can permit the identification of the specific effect of the chemical. A proper analysis based on the follow-up of the specific peak intensities or areas can permit the determination of the chemical concentration. The analysis can be focused on the salt direct effect, on analysis of the salt signature, or on its indirect effect by the analysis of the water signature.

As presented earlier in this chapter, the peak intensity is linked to the concentration. Peak intensity measurements are thus used in most quantitative analyses. However, the absolute intensity of a Raman spectrum can vary considerably from one instrument to another as it is also sensitive to changes in instrumental resolution, calibration and signal/noise ratio, etc. Thereby, it is necessary to use integrated intensity, a measure of the total intensity of the band, which is much less sensitive to instrumental resolution [4].

4.1. Application to winter maintenance aqueous solutions

The application example will concern aqueous solutions used in winter maintenance where different anti-icing chemicals are spread on runways in order to maintain a proper grip.[3] This is a commonly employed technique to avoid the occurrence of ice or to generate its melting [38, 39]. The products used are mainly salts, heard as ionic compounds composed of cations and anions. Each of these salts contains an "active compound" permitting to lower the freezing temperature of the liquid present on the road surface [40]. The most commonly employed anti-icing product for road winter maintenance is the NaCl (in France, up to 99% of the cases). The active compounds of the products applied on the airport surfaces are mainly acetates and formates of sodium or potassium for corrosion reasons [41].

4.1.1. Concentration determination

For the quantification of the molecules of interest, the anti-icing active compounds, it is necessary to construct appropriate calibration curves. For that purpose, solutions with well known composition and concentration of each active compound are prepared and analyzed spectroscopically. The first step is to identify the spectral signature of an analyzed compound. Raman spectra of the different salts used in winter maintenance are presented on **Figure 8**. The more complex the chemical composition, the richer is its Raman spectrum;

3 The quantification of the active compound of these products is necessary for both the characterization of the nature of the commercial products and the optimization of the quantities applied on airport areas.

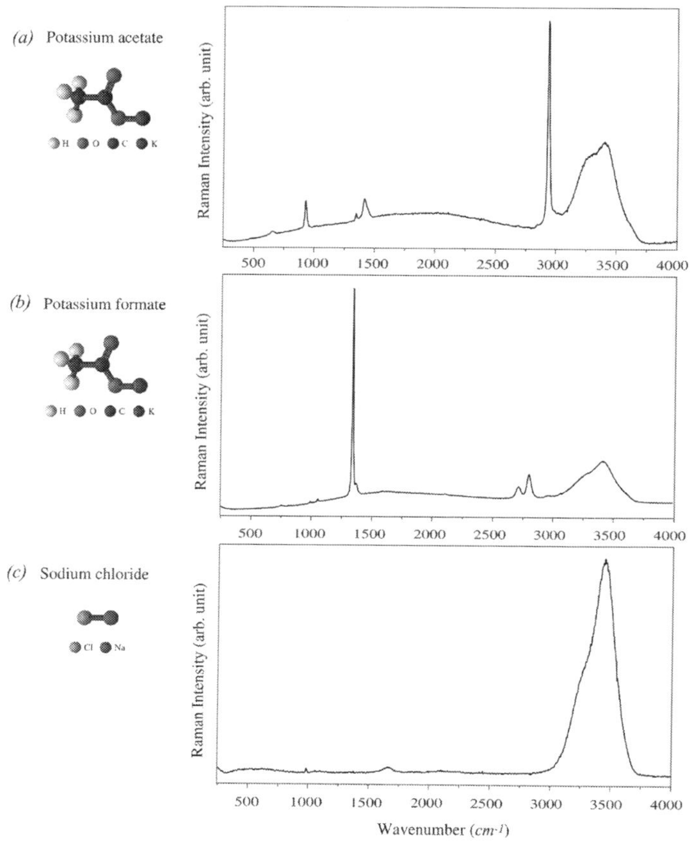

Figure 8. Raman spectra of aqueous solutions of the salts used in winter maintenance: potassium acetate (a), potassium formate (b), and sodium chloride (c) obtained with a 532-nm excitation wavelength.

The Raman spectrum of potassium acetate presents peaks corresponding to the vibrations of O—C—O bending, C—C stretching, CH_3 bending, C—O stretching and CH_3 stretching [42–45]. For potassium formate, the peaks corresponding to O—C—O and C—H bending, as well as C—O and C—H stretching [46–48] are present. The sodium chloride spectrum, on the other hand, presents only the peaks characteristic of the water, O—H bending and O—H stretching. Indeed, chemicals that present only monoatomic ions once dissolved in water (like NaCl gives [Na^+] and [Cl^-]), do not possess specific peaks, as they do not have bonds anymore [49, 50].

Clearly, it is not possible to use the same analysis method for the elaboration of the calibration curves for all these species. We then consider two cases: on the one hand, the case when the salt presents specific peaks, making it possible to detect its presence and concentration directly, and, on the other hand, where the salt does not present any specific peak and it is only possible to detect and quantify it indirectly.

Furthermore, for a better baseline correction, it is possible to focus the study on only one part of the spectrum. In the case of winter maintenance salts presented below, the study considered the region above 2500 cm^{-1}.

4.1.1.1. Direct concentration determination: CH_3COOK and CHOOK

For an optimized calibration curve, the analysis of aqueous solutions covering a large scale of concentrations is necessary. For the potassium acetate and formate, solutions with a weight percent up to 65 and 60%, respectively, were analyzed (**Figure 9**). The weight percent (wt%) is defined as the ratio of the mass of the salt dissolved m_{salt} and the mass of the solution sample m_{sample}.

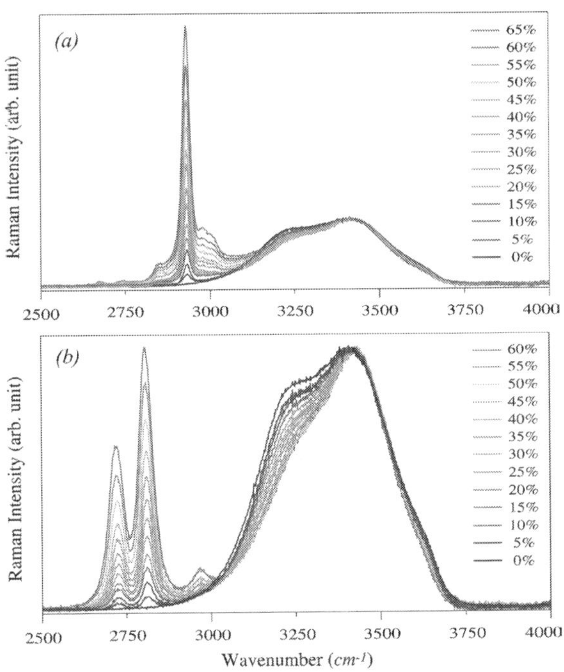

Figure 9. Normalized Raman spectra of aqueous solutions of potassium acetate (a) and potassium formate (b) with a weight percent between 0 and 65% and 0 and 60%, respectively. Spectra obtained with a 532-nm laser at 100 mW.

In **Figure 9** are presented spectra of potassium acetate and potassium formate solutions at different concentrations/weight percent in the 2500–4000 cm^{-1} spectral region. This region contains, for both salts, a salt specific peak, as well as the O—H stretching band. As expected, in both cases, the intensity of specific peaks increase with the concentration. A proper signal treatment permits to extract the effect of the active compound on the Raman spectra and to elaborate appropriate calibration curves. A calibration curve defines the relationship between an analytical signal produced by the analyte and its concentration.

In the simplest case, the calibration curve is linear and a simple linear regression permits to fit the analytical signal to the concentration. Most generally, however, the analysis is more complex since the calibration curves are often affected by overlapping bands, additional interfering components. The resultant calibration curves will often be non-linear [5].

On the whole, for the elaboration of a calibration curve, the calculation of the peak intensity as a function of the concentration can be used, however, it is important to ensure that the peak characteristics that will be calculated will be affected only by the compound itself. As mentioned before, in order to avoid the bias of spectral intensity caused by the experimental setup, it is recommended to use integrated intensity. Furthermore, experimental conditions, such as ambient light, can also influence the spectral intensity and even the integrated intensity. One way of overcoming all these difficulties and to maintain only the compound concentration influence is to use a ratio of integrated intensities [28].

Hence, for the elaboration of the calibration curves of potassium acetate and formate, a spectral marker S_D is defined as a ratio of an integrated intensity of a specific peak ΔI_{salt}, over the integrated intensity of the O—H asymmetric stretching vibrations ΔI_{OH}. For the potassium acetate, the integrated intensity 2801–3001 cm^{-1} including the CH_3 stretching band was chosen as ΔI_{salt}, and for potassium acetate ΔI_{salt} was defined as the integrated intensity of the 2725–2809 cm^{-1} region including the C—H symmetric stretch. For both cases, ΔI_{OH} was defined as the integrated intensity of the water 3324–3486 cm^{-1} spectral region. The thus obtained calibration curves are presented in **Figure 10**.

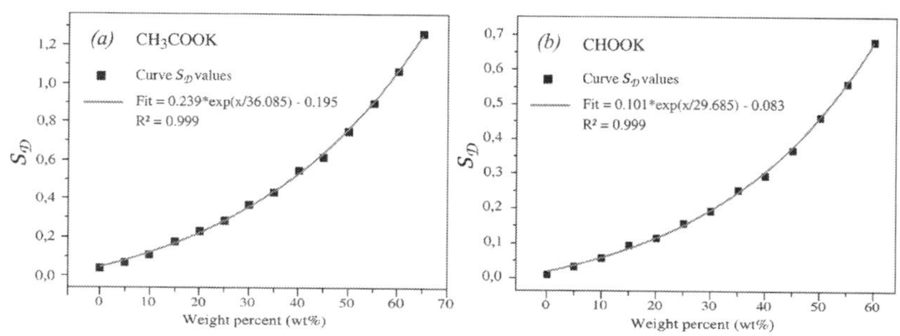

Figure 10. Calibration curves obtained for potassium acetate (a) and potassium formate (b). Both curves present an R^2 of 0.999.

In this particular case, there is a slight overlapping of the bands chosen, as there is a small contribution of the O—H band in the region of the C—H band. As a result, the resultant

http://dx.doi.org/10.5772/64550

calibrations curves obtained for the potassium acetate and potassium formate presents an exponential evolution.

These calibration curves can then permit not only to identify the presence but also to quantify these species in an aqueous solution. The precision of the method will depend on the quality of Raman spectra obtained. In general, with the 532 nm laser excitation and a 60 second accumulation time, it is possible to obtain a good signal/noise ratio permitting the quantification of these species with an 1% uncertainty. Depending on the application aimed, it is possible to diminish even more the uncertainty by analyzing more aqueous solutions in order to have more points for the calibration curves. For the winter maintenance application, however, the uncertainties aimed are rather high, 5%, and then this method is well-fitted just as it is shown here.

4.1.1.2. Indirect concentration determination: NaCl

The concentration determination by a direct analysis of a chemical specific peak is immediate and will, logically, offer better results. However, chemicals that dissolve in monoatomic ions do not possess specific peaks, as they do not have bonds [49, 50]. In order to detect their presence in water and to quantify them, it is necessary to analyze the water signature and deduce information on the chemical through their influence on the water structure. A typical example of this behavior is sodium chloride, NaCl.

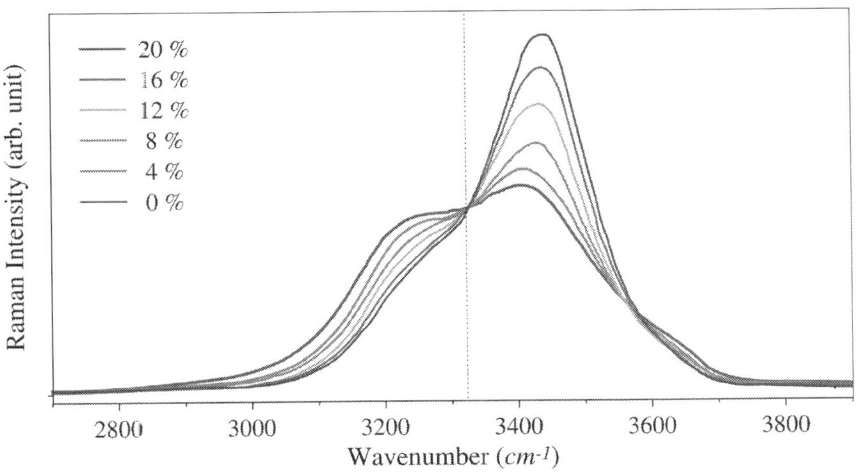

Figure 11. O—H stretching region of Raman spectra of NaCl aqueous solutions in a weight percent up to 20%.

Figure 11 shows the influence of NaCl concentration dissolved in water on its Raman spectrum. Two main effects can be underlined: NaCl affects both the O—H symmetric and O—H asymmetric stretch. The concentration increase enhances the morphological changes of the O—H stretching band, diminishing the O—H symmetric and increasing the O—H asymmetric stretching vibrations [51], leading to a spectrum shift towards higher wavenumbers. This can be considered as a direct result of the dissolution of NaCl in water, which provokes the decrease of the number of hydrogen bonds in the intermolecular structure [50, 52, 53].

It is then possible to use the ratio of the asymmetric and symmetric O—H stretching vibrations as a spectral marker for the elaboration of the calibration curve [54]. The O—H stretching band was then divided into two parts, 3325–3650 cm^{-1} as representative of the O—H asymmetric stretching vibrations, and 3000–3325 cm^{-1} as representative of the O—H symmetric stretching vibrations.

After an elaboration of a calibration curve, a set of blind tests should be performed in order to verify the uncertainty of this indirect way of concentration determination. The plot presenting the calculated concentration via the experimental S_D versus the theoretically calculated concentration is presented in **Figure 12**.

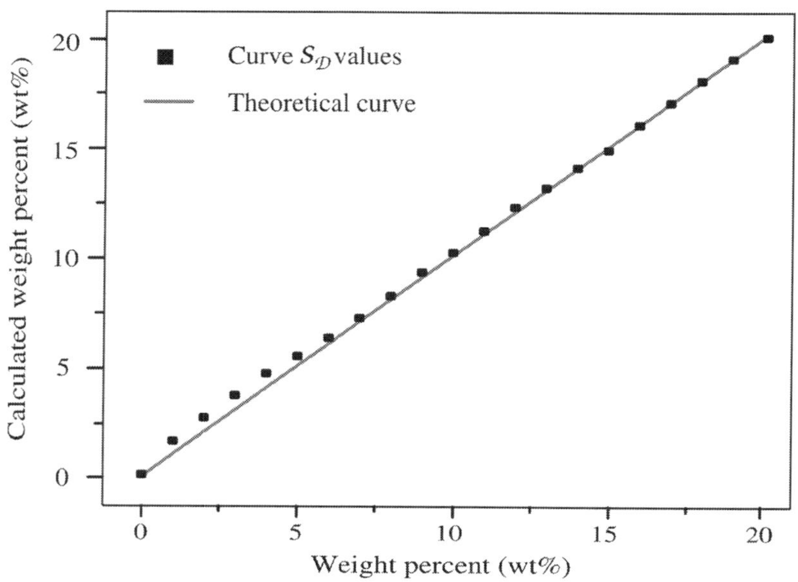

Figure 12. Comparison of calculated weight percent (represented by the experimental points in black) and weight percent (represented by the curve in red).

As shown in the **Figure 12**, even an indirect way of concentration determination can offer good precision, especially for more concentrated solutions; For this specific case, the standard deviation is of 0.25% for weight percentages over 7 and of 0.75% for weight percentages up to 5%.

To sum up, this approach can also be applied to all salts that dissolve into monoatomic ions, since in their dissolved form they interact with water, and these interactions are detectable and quantifiable by the means of the Raman spectra of water. Hence, similar studies even in applications other than the winter maintenance domain were also conducted, for example on the effect of different alkali halide on the water structure [50, 51]. These studies showed that the influence of dissolved salts on water structure depends on the size and the charge of the ions, as well as the strength they form with the O—H complex.

It is also possible to analyze the O—H stretching band for the quantification of dissolved salts in solutions where several salts are present. For instance, quantification of NaCl with a 0.14 wt % error was realized in a solution of NaCl and NaF at 2.8 wt% [55].

4.1.2. Phase diagram elaboration

The method described in Section 3.3 can also be applied to aqueous solutions where the main morphological change appearing with the phase change concerns the O—H stretching band. It is then possible to apply exactly the same spectral marker as for water for the detection of the phase transition, that is to say a ratio of integrated intensities of the asymmetric and the symmetric part of the O—H band.

The analysis of a set of aqueous solutions at different concentrations in a large range of temperature can therefore permit the construction of the phase diagram which has a capital importance in the winter maintenance domain. With a large temperature range, it is possible to detect the different phase transitions. Indeed, in mixtures such as aqueous solutions, upon cooling down, each component will solidify at a different temperature. The first phase transition will then be a transition from a liquid to a mixture of liquid and solid phases. For binary systems, where only two components are present, the second phase transition will then transform the solid-liquid mixture into an entirely solid. For illustration purpose, the NaCl-H_2O phase diagram is presented **Figure 13**.

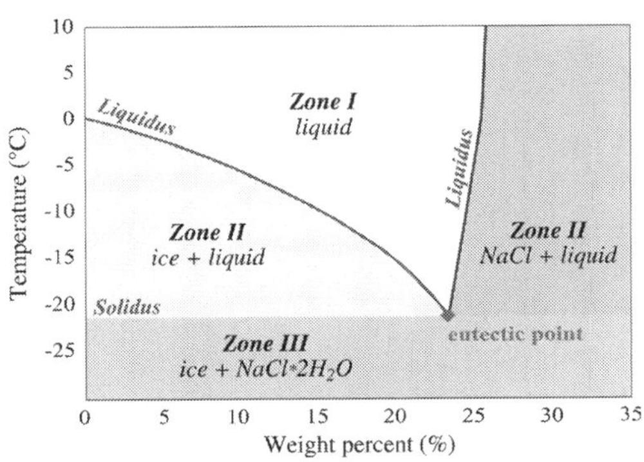

Figure 13. Phase diagram of the NaCl-H_2O binary system.

Upon cooling down a mixture of NaCl and water, the ice will be formed below a certain temperature of the so-called *liquidus* curve, this temperature being dependent on the brine concentration. During a certain temperature range, the mixture will be then composed of ice and liquid brine (Zone II). Upon further cooling down, at some point the entire mixture will solidify into ice and a hydrated form of NaCl (Zone III). This second phase transition will occur at temperatures below the *solidus* line, which is located at the eutectic temperature. The concentration of the eutectic point permits to determine the precise hydrated form of salt.

Such a phase diagram can be built experimentally by applying the spectroscopic method described before to any aqueous solutions [56].

4.2. Application to water pollution

Another application example could be the detection of water pollutants where there is an increasing need for a technique that could permit an *on-site* detection of the pollution sources. Many pollutant families are identified as being potentially very harmful for the environment and thus it is important to survey. As an example, we can cite the pollutants coming from the agricultural activities, such as the fertilizers, nitrates, or phosphates, or some drugs, such as hormones, which can have particularly important impacts on the environmental media.

In that objective, many studies have investigated the possibility to use optical methods, and more specifically RS for the detection of pollutants in water media [57, 58]. The method described earlier in this chapter, based on the calculation of ratios of pollutant-specific peak and water peak, can then also be applied to water pollutants [59, 60].

Author details

Ivana Durickovic

Address all correspondence to: ivana.durickovic@cerema.fr

Centre for Studies and Expertise on Risks, Environment, Mobility, and Urban and Country Planning Territorial Division for Eastern Regions – Laboratory of Nancy, Tomblaine, France

References

[1] McMillan P. Vibrational spectroscopy in the mineral sciences. In: Kieffer SW, Navrotsky A, editors. Microscopic to Macroscopic: Atomic Environments to Mineral Thermodynamics. Volume 14. Washington: Mineralogical Society of America; 1985, p. 9-64.

[2] Bhadekar R, Pote S, Tale V, Nirichan B. Mint: Developments in analytical methods for detection of pesticides in environmental samples. American Journal of Analytical Chemistry. 2011;2:1-15. DOI: 10.4236/ajac.2011.228118.

[3] Storey MV, van der Gaag B, Burns BP. Mint: Advances in on-line drinking water quality monitoring and early warning systems. Water Research. 2011;45:741-747.

[4] AU: Please provide location details for Refs. [4, 5, 38].Colthup NB, Daly LH, Wiberly SE. Introduction to Infrared and Raman Spectroscopy. 3rd ed., Academic Press; 1990.

[5] Larkin PJ. IR and Raman Spectroscopy: Principles and Spectral Interpretation. Elsevier; 2011. 535 p.

[6] AU: Please provide the journal title in English and also provide complete details for Ref. [6].Colomban P. Raman imaging of materials and heterogeneous devises. In: Techniques de l'Ingénieur. 2002. RE-5, p. 13

[7] Baia M, Astilean S, Illiescu T. Raman and SERS Investigations of Pharmaceuticals. Berlin: Springer; 2008.

[8] Claverie R, Fontana MD, Durickovic I, Bourson P, Marchetti M, Chassot JM. Mint: Optical sensor for characterizing the phase transition in salted solutions. Sensors. 2010;10:3815-3823. DOI: 10.3390/s100403815.

[9] Lewis IR, Edwards HGM, editors. Handbook of Raman Spectroscopy: From the Research Laboratory to the Process Line. New York: Marcel Dekker; 2001.

[10] AU: Please provide the journal title in English for Ref. [10].Barbillat J. Spectrométrie Raman. Techniques de l'Ingénieur. 1999;9:2865-1-2865-31.

[11] Walrafen GE, Blatz LA. Mint: Weak Raman bands for water. The Journal of Chemical Physics. 1973;59:2646-2650. DOI: 10.1063/1.1680382.

[12] Auer BM, Skinner JL. Mint: IR and Raman spectra of liquid water: Theory and interpretation. The Journal of Chemical Physics. 2008;128:224511-1224511-12. DOI: 10.1063/1.2925258.

[13] Carey MD, Korenowski GM. Mint: Measurement of the Raman spectrum of liquid water. Journal of Chemical Physics. 1998;108:2669-2675. DOI: 10.1063/1.475659.

[14] Eisenberg D, Kauzmann W. The water molecule. In: Pauling L. The Structure and Properties of Water. London: Oxford University Press; 1969. p. 1-34.

[15] Pimentel GC, McClellan AL. The Hydrogen Bond. San Francisco: Freeman; 1960

[16] Shriver P., Atkins W. Inorganic Chemistry. 3rd ed. Oxford: Oxford University press; 1999. 706 p

[17] Max JJ, Chapados C. Mint: Isotope effects in liquid water by infrared spectroscopy. III. H_2O and D_2O spectra from 6000 to 0 cm^{-1}. Journal of Chemical Physics. 2009;131:4626. DOI: 10.1063/1.1448286.

[18] Sun G. Mint: The Raman OH stretching bands of liquid water. Vibrational Spectroscopy. 2009;51:213-217. DOI: 10.1016/j.vibspec.2009.05.002.

[19] Pershin SM. Mint: Structure of the Raman band of the OH stretching vibrations of water and its evolution in a field of second harmonic pulses of a Nd:YAG laser. Molecular Spectroscopy. 2004;96:811-815. DOI: 10.1134/1.1771411.

[20] Bunkin AF, Pershin SM, Rashkovich LN. Mint: Changes in the Raman spectrum of OH stretching vibrations of water in an ultrasonic cavitation field. Molecular Spectroscopy. 2004;96:512-514. DOI: 10.1134/1.1719137.

[21] Kargovsky AV. Mint: On temperature dependence of the valence band in the Raman spectrum of liquid water. Laser Physics Letters. 2006;3:567-572. DOI: 10.1002/lapl.200610063.

[22] Laubereau A., Laenen R. Ultrafast coherent Raman and infrared spectroscopy of liquid systems. In: Fayer MD, editor. Ultrafast Infrared and Raman Spectroscopy. New York: Marcel Dekker; 2001. p. 1-80.

[23] Skinner JL, Auer BM, Lin YS. Mint: Vibrational line shapes, spectral diffusion, and hydrogen bonding in liquid water. Advances in Chemical Physics. 2009;142:59-103. DOI: 10.1002/9780470475935.ch2

[24] Risovic D, Furic K. Mint: Comparison of Raman spectroscopic methods for the determination of supercooled and liquid water temperature. Journal of Raman Spectroscopy. 2005;36:771-776. DOI: 10.1002/jrs.1359

[25] Walrafen G. Mint: Raman spectral studies of the effects of temperature on water structure. Journal of Chemical Physics. 1967;47:114-126. DOI: 10.1063/1.1711834.

[26] Scherer JR, Go MK, Kint S. Mint: Raman spectra and structure of water from −10 to 90 deg. The Journal of Physical Chemistry. 1974;78:1304-1313. DOI: 10.1021/j100606a013.

[27] Chumaevskii NA, Rodnikova MN. Mint: Some peculiarities of liquid water structure. Journal of Molecular Structure. 2003;106:167-177. DOI: 10.1016/S0167-7322(03)00105-3.

[28] Durickovic I. Study of the residual salinity by Raman spectroscopy [thesis]. Metz: University of Paul Verlaine – Metz; 2008.

[29] Sun Q. Mint: The single donator-single acceptor hydrogen bonding structure in water probed by Raman spectroscopy. Journal of Chemical Physics. 2010;132:054507. DOI: 10.1063/1.3308496.

[30] Walrafen GE. Raman and infrared spectral investigations of water structure. In: Franks F., editor. The Physics and Physical Chemistry of Water. New York: Springer; 1972. p. 151-214. DOI: 10.1007/978-1-4684-8334-5_5.

[31] Walrafen GE. Mint: Effects of equilibrium H-bond distance and angle changes on Raman intensities from water. Journal of Chemical Physics. 2004;120:4868-4876. DOI: 10.1063/1.1640568.

[32] Rey R, Moller KB, Hynes JT. Mint: Hydrogen bond dynamics in water and ultrafast infrared spectroscopy. Journal of Physical Chemistry A. 2002;106:11993-11996. DOI: 10.1021/jp026419o.

[33] Schmidt DA, Miki K. Mint: Structural correlations in liquid water: A new interpretation of IR spectroscopy. Journal of Physical Chemistry A. 2007;111:10119-10122. DOI: 10.1021/jp074737n.

[34] Kitadai N, Sawai T, Tonoue R, Nakashima S, Katsura M, Fukushi K. Mint: Effects of ions on the OH stretching band of water as revealed by ATR-IR spectroscopy. Journal of Solution Chemistry. 2014;43:1055-1077. DOI: 10.1007/s10953-014-0193-0.

[35] Beccucci M, Cavalieri S, Eramo R, Fini L, Materazzi M. Mint: Accuracy of remote sensing of water temperature by Raman spectroscopy. Applied Optics. 1999;38:928-931. DOI: 10.1364/AO.38.000928.

[36] Praprotnik M, Janezic D, Mavri J. Mint: Temperature dependence of water vibrational spectrum: A molecular dynamics simulation study. Journal of Physical Chemistry A. 2004;108:11056-11062. DOI: 10.1021/jp046158d

[37] Durickovic I, Claverie R, Bourson P, Marchetti M, Chassot JM, Fontana MD. Mint: Water-ice phase transition probed by Raman spectroscopy. Journal of Raman Spectroscopy. 2011;42:1408-1412. DOI: 10.1002/jrs.2841

[38] U.S. Department of Transportation. Manual of Practice for an Effective Anti-icing Program: A Guide for Highway Winter Maintenance Personnel. Federal Highway Administration; 1996.

[39] Switzenbaum MS, Veltman S, Schoenberg T, Durand CM. Best Management Practices for Airport Deicing Stormwater. Amherst: Department of Civil and Environmental Engineering, University of Massachusetts; 1999. 57 p.

[40] National Cooperative Highway Research Program. Snow and Ice Control: Guidelines for Materials and Methods. Washington: Transportation Research Board; 2004.

[41] Airport Cooperative Research Program. Impact of Airport Pavement Deicing Products on Aircraft and Airfield Infrastructure. Washington: Transportation Research Board; 2008.

[42] Frost RL, Kloprogge JT. Mint: Raman spectroscopy of the acetates of sodium, potassium and magnesium at liquid nitrogen temperature. Journal of Molecular Structure. 2000;526:131-141. DOI: 10.1016/S0022-2860(00)00460-9.

[43] Yang RD, Tripathy S, Li Y, Sue HJ. Mint: Photoluminescence and micro-Raman scattering in ZnO nanoparticles: The influence of acetate adsorption. Chemical Physics Letters. 2005;411:150-154. DOI: 10.1016/j.cplett.2005.05.125

[44] Tse WS, Chiang PY, Lin SJ. Mint: A Raman spectral study of the phase transitions in crystalline potassium acetate. Chinese Journal of Physics. 1986;24:63-68.

[45] Musumeci AW, Frost RL, Waclawik ER. Mint: A spectroscopic study of the mineral paceite (calcium acetate). Spectrochimica Acta Part A: Molecular and Biomolecular Spectroscopy. 2007;67:649-661. DOI: 10.1016/j.saa.2006.07.045.

[46] AU: Please provide complete details for Ref. [46].Poulston S, Holroyd RP, Bowker M, Parker SF, Mitchell PCH. Mint: Molecular spectroscopy of formates on copper surfaces. Scientific Highlights.

[47] Robinet L, Eremin K, Cobo des Arco B, Gibson LT. Mint: A Raman spectroscopic study of pollution-induced glass deterioration. Journal of Raman Spectroscopy. 2004;35:662-670. DOI: 10.1002/jrs.1133.

[48] Mitchell PCH, Holroyd RP, Poulston S, Bowker M, Parker SF. Mint: Inelastic neutron scattering of model compounds for surface formates. Potassium copper formate and formic acid. Journal of Chemical Society, Faraday Transitions. 1997;93:3569-3575.

[49] Bakker RJ. Mint: Raman spectra of fluid and crystal mixtures in the systems H2O, H2O-NaCl and H2O-MgCl2 at low temperatures: Applications to fluid-inclusion research. The Canadian Mineralogist. 2004;45:1283-1314.

[50] Terpstra P, Combes D, Zwick A. Mint: Effect of salts on dynamics of water: A Raman spectroscopy study. Journal of Chemical Physics. 1990;92:65-70.

[51] Schultz JW, Hornig DF. The effect of dissolved alkali halides on the Raman spectrum of water. Journal of Physical Chemistry. 1961;65:2131-2138. DOI: 10.1021/j100829a005.

[52] Hibben JH. Mint: The Raman spectra of water, aqueous solutions and ice. Journal of Chemical Physics. 1937;5:166-172. DOI: 10.1063/1.1750001.

[53] Rull F. Mint: Structural investigation of water and aqueous solutions by Raman spectroscopy. Pure Applied Chemistry. 2002;74:1859-1870. DOI: 10.1351/pac200274101859.

[54] Durickovic I, Marchetti M, Claverie R, Bourson P, Chassot JM, Fontana MD. Mint: Experimental study of NaCl aqueous solutions by Raman spectroscopy: Towards a new optical sensor. Applied Spectroscopy. 2010;64:853-857.

[55] Burikov SA, Dolenko TA, Fadeev VV, Sugonyaev AV. Mint: Identification of inorganic salts and determination of their concentrations in aqueous solutions based on the valence Raman band of water using artificial neural networks. Applied Problems. 2007;17:554-559. DOI: 10.1134/S1054661807040141.

[56] Durickovic I, Thiebaud L, Bourson P, Kauffmann T, Marchetti M. Mint: Spectroscopic characterization of urea aqueous solutions: Experimental phase diagram of the urea-water binary system. Applied Spectroscopy. 2013;67:1205-1209. DOI: 10.1366/12-06870.

[57] Li Z, Deen MJ, Kumar S, Selvaganapathy PR. Mint: Raman spectroscopy for in-line water quality monitoring – instrumentation and potential. Sensors. 2014;14:17275-17303. DOI: 10.3390/s140917275.

[58] Braunlich G, Gamer G, Petty MS. Mint: Detection of pollutants in water by Raman spectroscopy. Water Research. 1973;7:1643-1647. DOI: 10.1016/0043-1354(73)90133-4.

[59] Ben Mabrouk K, Kauffmann TH, Fontana MD. Mint: Abilities of Raman sensor to probe pollutants in water. Journal of Physics: Conference Series. 2013;450:012014. DOI: 10.1088/1742-6596/450/1/012014.

[60] Durickovic I, Marchetti M. Mint: Raman spectroscopy as a polyvalent alternative for water pollution detection. IET Science, Measurement and Technology. 2014;8:122-128. DOI: 10.1049/iet-smt.2013.0143.